T0231647

CONVECTIVE
FLOW
BOILING

Supporting Sponsors

The Engineering Foundation
National Science Foundation

Cooperating Sponsors

Heat Transfer Division
American Society of Mechanical Engineers

Heat Transfer and Energy Conversion Division
American Institute of Chemical Engineers

CONVECTIVE FLOW BOILING

Proceedings of Convective Flow Boiling, An International Conference
held at the Banff Center for Conferences, Banff, Alberta, Canada
April 30 – May 5, 1995

EDITOR
JOHN C. CHEN
LEHIGH UNIVERSITY
Bethlehem, Pennsylvania, USA

ASSOCIATE EDITORS
Yasunobu Fujita
Kyushi University
Fukuka, Japan

Franz Mayinger
Technical University of Munich
Germany

Ralph Nelson
Los Alamos National Laboratory
New Mexico, USA

CRC Press
Taylor & Francis Group
Boca Raton London New York

CRC Press is an imprint of the
Taylor & Francis Group, an **informa** business

A TAYLOR & FRANCIS BOOK

CONVECTIVE FLOW BOILING

First published 1996 by Taylor & Francis

Published 2019 by CRC Press
Taylor & Francis Group
6000 Broken Sound Parkway NW, Suite 300
Boca Raton, FL 33487-2742

© 1996 by Taylor & Francis Group, LLC
CRC Press is an imprint of the Taylor & Francis Group, an informa business

No claim to original U.S. Government works

ISBN 13: 978-1-56032-507-9 (hbk)

Visit the Taylor & Francis Web site at
http://www.taylorandfrancis.com

and the CRC Press Web site at
http://www.crcpress.com

A CIP catalog record for this book is available from the British Library.

Library of Congress Cataloging-in-Publication Data

International Conference on Convective Flow Boiling (1st : 1995 :
  Banff, Atla.)
    Convective flow boiling: proceedings of Convective Flow Boiling,
  an International Conference held at the Banff Center for
  Conferences, Banff, Alberta, Canada, April 30–May 5, 1995 / editor,
  John C. Chen; associate editors, Yasunobu Jukita . . . [et al.].
      p. cm.
    Includes index.
    1. Heat—Transmission—Congresses. Ebullition—Congresses.
  I. Chen, John C., date. II. Title.
  TJ260.I574   1995
  621.402'2—dc20
  ISBN 1-56032-507-0 (cloth)

                                                          96-8655
                                                          CIP

CONTENTS

CRITICAL HEAT FLUX AND TRANSITION-FILM-BOILING

Contents

MULTICOMPONENT BOILING

SPECIAL SYSTEMS AND EFFECTS

INDUSTRIAL APPLICATIONS AND DEVELOPMENTS

PREFACE

As one of the most efficient modes of heat transfer, boiling finds wide application in industry. Examples include steam generators in power plants, reboilers in chemical plants, evaporators in refrigeration systems, two-phase heat exchangers in LNG plants, high-flux coolers for electronic components, and many others. Boiling is an extremely complex phenomenon, involving simultaneous heat and mass transfer across liquid-vapor, solid-liquid, and solid-vapor interfaces. In spite of considerable research in the last fifty years, designs of boiling systems still depend extensively on empirical inputs. The pursuit of mechanistic understanding and first-principle models is a continuing challenge that has both intellectual and pragmatic goals.

This book comprises selected papers from the First International Conference on Convective Flow Boiling. The purpose of the conference was to examine state-of-science and recent developments in technology of flow boiling, i.e., boiling systems which are affected by convective flows. A total of fifty-five papers are included, representing the work of experts from fifteen countries. The volume is organized as a section of keynote papers and six sections of technical topical areas:

Keynote Papers
Boiling Inception and Subcooled Boiling
Saturated Convective Boiling
Critical Heat Flux and Transition-Film Boiling
Multicomponent Boiling
Special Systems and Effects
Industrial Applications and Developments

Together these papers represent a benchmark in continuing development of convective-boiling technology. It is clear that much has been learned, but much more yet remains to be learned and to be applied.

We wish to express our gratitude to all authors and participants, noting with special appreciation the extra efforts of the session chairs, who also served as coordination reviewers for the papers:

S. G. Bankoff	D. Gorenflo
A. Bar-Cohen	R. Nelson
A. E. Bergles	J. Palen
V. P. Carey	M. Shoji
G. Celata	L. W. Swanso
K. J. Cornwell	J. Thome
D. Gibbons	

We also wish to acknowledge Charles Freiman, Barbara Hickernell, and Dr. Frank Schmidt of the Engineering Foundation for their sustained encouragement and help in organizing the Conference. Lisa Ehmer and Carolyn Ormes of Taylor & Francis deserve our thanks for their contribution toward the publication of this book. A special note of appreciation is due to Ms. Elaine Correll, Secretary for the Institute of Thermo-Fluid Engineering and Science at Lehigh University, for personally processing all manuscripts and dealing with hundreds of letters, telephone calls, faxes, and e-mail messages that made possible the conference and this book.

Finally, we acknowledge with appreciation the financial support of the Engineering Foundation and the National Science Foundation, as well as the programmatic support of the Heat Transfer Division of the American Society of Mechanical Engineers (ASME) and the Heat Transfer and Energy Conversion Division of the American Institute of Chemical Engineers (AIChE).

Editor
John C. Chen

Associate Editors
Yasunobu Fujita
Franz Mayinger
Ralph Nelson

KEYNOTE PAPERS

PHEOMENOLOGICAL ISSUES IN FORCED CONVECTIVE BOILING

Geoffrey F. Hewitt

Department of Chemical Engineering & Chemical Technology
Imperial College of Science, Technology & Medicine
London, England

ABSTRACT

Some problems in forced convective boiling are addressed with particular reference to the annular flow regime. First, correlation approaches to critical heat flux are discussed and their inadequacies demonstrated. The importance of modelling is emphasised and more recent developments in modelling of annular flow are reviewed with particular reference to entrainment and deposition mechanisms. Here, the modelling can be applied to any annular flow but is of particular interest in critical heat flux prediction. With this as a background, the traditional understanding of forced convective boiling heat transfer is reviewed and disentient viewpoints discussed. Experiments in equilibrium annular flow are described which aim to elucidate any differences between evaporation and condensation.

1. INTRODUCTION

Annular flow is arguably the most important regime of two-phase flow, occurring in a wide variety of industrial equipment and existing over a wide range of qualities and mass fluxes. In normal boiler operations, the onset of annular flow occurs at qualities above a few percent and phenomena within the regime are dominant in determining the heat transfer coefficient and critical heat flux in most boiling channels.

In what follows, the interlinked hydrodynamic, heat transfer and flux limitation phenomena will be discussed. Section 2 deals with the prediction of annular flows with particular reference to dryout prediction. In Section 3, the emphasis is shifted to the heat transfer processes in the liquid film in annular flow. The traditional picture is reviewed, challenges to this picture are reviewed and equilibrium annular flow experiments aimed at resolving the phenomenological issues are described.

A crucial phenomenon in annular flow is the existence of *disturbance waves* which are the source of liquid entrainment and whose influence on heat transfer is likely to be dominant. The influences and behaviour of disturbance waves has been recently investigated using computational fluid dynamics (CFD) techniques and this investigation is described in Section 4 below

2. DRYOUT (CRITICAL HEAT FLUX)

Dryout (or critical heat flux) is arguably the most important phenomenon in forced convective evaporation in annular flow. It sets the upper limit to the very good heat transfer behaviour which is associated with this flow regime and is usually to be avoided in practical applications, not only because of overheating and destruction of the heat transfer system but also because of the severe deposition and corrosion which is induced when the phenomenon is initiated. The normal way of dealing with dryout (CHF) is to use empirical correlations to predict its onset and to design the plant to lie within the restrictions imposed. However, there is a major difficulty. This is that the correlations often specifically embody misunderstandings of the nature of the phenomenon.

The two principal methods of correlating critical heat flux data are:

(1) Representation of the data in terms of the critical heat flux

($\dot{q}_{cr}$) as a function of quality (x, namely the fraction of

the mass flow which is in the form of vapour). $\dot{q}_{cr}$ decreases with increasing x.

(2) Representation and correlation of data in terms of the quality at dryout as a function of "boiling length", namely the length between the onset of bulk boiling and the point at which dryout occurs.

Neither approach to correlation has any fundamental justification; the first is more popular since it implies the ability to calculate dryout on a "local conditions" basis whereas the second is usually much better for the representation of the usual practical case of non-uniform heating (for uniformly heated tubes, then the two representations are interchangeable). The inadequacy of these correlations approaches is demonstrated in the results shown in Figure 1.

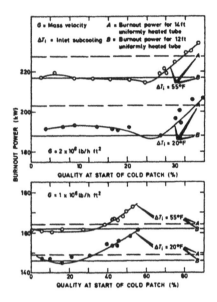

FIGURE 1: VARIATION OF BURNOUT POWER WITH POSITION OF COLD PATCH (BENNETT ET AL, 1967).

Here, dryout (or "burnout") heat flux was determined for a 4.26 m long 12.6 mm diameter tube which had an unheated zone, 0.61 m long (a "cold patch"), whose location could be moved along the tube. As will be seen, there is a systematic change in burnout power depending on the location of the cold patch but circumstances occur in which, as the patch is moved towards the end of the tube, the power can actually *be greater than the power for a uniformly heated tube without the cold patch*. This vitiates both of the correlation approaches and it is clear that a more phenomenological approach is needed to the prediction of dryout. A basis for this was provided in the early work at Harwell by making film flowrate measurements. Such measurements for uniformly heated tubes with cold patches near the inlet and near the outlet are illustrated in Figure 2. It is more instructive to plot the results in terms of entrained flow rate as a function of local quality and the results are plotted in this form in Figure 3. Note that dryout occurs when the entrained flow equals the total liquid flow. As will be seen, the effect of the cold patch is due to relaxation of the system towards equilibrium entrained fraction at which the entrainment and deposition rates are equal.

Even a relatively crude model of annular flow will give improvements in the phenomenological understanding and the prediction of dryout. The early models used in this area are summarised by Hewitt (1982) and, in recent years, there has been an attempt to address some of the weaknesses of these early

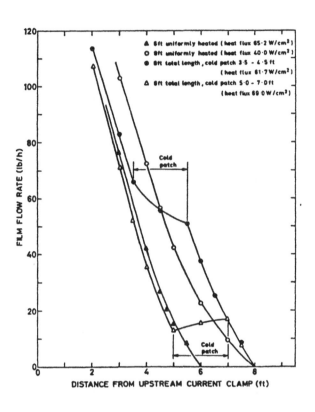

FIGURE 2: VARIATION OF FILM FLOW RATE WITH LENGTH FOR UNIFORMLY HEATED TUBES AND TUBES WITH UNHEATED ZONES ("COLD PATCHES") FOR THE CASES WHERE DRYOUT IS OCCURRING AT THE END OF THE TUBE (BENNETT ET AL, 1967).

models, particularly relating to the droplet deposition and entrainment relationships (Hewitt and Govan, 1990).

The fundamental problem to be solved in predicting dryout is estimating when the film flowrate approaches zero. Thus, the following equation is solved by integration along the channel:

$$\frac{d \dot{m}_{LF}}{dz} = \frac{4}{d_o} \left[D - E - \frac{\dot{q}}{h_{LG}} \right] \qquad (1)$$

where $\dot{m}_{LF}$ is the mass flux (related to the full tube diameter) of the liquid flowing in the film, z is the distance, d_o the tube diameter, D the deposition rate per unit peripheral area of the tube, E the entrainment rate, $\dot{q}$ the heat flux and h_{LG} the latent heat of vaporisation. D is often related to the droplet concentration in the gas core (calculated on a homogeneous basis) by the equation:

$$D = kC \qquad (2)$$

where k is a deposition mass transfer coefficient.

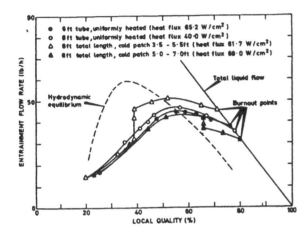

FIGURE 3: VARIATION OF ENTRAINED DROPLET FLOW WITH QUALITY (BENNETT ET AL, 1967).

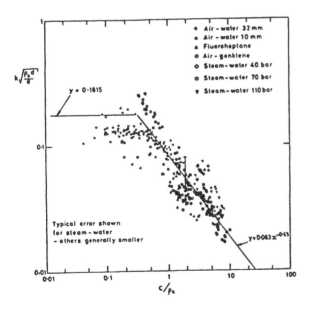

FIGURE 4: CORRELATION FOR DEPOSITION COEFFICIENT (GOVAN ET AL, 1988)

The processes of deposition and entrainment are extremely complex and it is probably impossible to represent them accurately. The diffusional behaviour implied by equation 2 is certainly not generally true. Thus, droplets may have a significant radial velocity at their time of creation and this influences the

deposition behaviour. Furthermore, the coefficient k may vary with C at high concentrations. However, even approximate correlations for deposition are helpful in annular flow predictions provided they take full account of concentration effects in particular. The same is true for entrainment correlations; entrainment does not begin until the liquid film flow exceeds a critical value m_{LF}. New correlations for deposition and entrainment are presented by Govan et al (1988) and are shown in Figures 4 and 5 respectively where σ is the surface tension, ρ_G and ρ_L the gas and liquid phase densities and m_G the gas mass flux.

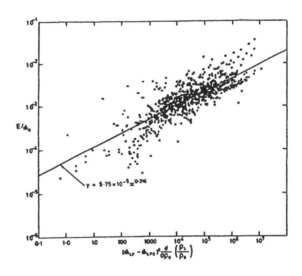

FIGURE 5: CORRELATION FOR ENTRAINMENT RATE (GOVAN ET AL, 1988).

Despite the obvious scatter in these correlations, their use has led to much improved predictions of a whole variety of situations relating to dryout and annular flow. These improved predictions are illustrated here by two examples:

(1) The cold patch data shown in Figure 3 can be predicted much more accurately with the new models than with the original Harwell annular flow model (HANA) as shown in Figure 6.

(2) Data for transient dryout has been investigated in detail (see Hewitt and Govan, 1990 and Govan, 1990). An example of the predictions here is that of the data of Moxan and Edwards (1967), who studied the time to dryout in flow in a vertical tube at

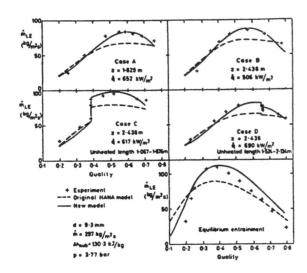

FIGURE 6: PREDICTION OF ENTRAINMENT FLOW FOR
TUBE WITH AND WITHOUT COLD PATCHES (HEWITT
AND GOVAN, 1990).

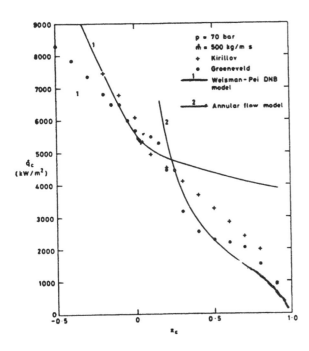

FIGURE 7: FULL RANGE MODEL FOR CHF (HEWITT AND
GOVAN, 1990).

70 bar, when the inlet flow was varied with time according to the expression:

Calculated and experimental times to dryout are shown in Table 1. With the new model, improved predictions of the time to dryout are shown. Perhaps the most interesting aspect of Table 1 is where the new model has been applied in a pseudo steady-state mode. This is equivalent to what is done in many nuclear reactor codes where local conditions dryout correlations are applied, quite incorrectly, to transient dryout. Fortunately, the pseudo steady-state approach gives much shorter times to dryout and is therefore conservative.

$$\dot{m}(t) = 786 + 1926 \exp(-t / 0.275) \qquad (3)$$

A question which is often asked about annular flow modelling of dryout is the lower limit at which it would apply. Obviously, in the subcooled region, the critical heat flux phenomenon still occurs but the mechanism would clearly be different (more analogous to film boiling than to film dryout). A curious feature is, of course, that the relationship between flux and quality for dryout (CHF) is a continuous one from the subcooled region into the quality region. In an attempt to produce a full-range model of dryout, Hewitt and Govan (1990) report taking the following steps:

TABLE 1
PREDICTION OF MOXON AND EDWARDS
TRANSIENT DRYOUT DATA (GOVAN, 1990)

Time to dryout (s)

Heat flux (kW/m²)	Sub-cooling (kJ/kg)	Experi-ment	Old Model (PWJI)	New Model (PWJIA)	New model pseudo steady-state
955	58.15	0.95	0.65	0.80	0.36
960	53.50	0.89	0.63	0.77	0.34
1155	46.52	0.40	0.34	0.37	0.08
1174	58.15	0.30	0.34	0.35	0.07

(1) They further developed the annular flow model so that it would apply at lower qualities. This included the improvement of the boundary condition for the onset of annular flow and taking account of the small but significant effect of heat flux on entrainment, particularly at lower qualities. For the first of these, the boundary condition was taken at a void fraction of 0.5 and the entrained fraction at this condition was calculated assuming equilibrium between entrainment and deposition. The effect of heat flux on entrainment was accounted for using a correlation developed by Milashenko et al (1989).

(2) Predictions for the film boiling type transition were obtained by modification of the model of Weisman and Pei (1983) to allow this model to be extended into the higher

quality region (it was originally developed for subcooled conditions and very low qualities).

The modified models are compared in Figure 7 with data taken from the tables of "standard CHF values" given by Kirillov et al (1989) and Groeneveld et al (1986). It is noteworthy that the used standard values themselves show discrepancies but in general, the agreement with the predictions is rather good. The governing mechanism for dryout would be the film boiling type at low qualities and the annular flow type at high qualities. Even at high qualities, if the local heat flux were sufficient, then a film boiling type CHF phenomenon would be expected as is obvious from Figure 7.

It will be seen from the above that there is now a reasonable phenomenological basis for interpretation of dryout (CHF) and this understanding is important in considering heat transfer aspects as we shall see below. However, there is scope for more work on deposition and entrainment processes and on the more widespread applications of the model.

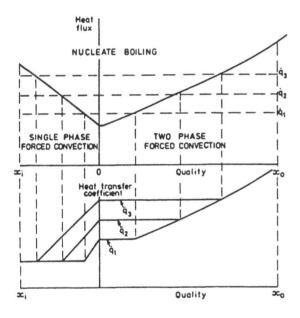

FIGURE 8: CONVENTIONAL REPRESENTATION OF REGIMES OF FORCED CONVECTION HEAT TRANSFER AND VARIATIONS OF HEAT TRANSFER COEFFICIENT WITH HEAT FLUX AND QUALITY.

3. FORCED CONVECTIVE BOILING HEAT TRANSFER

The conventional interpretation of heat transfer in forced convective boiling in a channel is illustrated in Figure 8. The upper part of the diagram shows the regions of heat transfer as a function of heat flux and quality and the lower part of the diagram shows how the heat transfer coefficient varies with the heat flux and quality. It has been conventionally assumed that the nucleate

boiling component of the heat transfer is surpressed as the quality increases and that, ultimately, the heat transfer is pure forced convective with no nucleate boiling. These concepts are embodied in the most widely used correlation for forced convective evaporation, namely that of Chen (1966).

The conventional picture illustrated in Figure 8 has been increasingly challenged in recent years. Particular forms of dissent have included:

(1) Many recent workers treat forced convective boiling as *either* nucleate boiling *or* forced convective evaporation, thus suggesting that either one process or the other is dominant and that the changeover is sharpe rather than gradual. An example of this approach is the work of Steiner and Taborek (1992).

(2) Mesler (1973) suggested that nucleate boiling was *enhanced* rather than *supressed* at high qualities. Here, he drew the analogy with boiling in static thin films where the heat transfer coefficient increases as the film gets thinner.

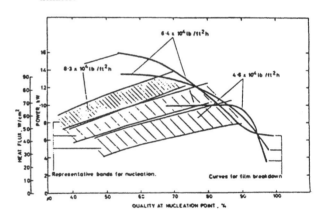

FIGURE 9: ONSET OF NUCLEATION IN ANNULAR FLOW (HEWITT ET AL 1965).

Although treating the forced convective boiling process as either nucleate boiling or forced convective evaporation may be convenient for design purposes, it is difficult to see how the mechanisms could switch suddenly in practice. Direct observations of the onset and suppresion of nucleate boiling in liquid films were made in early work at the Harwell Laboratory (Hewitt et al, 1965) and some of the results are illustrated in Figure 9.

The experiments were carried out with an annulus with a climbing film on the inside which could be clearly viewed through an outside glass tube, condensation on which was prevented by having a further outer glass tube through which a hot air stream was passed. Nucleation centres could clearly be seen in the liquid film and the regions of onset and suppresion of nucleate boiling could be established. As will seen from Figure 9, the heat flux required for nucleate boiling increases with increasing quality and mass flux, following the pattern expected from the conventional view of boiling mentioned above.

In response to the suggestion from Mesler (1973) the literature on forced convective boiling was reexamined and it had to be admitted that the direct evidence on forced convective boiling mechanisms left a great deal to be desired. In other words, it was not easy to directly refute the alternative mechanism. This led to a series of experiments, initially carried out collaboratively between Harwell and Oxford University, and more recently continued at Imperial College, London. There is a fundamental difficulty in designing objective experiments on forced convective boiling because of the effect of flow history. This point is illustrated in Figure 10 which plots entrained flow as a function of local quality with the heat flux increasing successively from curve B to curve C to curve D.

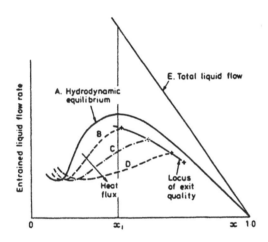

FIGURE 10: EFFFECT OF HEAT FLUX ON RELATION BETWEEN ENTRAINED LIQUID FLOW AND QUALITY FOR A FIXED MASS FLUX AND TUBE LENGTH.

The locus of entrained liquid flow rate as a function of exit quality is shown; ultimately, this curve would meet the line for total liquid flow (E) at the dryout point. However, let us consider the behaviour of the system at a given quality x_1. As will be seen, as the heat flux increases, the entrained liquid flowrate decreases and this implies, by difference, an increase in the liquid film flow. Thus, fixing the quality in forced convective boiling situation automatically gives a variation of film flow with heat flux. This implies gross changes in the hydrodynamic conditions (film thickness, interfacial wave configuration, interfacial shear stress and pressure gradient) and this provides a very serious challenge to objectivity in interpreting forced convective boiling processes. The approach adopted in the Harwell/Oxford/Imperial College experiments is illustrated schematically in Figure 11. Subcooled water was fed to the bottom of a 9.6 mm internal diameter tube and the section AB was heated by passing a large current through the tube walls. This generated a known quality at point B. There then followed an adiabatic section (BC) in which the hydrodynamic characteristics of the flow (entrained fraction, interfacial structure) reached equilibrium. Finally, there was a further section (CD) in which the heat transfer experiments were

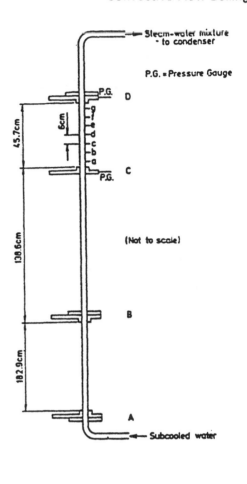

FIGURE 11: TEST SECTION FOR EQUILIBRIUM ANNULAR FLOW EXPERIMENT (AOUNALLAH ET AL, 1982).

carried out. In the first experiments with this system (Aounallah et al, 1982) the heat transfer section was a conventional one with Joule heating of the wall and thermocouples on the outside of the wall.

The behaviour of the system shown in Figure 11 with respect to the entrainment flow diagram is shown in Figure 12. The initial boiling section brings conditions to point B after which the adiabatic section (BC) leads to hydrodynamic equilibrium being reached. The next section (CD) then leads to a shift in this equilibrium but measurements made near the bottom end of the heat transfer section would be expected to conform reasonably closely to equilibrium conditions as shown.

Figure 13 shows results obtained from the first experiments using the above system (Aounallah et al, 1982). As will be seen, the heat flux varies linearly with

wall superheat under these circumstances and this can be taken as indirect evidence that nucleate boiling is not occuring; for nuceate boiling, the relationship between heat flux and wall superheat is

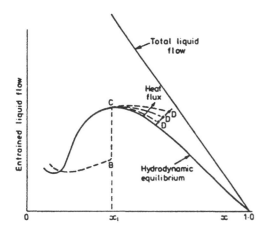

FIGURE 12: SCHEMATIC REPRESENTATION OF
EXPERIMENTAL SCHEME IN TERMS OF ENTRAINMENT
FLOW DIAGRAM.

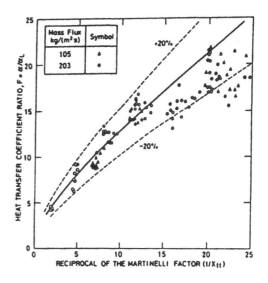

FIGURE 14: COMPARISON OF EQUILIBRIUM ANNULAR
FLOW DATA WITH CHEN CORRELATION (AOUNALLAH
ET AL, 1982).

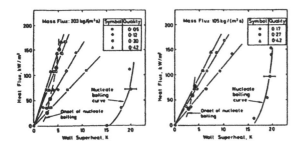

FIGURE 13: VARIATION OF HEAT FLUX WITH WALL
SUPERHEAT IN EQUILIBRIUM FLOW EXPERIMENTS
(AOUNALLAH ET AL, 1982).

The linearity of the flux/wall superheat relationship shown in
Figure 14 is consistent with the absence of conventional nucleate
boiling with wall nucleation. However, if the nucleation process
is governed in some way by the *hydrodynamics* of the system,
then the flux/superheat relationship could still be linear whilst
allowing bubble growth within the liquid film to play an
important role. Such *secondary* nucleation mechanisms were
suggested, again by Prof. R.B. Mesler, as having a potentially
important role. Clearly, the significance of this mechanism is
much more difficult to investigate. A procedure which was
evolved initially at Harwell and later at Imperial College was to
replace the electrically heated test section illustrated in Figure 11
with a test section which was fluid heated to allow both heating
and cooling, namely both evaporation and condensation. This
then allows the comparison of evaporation and condensation
coefficients; clearly, in the forced convective region, secondary
nucleation would only have a small influence on condensation
(the nuclei being condensed) whereas it would have a larger
influence on evaporation.

The test section used in the Harwell experiments by Chan
(1990) is illustrated in Figures 15 and 16. A single phase fluid
passed down an annulus around the tube carrying the two-phase
flow, the bottom end of this test section corresponding to point C
in Figure 11. The wall temperature was measured using
thermocouples brazed into the wall as illustrated in Figure 16 and
the fluid temperature in the annulus was measured using
thermocouples carefully centred between the two bounding
surfaces. For heating, glycerol was used (allowing a high
temperature whilst retaining the low pressure on the secondary
side) and for cooling, cold water was employed. The heat flux is
determined by measuring the differences between the

non-linear. Shown on Figure 11 is a curve for the onset of
nucleation calculated from the criterion of Davis and Anderson
(1966). This criterion assumes the availability of the full range of
nucleations sizes whereas nucleate boiling tests in the same test
section (carried out with the same test section in the slightly
subcooled region) indicated much higher wall superheats for
nucleation. The Davis and Anderson criterion indicated a critical
nucleation site size of around 10 microns whereas the nucleate
boiling tests indicated the critical size of about 1.7 microns.
Thus, it can be assumed that on the basis of the linearity and on
the basis of the observed superheats in nucleate boiling that there
was no wall nucleation. in these tests. This was taken to disprove
the original hypothesis of Mesler (1973) implying enhancement of
nucleate boiling as the film thins. The data from these
experiments were in good agreement with the Chen (1966)
correlation for forced convection as is illustrated in Figure 14.

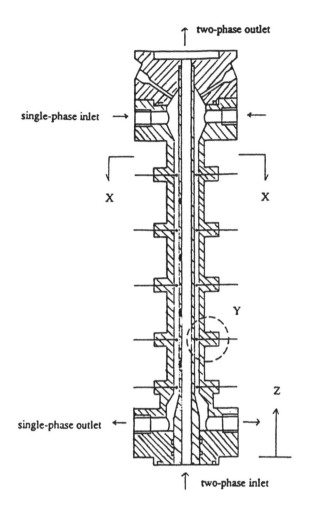

FIGURE 15: HEAT TRANSFER TEST SECTION (CHAN, 1990).

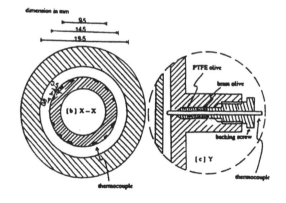

FIGURE 16: HEAT TRANSFER SECTION: CROSS SECTION AND CLOSEUP (CHAN 1990).

thermocouples in the annulus. The differences are rather small and this was the main problem with the measurements. Because of the high viscosity of glycerol, and because of the need for a turbulent flow in the annulus to mix the fluid there, the glycerol

velocities were higher than those for the water and the temperature changes along the annulus where therefore smaller. Thus, the accuracy of the experiment is limited. Nevertheless, these first experiments appeared to show significant differences between evaporation and condensation as shown in Figure 17

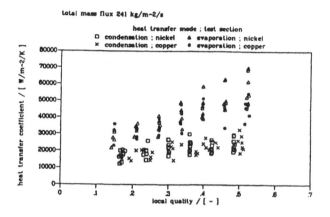

FIGURE 17: RESULTS FOR HEAT TRANSFER COEFFICIENT (CHAN, 1990).

As was stated above, there are problems in using glycerol as the secondary side heating fluid due to the small temperature differences obtained. Another problem is that the secondary side fluid must be changed in going from evaporation to condensation. The experiment has been reconstructed at Imperial College over the past year and the problem of the secondary side fluid alleviated by using sub-atmospheric pressure on the primary side and using somewhat pressurised water on the secondary side. This allows the water to be used for both heating and cooling and is clearly much more convenient. Although the first experiments appeared to confirm the differences between evaporation and condensation, a problem was discovered related to the steadiness of the flow of the secondary fluid, especially when water was used as the fluid. When this problem was corrected, the evaporation and condensation data show agreement! There is clear need for further investigation of this problem.

4. CFD STUDIES OF DISTURBANCE WAVES

The question arises of whether the basic forced convection coefficient can be predicted. Application of turbulent film theory based on average interfacial shear stress and mean film thickness tends to give results which are higher than the data so it is clear that a much more detailed examination is required. The techniques that are available using computational fluid dyanmics (CFD) offer the possibility of doing a more detailed calculation taking accout of the real interface.

CFD operates by obtaining numerical solutions for the conservation equations of mass, momentum and energy. Direct simulation of turbulent flows is possible only at very low Reynolds numbers and recourse normally has to be had to one form or another of turbulence model. In the investigation of annular flow described below, a low Reynolds number k-ε model has been employed and will now be briefly described. For a turbulent flow, the momentum conservation equation takes the form:

$$U_j \frac{\partial U_i}{\partial x_j} = -\frac{1}{\rho}\frac{\partial p}{\partial x_i} + \frac{\partial}{\partial x_i}\left(\nu \frac{\partial U_i}{\partial x_j} - \overline{u'_i u'_j}\right) \quad (4)$$

where U_i is the time averaged velocity in the direction x_i, ρ the fluid density, p the pressure, ν the kinematic viscosity and u'_i the instantaneous velocity of the fluid in direction x_i. The term $\overline{u'_i u'_j}$ appearing on the right hand side is due to turbulent motion and is called the Reynolds stress tensor. It acts to promote the diffusion of momentum and cannot be obtained by further time-averaging alone and has to be modelled. In the standard k-ε model, the turbulent motion is characterised by two quantities, namely k, the turbulent kinetic energy, and ε, the turbulent energy dissipation rate. These values are given by:

$$k = \frac{1}{2}\overline{u'_i u'_j} \quad (5)$$

$$\varepsilon = \nu \overline{\frac{\partial u'_i}{\partial x_j} \cdot \frac{\partial u'_j}{\partial x_i}} \quad (6)$$

The Reynolds stress tensor can be expressed in terms of mean velocity gradients and an effective turbulent kinematic viscosity ν_t as follows:

$$\overline{u'_i u'_j} = -\nu_t \left(\frac{\partial U_i}{\partial x_j} + \frac{\partial U_j}{\partial x_i}\right) \quad (7)$$

with ν_t being given by:

$$\nu_t = c_\mu \frac{k^2}{\varepsilon} \quad (8)$$

where c_μ is an empirical constant whose value has been determined from measurement as 0.09. Thus, if the distribution of k and ε is known, then the momentum equation can be solved

and the flow field calculated. In the k-ε model, transport equations are written for k and ε which have the form:

Advection = Production - Diffusion - Dissipation   (9)

and these equations for k and ε respectively, are as follows:

$$\frac{\partial k}{\partial t} + \frac{\partial (U_j k)}{\partial x_j} = \nu_t \left(\frac{\partial U_i}{\partial x_j} + \frac{\partial U_j}{\partial x_i}\right)\frac{\partial U_i}{\partial x_j}$$
$$+ \frac{\partial}{\partial x_j}\left(\frac{\nu}{\sigma_k}\frac{\partial k}{\partial x_j}\right) - \varepsilon \quad (10)$$

$$\frac{\partial \varepsilon}{\partial t} + \frac{\partial (U_j \varepsilon)}{\partial x_j} = c_1 \frac{\varepsilon}{k}\nu_t \left(\frac{\partial U_i}{\partial x_j} + \frac{\partial U_j}{\partial x_i}\right)\frac{\partial U_i}{\partial x_j}$$
$$+ \frac{\partial}{\partial x_j}\left(\frac{\varepsilon}{\sigma_\varepsilon}\frac{\partial \varepsilon}{\partial x_j}\right) - c_2 \frac{\varepsilon^2}{k} \quad (11)$$

where four new empirical constants have been introduced, namely $c_1 = 1.44$, $c_2 = 1.92$, $\sigma_k = 1.0$ and $\sigma_\varepsilon = 1.717$. The standard k-ε described above is valid only at high Reynolds numbers. Since the turbulent fluctuations are damped near a solid wall, there will be a region close to the wall where viscous effects are important. In this region, the local turbulence Reynolds number, defined by Jones and Launder (1972) as:

$$Re_t = \rho k^2 / \mu\varepsilon \quad (12)$$

where μ is the dynamic viscosity of the fluid, will be small and the high Reynolds number models are no longer applicable. For this reason, the standard k-ε model cannot be used to calculate the flow right up to the wall through the viscous sub-layer. Use is made therefore of empirical laws of sufficient generality (such as the Universal Velocity Profile) to connect the wall conditions such as shear stress and heat flux to the dependent variables just outside the viscous layer. This "wall function" approach is not always very satisfactory and special low Reynolds number k-ε models can be employed as proposed initially by Jones and Launder (1972). In this model, two of the empirical constants used in the original k-ε model (namely c_μ and c_2) are expressed (Launder and Sharma, 1974) as functions of the turbulence Reynolds number as follows:

$$c'_\mu = c_\mu \exp\left(-3.4 / (1 + Re_t / 50)^2\right) \qquad (13)$$

$$c'_2 = c_2 \left[1 - 0.3 \exp\left(-Re_t^2\right)\right] \qquad (14)$$

The transport equations for k and ε are also modified to the forms:

$$\frac{\partial k}{\partial t} + \frac{\partial(U_j k)}{\partial x_j} = \nu_t \left(\frac{\partial U_i}{\partial x_j} + \frac{\partial U_j}{\partial x_i}\right) \frac{\partial U_i}{\partial x_j}$$

$$+ \frac{\partial}{\partial x_j}\left[\left(\nu + \frac{\nu_t}{\sigma_k}\right)\frac{\partial k}{\partial x_j}\right] - \varepsilon - 2\nu\left(\frac{\partial k^{1/2}}{\partial x_i} \cdot \frac{\partial k^{1/2}}{\partial x_j}\right)$$

$$(15)$$

$$\frac{\partial \varepsilon}{\partial t} + \frac{\partial(U_j \varepsilon)}{\partial x_j} = c_1 \frac{\varepsilon}{k} \nu_t \left(\frac{\partial U_i}{\partial x_j} + \frac{\partial U_j}{\partial x_i}\right)\frac{\partial U_i}{\partial x_j}$$

$$+ \frac{\partial}{\partial x_j}\left[\left(\nu + \frac{\nu_t}{\sigma_\varepsilon}\right)\frac{\partial \varepsilon}{\partial x_j}\right]$$

$$- c'_2 \frac{\varepsilon^2}{k} + 2\nu\nu_t \left(\frac{\partial^2 U_i}{\partial x_j \partial x_i}\right)^2$$

$$(16)$$

The above equations are, of course, solved by finite difference methods; in the calculations described below, the CFDS-FLOW3D code was used which offers a number of options for the differencing scheme. In most of the calculations described here, the quadratic upwind differencing option (QUICK) was employed.

The occurance of large interfacial *disturbance waves* is a dominant feature in annular flow (see Hewitt and Hall-Taylor, 1970). Some characteristics of the waves are as follows:

(1) Disturbance waves occur above a critical liquid film Reynolds number (around 250); below this Reynolds number, only small ripples occur on the interface.

(2) The frequency of the disturbance waves falls with distance and reaches an asymptotic value at large distances. There is some evidence that the initial frequency is close to that of turbulent bursts. The spacing between the waves is random.

(3) Wave velocities are typically 3-10 m/s and the waves are typically 1-2 tube diameters in extent.

(4) Shear stress measurements indicate that the wall shear stress is high underneath the waves.

(5) Visualisation and entrainment measurements confirm that the waves are the main source of entrained droplets, entrainment being a vital feature of annular flow.

Although there is some suggestion that the waves are associated with turbulence inception phenomena within the liquid film, their source and nature is till somewhat mysterious. In the work summarised here (and described in more detail in Jayanti and Hewitt, 1994), attention was focussed on the flow in the liquid film, the influence of the gas phase being made manifest by its influnce on interfacial shear stress only. The influence of resolved normal stress (i.e. differences in net pressure force between the front and the back of the waves) were not taken into account. Further calculations by Wolf (1995), who calculated both the gas and liquid flow fields, indicated that this was a reasonable approximation.

A typical disturbance wave was chosen for study and had the following characteristics:

(1) The length of the disturbance wave region was taken as 20 mm followed by a substrate film region of 80 mm length. The disturbance wave was assumed to be of sinusoidal shape in its region of influence with a maximum applitude of 0.75 mm, namely five times the assumed substrate thickness of 0.15 mm.

(2) The film flow was assumed to be driven by air with a density of 2.15 kg/m³ flowing at a mean velocity of 50 m/s. The tube diameter was assumed to be 30 mm.

(3) The wave was maintained in a stationary position in the computational domain with the wall moving at the wave velocity. The wave velocity may be estimated as that condition for which the forces on the film are in equilibrium. This velocity was calculated at around 2.6 m/s for the results summarised here (the velocity varied slightly with the turbulence model chosen), in good agreement with the values measured in experiments.

As was stated above, it is necessary to specify reasonable values for interfacial shear. For the substrate region, analysis of data by Shearer and Nedderman (1965) for annular flow in the sub-critical (ripples only) regime produced the following correlation for the substrate region interfacial friction factor f_i:

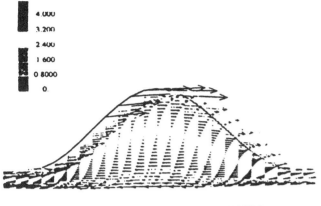

(A) VELOCITY VECTORS

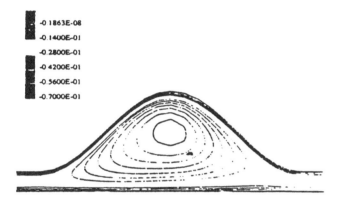

(B) STEAM FUNCTION CONTOURS

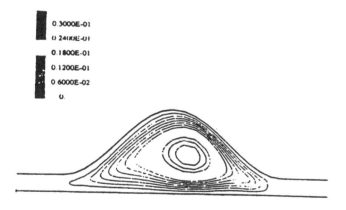

(C) TURBULENT VISCOSITY DISTRIBUTION

FIGURE 18: FLOW AND TURBULENCE FIELD IN A DISTURBANCE WAVE (JAYANTI AND HEWITT, 1994).

$$\frac{f_i}{f_G} = 0.856 + 0.00281 \, Re_G \, \frac{\delta_s}{D} \qquad (17)$$

where f_G is the gas phase friction factor, Re_G the gas phase Reynolds number, δ_s the substrate film thickness and D the tube diameter. For the disturbance wave region, the equivalent sand roughness height was taken to be five times the local film thickness; this gives reasonable agreement with overall pressure drop measurements but is, of course, only an approximate estimate. The interfacial friction factor was then calculated using the standard Colebrook-White formula as a function of position. From the interfacial friction factor, the interfacial shear stress was calculated for both regions by the equation:

$$\tau_i = \frac{1}{2} f_i \, \rho_G \, U_G^2 \qquad (18)$$

where ρ_G is the gas density and U_G the gas superficial velocity.

The results were obtained with both the standard k-ε model and with the low Reynolds number k-ε model (Jayanti and Hewitt, 1994). However, the low Reynolds number model was found to give much more physically realistic results (mainly because it does not need "wall functions") and only these will be given here. The outcome from the calculations on flows within the disturbance waves are illustrated in Figure 18 which shows the calculated velocity vector distributions, the stream function contours and the turbulent viscosity distributions. It should be borne in mind that, for presentational purposes, the wave height has been exaggerated (i.e. the scales in the vertical and horizontal directions are different). It should be remembered that the wave height is 0.75 mm whereas the zone occupied by the wave has an extent of 20 mm.

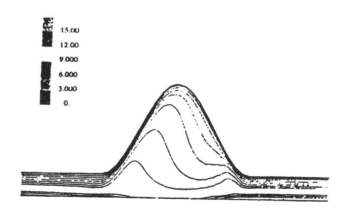

FIGURE 19: SIMULATION OF DISTURBANCE WAVE. TEMPERATURE DISTRIBUTION FOR A FIXED HEAT FLUX AND FIXED INTERFACE TEMPERATURE.

The first noticeable feature in the results is that of the recirculation pattern within the disturbance wave. This is evidenced by both the velocity vector plots (Figure 18a) and the stream function plots (Figure 18b). The standard k-ε model predicts that the flow is turbulent everywhere whereas the low Reynolds number k-ε model predicts that the flow is turbulent only within the disturbance wave region (see Figure 18c).

Using CFDS-FLOW3D, it is possible to predict the temperature distribution in the computational domain for a situation with heat transfer. The simulation chosen was that with a fixed wall heat flux and a fixed interfacial temperature (the latter would be true for evaporation or condensation of a pure vapour). The results are illustrated in Figure 19; note the distortion in the temperature profiles caused by the recirculation within the disturbance wave.

Perhaps the most interesting result from these calculations relates to average values for the heat transfer coefficient. For the low Reynolds number k-ε model a value of 6065 W/m^{2}K was obtained.

The calculations were also carried out assuming laminar flow in the substrate and disturbance wave and this gave a coefficient of 4505 W/m^{2}K. If the standard k-ε model is used for this two-dimensional calculation, a value of 37000 W/m^{2}K is obtained for the heat transfer coefficient. Classically, analysis of liquid film heat transfer has been carried out using one-dimensional analyses assuming a flat interface and taking the average value of interfacial shear stress. A calculation of this kind (see Hewitt and Jayanti, 1994) produces a value of 15650 W/m^{2}K. The actual values of heat transfer coefficient lie somewhere between the values for the low Reynolds number k-ε model and that predicted using the standard k-ε model.

5. CONCLUSION

The phenomena occuring in forced convective evaporation are extremely complex and the basis of understanding is certainly not completely secure! There is a clear need for more detailed studies of mechanisms particularly in the area of the relationships between forced convection and nucleate boiling. CFD provides a useful tool in illuminating the *two-dimensional* nature of the flow system but the results are very sensitive to the choice of turbulence model.

REFERENCES

Bennet, A.W., Hewitt, G.F., Kearsey, H.A., Keeys, R.K.F. and Pulling, D.J. 1967. "Studies of Burnout in Boiling Heat Transfer." Trans. Inst. Chem. Eng. 45, T139-T333.

Carroll, K. and Mesler, R.V., 1981. "Bubble Nucleation Studies. Part 2: Bubble Entrainment by Drop-formed Vortex Rings." AIChEJ 27, 851-853.

Chan, W.H.G.T., 1990. "Evaporation and Condensation in Annular Vertical Upward Flow of Water-Steam." Ph.D. Thesis, University of London (Imperial College).

Chen, J.C., 1966. "A Correlation for Boiling Heat Transfer to Saturated Fluids in Convective Flow." I & EC Process Design and Development 5, 3.

Esmailizadeh, L. and Mesler, R.B., 1986. "Bubble Entrainment with Drops." J. Coloid Interface Sci. 110, 561-574.

Govan, A.H., 1990. "Modelling of Vertical Annular and Dispersed Two-Phase Flows." Ph.D. Thesis, University of London (Imperial College).

Govan, A.H., Hewitt, G.F., Owen, D.G. and Burnett, G., 1989. "Wall Shear Stress Measurements in Vertical Air-Water Annular Two-Phase Flow." Int. J. Multiphase Flow 15, 307-325.

Groeneveld, D.C., Cheng, S.C. and Doan, T., 1986. "The CHF Look-Up Table, A Simple and Accurate Method for Predicting Critical Heat Flux." Heat Transfer Eng. 7, 46-62.

Hewitt, G.F., 1982. "Prediction of Pressure Drop in Annular Flow by Phenomenological Modelling." Section 2.2.4 of Handbook of Multiphase Systems (Ed. G. Hetsroni) McGraw-Hill Book Company, New York.

Hewitt, G.F. and Hall-Taylor, N.S., 1970. "Annular Two-Phase Flow." Pergamon Press, Oxford.

Hewitt, G.F. and Govan, A.H., 1990. "Phenomena and Prediction in Annular Flow." Invited Lecture, Symposium on Advances in Gas Liquid Flows, ASME, Dallas, November 1990 (FED-Vol. 99, HTD-Vol. 155, pp 41-56).

Hewitt, G.F., Jayanti, S. and Hope, C.B., 1990. "Stucture of Thin Liquid Films in Gas-Liquid Horizontal Flow." Int. J. Multiphase Flow 16, 951-957.

Hewitt, G.F. and Jayanti, S., 1994. "Hydrodynamics and Heat Transfer in Annular Gas-Liquid Flow: A CFD Study." Multiphase System Programme Report MPS/59 (Imperial College, London).

Jones, W.P. and Launder, B.E., 1972. "The Prediction of Laminarisation with a Two-Equation Model of Turbulence." Int. J. Heat Mass Transfer 15, 301-314.

Kirillov, P.L., Bobkov, V.P., Vinogradov, V.N., Denisov, V.S., Ivashkevitch, A.A., Katan, I.B., Paniutchev, E.I., Smogalev, I.P. and Sal'Nikova, O.B., 1989. "On Standard Critical Heat Flux Data for Round Tubes." Proc. 4th Int. Topical Meeting on Nuclear Reactor Thermal Hydrolics (NURETH-4), Karlsruhe, October 10th-13th, 1989, pp 103-108.

Milashenko, V.I., Nigmatulin, B.I., Petukhov, V.V. and Trubkin, N.I., 1989. "Burnout and Distribution of Liquid in Evaporative Channels of Various Lengths." Int. J. Multiphase Flow 15, 393-402.

Moxon, D. and Edwards, P.A., 1967. "Dryout During Flow and Power Transients." UKAEA Report No. AEEW-R553.

Sharer, C.J. and Nedderman, R.M., 1965. "Pressure Gradient and Liquid Film Thickness in Co-Current Upwards Flow of Gas-Liquid Mixtures: Application to Film Cooler Design." Chem. Eng. Sci. 20, 671-683.

Wolf, A., 1995. "Film Structure of the Vertical Annular Flow." Ph.D. Thesis, University of London (Imperial College).

ADVANCED EXPERIMENTAL METHODS

Franz Mayinger
Lehrstuhl A für Thermodynamik
Technische Universität München
München, Germany

ABSTRACT

The enormous advances in electronic data processing today enable the engineer to formulate fluiddynamic transport processes, especially also with phase change in a detailed mathematical, numerical way, and modern date aquisition systems make it possible to gain detailed insight into the nature of physical phenomena.

Modern optical methods and also impedance methods are especially well suited for experimental studies, because of their inertialess and non-invasive mode of operation. Possible applications of optical methods, and also impedance methods are briefly described with emphasis to heat transfer with phase change and also to fluiddynamics in two-phase flow. The examples demonstrate the capabilities and the importance of new measuring techniques.

1. INTRODUCTION

Flow boiling heat transfer is going along with very dynamic processes like nucleation, bubble growth, bubble collapse, phase interface interactions and mixing between the phases. These processes result in thermo- and fluiddynamic conditions, responsible for heat transfer and friction-loss for example. To improve mathematical modelling of transport processes, it is essential to know details of phase interactions as good as possible. A key for such insights are advanced experimental techniques, which work non-invasive and inertialess. Optical methods offer several possibilities to get a better knowledge of thermo-, fluiddynamic events in convective flow boiling.

However also non-optical methods like sensors, working on a capacitive basis, experienced a very remarkable development with respect to non-invasivity and signal forming of low inertia. Methods, based on the attenuation of ionising rays like X-rays or Gamma-rays do not interfere with the boiling flow also, however, they need a certain counting rate, which means, that they do not work inertialess.

The development of optical methods and also of capacitive techniques was supported by the availability of new electronic devices, which allow to reduce the data processing considerably, having been very time consuming in the past. So modern electronics, together with newly developed computer software, opened new prospects for optical techniques, applicable to convective flow boiling.

There are several examples for modern optical measuring techniques generally being of interest in heat- and mass transfer and in fluid-dynamics (Mayinger, 1994). These techniques are using various physical effects and find a broad field of applications. They are either working on a global basis or give only pointwise information. In heat- and mass transfer mainly temperature, concentration and velocity are of interest. For each physical property usually various optical techniques can be applied. Velocity in a droplet-spray for example can be measured by pulsed laser holography (global method) or by laser doppler anemometry (pointwise technique, LDA or PDA).

Holographic interferometry allows to determine the temperature field and by this gives precise data for evaluating heat transfer.

In addition one has to mention Particle Image-Velocimetry, which is mainly using light sheet illumination and which is competitive to double- or multi-pulse-holography.It is not possible to discuss all optical techniques known from the literature in detail here, and therefore emphasis is given to

- Holography
- Holographic Interferometry and
- Laser-Induced Fluorescence

In the field of non-optical techniques a brief discussion will be devoted to a non-invasive technique of the capacitive method, based on the dielectric constant of liquid and vapour. Examples of application will demonstrate the usefulness and the capability of these methods. Brief comments will also be given on computer-aided evaluation of the optical signals or the optical images.

2. HOLOGRAPHY

The general theory of holography is so comprehensive, that for a detailed description one must refer to the literature (Gabor 1951, Kiemle, Röss 1969, Mayinger 1994). Here only the principles, necessary for understanding the holographic measurement technique can be mentioned. In Fig. 1 the holographic 2-step image forming process of recording and reconstructing an arbitrary wave front is illustrated. The object is illuminated by a monochromatic light source and the reflected, scattered light falls directly onto a holographic plate. This object wave has a very complicated wave front. According to the principle of Huygens, one however can regard it to be the super-position of many elementary spherical waves. In order to simplify the matter, only one wave is drawn in Fig. 1. This wave is superimposed by a second one called reference wave. If both waves are mutually coherent, they will form a stable interference pattern, when they meet on a photographic plate. This system of fringes can therefore be recorded on a photographic emulsion. After the development, the plate is called "Hologram". The microscopic pattern (in general it consists of up to 3000 lines per mm) contains all information about the wave. The amplitude is recorded in the form of different contrast of the fringes, the phase is recorded in the spatial variations of the pattern.

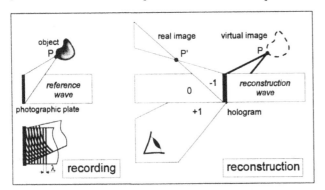

Figure 1. Principle of Holographic Two-Step Image-Forming Process.

If the plate, after chemical processing, is illuminated by a light beam, similar to the original reference-wave, the microscopic pattern acts like a diffraction grating with variable grating constant. The light transmitted, consists of a zero-order wave, travelling in the direction of the reconstructing beam, plus 2 first-order waves. One of these first-order waves travels in the same direction as the original object wave and has the same amplitude and phase distribution. This first-order wave produces a virtual image in front of the holographic plate, seen from the side of the incoming reference beam. The other wave goes in the opposite direction and creates a real image of the object behind the photographic plate. This real image can be

studied with various reconstruction devices, such as a microscope.

For conventional application of holography one can use laser emitting continuous light. The recording of very fast moving or changing objects - like in convective, two-phase flow - needs ultra-short exposure times, which can be achieved by using a pulsed laser, for example a ruby-laser with pulse durations of 20 - 30 ns. A holographic set-up, using a pulsed laser is shown in Fig. 2.

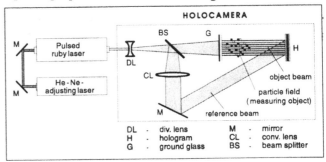

Figure 2. Holographic Set-Up for Ultra-Short Time Exposures with a Pulsed Laser.

A more sophisticated arrangement for recording pulsed laser holograms is presented by Fig. 3. In this arrangement the light, emitted by the pulsed ruby-laser, travels through a lense- and mirror system, where it is expanded, devided and guided through the measuring object and onto the holographic plate. This set-up is suitable for studying particle flow or phase distribution in multi-phase mixtures. It allows to visualise dispersed flow - like in post dry-out heat transfer with droplets not smaller than 10λ, whereby λ is the wave-length of the laser light.

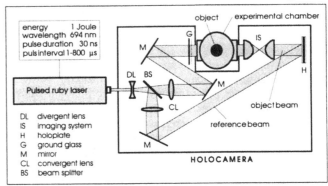

Figure. 3: Holographic Interferometer for Spray Analysis.

If the electronic system of the ruby-laser allows to emit more than one laser pulse within a very short period of time, sequences of the droplet- or spray-behaviour can be stored on the photographic emulsion of the same holographic plate, from which the velocity of the droplets, as well as their changes in size and geometric form can be evaluated. The evaluation however needs a very sophisticated and computerised procedure.

For evaluating the hologram, it first has to be reconstructed, as demonstrated in Fig. 4. After chemical processing the holographic plate is replaced in the old position and then illuminated by a continuously light-emitting helium-neon laser. If the holographic plate is replaced in the same orientation as during the recording process, one can look at it with the naked eye and one sees a virtual image of the droplet cloud or spray, exactly at the place where it was produced previously. For a quantitative evaluation one needs a closer examination via an enlarging lense or a microscope, which for example is connected with a video-camera. To do this, the holographic plate has to be turned by 180°, when positioning it to the old place. This is demonstrated in the lower part of Fig. 4. By illuminating with the reconstruction beam, a real image of the spray or droplet cloud is now produced, which has a three-dimensional extension.

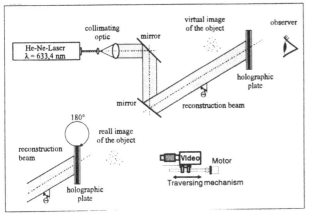

Figure 4. Optical Arrangement for the Reconstruction of Pulsed Laser Holograms.

An example of the arrangement of the video-camera, together with the evaluating components - computer, digitizer and monitor - is shown in Fig. 5.

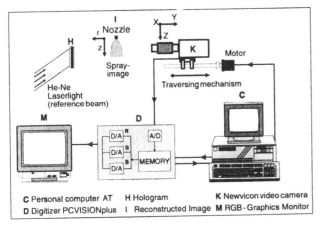

C Personal computer AT   H Hologram   K Newvicon video camera
D Digitizer PCVISIONplus   I Reconstructed Image   M RGB - Graphics Monitor

Figure 5. Digital Image-Processing System for the Evaluation of Pulsed Laser Holograms.

As mentioned before the reconstructed image is of three-dimensional extension and by using a lense of a long focus, only a very narrow area of the spray will be well focused. This has the advantage, that the spray cloud can

be evaluated plane by plane, either by adjusting the focal plain of the lense or by moving the camera forward or backward with a fixed focus of the lense. The latter procedure has the benefit, that the scale of the image, recorded in the video-camera, is not changed. That is the reason why it is preferably used in technical applications.

The very rapid development of computer technology made image analysis and image-processing applicable in many technical and scientific areas. Numerous problems of pattern-recognition, data handling of digitised pictures and computer graphics - formerly only reserved for computing centres - can now be solved on a personal computer. These modern techniques can also be applied to holograms, representing pictures of conditions in two-phase flow. The purpose of digital image-processing is to reflect the main features of a picture more clearly and informatively, than in the original and to judge the contents of an image quantitatively by employing pattern recognition algorithms. In the following some principles of digital image processing for evaluating pictures of droplet fields, obtained from holographic reconstructions and the corresponding set-up of a computer aided image-processing system, is briefly described. Basic steps necessary to digitise a picture are

- superposition of a two-dimensional grid
- with a simple black and white image
- quantization of the pixels in a grid box
- transfering the pixel pattern into a binary matrix.

The quality of a digitised picture depends on the mode of digitation i.e. on the sampling and on the quantization. For very high quality it is desirable, that the width of a grid box is of the magnitude of the grain of the photographic layer. A detailed description of the numerical technique to perform the processes for grey value pictures can be found in Gonzales and Wirtz (1977) and in Pavlides (1982). A good introduction was published by Haberäcker (1987).

If the pictures, to be evaluated, are obtained from holographic reconstructions by using a video-camera, it is necessary to scan them. With the procedure of scanning we have to distinguish between in-line holograms and off-axis holograms. In the optical arrangement for recording in-line holograms the optical access of the object-beam coincides with the reference-beam. Both beams also coincide with the optical axis of the scanning camera. To resolve fine details of the object, the scanning camera can be equipped with a micro-lens, which reveals only a small fraction of the whole picture. To scan the whole object part by part, a relative movement between hologram and camera has to be facilitated. Equipments are on the market, which allow to fix a holographic plate in a traversing mechanism, movable in three dimensions.

For the reconstruction of off-axis holograms a laser-beam is used, having the same angle of incidents at the holographic plate as the reference-beam at the moment of the holographic recording. The illumination of the holographic plate by the reconstruction beam reproduces the object-beam, which contains the spatial information of the object. Also this holographic reconstruction can be scanned by a video-camera. In this case however, the only possibility to realise the relative motion between reconstructed picture and video-camera is to move the video-camera. A movement of the photo-plate would change the illumination.

The processing of images obtained from holograms involves:

- the separation of the image from the background
- the identification of sharp focused image
- parts, measuring of their projected areas and
- the evaluation of their equivalent diameters or
- center points in respect to a reference frame.

These processing steps will be explained, using the example of a dispersed two-phase flow behind an injecting nozzle. By applying an average filter, the noisy background can be suppressed. There is software on the market, which can be used to do this.

After the suppression of the noisy background, the sharply focused parts of the image - in this case droplets - have to be identified. This can be realised by contouring these parts of the image. To explain this procedure, instead of droplets more simple objects are chosen here, namely glasspearls adhered to a thin wire. Fig. 6 presents two reproductions of a hologram, taken from these glasspearls, where the first one was well focused, and the second one was 1 mm out of focus. The histograms in Fig. 6 represent the grey value along the line, which cuts the contour of the second pearl.

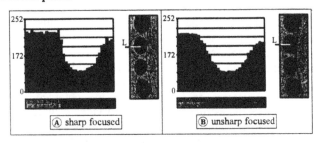

Figure 6.Histogram of Grey Value of a Sharply Focused and an Unsharply Focussed Image.

From the observation of both images and their corresponding histograms it is obvious, that sharply focused contours deliver a strong gradient of grey values and unsharply focused ones show a smooth transition of grey values. The decision about the sharpness of a contour is made by a digital image processing system. For this procedure many so called "out of focus"-algorithms have been developed and are available from literature (Ligthart and Groen 1982). Based on the different gradients of the grey values, a Sobel-operator enhances strong gradients, while making images with weak gradients disappear. The last step is to eliminate all images, which are not well focused. This can be done by allowing only pixels to remain in the image, which have a grey gradient to their neighbours

above a certain value. For details reference is made to the work by Chávez and Mayinger (1992).

An example of a computerised reproduction of the veil and the droplet swarm originating from a nozzle is shown in the Fig. 7 and Fig. 8. They convey an impression of how the information of the hologram can be improved by computerised evaluation. While the photograph on the left side of Fig. 7 shows only the shape of the veil, the computerised picture on the right side clearly presents the thickness of the liquid film of the veil and also shows its wavy nature. In addition the droplets, operating from the lower end of the veil are clearly defined.

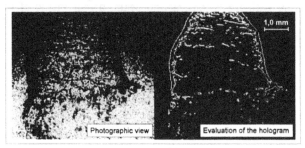

Figure 7.Veil and Spray Flow Behind a Nozzle: Photographic View and Evaluation of the Hologram.

Fig. 8 presents the sequence of reproduced "pictures" by this computer-aided image-processing. The upper row in this figure shows the region of the spray near the nozzle and the lower one, a region further downstream, where the veil is already disintegrated into a droplet swarm. By applying specially developed algorithms (Chávez 1991), the cross-section area, the diameter and the concentration of the droplets can be determined.

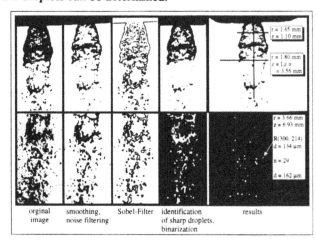

Figure 8. Steps of Image-Processing of a Hologram.
Upper Row: Spray Near the Nozzle.
Lower Row: Spray Downstream the Nozzle.

The figure gives an impression how the numerical procedure changes the original photographic picture into a computerised one, which contains only information of particles, being exactly in the focus plane, a slice, which in this case is thinner than 0,5 mm.

If two exposures of a moving particle collective are illuminated onto the same holographic plate within a short time, the velocity of the particles can be determined from such a hologram (Chávez and Mayinger 1992). For simplicity sake let us assume, that the particles are only moving in a two-dimensional way i.e. within the focused layer of the holographic reproduction.

Now a Fourier analysis is applied. To do this, the computer starts a process in which each image of the droplet is connected with all other ones and where the distance, as well as the angle with reference to the nozzle access between two "images" is determined. The Fourier analysis converts the spatial distribution into an normalised frequency distribution, as shown in Fig. 9 for the angle and for the distance of the connecting lines. The preferential angle appears as a peak in the Fourier diagram, showing the main direction and by this the direction of the movement of the particles. The same is done with the distance of the droplet "portraits". Now the preferential distance appears as a peak in the Fourier diagram. From that the average velocity can be calculated by using the time distance between the two exposures.

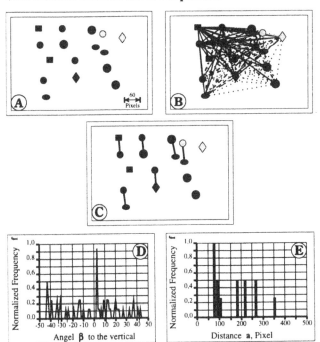

Figure 9. Steps for the Recognition of Couples of Spots, Corresponding to Two Successive Positions of Droplets.

If the movement of the particles is strongly three-dimensional, the procedure is much more complicated. In such a case electronically reproduced pictures of 3 or more focus planes have to be taken into account and the searching for the image of the second exposure of a particle has to be extended to all these planes being under consideration. So the Fourier analysis can be made in a three-dimensional way.

The results of a Fourier analysis in a two-dimensional way are presented in Fig. 10. The method is accurate enough to monitor even small influences on the droplet velocity, as they are for example produced by varying the pressure of the gas atmosphere into which the spray and by this the droplets are injected.

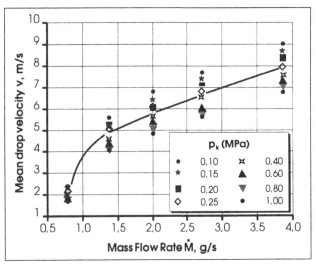

Figure 10. Mean Droplet Velocity in a Spray as a Function of the Flow-Rate at Different Ambient Pressures.

3. HOLOGRAPHIC INTERFEROMETRY

In flow boiling heat transfer, we are interested in heat flux densities, temperature fields and resulting from this in heat transfer coefficients. A very powerful method to record temperature fields and temperature gradients is the holographic interoferometry. The benefit of holographic interferometry compared to other interferometric methods - like Mach-Zehnder interferometry - is, that there is no need for a high optical quality of the optical components, because only relative changes of the object wave are recorded, and optical errors are automatically compensated by this interferometric method. On the other hand the monochromatic light, producing the wave-front, has to be very stable, therefore a laser of good coherence is needed as a light-source. There are many possibilities for arranging optical set-ups to form a holographic interferometer, which cannot be discussed here in detail. Reference is made to the literature (Mayinger 1994, Mayinger and Panknin 1974 and Panknin 1977).

A most commonly used arrangement for a holographic set-up is shown in Fig. 11. By means of a beam-splitter, the laser-beam is devided into an object and a reference-beam. Variable beam-splitters are especially useful for allowing easy adjustment of the intensities of the beams. Both beams are then expanded to parallel waves by a telescope, which consists of a microscope-lens and a collimating lens. The object wave passes through the test section, in which the temperature field is to be examined, whereas the reference wave directly falls onto the photographic plate.

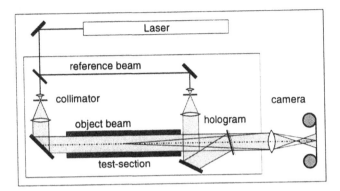

Figure 11. Optical Set-Up for Holographic Interferometry.

To obtain good reconstructions from a hologram, the inter-ference pattern between object and reference wave must be stable during the exposures of the plate. Therefore the optical components are mounted on a vibrationfree table. To avoid even those vibrations, which can be caused by water-cooled laser, it is sometimes necessary to place the laser away from the optical set-up.

The principle of the holographic interferometry is, that two waves are superimposed, which passed through the same test section at different moments. Therefore changes, which occur between the two recordings are interferometrically measured. For measuring the heat transfer for example, the recording of the first wave - the first exposure - is made, when all desired processes in the test section are in operation - fluid flow, pressure and mean temperature - but not that process - the heat transfer from an object, from a plate or from a bubble - which is of interest. There are in principle 2 techniques to record both exposures:

- the double exposure technique and
- the real time technique.

3.1 Recording Techniques for Holographic Interferometry

At first, the double exposure technique, which is very simple to use, will be briefly described. According to the holographic principle, several object waves can be recorded, one after the other on one and the same hologram. By illuminating with the reference wave, they are all simultaneously reconstructed. If they differ only slightly from one another, they will interfere micro-scopically. In the interference pattern, the differences between the object waves are discernible. This principle is used for the double exposure technique, illustrated in Fig. 12. In this illustration, the temperature distribution in a heated tube is chosen as an example. In a first exposure, the wave passing through the test section with constant temperature distribution is recorded. After recording this first exposure, the phenomena to be investigated is started. In this case, the temperature field is established by heating the wall of the tube. Now the incoming wave receives a continuous, additional phase-shift, due to the temperature changes. This resulting wave-front called the measuring-beam is recorded

on the same plate. After processing, the hologram is repositioned and illuminated by the reference-beam. Now both object waves are reconstructed simultaneously and will interfere. The interference picture can be observed or photographed. The main difference from classical interferometry is, that the object-beam is compared to itself. This allows a range of new applications, improvement and simplifications of optical interferometry, compared to conventional techniques. Since both waves pass through the same test section, any imperfections on the windows, mirrors and lenses are eliminated. Examinations even at very high pressure can be made, because the deformation of the windows can be compensated.

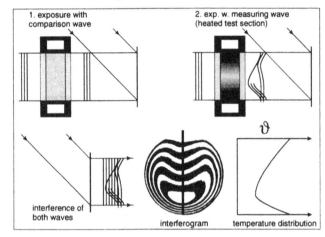

Figure 12. Principle of the Double Exposure Technique.

The method of double-exposure technique has the benefit, that it is simple to handle, however, it has the draw-back, that the experimentalist does not see, what he has recorded until the photographic plate is chemically processed, which usually takes 1/2 an hour. With the double-exposure technique, the investigated process cannot be continuously recorded, which is another disadvantage. Therefore a more sophisticated recording process for holographic interferometry was developed, the so-called real-time-method, which is illustrated in Fig. 13.

After the first exposure, by which the comparison wave is recorded and during which no heat transfer is going on in the test section, the hologram is developed and fixed. After this chemical process, the photographic plate is repositioned accurately to its old place, and the comparison wave can be reconstructed continuously by illuminating the photographic plate - the hologram - with the reference wave. This reconstructed wave, showing the situation without heat transfer in the test section, can now be superimposed onto the momentary object wave. If the object wave is not changed, compared to the situation before the chemical developing process, and if the hologram is precisely repositioned, no interference fringes will be seen on the hologram.

This indicator can be used for replacing the hologram exactly to its old place - within an accuracy of half a wavelength - by using micrometer and piezo quartz positioning devices.

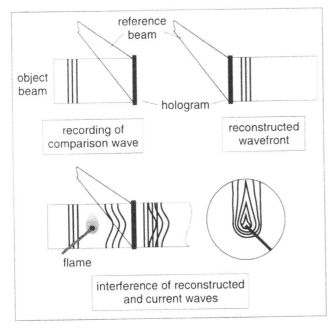

Figure 13. Principle of the Real-Time-Method.

After repositioning, the heat transfer process can be started. Due to the heat transport, a temperature field is formed in the fluid, and the object wave receives an additional phase-shift, when passing through this temperature field. Behind the hologram, both waves interfere with each other, and the changes of the interference pattern can be continuously observed or photographed.

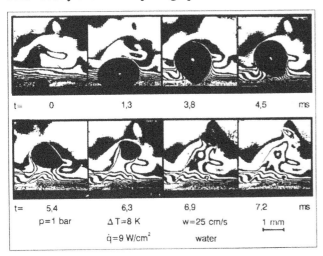

Figure 14. Detachment of a Single Bubble from a Heated Wall.

An example of such an holographic interferogram shows Fig. 14, which was taken from a simple flow boiling process. Slightly subcooled water was slowly flowing over a horizontal surface, which was electrically heated. Due to

the heating, nucleation started in the superheated boundary layer and a bubble was forming. In the first sequences of the interferograms, shown in Fig. 14, the bubble gets heat and by this also mass from the superheated boundary layer, and it grows rapidly in a few milliseconds. The black and white lines - called fringes - in these interferograms represent in a first approximation isotherms in the liquid, and the temperature difference between each isotherm is constant, as explained a little later. So one can read the fringes like a pattern of constant height - in this case temperature height - and, where the fringes are close together, there is a steep temperature gradient, and where they are far apart, a plateaux of almost constant temperature - small gradients - is demonstrated.

In Fig. 14 one can also see the limitation of this method, which is due to the fact, that the light, passing through the heated test section is not only shifted in phase, but is also deflected. So depending on the sign of the temperature gradient - positive or negative - the laser-beam is deflected either to the wall or from the wall. Therefore in a thin zone very near to the wall, we get no information and we can only see a grey pattern without interference fringes.

Figure 15. Detachment and Recondensation of Bubbles Under Various Boundary Conditions.

Following the sequences, presented in Fig. 14, we realise, that the vapour bubble, after penetrating from the superheated and saturated layer near the wall into the sub-cooled bulk of the liquid, starts to condense again and disappears within approximately 7 ms. The boundary layer is highly inhomogenous and is disturbed by the growing and by the buoyancy induced rising of the bubble.

Boiling on a heated surface is a statistical process and nucleation as well as recondensation of a forming bubble can not at all be described by a simple theory. This is demonstrated in Fig. 15. There also a sequence of interferograms is presented, covering a total time span of approximately 5 ms. Under a relatively thick boundary layer with flat temperature distribution, a bubble starts to form (0,3 ms), grows and recondenses again after reaching the subcooled bulk of the fluid, which in this case is also water. Bubble growth and recondensation is finished after 4,7 ms.

Another bubble also at the moment of 0,3 ms starts to grow under a position of the boundary layer with high temperature gradient. Due to the high temperature gradient, this bubble grows explosively within 0,3 ms, moves very fast into the subcooled bulk and condenses there as rapidly as it was grown. Its total lifespan is less than a half of that of the first bubble. For getting quantitative data of the temperature distribution and also for derivating the heat transfer coefficient, we have to perform an evaluation procedure according to the optical laws. This evaluation procedure will be briefly described in the following:

3.2 Evaluation of the Interferograms

The evaluation of a holographic interferogram, made in an optical set-up with parallel object wave, is very similar to the evaluation of interference patterns recorded in a Mach-Zehnder interferometer (Mayinger 1994, Mayinger, Panknin 1974). Therefore only the basic equations will be given. In Mach-Zehnder interferometry the wavefront, which is distorted by the object in the test section, is compared to a plane wave. In holography the object waves, passing through the test section at different times are superposed, and therefore reveal the changes in optical path-length between the two exposures. Expressed in multiples S of a wavelength, this change is calculated to

$$S(x,y) \cdot l = l \left[n(x,y)_2 - n(x,y)_1 \right] \qquad (1)$$

l is the length of the test section, in which the refractive index is varied, because of temperature or concentration changes. The refractive index distribution $n(x,y)$ during the recording of the two waves is assumed to be two-dimensional (no variation in light direction). Equation (1) shows, that initially only local variations can be determined. Only if the distribution of the refractive index $n(x,y)_1$ during the recording of the comparison wave is known, absolute values can be obtained. Therefore one usually establishes a constant refractive index field (constant temperature) while recording the comparison wave.

$$S(x,y) \cdot \lambda = l \left[n(x,y)_2 - n_\infty \right] \qquad (2)$$

To obtain absolute values for the temperature field, the temperature at one point of the cross section has to be determined by thermocouple measurements. This is usually done in the undisturbed region or at the wall of the test chamber. Equation (2) is the equation of ideal interferometry. It was assumed, that the light ray propagated in a straight line. Passing through a boundary layer, the light rays, however, are deflected because of refractive index gradients.

The light deflection can be converted into an additional phase shift ΔS, if a linear distribution of the refractive index is assumed to be within this small area.

$$\Delta S = \frac{n_0 \cdot \lambda \cdot l}{12 \cdot b^2} \qquad (3)$$

b is the fringe width, n_0 the average refractive index.

In many applications an ideal two dimensional field cannot be found. Often the boundary layer extends over the ends of the heated wall, or there are entrance effects or temperature variations along the path of the light beam (axial flow in the test section). Therefore only integrate values are obtained. Having corrected the interferogram, the obtained refractive index field can be converted into a density field. The relation is given by the Lorentz-Lorenz-formula, where N is the molar refractivity and M the molecular weight.

$$\frac{n^2 - 1}{n^2 + 2} \frac{1}{\rho} = \frac{N}{M} \qquad (4)$$

For gases with n=1, this reduces quite accurately to the Gladstone-Dale-equation.

$$\tfrac{2}{3}(n-1)\frac{1}{\rho} = \frac{N}{M} \qquad (5)$$

If there is only one component in the test section, and the pressure is kept constant, the density variations can only be caused by temperature changes. If the fluid is a gas, one can use the equation of state, in order to obtain the following formula, which relates the fringe shift to the temperature.

$$T(x,y) = \left[\frac{S(x,y) \, 2 \cdot \lambda \cdot R}{3 \, N \cdot p \cdot l} + \frac{1}{T_\infty} \right]^{-1} \qquad (6)$$

For liquids, the situation is more complicated. One has to go back to equation (1) or equation (2) respectively, and one has to use an equation from the literature, which correlates the refractive index with the temperature, being available in handbooks from thermodynamics.

Often local heat transfer coefficients are of special interest. In this case the temperature gradient at the wall is determined, and assuming a laminar boundary layer next to the wall, the heat transfer coefficient is obtained by :

$$\alpha = \frac{-k \cdot \left(\dfrac{dT}{dy} \right)_w}{T_w - T_\infty} \qquad (7)$$

3.3 Finite Fringe Method

With very high heat transfer coefficients, the boundary layer at a heat-transferring surface becomes very thin, down to a few hundreds of a millimetre. In this case it is difficult to evaluate the interference pattern, if it is registered with the procedure, described up to now. A slightly altered method, the so called "finite fringe method" offers some benefits. In this method, after the reference hologram was produced, a pattern of parallel interference fringes is created by tilting the mirror in the reference wave of Fig. 16, or by moving the holographic plate there within a few wave-lengths. The direction of the pattern can be selected as one likes, and it only depends on the direction of the movement of the mirror or of the holographic plate. By imposing a temperature field, due to the heat transport process, this pattern of the parallel interference fringes is then distorted. The distortion or deflection of each fringe from its original parallel direction is a measure for the temperature gradient at this spot and allows to deduce the heat flux and by this the heat transfer coefficient.

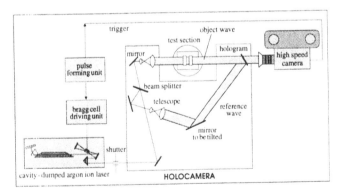

Figure 16. Finite Fringe Method for Holographic Interferometry.

In Fig. 17 this method is applied for monitoring the boundary layer and by this also the heat transfer at the phase interface at a bubble, filled with saturated vapour and condensing in a subcooled liquid of the same substance. The velocity of the condensation and by this the movement of the phase interface can be controlled by the heat transfer process or by inertia forces. The left in-

terferogram in Fig. 17 demonstrates the situation at the phase-interface between vapour and liquid, when the heat transfer dominates and the right one gives an impression of inertia controlled condensation. In the latter case, the heat transfer coefficient cannot be measured by the holographic interferometry, because there is no laminar boundary layer at the phase interface.

Figure 17. Phase Interface Boundary Layers with Heat Transfer- and Inertia Controlled Condensation.

The interferogram on the left-hand side has to be interpreted in such a way, that in these areas, where the interference fringes have a constant gradient (parallel lines), resulting from tilting the mirror or moving the holographic plate, there is a homogeneous temperature field. When the temperature changes, due to heat transfer, these parallel lines are deflected and the change in the original direction - the deflection of the fringe - can be used for evaluating the local and instantaneous heat transfer coefficient. For details of this evaluation process, reference is made to the literature (Nordmann, Mayinger 1981 and Chen, Nordmann, Mayinger 1991).

A sequence of evaluations of such interferograms is presented in Fig. 18. The sequence of these interferograms was taken by combining the holographic interferometry with the high-speed cinematography. The method allows instantaneous and local measurements of the heat transfer coefficient with good accuracy.

Saturated steam was blown through a capillary into slightly subcooled water. A bubble is formed at the outlet of the capillary and the condensation process starts immediately. For a short period - approximately 15 ms - the condensation at the phase interface reduces the volume of the bubble to a larger extent, than steam can be fed into the bubble via the flow through the capillary. Therefore in the scene 2 (approximately 10 ms after the opticalmonitoring started) the bubble takes a "squatting position" and the heat transfer at the top of the bubble is strongly reduced by the moving down of the layer of saturated liquid formed before and now insulating the steam from the subcooled bulk. Afterwards the steam flow through the capillary can overcome the rate of the condensed volume again, the bubble grows, separates from the nozzle and the heat transfer coefficient is increased.

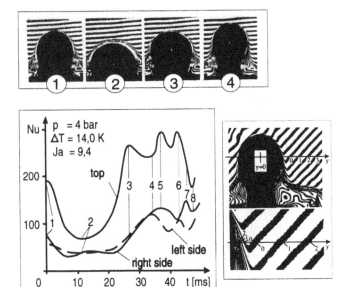

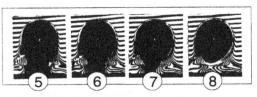

Figure 18. Heat Transfer at the Phase-Interface of a Vapour Bubble, Condensing in a Subcooled Liquid, Deduced from a Sequence of Interferograms.

4. LASER-INDUCED FLUORESCENCE

Light as sensor can provide several informations and not only the refractive index and the phase-shift can be used to get information about the distribution of temperature in a substance or of concentration in a mixture. Besides the phase-shift, the effect of scattering is most commonly used to get information about the physical and chemical conditions in a liquid or in a gas. Scattering methods are for example

- Raman-scattering
- Rayleigh-scattering
- Mie-scattering
- Bragg-scattering or
- Brillouin-scattering

It should be briefly mentioned, that Mie-scattering is recently used for flow visualisation in the particle-image-velocimetry (PIV). For using this method, a laser beam is formed into a very thin light-sheet, illuminating a plane within the volume of interest. For one measurement, 2 consecutive laser pulses are fired within a very short time interval, and the radiation, scattered by the particles in the illuminated area, is recorded two-dimensionally by a camera. The diameter of scattering particles usually is between 3 μm and 300 μm.

Here only the laser-induced fluorescence will be discussed a little more in detail. When applying this method,

the molecule under consideration absorbs one photon of the incoming laser-light. The energy of the photon must be equal to the energetic difference of two energy levels, i.e. the original level in the ground electronic state and a corresponding level in the first electronic state. The situation is principally explained in Fig. 19.

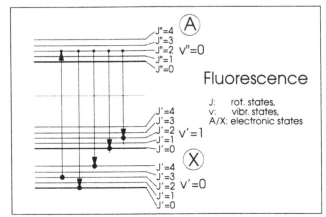

Figure 19. Energy Diagrams for Laser-Induced Fluorescence.

Since the energy differences involved are discrete and specific for each species, the frequency of the laser light has to be chosen in accordance with the molecule of interest. The states in the upper electronic levels are semi-stable with extremely short life-times and, therefore, soon after the transition to the upper state, the excited molecule drops back to the stable energy level within the ground electronic state, emitting a photon. In the most common case, the mean part of the fluorescence occurs at the same wave-length as the incoming light. Therefore it is sometimes difficult to distinguish between primary (incoming) radiation and the radiation by fluorescence.

In competition to this laser-induced fluorescence (LIF) a somewhat revised method was developed, the so called laser-induced predissociated fluorescence (LIPF). Here the excitation transition is chosen such, that predissociation at the upper state occurs at a high rate, and the wave-length of the fluorescent light is different from that of the incoming light for one and the same molecule. A good overview and detailed information about the before mentioned fluorescence technique can be obtained from the literature (e.g. Andresen 1988/1990/1991, Hanson 1990).

LIF and LIPF can be used to measure local concentrations, especially in chemical reactions, but also in newer times, it is applied to evaporation processes. An arrangement of optical components for performing LIF or LIPF measurements is shown exemplary in Fig. 20. A laser-beam electronically excites the molecules of interest. The author of the present paper was studying the concentration fields in combustion processes by LIF (Haibel, Mayinger, Strube 1993). Fig. 20 demonstrates how the laser-beam, originating from an EXCIMER-laser is expanded and deformed into a thin light-sheet with a height of approximately 5 cm and a thickness less than 0,7 mm. This light-sheet travels through a narrow quartz-window into the reaction zone, where the fluorescence is produced. The fluorescense is observed and recorded in perpendicular direction to this light-sheet with the aid of a CCD-camera, which is in this case intensified in the same wave-length as the laser produces (308 nm). The video signal of the camera is processed in a image evaluating unit and transformed into pseudo-colour pictures. These pictures are recorded by a SVHS-video-recorder, working in an analog way. The timewise co-ordination of laser and camera is done via a triggering unit, which is synchronised by the video-camera.

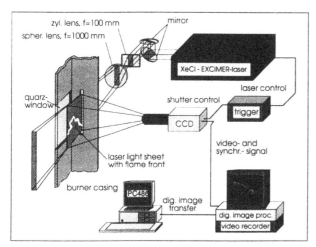

Figure 20. Optical Set-Up for Laser-Induced Fluorescence.

The chemical reaction in a flame is much more unstable and dynamic, than one would estimate by looking to the flame with the naked eye. This is also true for steady state combustion chambers. If we monitor a hydrogen flame in a simple combustion chamber, as shown in the upper picture of Fig. 21, without using the ultra-short time registration technique, we see a flame, corresponding to our usual imagination. This upper fluorescence picture in Fig. 21 integrates 50 single shots, ave-raging the distribution of the OH-radical fluorescent due to the excitation by the laser-sheet. Due to the fact, that the OH-radical is the most active product in the reaction-genetics of hydrogen combustion, the grey tunes in this figure mediate the local reaction activity. One realises zones of high reaction rates - black and dark grey areas - downstream and a little upwards of the spot where the mixture of hydrogen and air is injected into the combustion chamber.

The apparently quiet picture of the flame changes drastically, if the recording time is reduced to very short periods - down to 17 ns. Two ultrashort shots of the flame situation are shown in the lower pictures of Fig. 21. In the upper one of these two pictures, the flame is separating from the injection spot of the burner and inspite of a homogeneous mixture of hydrogen and air, there is no reaction at this spot for that moment. Such non-reacting zones can be observed - as shown in the lowest picture of Fig. 21 - over a long flowpath of the flame.

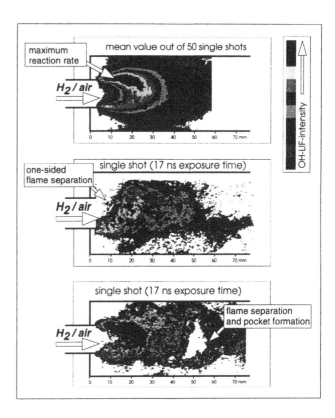

Figure 21. Structure of Premixed Hydrogen Flames in a Burner.

There are several papers in the literature, describing the use of laser-induced fluorescence for measuring vapourisation dynamics (Bazile, Stepowski, 1994, Senda, Fukami, Tanabe, Fujimoto, 1992 and Melton, Verdieck, 1985). Most of these authors use the exciplex fluorescence method.

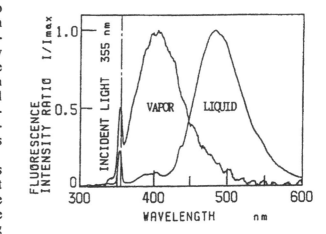

Figure 22. Frequence Spectra in Vapour and Liquid with Laser-Induced Fluorescence (Senda et al).

In this method, liquid additives - tracers are doped into the liquid to be evaporated and are assumed to be co-evaporative with it. In addition a second tracer can be added to the liquid in a very low concentration, which is not evaporating. So 2 spectra can be observed in the vapour-liquid mixture. One, which results from the fluorescense in the vapour and another one, originating from the fluorescence in the liquid. Bazile and Stepowski 1994 used the organic dye Rhodamin 6G, which does not vapourise for visualising a liquid/fuel spray. Senda e.a. 1992 reports on the visualisation of evaporative diesel spray by the exciplex fluorescence method. A TMPD monomer is used as a tracer for visualising the vapour phase. Fig. 22, taken from a paper by Senda e.a. (1992), shows fluorescence spectra of a liquid and a vapour phase. The spectra were measured by a multi-channel analyser.

5. A NOVEL IMPEDANCE METHOD

Impedance methods are widely applied for measuring the volumetric concentrations in multi-phase flows. They are based on different electrical properties (permittivity and conductivity) of the flow components and their effect on the measured impedance (capacitance and conductance) of an appropriate sensor. Although the impedance method offers a number of advantages, like instantaneous answers, its sensitivity to the flow pattern sometimes limits the range of application. Many different electrode designs have been developed, trying to minimise this limitation. Widely used arrangements are parallel plate electrodes (Auracher 1985), ring electrodes (Özgü 1973) or helical electrodes (Geraets 1988). For the generally occurring flow conditions, a number of analytical formulae are known, which connect the permittivity ε of a mixture of two fluids with their volumetric concentration ratio. These models are based on general conditions, regarding the shape of the dispersed particles and are often limited to relatively small maximum concentrations. The most popular models are those by Maxwell. The sensitivity of the impedance to the distribution pattern of the components - liquid and vapour - within the sensor leads to different calibration curves for each flow regime. In many practical flow situations, these curves differ very much.

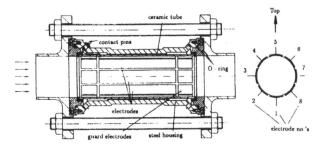

Figure 23. Impedance Sensor for Void Measurements.

To overcome the insufficiencies mentioned above, a new approach is made for measuring void fraction and flow pattern in multi-phase flows (Klug, Mayinger 1994/1992). A non-intrusive impedance probe, consisting of eight surface-plate electrodes implemented into the inner side of a tube (see Fig. 23).

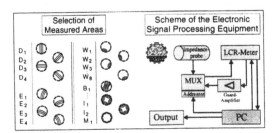

Figure 24. Examples of Measuring Fields.

With this probe, the impedance between different combinations of electrodes - the so called measuring fields - is measured, as demonstrated in Fig. 24 For every measuring field, the impedance - as an integral parameter - is determined by the distribution of the phases within the whole sensing volume of the probe. However, the individual domains of the sensing volume make different contributions to the total amount of the flow-influenced probe impedance. Therefore characteristic contribution patterns for the spatial sensitivity (Bair, Oakley 1992) can be observed, which allows to classify the multitude of measuring fields into several groups. Each group consists of a certain number of fields, which depends on the degree of field symmetry. One can distinguish between diametrical fields (D), eccentric fields (E), wall fields (W), large fields (B), integral fields (I) and Maltese-cross-shaped fields (M). For each field and for each flow pattern, one can make a theoretical calculation, and one can compare the theoretical prediction with the measured one. This comparison is a kind of calibration.

For practical use, where the flow pattern and also the void fraction are unknown, one can switch through all these mentioned field types and by comparing the measured signals with the calibration data, one finds a minimum of deviation for a certain field group and a certain flow pattern between actual measured data and calibration values. From that, one knows which flow pattern exists during the actual measurement, and one can choose the right correlation for evaluating the impedance signals to get the local volumetric void fraction.

This method is briefly sketched in Fig. 25 for 3 different flow patterns, namely stratified flow, annular flow and inversed annular flow (vapour outside and liquid in the bulk). The good performance of this reconstruction technique for the flow regimes, stratified flow, bubble flow and annular flow is shown in Fig. 26. For each example, the different measuring fields are indicated, which were used to reconstruct the data.

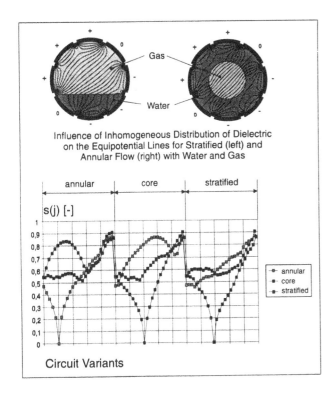

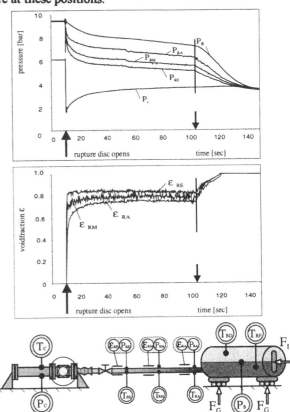

Figure 25. Calculated Derivations for 3 Cases, by Using 27 Measuring Fields.

the pipe is shown. The lower part of this figure demonstrates clearly, that the impedance probe accurately monitors the instantaneous void fraction of the vapour-liquid-mixture at these positions.

Figure 27. Pressure- and Void Distribution in a Pipe During Blow-Down.

Of course the same result would have been also obtainable by using a tomographic method, instead of the calibration procedure. However, the data handling with the tomographic method is much more time consuming, than that of the described procedure.

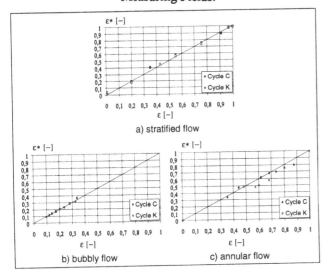

Figure 26. Calibration Tests with Impedance Sensor.

Finally in Fig. 27 a practical application of this method is presented. Three impedance probes were installed in a 2 m long pipe, through which a nonsteady, two-phase flow mixture was flowing during a blow-down process. A pressure vessel, filled partially with propane, was depressurised, feeding a liquid-vapour-mixture - due to flashing - into the pipe and flowing, via an orifice into a low pressure chamber. In the upper part of Fig. 27, the temporal course of the pressure at various positions along

6. CONCLUDING REMARKS

Optical methods and also other instantaneous and non-invasive techniques are expected to experience a powerful revival, due to three reasons:

Sophisticated theoretical treatment of heat transfer with phase change and two-phase flow with large computer codes needs very detailed information about temperature distribution and flow pattern for assessing and improving the physical models, used in the codes and for verifying such codes. An optically determined pattern of isotherms and an non-invasive impedance based measurement of flow pattern is a very stringent touchstone for the reliability and accuracy of a code.

Modern developments in power- and process engineering make transient situations more and more interesting, especially for controlling procedures and safety deliberations. Non-invasive and inertialess measuring techniques are a great help for getting good informations.

A former drawback of the image forming, optical measuring techniques and for the non-invasive impedance techniques, namely the laborious and time-consuming evaluation does not exist any more. Even a personal computer is good enough for evaluating a hologram or an interferogram within a few seconds, a process, that took several hours in the past. The costs of such an evaluating equipment are relatively moderate.

A new symbiosis could come into being between theorists and experimentalists, working in heat transfer and in multi-phase flow.

REFERENCES

Andresen, P., Bath, A., Gröger, W., Lülf, H.W., Meijer. G., ter Meulen, J.J., 1988, Laser-Induced Fluorescence with tunable Laser as a possible method for instantaneous temperature field measurements at high pressures: Check with an athmospheric flame, Appl. Optics. 27: 365, 1988.

Andresen, P., Meijer, G., Schlüter, H., Voges, H., Koch, A., Hentschel, W., Oppermann, W., Rothe, E., (1990), Fluorescence imaging inside an internal combustion engine using tunable eximer lasers, Appl. Optics. 29:2392, June 1990.

Andresen, P., Wolff, D., G., Schlüter, H., Voges, H., Koch, A., Hentschel, W., Oppermann, W., Rothe, E., (1990), Identification and imaging of OH and O_2 in an automobile engine using a tunable KrF eximer laser, Appl. Optics, 1991.

Auracher, H., and Daubert, J., 2nd Int. Conf. on Multi-Phase Flow, London, 1985.

Bair, M.S., and Oakley, J.P., 1st Meeting European Concerted Action on Process Tomography, Manchester, March 26-29, 1992 (1992).

Bazile, R., Stepowski, D., Measurements of the vapourisation dynamics in the development zone of a burning spray by planar laser-induced fluorescence and Raman scattering, Experiments in Fluids 16,171-180 (1994).

Chàvez, A., Holografische Untersuchung an Einspritzstrahlen - Fluiddynamik und Wärmeübergang durch Kondensation, Diss. Techn. Univ. München 1991.

Chàvez, A., Mayinger, F., (1992), Measurement of direct-contact condensation of pure saturated vapour on a injection spray by applying pulsed laser holography, Int. J. Heat Mass Transfer, Vol. 35 No. 3, pp 691-702, 1992.

Chen, Y.M., Mayinger, F., Nordmann, D., Heat Transfer at the phase-interface phenomena in Multiphase Flow. Eds.: Hewitt, G.F., et al. New York: Hemisphere, 1991, S.433-442.

Gabor, D., A New Microscopic Principle, Nature 161, 777 (1948), Microscopy by Reconstructed Wavefronts, Proc. Roy. Soc. A 197, 454 (1949), Microscopy by Reconstructed Wavefronts II, Proc. Phys. Soc. B 64, 449 (1951).

Geraets, J.M., and J.C. Int. J. Multiphase flow 14, No. 3 (1988) 305-320.

Gonzales, R.C., Wintz, P. (1977), Digital image processing, Addisson-Wesley, Massachussetts.

Haibel, M., Mayinger, F., Strube, G., Application of non-intrusive diagnostic methods to sub- and supersonic H_2-air combustion, 3rd intl. Symposium on special topics in chemical propulsion: non-intrusive combustion diagnostics, Scheveningen, NL, 1993.

Hanson, R.K., Seitzmann, J.M., Paul, P.H. (1990), Planar fluorescence imaging of combustion gases, Applied Physics, Vol. B. 50.

Kiemle, H. and Röss, D., Einführung in die Technik der Holographie, Akademische Verlagsgesellschaft, Frankfurt a.M. (1969).

Klug, F. and Mayinger, F., Proc. NURETH-5 Meeting, Salt Lake City, Sept. 21-24, 1992 (1992).

Klug, F., Mayinger, F., Impedance based flow reconstruction - A novel flow composition measuring technique for multi-phase-flows, Nuclear Engineering and Design, 146, 35-42 (1994).

Lighthart, G., Greon, C. (1982), A Comparison of different autofocus algorithms, IEEE Transactions, Vol. XX, pp 597-604.

Mayinger, F., Optical Measurements, Techniques and Application, Springer Verlag, Heidelberg, (1994).

Melton, L.A. and Verdieck, J.F., Vapour/Liquid Visualisation for Fuel Sprays, Combust. Sci. and Tech., 42, 217-222, (1985).

Nordmann, D. and Mayinger, F., Temperatur, Druck und Wärmetransport in der Umgebung kondensierender Blasen, VDI-Forschungsheft 605, 1981.

Özgü, M.R. and Chen, J.C., Rev. Sci. Instrum. 44, No. 12, (1973), 1714-1716.

Panknin, W., Mayinger, F., (1974), Holography in Heat and Mass Transfer, 5th Int. Heat Transfer Conference, VI, 28, Tokio.

Panknin, W., (1977), Eine holographische Zweiwellenlängen Interferometrie zur Messung überlagerter Temperatur- und Konzentrationsgrenzschichten, Diss. Universität Hannover.

Panknin, W., Mayinger, F., (1978), Anwendung der holographischen Zweiwellenlängen Interferometrie zur Messung überlagerter Temperatur- und Konzentrationsgrenzschichten, Verfahrenstechnik, Vol 12, No. 9, pp 582-589.

Pavlidis, T., (1982), Algorithms for graphics and image processing, Springer Verlag, Berlin.

Senda, J., Fukami, Y., Tanabe, Y., Fujimoto, H., Vizualization of Evaporative Diesel Spray Impinging Upon Wall Surface by Exciplex Fluorescence Method, Proc. of the Int. Congress and Exposition, Detroit, February 24-28, 1992, SAE.

CRITICAL HEAT FLUX MECHANISMS

Y. Katto

Department of Mechanical Engineering, Nihon University
Kanda-Surugadai, Chiyoda-ku
Tokyo 101, Japan

ABSTRACT

The gradually clarifying picture of the critical heat flux (CHF) is presented by explaining the outline of the CHF mechanism in fundamental boiling systems. First, the liberation from the long-lived restraint of the hydrodynamic instability CHF model is described. Then the CHF conditions in pool, external flow, and internal flow boiling are explained in a relatively unified manner as the limiting state of the macrolayer (or liquid sublayer) reducing the thickness or the mass flow rate with increasing heat flux, showing the need of the logically coordinative treatment of the Taylor and Helmholtz instabilities. Finally, considering the actual states on the heated surface near the CHF condition, the CHF is divided into two categories, i.e., the normal and the real CHF, and their relationship and characteristics are explained.

1 INTRODUCTION

Due to the complexity of the critical heat flux (CHF) phenomenon, complete clarification has not yet been offered on the CHF mechanism in the strict sense. However, the studies performed to date have developed our knowledge up to the stage offering a rather reliable perspective of the CHF mechanism. Hence, mainly based on the CHF studies during the last decade (see Katto 1994) and on the other associated studies, the gradually clarifying picture of the CHF in external and internal flow boiling is presented below.

In order to reserve the logical certainty as far as possible, and at the same time, to avoid the useless confusion, discussion is restricted to the CHF in fundamental boiling systems being expressable in the generalized form as far as possible. In addition, for the similar purpose, the author refers to the CHF in pool boiling as well, because the relationship (intimate or unintimate) between pool and flow boiling as to the CHF mechanism is one of the most important factors in clarifying the CHF phenomenon collectively.

2 CHF MECHANISMS

(1) Difficulties in the Study The difficulty encountered in the work of studying CHF mechanisms is considered as composed of the following items: (i) the difficulty in the idealized modeling of the two-phase fluid behavior at high heat fluxes near CHF, (ii) the difficulty involved in reliable measurements of the fluid state in the immediate vicinity of the heater surface (in general, an overall information on the whole area of the heater surface is required), and (iii) the difficulty coming from the overlooking of important key features associated with high-heat flux boiling and CHF, which is apt to form a quite different image of the CHF mechanism.

(2) Principles as the Basis of Discussion In dealing with the CHF mechanism under such complex conditions, it seems useful to take some steps so as to arrange the existing views of the CHF phenomenon in proper order. In the present paper, therefore, the following three principles are adopted:

<1> To deal with primary CHF models which may be of simple nature but are useful for throwing light on the framework of the CHF phenomenon. At the present stage of the CHF study, too much detailed analyses are not necessarily adequate because of the difficulty in examining the validity of many assumptions involved.

<2> To consider the CHF mechanisms not separately but rather synthetically through the various boiling modes, i.e., pool, external flow and internal flow boiling, in order to avoid the confused and incorrect views such as those in the case of the five blind men and the elephant.

<3> To adopt the convenient concept that CHF can be divided into 'normal' and 'real' CHF, as explained later in section 8.2, where the normal CHF is of hydrodynamic origin, while the real CHF is the CHF that is subject to the influence of the heater-side factors as well. In this paper, the normal CHF is discussed through sections 3 to 7, and the real CHF in section 8.

3 PIONEER STUDIES

A wide variety of ideas as to the CHF mechanisms in pool and flow boiling were cited in the review or similar papers published in the 1960s and 1970s (Gambill 1968a, 1968b; Macbeth 1968; Tong & Hewitt 1972; Bergles 1975, 1977, 1979; Hewitt 1978). Among those ideas, the following two models are particularly noticeable.

(1) The CHF Condition in High Quality Flow Boiling. The CHF condition found by Hewitt et al.(1965) for high quality flow boiling is regarded as the first (and perhaps the only one) clear result attained in the study of CHF mechanisms. It owes to the favorable situation such as that shown in Fig.1 (quoted from

Staniforth et al.1965), where the liquid film mass flow rate m can be measured with enough accuracy at the exit end of the heated channel as a function of heat flux q, suggesting the onset of CHF at the limiting state:

$$m \rightarrow 0$$

In other words, the difficulty to be encountered in the ordinary case of direct measurement of the CHF condition of m = 0 can be avoided by the foregoing limiting situation.

(2) The Hydrodynamic Instability CHF Model   After the success of Kutateladze (1952) in correlating the experimental CHF data for saturated pool boiling in a generalized form (cf. Fig.11 shown later) as

$$q_\infty/\varrho_G H_{fg})/[\sigma g(\varrho_L - \varrho_G)/\varrho_G^2]^{1/4} = \text{const.} \ (\approx 0.16) \qquad (1)$$

the hydrodynamic instability model, capable of deriving the above equation theoretically, was presented by Zuber (1959) for pool boiling on an infinite plate heater, and subsequently extended by Lienhard & Dhir (1973) to finite bodies (as to the history of this model, see Lienhard & Witte 1985, Lienhard 1988, and Lienhard 1994). Fig.2 quoted from Dhir & Lienhard (1974) represents the steady-state vapor escape passage (in the form of large and long vapor jets) assumed in this model.

(3) Solitary Nature of the Hydrodynamic Instability Model.   The above-mentioned hydrodynamic instability model is the first theoretical model for the CHF of DNB type, but has a nature of inapplicability to the analysis of CHF under convective flow or transient conditions. The CHF in the cylinder-crossflow system (cf. Fig.14) and that in the disk-jet system (cf. Fig.13a) were analyzed by Lienhard & Eichhorn (1976, 1979) with "mechanical energy stability criterion (MESC)". However, MESC, which depends on a scalar quantity balance of energy, does not seem to be equivalent to the dynamic criterion of the hydrodynamic instability at the vapor/liquid interface. Early in the 1980s, Hewitt (1982) wrote "it has become increasingly clear that pool boiling mechanisms have little direct relevance to burnout mechanism in forced convective systems".

4   ON THE HYDRODYNAMIC INSTABILITY MODEL

The hydrodynamic instability model has been established based on the Taylor (at the horizontal vapor/liquid interface) and the Helmholtz (at the vertical vapor-jet interface) instability conditions, which are written, respectively, as:

$$\lambda_{Td} = 2\pi\sqrt{3} \ [\sigma/g(\varrho_L - \varrho_G)]^{1/2} \qquad (2)$$

$$\lambda_H = 2\pi\sigma \ (\varrho_L + \varrho_G)/[(\varrho_L \varrho_G)(u_G - u_L)^2] \qquad (3)$$

In the case of Zuber's analysis for the 'infinite plate' in Fig.2, the Taylor wavelength λ_{Td} of equation (2) determines the side length of a square unit area, while the Helmholtz instability equation (3) is employed to determine the critical relative velocity $(u_G - u_L)$ between vapor and liquid flow perpendicular to the heater surface so as to give the critical value of heat flux q through the mass and energy balance relationship:

$$u_G - u_L = (q/\varrho_G H_{fg})(A_w/A_v)[1 + (\varrho_G/\varrho_L)/(A_w/A_v - 1)] \qquad (4)$$

where A_v is the vapor jet base area, and the fraction of A_v on the heater surface A_v/A_w is evaluated from the pitch and diameter of vapor jet on the infinite plate (Fig.2). Extension to finite bodies by Lienhard & Dhir (1973) was made by modifying the above-mentioned geometrical sizes under assumptions of $\varrho_G \ll \varrho_L$ and $u_L \ll u_G$ implicit in their analysis.

4.1   Postulations in the Instability Model   It is true that no complete models can exist at the present stage of the CHF study, but the following two debatable points in the hydro-

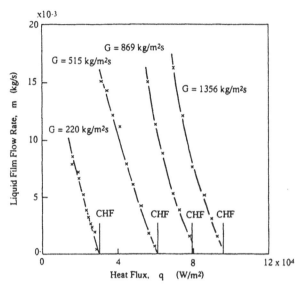

Fig.1   Liquid film mass flow rate vs. heat flux for various mass velocities (from Staniforth et al.1965).

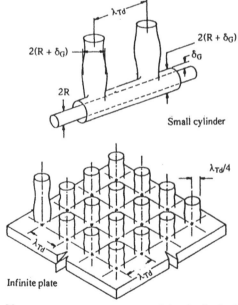

Fig.2   Vapor escape passage assumed in the hydrodynamic instability model (from Dhir & Lienhard 1974).

dynamic instability model are rather serious.

(1) No gravity effects on the vapor escape flow.   The cylindrical interface of the long vapor column assumed in the bulk region (see Fig.2) is subject to the pressure difference due to the weight difference between the vapor and the surrounding liquid, and hence, unless the sufficient pressure is preserved within the vapor column, the radial movement of liquid toward the vapor column cannot be prevented. According to the Zuber model, however, the thrust pressure exerted by the vapor jet on its base area is calculated to be only 1.1 mmAq at the CHF condition for water boiling at 1 atm, for example. Actual measurement of the pressure exerted on the high-heat flux heater surface has been conducted by Katto & Kikuchi (1972) with the apparatus of Fig.3, and it proves the pressure to be lower than the surrounding pressure at the heater surface level. These facts suggest the nonreality of the assumption of the steady-state

columnar vapor escape flow, and the solitary nature of this model [as mentioned in section 3-(3)] is considered to come from the postulation of such imaginary flow state.

(2) Logical inconsistency. The Taylor and Helmholtz instabilities are the phenomena of the same roots only having a difference in the acting forces, i.e., the gravity and the inertia. In the hydrodynamic instability model, however, equation (2) is used to determine the instability wavelength λ_{Td} for a given magnitude of g, whereas equation (3) is used to determine the relative velocity $(u_G - u_L)$, and moreover, this is done by employing the magnitude of λ_H that is determined elsewhere independent of the instability condition of equation (3). This is a strange situation because it means that the Helmholtz instability condition, equation (3), works only at one specific magnitude of $(u_G - u_L)$.

4.2 Vapor Escape Behavior under Gravity Effects
Discussion in section 4.1-(1) suggests the appearance of the periodic vapor escape behavior with radial liquid movement. In fact, Davidson & Schüler (1960) have observed that a pair of bubbles, in spherical and cylindrical shape, are formed successively from a continuous air flow discharged into water through an orifice. Katto & Otokuni (1994) have shown successive states of the vapor generated near the CHF condition on a horizontal and a vertical, uniformly heated rectangular heater (10x100 mm), respectively, observed through the simulation of air-water system for water boiling at 1 atm. In the case of the horizontal heater (Fig.4), one can notice the division of the heater surface into three unit areas due to the Taylor instability, and the successive formation of a roughly spherical vapor clot followed by a cylindrical one. Next, Fig.5 (from Haramura 1989a) represents the successive states of the vapor escape behavior near the CHF condition from a horizontal, thin cylindrical heater of 0.69 mm in diameter kept at a constant temperature with the help of computer. The same characteristics as in Fig.4 can be observed for the formation of vapor clots in Fig.5 also.

4.3 Mechanism of Vapor Clot Formation
As has been seen in Figs.4 and 5, the actual formation of vapor clots on the heater surface is fairly complicated. Accordingly its detailed theoretical analysis is difficult, but intelligible approximate explanations can be offered as follows;
<1>. A roughly spherical vapor clot appears first, then a cylindrical one follows (section 4.2); and this conjugate vapor clot formation is attributable to the circumstance that the preceding vapor clot is formed rather independently, while the behavior of the following vapor clot is strongly affected by the induced flow generated by the rising motion of the preceding vapor clot. [N.B. The term "vapor mushroom" or "mushroom shaped bubble" is often used recently in the study of CHF, but this is liable to lead to misunderstanding of the vapor clot formation].
<2>. The successive formation of vapor clots from the 'continuous' vapor generation on the heater surface is caused through the balance between the volume increase due to accumulation of vapor and the rising motion due to buoyancy. For a spherical vapor clot illustrated in the top figure of Fig.6, the equation of motion is written as (Katto & Yokoya 1976):

$$d\left[\{(11/16)\varrho_L + \varrho_G\}V(t)(ds/dt)\right]/dt = (\varrho_L - \varrho_G)V(t)g \quad (5)$$

where V(t) and s(t) are the instantaneous vapor clot volume and center height from the heater surface, respectively, at time t. In the case of V(t) = const.t (i.e., constant heat flux), for example, the vapor clot center height s(t) derived from equation (5), and the instantaneous vapor clot radius $r(t) = [3V(t)/4\pi]^{1/3}$, change with time as represented in Fig.6.

The two lines s(t) and r(t) intersect at the point C,

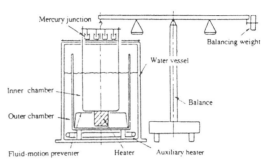

Fig.3   Experimental apparatus to measure the pressure on the heater surface in pool boiling (from Katto & Kikuchi 1972).

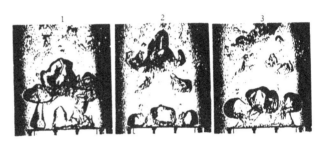

Fig.4   Vapor escape configurations at interval of 16.7 ms near the CHF condition on a rectangular horizontal surface (10 x 100 mm) (from Katto & Otokuni 1994).

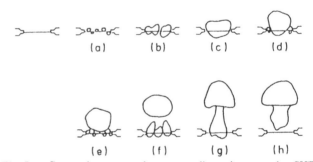

Fig.5   Successive states of vapor configuration near the CHF condition in pool boiling on a 0.69-mm dia, horizontal heater kept at a constant temperature (from Haramura 1989a).

when the separation from the heater surface occurs; and accordingly, the time τ shown in Fig.6 is called the hovering period (or residence time) of the vapor clot on the heater surface. [N.B. Equation (5) suggests that the increase of g is followed by the increase of s(t) in magnitude, leading to the decrease of the hovering period τ (see Fig.6)].

5   THE MACROLAYER DRYOUT MODEL

5.1 Framework of Macrolayer Dryout Model
The macrolayer dryout model illustrated in Fig.7(a) for pool boiling has been made on the basis of (i) the successive formation of vapor clots mentioned in section 4.2, and (ii) the existence of the macrolayer underneath the vapor clot confirmed by various methods (Katto & Yokoya 1968, Yu & Mesler 1977, and the experiments mentioned later in section 5.2). If the inflow of bulk liquid into the macrolayer is prevented by the overlying and neighboring vapor clots, then the instantaneous thickness of the macrolayer δ reduces with time due to boiling or evaporation, and as heat flux q is increased, it brings about the limiting state:

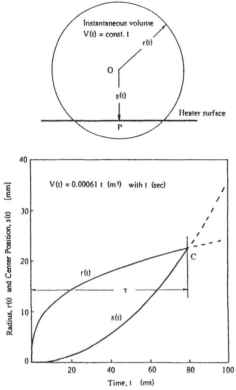

Fig.6   A simple model to analyze the behavior of a vapor clot growing and rising in liquid pool, together with an example of the analytical result.

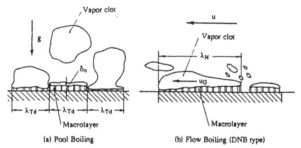

Fig.7   Behavior of the vapor clot and macrolayer assumed in the macrolayer dryout model.

$$\delta \rightarrow 0$$

at the end of the hovering period τ, when CHF occurs. Then, the critical heat flux for saturated boiling q_{co} is predicted by

$$q_{co} = j_L \, \delta_0 \, \varrho_L \, H_{fg} / \tau \qquad (6)$$

where j_L is the liquid volume fraction in the macrolayer (which has numerous vapor escape passages in it as mentioned later in section 5.3-(3)), and δ_0 is the initial thickness of macrolayer. [N.B. If equation (5) is used to evaluate τ in case of the infinite plate, V(t) at $q = q_{co}$ is written as $V(t) = (q_{co}/ \varrho_G H_{fg}) \, \lambda_{Td}^2 \cdot t$ with λ_{Td} given by equation (2), and thereby equation (6) yields the result in the form of equation (1) (see Haramura & Katto 1983)].

5.2 Fully-Developed Nucleate Boiling

(1) Characteristics of fully-developed nucleate boiling heat transfer. As has already been mentioned elsewhere (Katto 1992), nucleate boiling heat transfer at high heat flux is almost

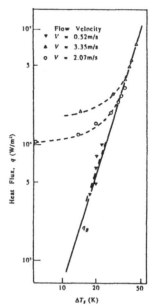

Fig.8   Flow boiling heat transfer in a tube for different flow velocities (from Bergles & Rohsenow 1964).

insensitive to various external conditions, such as heater orientation, gravity, bulk liquid flow state, and many others. As an example, Fig.8 quoted from Bergles & Rohsenow (1964) shows that if heat flux is increased sufficiently, the effects of forced convection on the heat transfer in tubes disappears, and consequently, forced flow boiling curves coincide approximately with the extrapolated line of the pool boiling curve q_B. These facts suggest that the macrolayer is characterized by the self-regulating nature dominating the foregoing characteristic heat transfer in fully-developed nucleate boiling.

(2) Relations of CHF between pool and forced flow boiling. Due to the self-regulating nature mentioned above, it is presumed that the macrolayer illustrated in Fig.7(a) can appear in other types of boiling at high heat fluxes also preserving nearly the same capacity as that observed in pool boiling. Accordingly, the solitary nature of CHF in pool boiling mentioned in section 3.3 is now removed, making it possible to assume the existence of relationship between pool and flow boiling in such a way as that illustrated in Fig.7(a) and (b), where λ_{Td} and λ_H are the Taylor and Helmholtz instability wavelengths given by equations (2) and (3), respectively.

5.3 Initial (or Maximum) Thickness of Macrolayer

(1) Experimental data. Considering the complicated behavior of vapor clots (see Figs.4 and 5) as well as the possible effects of bulk liquid entry into the macrolayer from the circumference of a finite heater employed usually in experiments, it poses some problems as to how to define and measure the initial thickness of macrolayer δ_0. Nevertheless, Fig.9 shows a rather orderly nature in the existing experimental data of the macrolayer thickness δ_0 measured in pool boiling of water at 1 atm on disk heaters by Gaertner (1965), Bhat et al.(1986), Shoji & Kuroki (1994) as well as on a thin cylindrical heater by Haramura (1989b). Meanwhile, Fig.10 represents the data of δ_0 measured by Rajvanshi et al.(1992) for pool boiling of six different fluids at 1 atm on a disk heater. Comparison of the above-mentioned data with a generalized correlation, equation (8), will be described later.

(2) The macrolayer formation mechanism. Several studies associated with the macrolayer formation mechanism

have so far been published (Haramura & Katto 1983, Bhat et al.1983, Chappidi et al.1991, Sadasivan et al.1992, Kumada & Sakashita 1992, and Shoji & Kuroki 1994), but many unsolved problems or unknown matters still remain. In these studies, however, two generalized correlations of δ_0 have been presented: one derived by Kumada & Sakashita (1992) through a semi-empirical analysis based on the lateral bubble coalescence mechanism on the heater surface, and another by Haramura & Katto (1983) considering the Helmholtz instability at the vapor/liquid interface of very small vapor stems involved in the macrolayer. The former correlation has a quite involved form, and includes the effects of g being inconsistent with the insensitivity to gravity as described in section 5.2-(1). Meanwhile, Haramura & Katto's (1983) process of deriving the generalized correlation is noteworthy in dealing with the Helmholtz instability equation (3) based on the principle different from that of the hydrodynamic instability model [section 4.1-(2)] as explained below.

(3) The correlation of δ_0 derived from the vapor stem stability. Equations (3) and (4) are applicable to the two-phase counterflow along the interface of the vapor stem involved in the macrolayer as well (Haramura & Katto 1983), which gives the Helmholtz instability wavelength λ_H for 'any' magnitude of heat flux q. Then it is presumable that the vapor stem (which bottom end is fixed firmly at a nucleation site on the heater surface) can maintain 'stable' condition if the macrolayer thickness δ_0 (that is equal to the height of the vapor stem) is shorter than λ_H at least. Consequently, the initial (or maximum) thickness of the macrolayer δ_0 is written as:

$$\delta_0 = \xi \lambda_H \qquad (7)$$

where ξ is a constant less than unity. Now substituting λ_H of equation (3) [along with equation (4)] into λ_H in equation (7), and considering the fraction of the vapor stem base area on the heater surface $A_v/A_w \ll 1$ in equation (4) for the highly wettable surfaces (cf. section 8.2), it yields

$$\delta_0 = \xi \lambda_H = \xi \, 2\pi\sigma[(\varrho_L + \varrho_G)/\varrho_L\varrho_G](A_v/A_w)^2(\varrho_G H_{fg}/q)^2 \quad (8)$$

Then, Haramura & Katto (1983) have assumed $\xi = 1/4$ tentatively, and determined A_v/A_w to be $A_v/A_w = 0.0584(\varrho_G/\varrho_L)^{0.2}$ so that q_{∞} of equation (6) (where the relation of $j_L = 1 - A_v/A_w$ is assumed because of the existence of vapor stems) can agree with q_{∞} of equation (1) with const. = 0.131 for the CHF on the infinite plate heater. Later, Rajvanshi et al,(1992) reported that equation (8) with $\xi = 1/2$ (instead of 1/4) agree with their experimental data of δ_0 for various kinds of fluids as shown in Fig.10. The broken line in Fig.9 is also the prediction of equation (8) with $\xi = 1/2$ showing a fairly good agreement with the experimental data.

6 THE CHF IN EXTERNAL FLOW BOILING

6.1 Mechanism In View of Dimensional Analysis

(1) Dimensional analysis. The normal CHF [see section 2-(2)] in saturated boiling is a phenomenon associated with the hydrodynamic states, which can be represented in terms of the following quantities, (i) velocity: the vapor velocity $q_{co}/\varrho_G H_{fg}$ and the liquid velocity u; (ii) force: the inertial forces (ϱ_G and ϱ_L), the viscous forces (μ_G and μ_L), the surface tension σ, and the buoyancy $g(\varrho_L - \varrho_G)$; and (iii) length: the heater length L. Then, the dimensional analysis of these quantities yields a general relationship as follows:

$$q_{co}/\varrho_G H_{fg} u = f(\, \varrho_G/\varrho_L, \; \varrho_L u^2 L/\sigma, \; g(\varrho_L - \varrho_G)L/\varrho_L u^2,$$
$$\mu_G/\mu_L, \; \varrho_L uL/\mu_L) \qquad (9)$$

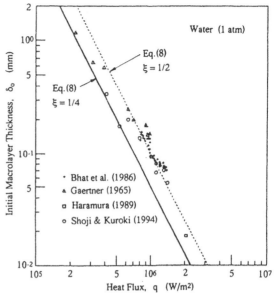

Fig.9 Initial macrolayer thickness measured in pool boiling of water at 1 atm, with the prediction of equation (8).

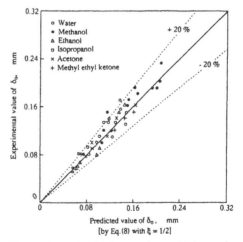

Fig.10 Comparison of initial macrolayer thickness δ_0 measured for various fluids with the prediction of equation (8) (from Rajvanshi et al.1982).

From equation (9), the following several specific expressions can be readily derived :

<1>. Pool boiling: by eliminating u among dimensionless groups in equation (9)

$$(q_{co}/\varrho_G H_{fg})/[\sigma g(\varrho_L - \varrho_G)/\varrho_G^2]^{1/4} = f(\, \varrho_G/\varrho_L, \; L',$$
$$[\sigma/g(\varrho_L - \varrho_G)]^{1/2}/(\mu_L^2/\sigma\varrho_L)) \qquad (9\text{-}1)$$

<2>. External flow boiling: by eliminating the term related to g in equation (9)

$$q_{co}/G H_{fg} = f(\, \varrho_G/\varrho_L, \; G^2L/\sigma\varrho_L, \; \mu_G/\mu_L, \; uL/(\mu_L/\varrho_L)) \qquad (9\text{-}2)$$

<3>. Intermediate region between 1 and 2: Equation (9) can be rewritten as follows:

$$(q_{co}/\varrho_G H_{fg})/[\sigma g(\varrho_L - \varrho_G)/\varrho_G^2]^{1/4} = f(\, \varrho_G/\mu_L, \; L', \; u') \qquad (9\text{-}3a)$$

or

$$q_{\infty}/GH_{fg} = f(\varrho_G/\varrho_L, \quad G^2L/\sigma\varrho_L,$$
$$(1 - \varrho_G/\varrho_L)/(gL\varrho_L^2/G^2)) \qquad (9\text{-}3b)$$

where $L' = L/[\sigma/g(\varrho_L - \varrho_G)]^{1/2}$, $u' = u/[\sigma g(\varrho_L - \varrho_G)/\varrho_G^2]^{1/4}$, and $G = \varrho_L u$. In the above two expressions, the effects of viscosities have been eliminated , for the sake of simplicity,

<u>(2) Comparison of equation (9) with empirical CHF correlations.</u> First, with respect to equation (9-1) for the CHF in pool boiling, the famous result of Fig.11 obtained by Borishanskii (1956) for water and five organic fluids show that the effects of viscosities are minor for ordinary fluids. Then, under this condition of viscosity, the existing empirical correlations such as equation (1) and those of Lienhard & Dhir (1973) for finite bodies, all take the form of equation (9-1). Second, with respect to the external flow boiling, the existing empirical correlations (Lienhard & Eichhorn 1986, Sadasivan & Lienhard 1987, Katto et al.1987 for the cylinder-crossflow system; Sharan & Lienhard 1985, Katto & Yokoya 1988 for the disk-jet system; and others) all take the form of either equation (9-2) or equation (9-3b) ignoring the viscosity term.

The above-mentioned empirical facts suggest that (i) the normal CHF in pool and external flow boiling is certainly of hydrodynamic origin, and that (ii) the effects of viscosity on CHF can be ignored approximately for ordinary fluids.

6.2 Boiling Systems With Simple Liquid Supply Conditions

Fig.12 represents the flow configuration observed by Katto & Ishii (1978) for boiling at 1 atm and 3.38 MW/m^2 near CHF on a rectangular heater with a plane jet of saturated water flowing out of a thin rectangular nozzle. In this case, a sheet-like liquid flow is formed being separated from the macrolayer maintained on the heater surface. In the two axially symmetrical boiling systems of Fig.13 (a) and (b) (for which experimental studies have been conducted by Monde & Katto 1978, Katto & Shimizu 1979, and many others on the former, and by Inamura 1982 on the latter), liquid flow separation from the macrolayer occurs with the same mechanism as before, but the splash of liquid appears because of the radially spreading flow.

As for saturated boiling on a cylindrical heater in a cross flow, Vliet & Leppart (1964) found the formation of a vapor cavity in the wake region (see Fig.14), and Sadasivan & Lienhard (1991) reported the state of Fig.15 for the vapor escape pattern along the curved heater surface, suggesting that the heater surface is exposed to liquid flow in a restricted narrow region along the stagnation line .

6.3 Approximate Analyses of the CHF

(1) A simple model making no use of empirical constants. In order to throw light on the CHF mechanism in the boiling systems mentioned in section 6.2, let us consider a quite simple liquid supply model making no use of empirical constants, based on the following three simplifications:

<1>. The macrolayer maintained on the heater surface has the thickness δ_0 of equation (8) at the upstream end due to the self-regulating nature of the macrolayer [see section 5.2-(1)].

<2>. The macrolayer is provided with liquid at the upstream end with velocity u (i.e., the nozzle exit velocity in Figs.12 and 13, and the approaching velocity of liquid in Fig.14).

<3>. Except the case of Fig.13 (b) where the gravity effects help the liquid film to flow along the heater surface, the heater length is sufficiently small, so that the liquid entering the macrolayer can wet readily over the heater surface. [N.B. Most of the available data of CHF in external flow boiling were obtained for boiling on comparatively small heaters].

Then, as heat flux q is increased under these conditions, it leads to the limit condition of heat flux (i.e., CHF) as:

$$q_{co} = \Pi \, \delta_0 u \, \varrho_L \, H_{fg}/A_w \qquad (10)$$

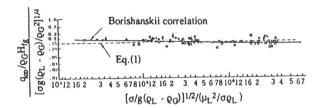

Fig.11   Correlation of CHF data for various fluids in pool boiling (from Borishansky 1956).

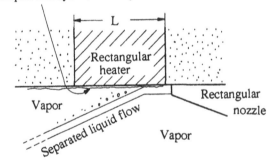

(a) Experimental observation (q = 3.38 x 10^6 W/m^2)

Fig.12 Boiling at 3.38 MW/m^2 on the rectangular heater of L = 10 mm with a plane jet of saturated water flowing out of a thin rectangular nozzle (from Katto & Ishii 1978).

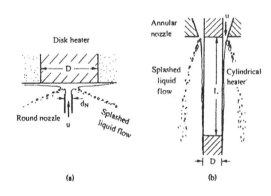

Fig.13 High heat flux boiling near CHF on heaters with a saturated liquid jet and their flow configurations: (a) a downward facing disk heater; (b) a vertical cylindrical heater.

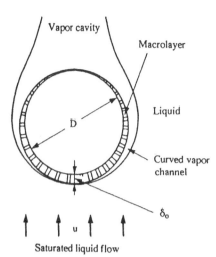

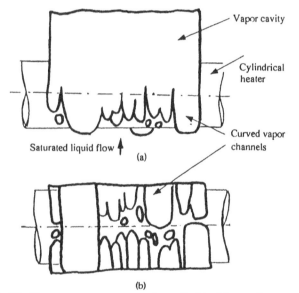

Fig.14 High heat flux boiling near CHF condition on a cylindrical heater in a cross flow of saturated liquid.

Fig.15 Vapor escape pattern on the cylindrical heater in a cross flow: (a) elevation view; (b) bottom view (from Sadasivan & Lienhard 1991).

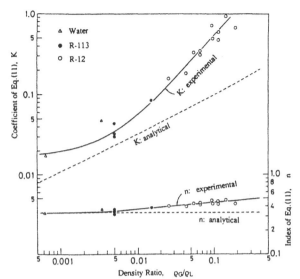

Fig.16 Correlation of CHF data in the disk-jet system of Fig.13(a) compared with the prediction of equation (11) in the wide range of ϱ_G/ϱ_L (from Katto & Yokoya 1988).

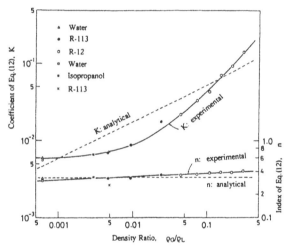

Fig.17 Correlation of CHF data in the cylinder-crossflow system of FIg.14 compared with the prediction of equation (12) in the wide range of ϱ_G/ϱ_L (from Katto et al. 1987).

where Π is the border length at the upstream end of the macrolayer, and A_w is the heater surface area.

    (2) Comparison of equation (10) with experimental data. CHF values predicted by equation (10) are compared with experimental results for two typical cases: Fig.13(a) and Fig.14.

    First, as for the disk-jet system of Fig.13(a), equations (10) and (8) with $\xi = 1/4$ result in

$$q_{co}/\varrho_L u H_{fg} = K \left[\sigma / \{ \varrho_L u^2 (D - d_N)(1 + D/d_N) \} \right]^n \qquad (11)$$

with $K = 0.278 (\varrho_G/\varrho_L)^{0.467} [1 + (\varrho_G/\varrho_L)]^{1/3}$, and $n = 1/3$. These analytical values of K and n are represented by broken lines in Fig.16, respectively. Meanwhile, Katto & Yokoya (1988) examined the available experimental q_{co} data, providing empirical correlations of K and n as represented by solid lines in Fig.16.

    Next, as for the cylinder-crossflow system of Fig.14, equations (10) and (8) with $\xi = 1/4$ yield

$$q_{co}/\varrho_L u H_{fg} = K \ (\sigma/\varrho_L u^2 D)^n \qquad (12)$$

with $K = 0.151 (\varrho_G/\varrho_L)^{0.467} [1 + (\varrho_G/\varrho_L)]^{1/3}$, and $n = 1/3$. Similarly to before, these analytical values of K and n are compared with the experimental correlations of K and n (see Katto et al.1987) in Fig.17.

    The CHF models assumed above are quite simple, so the noticeable difference between the analytical and experimental values of K and n in Figs.16 and 17 is unavoidable. However, it is also true that these CHF models, though employing no empirical constants, can outline the characteristic trend of K and n over a wide range of ϱ_G/ϱ_L (see Figs.16 and 17); and this seems to suggest at least that the foregoing simple model be the first approximation of the CHF concerned.

6.4 CHF in Other Boiling Systems There are many other external flow boiling systems, of course, a few of which are noted below.

<1>.   If the heater surface is upward-facing in Fig.12 and Fig.13(a), the depletion of the macrolayer due to evaporation is replenished under certain conditions by either the deposition of droplets from the separated liquid flow or the reattachment of the separated liquid flow to the heater surface.   Bain et al.(1984) attempted an analysis of the CHF in such a boiling system.

<2>.   When a heater surface is flush-mounted on a wall, along which bulk liquid flows (see Fig.7b), the liquid and vapor behavior is not simple, and so it is rather difficult to analyze the detailed mechanism of CHF. Experimental studies in such boiling systems have been conducted by Katto & Kurata (1980), Yagov & Puzin (1984), and Mudawar & Maddox (1989).

<3>.   As for the cylinder-crossflow system, if the cylinder diameter is fairly large, the fluid behavior may be considerably different from the simple pattern assumed in section 6.3-(1). Meyer et al.(1986) conducted an experimental study of the CHF on 25.4-mm. dia. horizontal cylinder placed perpendicularly to a horizontal uniform liquid flow.

7   THE CHF IN INTERNAL FLOW BOILING

7.1 Generalized Correlations of the CHF in Tubes
In the 1960s and 70s, various dimensional, empirical CHF correlations for water flow boiling were presented, and at the same time, studies were conducted on the laws of relating the CHF values for a prototype fluid to those for a model fluid. Then, in the subsequent decade (the 1980s), the three different types of "generalized" CHF correlations were presented as follows, covering wide ranges of the fluid and other conditions:
<1>.Tabular type correlation: developed by Groeneveld et al. (1986) with a table listing the values of $q_{c,8}/GH_{fg}$ as a function of ϱ_G/ϱ_L, $(G^2 d/\sigma \varrho_L)^{1/2}$, and x as:

$$q_{c,8}/GH_{fg} = f(\varrho_G/\varrho_L, \ \sqrt{G^2 d/\sigma \varrho_L}, \ x)$$

where $q_{c,8}$ is the CHF value for 8-mm dia. tube, and x is the local quality at the CHF condition. For diameter d other than 8 mm, the empirical relationship: $q_{c,d} = q_{c,8}(d/8)^{1/3}$ is assumed.
<2>.Graphic type correlation: developed by Shah (1987) with three graphs representing q_c/GH_{fg} as a function of Y, L/d, and x_i (for the region called UCC), or a function of Y, L/d, p_r and x (for the region called LCC), where

$$Y = (Gdc_{pL}/k_L)(\varrho_L^2 gd/G^2)^{-0.4} (\mu_L/\mu_G)^{0.6},$$

p_r is the reduced pressure, x_i is the inlet quality, and x is the local quality at the CHF condition.
<3>. Formula type correlation: developed by Katto & Ohno (1984) employing five equations in the form of

$$q_c/GH_{fg} = f(\varrho_G/\varrho_L, \ \sigma \varrho_L/G^2 L, \ L/d, \ \Delta H_i/H_{fg}),$$

respectively, where ΔH_i is the inlet subcooling enthalpy. This correlation is characterized by the positive use of Weber number $G^2 L/\sigma \varrho_L$ based on the tube length L instead of $G^2 d/\sigma \varrho_L$ based on the tube diameter d.

Among the foregoing three correlations, correlations 1 and 3 take the effects of σ(surface tension) into consideration, but neglect the effects of μ_L and μ_G (viscosity). On the contrary, correlation 2 considers μ_L and μ_G, but neglect the effects of σ. Hence, it is difficult to derive something useful from these generalized correlations to throw light on the CHF mechanism; but the foregoing success of correlating CHF data seems to suggest that the CHF in internal flow boiling in tubes is not so much confused as rejecting the analysis of the mechanism.

7.2 CHF in High Quality (Annular Flow) Region
(1) The CHF mechanism. As has been mentioned in section 3-(1), the CHF condition of the liquid film mass flow rate

$$m \rightarrow 0$$

and the physical process to attain this limiting state at the exit end of the heated tube was disclosed in the 1960s. Then the analytical procedure to predict the CHF value in a round tube was developed in the 1970s by Whalley et al.(1974), Würtz (1978), Saito et al.(1978), Levy et al.(1980), and others (cf. Hewitt 1978). The main mechanism here is the gradual decrease of the liquid film flow along the tube length due to the evaporation and the droplet entrainment, being delayed by the replenishment of the liquid film flow by the droplet deposition from the vapor core flow. When heat flux is high, the suppression of the droplet deposition due to vapor effusion from the liquid film, and the increment of the entrainment due to the bursting of bubbles in the liquid film can occur (as for the recent study of these items, see Milashenko et al.1989).

(2) The initial liquid film thickness at the starting point of annular flow. In the foregoing CHF prediction model, the calculation of the gradual decrease of liquid film flow along the tube length is started from the location at which the annular flow appears in the tube, where the liquid film has an initial thickness δ_1. This location is usually indicated in terms of certain critical values of void fraction ε_1, quality x_1, and some other adequate quantities associated with the onset of annular flow.

The CHF values predicted in such a procedure are represented by thick solid lines in Fig.18, showing that they begin to deviate noticeably from the experimental data as the tube length L_B decreases. The reduction of tube length results in the increase of CHF under fixed other conditions (see Fig.18), and it leads to the situation that the critical thickness of macrolayer δ_o of equation (8) is less than the above-mentioned δ_1 at the starting point of annular flow. In Fig.18, the broken line represents the CHF value predicted by replacing the initial liquid film thickness δ_1 by δ_o of equation (8); and Sugawara (1990) has adopted this concept into consideration in his CHF prediction code.

(3) A supplementary note. Fig.19 illustrates schematically the countercurrent two-phase flow boiling near the CHF condition in a uniformly heated vertical tube, closed at the bottom end, and open at the top end to an upper saturated liquid reservoir.   In this case, the following steady-state condition holds at any heat flux q lower than the CHF value:

$$m_G = m_L = \pi d \cdot z \cdot q / H_{fg} \qquad (13)$$

where m_G and m_L are the mass flow rates of vapor and liquid, respectively, through the cross-section at distance z from the bottom end. Both magnitudes of m_G and m_L change in proportion to z (from z = 0 to z = L), and the flow configuration such as illustrated in Fig.19 appears with the minimum liquid film thickness somewhere half way down the tube (Katto & Hirao 1991). If q is kept slightly higher than the limit condition of the steady-state countercurrent two-phase flow at the top end (Katto et al.1994), then the condition of equation (13) is no longer preserved, causing the repetition of the deficiency and recovery of the inflow of liquid at the top end, and thereby the rather slow, irregular up and down change of the wall temperature begins and continues near the point D in Fig.19. In other words, the CHF in this boiling system have quite different characteristics from those of the ordinary one.

7.3 CHF in Subcooled Flow Boiling
(1) CHF condition based on macrolayer dryout. For subcooled flow boiling in a uniformly heated tube, if heat flux is increased to high values near CHF, the state such as that illus-

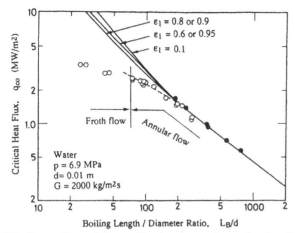

Fig.18 Comparison of the predicted and experimental value for the CHF in saturated flow boiling in a tube (Katto 1984).

Fig.19 Flow configuration near CHF condition in a uniformly heated vertical tube closed at the bottom end.

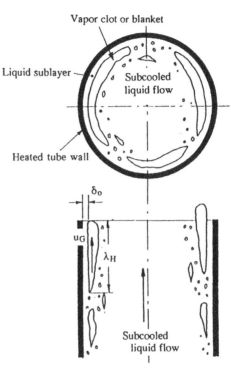

Fig.20 Flow configuration near CHF condition in subcooled flow boiling in tubes (from Katto 1990a).

[2] Following the same principle as that of the preceding Lee & Mudawar's study, Katto (1990a,b) developed a generalized CHF model applicable to not only water but also nonaqueous fluids (water, nitrogen, helium, R-11, R-12, and R-113). Then Katto (1992a) extended this model so as to cover the CHF of water boiling at low pressures also.
[3] Celata et al.(1994) developed an interesting model, making no use of empirical constants, yet being capable of predicting the CHF of water boiling in comparatively low pressure region. The next table compares the covering ranges of these models:

Model	Fluid	ϱ_G/ϱ_L	Subcooling	Void fraction
[1]	water	0.029 -0.231	$\Delta T_{sub} < 59$ K	$\varepsilon < 0.7$
[2]	general	0.0006-0.410	$\Delta T_{sub} < 117.5$K	$\varepsilon < 0.7$
[3]	water	0.0006- 0.063	$\Delta T_{sub} < 225$ K	$(25K < \Delta T_{sub})$

In addition, Fig.21 shows RMS errors of predicted CHF for the Celata (ENEA) and Katto models (quoted from Celata et al.1994)

(3) Similar and dissimilar points between the models.
In order to make equation (14) effective, it is necessary to know the magnitudes of λ_H, u_G, δ_o, and H_{fg}', respectively, and how to evaluate these values is an important problem being associated with the physical structure of each model.
<1>.Evaluation of λ_H and u_G: In all the foregoing three models, the vapor clot length (see the lower figure of Fig.20) is assumed to be given by the Helmholtz instability wavelength λ_H at the interface facing to the liquid sublayer, which is calculated by equation (3) under the assumption of $u_G \gg u_L$ with respect to the respective movements of the vapor clot and macrolayer, where the vapor clot velocity u_G is evaluated by considering the velocity distribution of the main stream in the tube under the assumption of homogeneous flow. Meanwhile, as to u_G, roughly speaking, the models [1] and [3] assume

$$u_G = U_{Lc} \qquad (15)$$

trated schematically in Fig.20 presumably appear near the tube exit end with thin vapor clot or blanket in the vicinity of the heated tube wall surface, holding a liquid sublayer (or macrolayer) between the vapor clot and the wall surface. Formally, this situation is similar to that of Fig.7(a), provided that the hovering period τ mentioned in section 5.1 is replaced by the passage time of the vapor clot τ_G during which the liquid sublayer is blanketed by the passing vapor clot; and if so, after equation (6) in pool boiling (putting $j_L = 1$ for simplicity), the CHF q_c is now predicted by

$$q_c = \delta_o \, \varrho_L \, H_{fg}'/\tau_G, \text{ with } \tau_G = \lambda_H/u_G \qquad (14)$$

where δ_o is the maximum thickness of the liquid sublayer, λ_H and u_G are the length and velocity of the vapor clot, respectively (see Fig.20), and H_{fg}' is the modified latent heat of evaporation including the sensible heating effect of subcooled liquid [see equation (19) mentioned later].
(2) Analytical models based on equation (14). Based on the concept of equation (14), the following somewhat different kinds of analytical CHF models have recently been proposed for subcooled flow boiling:
[1] Lee & Mudawar (1988) are the first in developing a mechanistic CHF model for subcooled water flow boiling based on the above-mentioned mechanism. Subsequently this model was implemented by Lin et al.(1989) with more theoretical basis for subcooled and low quality boiling.

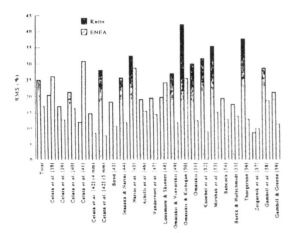

Fig.21   RMS errors of predicted CHF vs. single data set, for Celata (ENEA) and Katto models (from Celata et al.1994). The numbers in square blankets are the reference numbers used by Celata et al.(1994).

where U_{Lc} is the local velocity of the main stream at the centerline of the vapor clot. [N.B. The effects of gravity on the vapor clot motion can be neglected at high mass velocities]. Meanwhile, considering the complexity of the actual vapor clot behavior in the subcooled flow, the model [2] adopts the following expression:

$$u_G = k_v U_{L\delta} \qquad (16)$$

where $U_{L\delta}$ is the local velocity of the main stream at the level of the interface facing to the liquid sublayer, and k_v is the velocity coefficient correlated empirically in the following form:

$$k_v = f(Re, \varrho_G/\varrho_L, \varepsilon) \qquad (17)$$

where Re and ε are the Reynolds number and void fraction of the main steam, respectively.

<2>.Evaluation of δ_o: The above three models are quite different with each other in the procedure to evaluate δ_o. In the model [1], δ_o is determined based on the mechanistic analysis of the balance of forces exerting on the vapor clot. In the model [2], δ_o of equation (8) in pool boiling [section 5.3-(3)]:

$$\delta_o = \xi \, 2\pi\sigma[(\varrho_L + \varrho_G)/\varrho_L \varrho_G](A_v/A_w)^2(\varrho_G H_{fg}/q)^2$$

with $\xi = 1/4$ is used, only replacing q by q_e that is defined as the part of q used by evaporation [see equation (19) mentioned later]. Meanwhile. the model [3] considers the temperature distribution of the main stream in the tube under the assumption of homogeneous flow, determining the thickness y^* of the superheated layer, beyond which vapor clot cannot develop due to subcooling. Then, an available formula is used to evaluate the thickness of vapor clot D_b, when δ_o can be determined as:

$$\delta_o = y^* - D_b \qquad (18)$$

<3>.Treatment of conjugate heat flux: In subcooled flow boiling, heat flux q is formally expressed as

$$q = q_e + q_{sen} \qquad (19)$$

where q_e is the part of q used by latent heat of evaporation, and q_{sen} is that used by sensible energy heating. Then, in the model [1], H_{fg}' (i.e., latent heat of evaporation modified by considering q_{sen}) is used in equation (14), while in the model

[2], q_c of equation (14) is evaluated with H_{fg} instead of H_{fg}', then q_c thus evaluated is added to q_{sen}, yielding the magnitude of CHF q_c. On the other hand, in the model [3], q_c evaluated from equation (14) with H_{fg} instead of H_{fg}' is regarded as the CHF itself, because this model stands on the assumption that the existence of the vapor clot and the underlying liquid sublayer are restricted in the region of the superheated liquid layer.

<4>.On the Logical limitation of applicable range: In all the three models, the main stream in the tube is assumed to be homogeneous flow, and accordingly, the models [1] and [2] impose the limitation of void fraction $\varepsilon < 0.7$ as shown in the foregoing table in section 7.3-(2). Meanwhile, the model [3] does not impose this limit, but has the limit of $\Delta T_{sub} > 25K$ in order to avoid the excessive increase of the liquid sublayer thickness δ_o given by equation (18), where y^* increases up to the tube radius with decreasing ΔT_{sub}, while D_b does not change in magnitude.

As is clear from the above comparisons, three models [1], [2], and [3] differ delicately with each other, but it is of interest to note that they all accept equation (14) of the macrolayer dryout mechanism as the basis of CHF.

8   THE NORMAL AND REAL CHF

8.1 Actual States Near the Limit Condition
The CHF conditions defined as the limiting state such as m → 0 or δ → 0 have so far been explained for pool, external flow, and internal flow boiling, respectively.   But one must not forget that these are the 'idealized' CHF conditions assuming a 'uniform' reduction of the mass flow rate m or the thickness δ of the liquid sublayer.   Actually, however, such uniformity is deteriorated when the magnitude of m or δ becomes very small, leading to the appearance of dry spots and patches distributed irregularly over the heater surface; and this can have the connection with such the observed states as follow:
<1>.   As is well known, the surface temperature fluctuation appears near the CHF point (see Nishikawa et al.1972).
<2>.   Dry spots and patches are observed on the heater surface in pool boiling at the final stage of the macrolayer dryout (see Kirby & Westwater 1965, Katto & Yokoya 1968).
<3>.   In the high quality annular flow boiling in a tube, the CHF occurs at the film flow rate m > 0 instead of m = 0 (see Milashenko et al. 1989).

8.2 Two Categories of CHF
The characteristic situation of the CHF mentioned in section 8.1 suggests the existence of another field to be discussed, where the CHF is dominated not only by the hydrodynamic conditions but also by the heater-side factors. In this case, however, if the heater has sufficiently high values in wettability, thermal conductivity, etc., then the disorder and nonuniformity mentioned in the preceding section is eased or attenuated so that the influence of the heater-side factors on the CHF value can be very little.   Of course, systematic studies on this problem have not yet been performed enough, so it is difficult to draw a decisive final conclusion at the present stage; but it seems convenient at least to divide the CHF into two categories, i.e., the normal CHF and the real CHF, as follows:

(1) The normal CHF: This is defined as the idealized CHF of hydrodynamic origin, but nevertheless, very useful practically because it can be approximated by the CHF appearing on the heater wall with high wettability, high thermal conductivity, and sufficient wall thickness, etc. This is just a situation similar to that of the black body in thermal radiation.

(2) The real CHF: This is the CHF appearing under the influence of the hydrodynamic behavior as well as that of the heater-side factors.   In respect to the heater-side factors, the

'inside factors' and the 'surface factors' has quite different modes in influencing the CHF as explained below.

8.3 Heater Inside-Factor Effects The inside factors of the heater wall such as the wall thermal properties and thickness can exert their influence only through the unsteady-state heat conduction within the wall.

(1) Experimental information for pool boiling. Since very early days around 1960, not a few studies have been conducted on the heater inside factor effects; and very recently, relating to the high-level cooling problem of microelectronic components, active studies are still more going on by Bar-Cohen & McNeil (1992), Golobic & Bergles (1992), Carvalho & Bergles (1992), Watwe & Bar-Cohen (1994), and others. Fig.22 represents the comparatively old data correlated by Bar-Cohen & McNeil (1992) and Watwe & Bar-Cohen (1994), where δ_w is the wall thickness [m], and ϱ, c, and k are the density [kg/m^3], specific heat [J/kgK], and thermal conductivity [J/smK] of the wall material, respectively. Carvalho & Bergles (1992) performed thorough analyses of data obtained under various conditions suggesting that $\delta_w(\varrho ck)^{1/2}$ be the best parameter to correlate the CHF data affected by the inside factors.

Almost all of such studies on the inside factor effects (cf. Katto 1992b also) suggest that the heater inside factors can have the effects on the CHF only when the heater wall is very thin. For example, if $S < 20$ is tentatively assumed in Fig.22, it corresponds to $\delta_w < 0.4$ mm in the case of copper under ordinary conditions. In addition, as for δ_w of the heater wall subject to the inside factor effects, much lower value of $S < 3.3$ is suggested by Golobic & Bergles (1992).

(2) Experimental information for flow boiling. Studies of the inside factor effects on the CHF in flow boiling are relatively few. Collier (1972) referred to a few studies suggesting the possibility of the reduction in CHF value for tubes of wall thickness less than, say 0.4 mm. More recently, Del Valle M.(1983) experimented on the CHF in subcooled water boiling at 1 atm in a vertical channel of rectangular cross section (12 x 15 mm) electrically heated on one side wall (stainless steel), resulting in Fig.23 in the range of the mass velocity $G = 800 - 2000$ kg/m^{2}s and the quality at the CHF location $x = -0.12 - 0$ (negative values).

(3) Mechanism of the heater inside factor effects. Figs.22 and 23 show the similar trend that the heater inside effects can take place only when the heater wall is very thin (roughly, $\delta_w < 0.5 - 0.8$ mm). With respect to this problem, Watwe & Bar-Cohen (1994) conducted numerical analyses of the transient heat conduction in the heater wall near the dry spot during a hovering period of the vapor clot (cf. section 5.1), disclosing the important role of the heat conduction resistance in radial direction from the dry spot causing noticeable temperature rise at the dry spot. Similarly, Sadasivan et al. (1994) conducted numerical analyses of the transient conduction showing that the spacial distribution of vapor stems on the surface as well as the size distribution of vapor stems can affect the average surface temperature, and the interactions among vapor stems become greater as the wall thickness decreases.

(4) Supplementary notes.
<1> Lin & Westwater (1982) conducted the 'quenching test' of pool boiling, disclosing a fact that ϱck of the wall material exerts its effects on the CHF even when the heater wall is very thick. However, this case must be discriminated clearly from the above-mentioned case because of the different condition that, in the case of quenching (or transient) test, the whole region inside the heater wall is subject to the forced transient state.
<2> Not a few experimental studies (both steady-state and transient tests) on the basic CHF mechanism have so far been conducted with extremely thin wires paying no attension to the

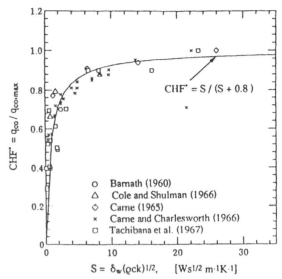

Fig.22   Correlation of CHF with $\delta_w(\varrho ck)^{1/2}$ (quoted from Watwe & Bar-Cohen 1994).

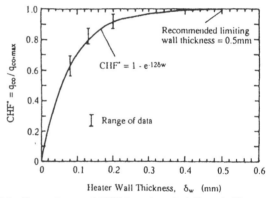

Fig.23   Dependence of CHF in subcooled flow boiling on the heater wall thickness (from Del Valle M. 1983).

heater inside factor effects. In addition to this, the nonuniform disturbance of the surface temperature is apt to occur in such cases. Accordingly, the use of thin wires may be undesirable for the basic study of CHF mechanisms.

8.4 Heater Surface-Factor Effects The surface conditions such as wettability have the ability of changing the fluid state in the immediate vicinity of the heater surface, and thereby the CHF value is changed. If the decrease of wettability causes the increase of the dry area fraction on the heater surface, there is a possibility of generating the same situation as that mentioned in section 8.3-(3) in more or less extended scale. In this case, however, when the magnitude of the quantity such as $\delta_w(\varrho ck)^{1/2}$ is high enough, the inside factor effects must disappear. Hence, the surface factor effects arising independent of the inside factor effects can be regarded as basically important factors.

(1) Experimental information for pool boiling. As has been mentioned in section 8.3-(4), experiments conducted with transient technique are inadequate to the study of the basic CHF mechanism; accordingly, only the typical CHF data obtained under 'steady-state' conditions are cited below.

Hahne & Diesselhorst (1978) experimented on the CHF in pool boiling of saturated water at 1 atm on 1.0 - 7.5 mm dia. horizontal cylindrical heaters of several different metals, showing the trend of the CHF decreasing as the contact angle increases. Meanwhile, Liaw & Dhir (1986) and Dhir (1992) obtained the results of Fig.24 for pool boiling on a 50-mm thick heater, showing the trend similar to that of the foregoing Hahne & Diesselhorst's results.

    With respect to the studies such as mentioned above, it must be noticed that most existing studies on the surface condition effects have so far been conducted with heaters considerably thicker than the limiting value of heater thickness (say, 0.5 - 0.8 mm) mentioned in section 8.3-(1).

    (2) Mechanism of the heater surface effects (Part 1). It is natural to presume that the reduction of the surface wettability (i.e., the increase of contact angle ϕ) leads to the increase of the dry area fraction A_v/A_w, and hence, the liquid volume covering the unit surface area, that is $j_L\delta_o$ of equation (6), decreases.

    Considering the change of δ_o in the term of $j_L\delta_o$, Pan & Lin (1990) conducted an analysis of transition boiling (including the CHF point) based on the idea that δ_o of equation (8) decreases with increasing the contact angle ϕ; but their assumption that the reduction of δ_o is caused by the reduction of A_v/A_w in equation (8), is unfortunately a trend reverse to the common sense that the reduction of A_v/A_w is brought about by the reduction of ϕ.

    Then recently Shoji & Kuroki (1994) proposed a simple macrolayer formation model based on the lateral bubble coalescence to give the following expression as to $j_L\delta_o$:

$$j_L\delta_o = (1 - A_v/A_w)\,\delta_o = N_a^{-1/2}\,f(\phi) \qquad (20)$$

where A_v/A_w and $f(\phi)$ are functions of ϕ, respectively, and N_a is the active nucleation site density. According to this model, A_v/A_w and $f(\phi)$ take the following values:

$\phi = 0°$	$10°$	$20°$	$30°$	$40°$	$45°$
$A_v/A_w = 0$	0.047	0.184	0.393	0.649	0.785
$1 - A_v/A_w = 1$	0.953	0.816	0.607	0.351	0.215
$f(\phi) = 0.139$	0.128	0.100	0.063	0.029	0.017

and, as to N_a, Shoji & Kuroki employ the following empirical correlation obtained from the experiment of Gaertner & Westwater (1960) for a specific condition of saturated water boiling at 1 atm:

$$N_a = 1.89 \times 10^{-7}\, q^{2.13}$$

where the units are $N_a[1/m^2]$ and $q[W/m^2]$, respectively. The CHF value q_{co} predicted by substituting $j_L\delta_o$ of equation (20) into equation (6) appears to agree approximately with the experimental data of Hahne & Diesselhorst etc. With respect to the variation of $j_L\delta_o$ vs. ϕ, the above table shows that the variation of j_L $(= 1 - A_v/A_w)$ is greater than that of δ_o which is proportional to $f(\phi)\,/\,(1 - A_v/A_w)$ [N.B. As for the effect of ϕ on N_a, see section 8.4-(4) mentioned later].

    According to the above table, A_v/A_w (i.e., dry area fraction) is so high as 0.649 for $\phi = 40°$, for example; and it raises a question as to whether or not the normal CHF condition of equation (6) can be assumed under such a state of the heater surface. In order to make it clear, reliable information is needed as to the detailed state of liquid in contact with the ill-wettable heater surface.

    (3) Mechanism of the heater surface effects (Part 2). As has been mentioned at the beginning of the present section 8.4, there is a possibility of generating the situation similar to that mentioned in section 8.3-(3). With respect to this problem, it is of interest to refer to the hot-spot-controlled CHF model proposed by Unal et al.(1992) based on the postulation that the CHF is basically caused by the critical liquid-solid contact temperature, i.e., the high temperature-rise at the dry

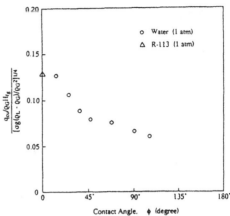

Fig.24   Dependence of CHF in pool boiling on the contact angle (from Dhir 1992).

patch region so as to prevent the liquid rewetting. This is a phenomenon closely related to the transient heat conduction within the heater wall, so it needs to obtain clear experimental evidence as to whether there is any influence of the heater surface conditions (such as contact angle) on, for example, the limiting magnitude of $S = \delta_w(\varrho ck)^{1/2}$ mentioned in section 8.3-(1). It is added that almost all the existing studies of the heater surface effects on the CHF have been conducted with quite thick heater walls paying no attention to such a limiting magnitude of S. In addition, one must not forget the fact that according to the studies conducted on the LHF of film boiling or the Leidenfrost phenomenon, the critical liquid-solid contact temperature does not seem a simply determinable property.

    (4) Active nucleation site density (including subcooled flow boiling). Taking an ingenious way, Wang & Dhir (1983) measured the active nucleation site density N_a as a function of the heat flux q in saturated pool boiling of water at 1 atm for three different contact angles $\phi = 15°$, $35°$, and $90°$. Their data are represented in Fig.25 together with the pioneer data of Gaertner & Westwater (1960), showing (i) that $N_a \propto q^{2.0}$ approximately, (ii) that N_a is a function of not only q but also ϕ, and (iii) that the heat transfer per an active site reduces with increasing the contact angle ϕ.

    As for subcooled flow boiling, there are experimental data of N_a measured by Del Valle M. & Kenning (1985) for water boiling at 1 atm: Fig.26 shows the spacial distribution of active nucleation sites on the heater surface at q = 95% of CHF (where CHF is 4.92 MW/m²), and Fig.27 represents four data points of N_a reported in their paper, showing the relation of $N_a \propto q^{1.43}$. As is noticed from the comparison with a part of the pool boiling data entered in Fig.27, quantitative difference between Fig.25 and Fig.27 seems very great even if the situation of equation (19) is taken into account. [N.B. Zeng & Klausner 1993 have measured N_a for high-quality flow boiling of R-113]

    (5) Possible difference between pool and flow boiling. As for flow boiling, few studies have been conducted to date on the effects of the heater-side factors on the CHF, so it is almost impossible to say something reliable at the present stage about the difference between pool and flow boiling. However, it may probably be of use to add the following comments:

<1>.   Reduction of the surface factor effects. If the forced convective flow has a power of smoothing the nonuniform state in the immediate vicinity of the heater surface, it should reduce the heater surface effects.

<2>.   Suppression of the temperature rise at dry area. In explaining the heater inside factor effects in pool boiling, the critical rewetting temperature at the dryout area, beyond which liquid can nolonger contact with the area, is assumed as the CHF condition (Tachibana et al.1967, Unal et al.1992, Watwe & Bar-

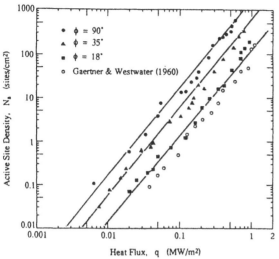

Fig.25 Active nucleation site density vs. heat flux in water pool boiling at 1 atm measured for three different contact angles (from Wang & Dhir 1983).

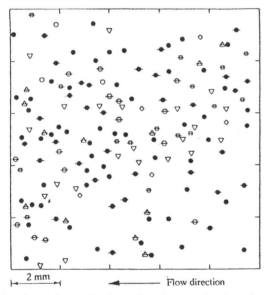

Fig.26 Spacial distribution of active nucleation sites in subcooled flow boiling of water at 1 atm at 95% of CHF (from Del Valle M. & Kenning 1985). The difference of symbols is concerned with the history of each site.

Cohen 1994, etc,). Meanwhile, Ueda & Kim (1986) observed the wall temperature variation and the vapor clot behavior near the surface in subcooled flow boiling, drawing the conclusion that the CHF seems to be initiated when the wall temperature 'rise' due to the partial disruption of the liquid sublayer underneath the vapor clot becomes greater than the temperature 'fall' resulting from the quenching caused by the following forced liquid flow.

9 SUPPLEMENTARY NOTES ON CHF
(1) CHF Mechanism Expressed in More General Form.
Selecting fundamental boiling systems, the basic CHF mechanisms have so far been explained in a unified manner as far as possible throughout pool, external flow and internal flow boil-

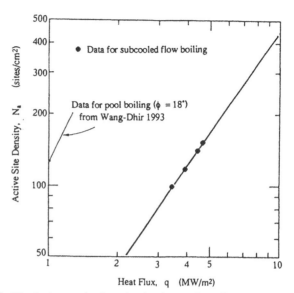

Fig.27 Active nucleation site density vs. heat flux in subcooled flow boiling of water at 1 atm based on the data of Del Valle M. & Kenning (1985).

ing. However, there are many other boiling systems with complex conditions of the heater geometry, the heater orientation, the flow pattern, the magnitude of pressure, flow rate, gravity, transient state, and so on (for example, see the study of Mudawwar et al.1985), where the CHF is generated by various different mechanisms in appearance.

Therefore, it is of use to extend the basic concept of the CHF mechanism mentioned so far up to the concept in more general form, with respect to the normal CHF at least; and that would be written as follows: the normal CHF occurs at the limiting state of the balance among (i) the liquid supply or replenishment to the heater surface, (ii) the liquid conservation or adhesion on the surface, and (iii) the liquid dissipation or removal from the surface caused by various processes.

On the other hand, as for the real CHF, it can be said that its studies have just started with many aspects to be clarified in the future. Accordingly it seems likely that we must wait a while before ataining a more systematic and reliable perspective.

(2) Trigger CHF Models. Though not included so far in this paper in order to keep a unified line as far as possible, there is another stream of studies with a common concept that the CHF is a phenomenon caused directly by a certain trigger situation. Several trigger states have so far been assumed such as (i) the hydrodynamic instability at the vapor/liquid interface (Zuber 1959, Kandula 1990), (ii) the boundary layer separation or blowoff due to the vapor effusion from the heater surface in flow boiling (Kutateladze & Leont'ev 1966, Kandula 1990), (iii) the critical bubbly layer condition near the heater surface in subcooled and low quality flow boiling (initiated by Weisman & Pei 1983 and followed by several papers extending the conditions), and (v) some others. All of these trigger states have a common characteristic feature of being associated with the limit of the 'steady-state' flow conditions.

10 CONCLUSIONS
There are many aspects of the CHF mechanisms that still require clarification, and further studies are expected to be conducted so as to solve these problems steadily. At the same time, however, it is also necessary to have a perspective of the

CHF mechanism at the present, and this paper has been written for this purpose.

NOMENCLATURE

A_v = dry area (or vapor jet base area) on heater surface
A_w = heater surface area
c = specific heat of wall material
c_L = specific heat of liquid
D = diameter of disk or cylindrical heater
D_b = thickness of vapor clot
d = diameter of tube
d_N = nozzle diameter
G = mass velocity; $[G = u\varrho_L$ for external flow]
g = gravitational acceleration
H = height
H_{fg} = latent heat of evaporation; $[H_{fg}'$: see equation (14)]
j_L = volume fraction of liquid in macrolayer
k = thermal conductivity of wall material
k_L = thermal conductivity of liquid
k_v = velocity coefficient
L = heater or tube length
$L' = L/[\sigma/g(\varrho_L - \varrho_G)]^{1/2}$
L_B = boiling length (tube length between zero quality section to exit end at CHF)
m = liquid film mass flow rate
N_a = active nucleation site density
p = pressure; $[p_r$ = reduced pressure]
q = heat flux
q_B = heat flux for pool boiling
q_c = critical heat flux; $[q_{co} = q_c$ for saturated boiling]
q_e = fraction of q used by evaporation
q_{sen} = fraction of q used by sensible energy heating
R = radius
Re = Reynolds number of flow in tube
$r(t)$ = instantaneous radius of spherical vapor clot
$s(t)$ = instantaneous height of vapor clot center
T = temperature; $[\Delta T_s$ = superheat of heater; ΔT_{sub} = subcooling of liquid]
t = time
U_{Lc} = local velocity of main stream at the vapor clot centerline
$U_{L\delta}$ = local velocity of main stream at distance δ from wall
u = liquid velocity in external flow boiling
$u' = u/[\sigma g(\varrho_L - \varrho_G)/\varrho_G^2]^{1/4}$
u_L = liquid velocity
u_G = vapor velocity, or passing velocity of vapor clot
$V(t)$ = instantaneous volume of spherical vapor clot
W = width
x = quality
y^* = thickness of superheated liquid layer
δ = instantaneous thickness of macrolayer
δ_0 = initial (or critical) thickness of macrolayer
δ_G = vapor film thickness
δ_w = wall thickness
ε = void fraction; $[\varepsilon_1 = \varepsilon$ at the onset of annular flow]
λ_H = the Helmholtz instability wavelength
λ_{Td} = the most dangerous Taylor instability wavelength
μ_G = dynamic viscosity of vapor
μ_L = dynamic viscosity of liquid
ϱ_G = vapor density
ϱ_L = liquid density
ϱ = density of wall material
σ = surface tension
τ = hovering period of vapor clot
τ_G = passage time of vapor clot
ϕ = liquid-solid contact angle

REFERENCES

Bains, R.P., El Masri, M.A. & Rohsenow, W.M. 1984 Critical heat flux in flowing liquid films. *Int. J. Heat Mass Transfer* **27**, 1623-1629.

Bar-Cohen, A. & McNeil, A. 1992 Parametric effects on pool boiling critical heat flux in dielectric liquids. *Pool and External Boiling* (Edited by Dhir, V.K. & Bergles, A.E.), pp.171-175. ASME, New York.

Bergles, A.E. 1975 Burnout in boiling heat transfer. Part I: Pool-boiling systems. *Nucl. Safety* **16**, 29-42.

Bergles, A.E. 1977 Burnout in boiling heat transfer. Part II: Subcooled and low quality forced- convection systems. *Nucl. Safety* **18**, 154-167.

Bergles, A.E. 1979 Burnout in boiling heat transfer. Part III: High-quality forced- convection systems. *Nucl. Safety* **20**, 671-689.

Bergles, A.E. & Rohsenow, W.M. 1964 The determination of forced-convection surface-boiling heat transfer. *Trans. ASME, J. Heat Transfer* **86**, 365-372.

Bernath, L. 1960 A theory of local boiling burnout and its application to existing data. *Chem. Engng Prog. Symp. Ser.* **56**, 95-116.

Bhat, A.M., Prakash, R. & Saini, J.S. 1983 On the mechanism of macrolayer formation in nucleate pool boiling at high heat flux. *Int. J. Heat Mass Transfer* **26**, 735-740.

Bhat, A.M., Saini, J.S. & Prakash, R. 1986 Role of macrolayer evaporation in pool boiling at high heat flux. *Int. J. Heat Mass Transfer* **29**, 1953-1961.

Borishanskii, V.M. 1956 An equation generalizing experimental data on the cessation of bubble boiling in a large volume of liquid. *Zhurn. Teckh. Fiz.* **26**, 452-456. [Translated in *Soviet Physics-Technical Physics* **1**(2), p.438].

Carne, M. & Charlesworth, D.H. 1966 Thermal conduction effects on the critical heat flux in pool boiling. *Chem. Eng. Prog. Symp. Ser.* **62**(64), 24-34.

Carne, M. 1965 Some effects of test section geometry in saturated pool boiling on the critical heat flux for some organic fluids and liquid mixtures. *Chem. Eng. Prog. Symp. Ser.* **61**(59), 281-289.

Carvalho, R.D.M. & Bergles, A.E. 1992 The effects of the heater thermal conductance/capacitance on the pool boiling critical heat flux. *Pool and External Boiling* (Edited by Dhir, V.K. & Bergles, A.E.), pp.203-211. ASME, New York.

Celata, G.P., Cumo, M., Mariani, A., Simoncini, M. & Zummo, G. 1994 Rationalization of existing mechanistic models for the prediction of water subcooled flow boiling critical heat flux. *Int. J. Heat Mass Transfer* **37**(Suppl.1), 347-360.

Chappidi, P.R., Unal, C., Pasamehmetoglu, K.O. & Nelson, R.A. 1991 On the relationship between the macrolayer thickness and the vapor-stem diameter in the high-heat- flux, pool nucleate boiling region. *Int. Comm. Heat Mass Transfer* **18**, 195-205.

Cole, R. & Shulman, H.L. 1966 Critical heat flux values at sub-atmospheric pressure. *Chem. Eng. Sci.* **21**, 723-724.

Collier, J.C. 1972 *Convective Boiling and Condensation*, p.260. McGraw-Hill, New York.

Davidson, J.F. & Schüler, B.O.G. 1960 Bubble formation at an orifice in an inviscid liquid. *Trans. I.Chem.E.* **38**, 335-342.

Del Valle M., V.H. 1983 An experimental study of critical heat flux in subcooled flow boiling at low pressure including the effect of wall thickness. *ASME-JSME Thermal Engng Joint Conf. Proc.*, Vol.1, 143-150.

Del Valle M., V.H. & Kenning, D.B.R. 1985 Subcooled flow boiling at high heat flux. *Int. J. Heat Mass Transfer* **28**, 1907-1920.

Dhir, V.K. 1992 Some observations from maximum heat flux data obtained on surfaces having different degrees of wettability. *Pool and External Boiling* (Edited by Dhir, V.K. & Bergles, A.E.), pp.185-192. ASME, New York.

Dhir, V.K. & Lienhard, J.H. 1974 Peak pool boiling heat flux in viscous liquids. *Trans. ASME, J. Heat Transfer* **96**, 71-78.

Gaertner, R.F. 1965 Photographic study of nucleate pool boiling on a horizontal surface. *Trans.ASME J. Heat Transfer* **87**, 17-29.

Gaertner, R.F. & Westwater, J.W. 1960 Population of active sites in nucleate boiling heat transfer. *Chem. Eng. Symp. Ser.* **56**, 39-48.

Gambill, W.R. 1968a Burnout in boiling heat transfer. Part I: Pool-boiling systems. *Nucl. Safety* **9**, 351-362.

Gambill. W.R. 1968b Burnout in boiling heat transfer. Part II: Subcooled forced- convection systems. *Nucl. Safety* **9**, 467-480.

Golobic, I. & Bergles, A.E. 1992 Effects of thermal properties and thickness of horizontal vertically oriented ribbon heaters on the pool boiling critical heat flux. *Pool and External Boiling* (Edited by Dhir, V.K. & Bergles, A.E.), pp.213-218. ASME, New York.

Groeneveld, D.C., Kiameh, B.P. & Cheng, S.C. 1986 Prediction of critical heat flux (CHF) for non-aqueous fluids in forced convective boiling. *Proc. 8th Int. Heat Transfer Conf.* Vol.5, 2209-2214.

Hahne, E. & Diesselhorst, T. 1978 Hydrodynamic and surface effects on the peak heat flux in pool boiling. *Proc. 6th Int. Heat Transfer Conf.*, 1, 209-214.

Haramura, Y. 1989a Characteristics of pool boiling heat transfer in the vicinity of the critical heat flux (relations between bubble motion and heat flux fluctuations). *Heat Transfer---- Jap. Res.* **18** (3), 18-31.

Haramura, Y. 1989b Measurement of thickness of liquid film formed on a heated surface in high heat flux saturation boiling (1st Report, A method to measure the thickness of the liquid film and the results in pool boiling). *Trans.JSME* **55B**, 1392-1306.

Haramura, Y. & Katto, Y. 1983 A new hydrodynamic model of critical heat flux, applicable widely to both pool and forced convection boiling on submerged bodies in saturated liquids. *Int. J. Heat Mass Transfer* **26**, 389-399.

Hewitt, G.F., Kearsey, H.A., Lacey, P.M.C. & Pulling, D.J. 1965 Burnout and nucleation in climbing film flow. *Int. J. Heat Mass Transfer* **8**, 793-814.

Hewitt, G.F. 1978 Critical heat flux in flow boiling. *Proc. 6th Int. Heat Transfer Conf.* **6**, 143-171.

Hewitt, G.F. 1982 Burnout . In *Handbook of Multiphase Systems* (Edited by Hetsroni, G.), pp.(6)66-(6)141. Hemisphere, Washington, D.C.

Inamura, H. 1982 Critical heat flux on a uniformly heated vertical cylinder cooled by a liquid film flow. MSc. Thesis, Dept. of Mech. Eng., University of Tokyo.

Kandula, M. 1990 Mechanisms and predictions of burnout in flow boiling over heated surface with an impinging jet. *Int. J. Heat Mass Transfer* **33**, 1795-1803.

Katto, Y. 1984 Prediction of critical heat flux for annular flow in tubes taking into account the critical liquid film thickness concept. *Int. J. Heat Mass Transfer* **27**, 883-891.

Katto, Y. 1990a A physical approach to critical heat flux of subcooled flow boiling in round tubes. *Int. J. Heat Mass Transfer* **33**, 611-620.

Katto, Y. 1990b Prediction of critical heat flux of subcooled flow boiling in round tubes. *Int. J. Heat Mass Transfer* **33**, 1921-1928.

Katto, Y. 1992a A prediction model of subcooled water flow boiling CHF for pressure in the range 0.1-20 MPa. *Int. J. Heat Mass Transfer* **35**, 1115-1123.

Katto, Y. 1992b Critical heat flux in pool boiling. *Pool and External Boiling* (Edited by Dhir, V.K. & Bergles,

A.E.), pp.151-164. ASME, New York.

Katto, Y. 1994 Critical heat flux. *Int. J. Multiphase Flow* **20**(Suppl.), 53-90.

Katto, Y. & Hirao, T. 1991 Critical heat flux of counter-flow boiling in a uniformly heated vertical tube with a closed bottom. *Int. J. Heat Mass Transfer* **34**, 993-1001.

Katto, Y. & Ishii, K. 1978 Burnout in a high heat flux boiling system with a forced supply of liquid through a plane jet. *Proc. 6th Int. Heat Transfer Conf.* **1**, 435-440.

Katto, Y. & Kikuchi, K. 1972 Study of forces acting on a heated surface in nucleate boiling at high heat fluxes. *Heat Transfer-Jap. Res.* **1**(3), 34-46.

Katto, Y. & Kurata, C. 1980 Critical heat flux of saturated convective boiling on uniformly heated plates in a parallel flow. *Int. J. Multiphase Flow* **6**, 575-582.

Katto, Y. & Ohno, H. 1984 An improved version of the generalized correlation of critical heat flux for the forced convective boiling in uniformly heated vertical tubes. *Int. J. Heat Mass Transfer* **27**, 1641-1648.

Katto, Y. & Otokuni, S. 1994 Behavior of vapor masses on a vertical flat surface of comparatively large height near critical heat flux condition in saturated pool boiling. *Int. J. Heat Mass Transfer* **37**(Suppl.1), 255-263.

Katto, Y. & Shimizu, M. 1979 Upper limit of CHF in the saturated forced convection boiling on a heated disk with a small impinging jet. *Trans. ASME J. Heat Transfer* **101**, 265-269.

Katto, Y. & Yokoya, S 1968 Principal mechanism of boiling crisis in pool boiling. *Int. J. Heat Mass Transfer* **11**, 993-1002.

Katto, Y. & Yokoya, S. 1976 Behavior of a vapor mass in saturated nucleate and transition boiling. *Heat Transfer-Jap. Res.* **5**(2), 45-65.

Katto, Y. & Yokoya, S. 1988 Critical heat flux on a disk heater cooled by a circular jet of saturated liquid impinging at the center. *Int. J. Heat Mass Transfer* **31**, 219-227.

Katto, Y., Sugiyama, K. & Fujita, M. 1994 An experimental and analytical study of CHF phenomenon appearing in a bottom-closed vertical tube for vapor/liquid density ratio in the range of 0.000624-0.136. *Proc. 10th Int. Heat Transfer Conf.* **7**, 461-466.

Katto, Y., Yokoya, S., Miake, S. & Taniguchi, M. 1987 Critical heat flux on a uniformly heated cylinder in a cross flow of saturated liquid over a very wide range of vapor-to-liquid density ratio. *Int. J. Heat Mass Transfer* **30**, 1971-1977.

Kirby, D.B. & Westwater, J.W. 1965 Bubble and vapor behavior on a heated horizontal plate during pool boiling near burnout, *Chem. Eng. Prog. Symp. Ser.* **61**(57), 238- 248.

Kumada, T. & Sakashita, H. 1992 Proposed model for Kutateladze correlation and new correlation of CHF. *Pool and External Boiling* (Edited by Dhir, V.K. & Bergles, A.E.), pp.177-183. ASME, New York.

Kutateladze, S.S. 1952 In *Teploperedacha pri Kondensatsii i Kipenti*. Mashgiz, Moscow (Also *Heat Transfer in Condensation and Boiling* ,2nd Ed. 1959, AEC, Rept-tr-3770, pp.95-113).

Kutateladze, S.S. & Leont'ev, A.I. 1966 Some applications of the asymptotic theory of the turbulent boundary layer. *Proc. 3rd Int. Heat Transfer Conf.* **III**, 1-7.

Lee, C.H. & Mudawar, I. 1988 A mechanistic critical heat flux model for subcooled flow boiling based on local bulk flow conditions. *Int. J. Multiphase Flow* **14**, 711-728.

Levy, S., Healzer, J.M. & Abdollahian, D. 1980 Prediction of critical heat flux for annular flow in vertical tubes. NP-1619, EPRI, Palo Alto, California.

Liaw, S.-P. & Dhir, V.K. 1986 Effect of surface wettability on transition boiling heat transfer from a vertical surface. *Proc. 8th Int. Heat Transfer Conf.* Vol.4, 2031-2036.

Lienhard, J.H. 1988 Burnout on cylinders. *Trans.*

ASME, *J. Heat Transfer* **110**, 1271-1286 (Correction in *Trans. ASME, J. Heat Transfer* 1991,**113**, 548).

Lienhard, J.H. 1994 The snares of pool boiling research: Putting our history to use. *Proc. 10th Int. Heat Transfer Conf.* **1**, 333-348.

Lienhard, J.H. & Dhir, V.K. 1973 Hydrodynamic prediction of peak pool-boiling heat fluxes from finite bodies. *Trans. ASME, J. Heat Transfer* **95**, 152-158.

Lienhard, J.H. & Eichhorn, R. 1976 Peak boiling heat flux on cylinders in a cross flow. *Int. J. Heat Mass Transfer* **19**, 1135-1142.

Lienhard, J.H. & Eichhorn, R. 1979 On predicting boiling burnout for heaters cooled by liquid jets. *Int. J. Heat Mass Transfer* **22**, 774-776.

Lienhard, J.H. & Witte, L.C. 1985 An historic review of the hydrodynamic history of boiling. *Reviews in Chem. Engng* **3** (No.3 & 4), 187-280.

Lin, W.-S., Lee, C.-H. & Pei, B.-S. 1989 An improved theoretical critical heat flux model for low-quality flow. *Nucl. Technol.* **88**, 294-306.

Lin, D.Y. & Westwater, J.W. 1982 Effect of metal thermal properties on boiling curves obtained by the quenching method. *Proc. 7th Int. Heat Transfer Conf.* Vol.4, 155-160.

Macbeth, R.W. 1968 The burnout phenomenon in forced-convection boiling. In *Advances in Chemical Engineers*, Vol.7, pp.207-293. Academic Press, New York.

Meyer, G., Gaddis, E.S. & Vogelpohl, A. 1986 Critical heat flux on a cylinder of large diameter in a cross flow. *Proc. 8th Int. Heat Transfer Conf.* Vol.5, 2125-2130.

Milashenko, V.I., Nigmatulin, B.I., Petukhov, V.V. & Trubkin, N.I. 1989 Burnout and distribution of liquid in evaporative channels of various lengths. *Int. J. Multiphase Flow* **15**, 393-401.

Monde, M. & Katto, Y. 1978 Burnout in a high heat-flux boiling system with an impinging jet. *Int. J. Heat Mass Transfer* **21**, 295-305.

Mudawar, I. & Maddox, D.E. 1989 Critical heat flux in subcooled flow boiling of fluorocarbon liquid on a simulated electronic chip in a vertical rectangular channel. *Int. J. Heat Mass Transfer* **32**, 379-394.

Mudawwar, I.A., El-Masri, M.A., Wu, C.S. & Ausman-Mudawwar, J.R. 1985 Boiling heat transfer and critical heat flux in high-speed rotating liquid films. Int. J. Heat Mass Transfer **28**, 795-806.

Nishikawa, K., Fujii, T. & Honda, H. 1972 Experimental study on the mechanism of transition heat transfer. *Bulletin JSME* **15**, 93-103.

Pan, C. & Lin, T.L. 1990 A model for surface wettability effect on transition boiling heat transfer. *Proc. 9th Int. Heat Transfer Conf.* Vol.2, 147-152.

Rajvanshi, A.K., Saini, J.S. & Prakashi, R. 1992 Investigation of macrolayer thickness in nucleate pool boiling at high heat flux. *Int. J. Heat Mass Transfer* **35**, 343-350.

Sadasivan, P. & Lienhard, J.H. 1987 Burnout of cylinders in flow boiling: The role of gravity influences on the vapor plume. ASME/AIChE Heat Transfer Conf. [See Lienhard, J.H. 1988 Burnout on cylinders. *Trans.ASME, J. Heat Transfer* **110**, 1271-1286].

Sadasivan, P. & Lienhard, J.H. 1991 Burnout of cylinders in flow boiling: surface wettability effects in gravity-uninfluenced flows. *ASME/JSME Therm. Engng Proc.*, Vol.2, xli-1.

Sadasivan, P., Unal, C. & Nelson, R. 1994 Pool boiling on the thin heaters: the role of thermal interaction among vapor stems. *Proc. 10th Int. Heat Transfer Conf.* **5**, 135-140.

Sadasivan, P., Chappidi, P.R., Unal, C. & Nelson, R.A. 1992 Possible mechanisms of macrolayer formation. *Int. Comm. Heat Mass Transfer* **19**, 801-815.

Saito, T, Hughes, D. & Carton, M.W. 1978 Multi-fluid modeling of annular two- phase flow. *Nucl. Eng. and Design* **50**, 225-271.

Shah, M.M. 1987 Improved general correlation for critical heat flux during upflow in uniformly heatedvertical tubes. *Int. J. Heat Fluid Flow* **8**, 326-335.

Sharan, A. & Lienhard, J.H. 1985 On predicting burnout in the jet-disk configuration. *Trans. ASME, J. Heat Transfer* **107**, 398-401.

Shoji, M. & Kuroki, H. 1994 A model of macrolayer formation in pool boiling. *Proc. 10th Int. Heat Transfer Conf.* **5**, 147-152.

Staniforth, R., Stevens, G.F. & Wood, R.W. 1965 An experimental investigation into the relationship between burnout and film flow-rate in a uniformly heated round tube. UKAEA, AEEW-R 430.

Sugawara, S. 1990 Analytical prediction of CHF by FIDAS code based on three-fluid and film-dryout model. *J. Nucl. Sci. Tech.* **27** 12-29.

Tachibana, F., Akiyama, M, & Kawamura, H. 1967 Non-hydrodynamic aspects of pool boiling burnout. *J. Nucl. Sci. Tech.* **4**, 121-130.

Tong, L.S. & Hewitt, G.F. 1972 Overall viewpoint of flow boiling CHF mechanisms. ASME-Paper 72-HT-54.

Ueda, T. & Kim, K.-K. 1986 Heat transfer characteristics during critical heat flux condition in a subcooled flow boiling system. *Proc. 8th Int. Heat Transfer Conf.* Vol.5, 2203- 2208.

Unal, C., Sadasivan, P. & Nelson, R.A. 1992 On the hot-spot-controlled critical heat flux mechanism in pool boiling of saturated fluids. In *Pool and External Flow Boiling* (Edited by Dhir, V.K. & Bergles, A.E.), pp.151-164. ASME, New York.

Vliet, G.C. & Leppert, G. 1964 Critical heat flux for nearly saturated water flowing normal to a cylinder. *Trans.ASME J. Heat Transfer* **86**, 59-67.

Wang, C.H. & Dhir, V.K. 1993 Effect of surface wettability on active nucleation site density during pool boiling of water on a vertical surface. *Trans. ASME, J. Heat Transfer* **115**, 659-669.

Watwe, A.A. & Bar-Cohen, A. 1994 The role of thickness and thermal diffusivity in pool boiling CHF in highly-wetting liquids. *Proc. 10th Int. Heat Transfer Conf.* **5**, 183-188.

Weisman, J. & Pei, B.S. 1983 Prediction of critical heat flux in flow burnout at low qualities. *Int. J. Heat Mass Transfer* **36**, 1463-1477.

Whalley, P.B., Hutchinson, P. & Hewitt, G.F. 1974 The calculation of critical heat flux in forced convection boiling. *Proc. 5th Int. Heat Transfer Conf.* **IV**, 290-294.

Würtz, J. 1978 An experimental and theoretical investigation of annular steam-water flow in tubes and annuli at 30 - 90 bar. Risφ Report No.372, Risφ National Lab., Denmark.

Yagov, V.V. & Puzin, V.A. 1984 Critical heat flux in forced-convection boiling of refrigerant-12 under conditions of local heat sources. *Heat Transfer---Sov. Res.* **16**(4), 47-52.

Yu, C.-L. & Mesler, R.B. 1977 A study of nucleate boiling near the peak heat flux through measurement of transient surface. *Int. J. Heat Mass Transfer* **20**, 827-840.

Zeng, L.Z. & Krausner, J.F. 1993 Nucleation site density in forced convection boiling, *Trans. ASME, J. Heat Transfer* **115**, 215-221.

Zuber, N. 1959 The critical heat flux in boiling from a horizontal surface. In *Hydrodynamic Aspects of Boiling Heat Transfer.* AECU-4439, pp.95-113.

CONVECTIVE BOILING APPLICATIONS IN SHELL-AND-TUBE HEAT EXCHANGERS

L. W. Swanson and J. W. Palen
Heat Transfer Research, Inc.
College Station, Texas

ABSTRACT

Convective boiling in shell-and-tube heat exchangers plays a vital role in many industrial heat transfer processes. Unfortunately, most convective boiling correlations found in the open literature cannot be used to design industrial vaporizers because of the complex nature of bundle geometries and the fluids processed. This situation is difficult to remedy because industrial-scale experimentation is very expensive.

This paper addresses shellside convective boiling phenomena in shell-and-tube heat exchangers. Shellside convective boiling calculations are an important design element of thermosiphon reboilers, feed-effluent exchangers, and flooded refrigeration evaporators. The problems often encountered during design and operation of these exchangers are summarized. Fundamental heat transfer methods are discussed, including basic relationships for both pure components and mixtures, and critical-heat-flux (CHF) phenomena such as departure from nucleate boiling (DNB) and annular mist flow. Practical examples are given to demonstrate how various process variables affect local heat transfer phenomena in industrial shell-and-tube heat exchangers. Finally, some of the vital research needs of these exchangers are outlined based on the shortcomings or lack of adequate design methods.

INTRODUCTION

Process Applications

All industrial processes that involve shellside vaporization in shell-and-tube heat exchangers require application of convective boiling techniques for proper design. The most common applications are horizontal thermosiphon reboilers, feed-effluent exchangers, feed vaporizers, and refrigeration system vaporizers. The kettle reboiler, often thought to be a pool boiling device, also operates in the convective boiling mode due to internal recirculation. These various applications have different internal baffle geometries, determined by the amount of available pressure drop. The design methods, therefore, must account for geometrically induced complications such as windowflow, bypassing, leakage, directional momentum changes, and ideal crossflow.

General Status of Present Design Methods

Most methods available in the literature apply to crossflow of single-component fluids over moderate pressure ranges for wet-wall conditions. In the chemical industries, however, almost all fluids processed are mixtures, containing a wide range of molecular sizes to be boiled or condensed. The literature does not contain any reliable procedures for calculating the heat transfer coefficients for such mixtures. The lack of boiling and condensing data both for deep vacuum and near critical pressure causes great uncertainty in design methods. For example, the Nusselt solution for condensation gives a zero heat transfer coefficient at the critical pressure, while the accepted methods for nucleate boiling give an infinite heat transfer coefficient at the critical pressure. Both results are unlikely, but verification of both is lacking. In deep vacuum, the mean temperature difference (MTD) for boiling and condensation is closely linked with the effect of two-phase pressure drop on the local saturation temperature. Phase-change pressure drop in vacuum can be dominated by momentum changes. Momentum changes depend on the vapor-liquid slip ratio, an empirically determined factor, which has never been evaluated for phase change under vacuum where the vapor-liquid density ratio is extreme.

Very few ideal tube banks exist in industrial heat

exchangers. The effects of fluid expansion, contraction, flow reversals, bypassing, and leakages are not known for two-phase flow.

Although most vaporizers are designed for wet-wall conditions, many design situations require vaporization with high-temperature heating mediums, and/or vaporization to superheated vapor. Only limited methods are currently available for calculating annular mist-flow or film boiling heat transfer coefficients in bundles and recognizing when these regimes exist.

In general, more studies addressing very non-ideal but very real process conditions are needed. This paper focuses on some of these conditions and their implications on heat exchanger design.

FLOW REGIMES AND HEAT TRANSFER MECHANISMS

Shellside convective boiling in shell-and-tube heat exchangers is a very complicated process because of the complex geometry of tube bundles and the variety of flow regimes and heat transfer mechanisms that can exist in the exchanger. Some of the heat transfer mechanisms corresponding to the wet-wall flow regimes include subcooled boiling, nucleate boiling, two-phase forced convection, and natural convection. Typical mechanisms for partial wet-wall and dry-wall flow regimes include transition boiling, film boiling, annular mist flow, mist flow, and sensible vapor cooling. To further complicate matters, flow stratification can occur under gravity dominated conditions causing partial dry out in the tube bundle. The heat exchanger's orientation, horizontal or vertical, also has an impact on the shellside flow regimes and heat transfer mechanisms.

This section discusses the basic pure component relationships for shellside heat transfer including nucleate boiling, natural convection, convective boiling, subcooled boiling, transition/film boiling, and annular mist-flow heat transfer. We conclude with a discussion of mixture correction factors for nucleate boiling and forced convection evaporation.

Pure Components

Wet-Wall Mechanisms. The derivation of a fundamental expression for the convective boiling heat transfer coefficient begins with a simple energy balance that accounts for the contributing modes of heat transfer (i.e., nucleate boiling, natural convection, forced convection).

$$q_b = q_{nb} + q_{nc} + q_{fc} \qquad (1)$$

Assuming that the temperature difference between the wall and the bulk fluid is equal to the wall superheat (subcooled boiling is an exception), and that this driving force applies to the various heat transfer mechanisms in eqn. (1), the

boiling heat transfer coefficient becomes

$$h_b = \alpha h_{nb} + h_{nc} + h_{fc} \qquad (2)$$

where α is the nucleate boiling suppression factor bounded by $0 \le \alpha \le 1$. Eqn. (2) is often referred to as a Chen-type correlation (Chen [1966]), where h_{nb} is the nucleate boiling heat transfer coefficient, h_{nc} is the natural convection heat transfer coefficient, and h_{fc} is the forced convection heat transfer coefficient. When the total heat flux is specified, h_{nc} must be calculated iteratively because it is a function of the temperature driving force, $T_w - T_b$, which is not known a priori. Depending on how the nucleate boiling suppression factor is defined, the nucleate boiling heat transfer coefficient can be based on the total heat flux or the nucleate boiling heat flux. For example, if h_{nb} is based on the nucleate boiling heat flux, it has to be calculated iteratively. The forced convection heat transfer coefficient accounts for heat transfer phenomena characterizing two-phase flow fields.

The most convenient way to apply eqn. (2) to shellside boiling is to base the nucleate boiling heat transfer coefficient on a correlation obtained from pool boiling experiments for single tubes or a tube within a bundle. This is logical because the nucleate boiling mechanism is a local phenomenon not greatly dependent on geometry. The forced convection heat transfer coefficient is dependent on many variables (e.g., bundle geometry, bundle orientation, two-phase flow regime, and flow direction), making two-phase flow data more difficult to generalize.

The nucleate boiling heat transfer coefficient in eqn. (2) can be written as

$$h_{nb} = h_{nbl} F_e F_s F_c \qquad (3)$$

The above expression is written in terms of the single-tube nucleate boiling heat transfer coefficient (h_{nbl}), fin efficiency factor (F_e), surface factor (F_s), and nucleate boiling mixture correction factor (F_c). A comprehensive list of single-tube nucleate boiling heat transfer correlations based on a wide range of pool boiling data has been compiled and reviewed by Dhir [1991] and Carey [1992]. Commonly used single-tube nucleate boiling correlations are those of Forster and Zuber [1955], Rohsenow [1962], and Stephan and Abdelsalam [1980]. The fin efficiency factor accounts for the decrease in heat transfer due the temperature gradient in the external fins. The surface factor corrects the heat transfer coefficient for enhancements due to external fins and corrects for the increase or decrease in nucleation site density for specialized outer tube surfaces (e.g., enhanced tubes, polished tubes). The nucleate boiling mixture correction factor accounts for mass transfer effects and vapor-liquid equilibrium (VLE) phenomena existing in fluid mixtures undergoing vaporization.

The nucleate boiling heat transfer coefficient in eqn. (2) is multiplied by a nucleate boiling suppression factor. Nucleate boiling suppression occurs when the viscous sub-layer in the turbulent boundary layer decreases in thickness as the liquid flow rate increases. As the viscous sub-layer decreases in thickness, the temperature gradient at the wall increases and the growing bubbles are exposed to colder fluid, which suppresses the number of active nucleation sites (e.g., see Klausner et al. [1993], Zeng and Klausner [1993]). This phenomenon was originally proposed by Chen [1966] for convective boiling in tubes and is discussed by Bennett and Chen [1980], Bennett et al. [1980], and Steiner and Taborek [1992]. Bennett et al. [1980] modified Chen's suppression factor and applied it to shellside convective boiling.

$$\alpha = \left(\frac{k_l}{h_{fc}Y_0}\right)\left[1 - \exp\left(-\frac{h_{fc}Y_0}{k_l}\right)\right] \qquad (4)$$

where

$$Y_0 = 0.041\left[\frac{\sigma}{g(\rho_l - \rho_v)}\right]^{0.5} \qquad (5)$$

Jensen and Hsu [1988] used the Bennet et al. suppression factor to develop a convective boiling model for crossflow in tube bundles. They concluded that a Chen-type correlation [i.e., eqn. (2)] applies to crossflow in tube bundles, but better predictive methods for the two-phase correction factor and nucleate boiling suppression factor are required.

After a thorough evaluation of the Bennett et al. nucleate boiling suppression factor, Webb and Gupte [1992] recommended that some type of asymptotic superposition of the contributing heat transfer modes be used instead of linear superposition with suppression. The general form for asymptotic superposition is

$$h_b = \left(h_{nb}^n + h_{nc}^n + h_{fc}^n\right)^{1/n} \qquad (6)$$

This approach was originally proposed for the boiling heat transfer coefficient by Kutateladse [1961]. Applying asymptotic superposition of the heat transfer modes with $n > 1$ imposes artificial nucleate boiling suppression by forcing a more abrupt transition between pure nucleate boiling and pure forced convective boiling. Figures 1a and 1b in Webb and Gupte's paper [1992] show this explicitly. This functional form has no formal theoretical basis, and the nucleate boiling heat transfer coefficient has to be determined iteratively because the nucleate boiling heat flux is not known a priori. In addition, the asymptotic form does not satisfy the energy balance given by eqn. (1) when the various heat transfer modes or coefficients are based on the same temperature driving force.

An expression for the natural convection heat transfer coefficient is also required in eqn. (2). The natural convection heat transfer coefficient is written as a function of the liquid-thermal expansion coefficient and temperature difference between the wall and bulk fluid. A correlation commonly used for natural convection heat transfer on single cylinders can be found in McAdams [1954]. Note that single-tube correlations for natural convection heat transfer do not account the bundle's effect on natural convection. Tube bundles can enhance the amount of natural convection heat transfer over that predicted by single-tube correlations. Fortunately, natural convection heat transfer is important only for low heat fluxes when the forced convection flow rate is low.

The forced convection heat transfer coefficient in eqn. (2) accounts for heat transfer enhancement due to the two-phase nature of the flow field.

$$h_{fc} = h_l F_{tp} F_e F_{cc} \qquad (7)$$

where h_l is the liquid-phase heat transfer coefficient. F_{tp}, F_e, and F_{cc} are the two-phase correction factor, fin efficiency factor, and mixture correction factor respectively. The forced convection mixture correction factor accounts for mass transfer effects that occur at the vapor-liquid interface of a fluid mixture during evaporation. The functional form for liquid-phase heat transfer coefficient depends on whether the flow is in the baffle window (longitudinal flow) or in crossflow. These coefficients are often written in terms of j-factors and (a) hydraulic diameter for flow in the baffle window or (b) tube-bundle characteristic dimensions for crossflow.

The two-phase correction factor for heat transfer is based on the analogy between two-phase momentum and heat transfer posed by Chen [1966]. The frictional correction factor for the pressure drop commonly used in the analogy is the functional form given by Lockhart and Martinelli [1949]. Ishihara et al. [1977], Palen and Yang [1983], and Dowlati et al. [1993] have successfully applied these relationships to shellside pressure drop for different types of heat exchangers with a variety of two-phase flow regimes. Palen and Yang [1983] found that the two-phase correction factor for heat transfer can be correlated in terms of the two-phase frictional correction factor (ϕ_l) as

$$F_{tp} = \left(\phi_l^2\right)^{m_1} \qquad (8)$$

where

$$\phi_l^2 = 1.0 + \frac{C_A}{X_{tt}} + \frac{1}{X_{tt}^2} \qquad (9)$$

$$C_A = f\left(y, \frac{\rho_v}{\rho_l}, \frac{\mu_l}{\mu_v}, C_{TD}\right) \qquad (10)$$

$$X_{tt} = \left(\frac{1-y}{y}\right)^{0.9}\left(\frac{\rho_v}{\rho_l}\right)^{0.5}\left(\frac{\mu_l}{\mu_v}\right)^{0.1} \qquad (11)$$

$$C_{TD} = g\rho_{tph}\frac{(\rho_l - \rho_v)\, D_0}{G^2} \qquad (12)$$

The exponent m_l in eqn. (8) is a function of the vapor/liquid density ratio. The Taitel-Dukler flow regime parameter [1976], C_{TD}, is included in the coefficient C_A to account for the effect of flow regime on the two-phase frictional correction factor.

In general, eqns. (1-12) were developed for convective boiling when the bulk liquid is saturated; however, for many process applications, liquid enters the exchanger with some degree of subcooling.

With minor modifications, eqns. (1-12) can be used to describe subcooled convective boiling. When the liquid is subcooled, heat transfer only occurs by sensible heating if the wall temperature is less than the liquid saturation temperature. In this case, only the liquid-phase sensible heating terms (h_{nc} and h_{fc}) are retained in eqn. (2). Partial or fully-developed nucleate boiling occurs when the wall temperature is greater than the saturation temperature of the liquid, and the associated superheat is large enough to activate nucleation sites on the tube wall. The formation, motion, and subsequent condensation of the bubbles enhance heat transfer over that for liquid-phase sensible heating. Under these conditions, the temperature driving force between the wall and the bulk liquid is not equal to the wall superheat. The energy balance given by eqn. (1) requires

$$h_b = \alpha\, h_{nb}\frac{T_w - T_{sat}}{T_w - T_b} + h_{nc} + h_{fc} \qquad (13)$$

Note that eqn. (2) is recovered when $T_b = T_{sat}$ and the

overall temperature driving force becomes the wall superheat. Refer to Dhir [1991] or Carey [1992] for a comprehensive review of subcooled boiling heat transfer for simple geometries. Unfortunately, very few studies of subcooled boiling in tube bundles can be found in the open literature (e.g., Jensen and Hsu [1988]).

Partial Dry-Wall and Dry-Wall Mechanisms. As the heat flux increases during convective boiling, eventually the outer tube wall begins to dry out and the critical heat flux is reached. Jensen and Tang [1994] conducted two-phase CHF experiments for ideal crossflow in staggered and in-line tube bundles and developed correlations for the CHF condition. Experimental data were taken at pressures of 150 kPa and 500 kPa using Freon 113 as the working fluid. Jensen and Tang defined three regions according to hypothesized mechanisms based on flow observations governing the CHF process (e.g., Leroux and Jensen [1992]). Region 1 accounts for the bubbly flow DNB mechanism preceding transition/film boiling in pool boiling experiments. Region 2 occurs when the vapor quality of the fluid is sufficiently high to shear droplets off the vapor-liquid interface while partially retaining the transition or film boiling nature of the flow field. Region 3 develops as the vapor quality increases further forming a flow field similar to the annular mist-flow regime characteristic of convective boiling inside tubes. The fraction of the outer tube wall that is wet for the annular mist mechanism in the tube bundle depends on process conditions. Eventually, the vapor quality becomes high enough to dry out the wall completely and sensible heat transfer in the vapor phase becomes the predominant mechanism. Figure 1 shows the Jensen-Tang (JT) flow regime map for CHF regions 1, 2, and 3.

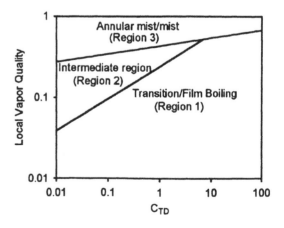

Figure 1. Jensen-Tang CHF Flow Regime Map [1994].

The JT CHF flow regime map is based on data for one working fluid over a fairly narrow range of pressures and requires further experimental verification. The boundaries of the map may actually vary as a function of the vapor/liquid density ratio or reduced pressure. For example, under vacuum conditions ($\rho_v/\rho_l \to 0$), the annular mist region (region 3) may expand, while for near critical pressure conditions ($\rho_v/\rho_l \to 1$), the transition/film boiling region may expand.

The CHF criteria given by Jensen and Tang for a staggered tube bundle are based on the vapor quality and the Taitel-Dukler flow regime parameter [eqn. (12)].

Transition line between regions 1 and 2

$$y_{12} = 0.242 C_{TD}^{0.396} \qquad (14)$$

Transition line between regions 2 and 3

$$y_{23} = 0.432 C_{TD}^{0.098} \qquad (15)$$

The corresponding values for the convective boiling critical heat flux in a staggered tube bundle are (Jensen and Tang [1994])

Region 1 ($y \le y_{12}$)

$$q_1 = q_{max}\exp\left(-0.0322 - \frac{10.1}{\Psi^{0.585}}\right) \qquad (16)$$

where

$$\Psi = D_0\left(\frac{\rho_{tph}}{\mu_l}\right)\left[\frac{\sigma g(\rho_l - \rho_v)}{\rho_l^2}\right]^{1/4} \qquad (17)$$

Region 3 ($y \ge y_{23}$)

$$q_3 = 1.97 x 10^{-5} G\lambda C_{TD} Re^{-0.858} \qquad (18)$$

For region 2 ($y_{12} < y < y_{23}$), the critical heat flux is determined by evaluating q_1 and q_3 at their respective transition qualities (given by eqns. (14) and (15)) and applying linear proration

$$q_2 = q_{1,tr} + \left(q_{3,tr} - q_{1,tr}\right)\left(\frac{y - y_{12}}{y_{23} - y_{12}}\right) \qquad (19)$$

Expressions for the single-tube DNB critical heat flux in eqn. (16) can be obtained from Lienhard [1988] or Carey [1992]. The exponential multiplier in eqn. (16) should be used with caution because earlier studies have shown that the DNB critical heat flux for large tube bundles ($A/\pi D_b L > 3$) is strongly dependent on the bundle geometry (e.g. see Palen and Small [1964]). The geometrical effect on the DNB critical heat flux is not accounted for in eqn. (16) because the geometrical parameter given above for the ideal tube bank tested by Jensen and Tang is less than 3. Secondly, under conditions outside the range of the Jensen-Tang data, the parameter Ψ can produce physically unrealistic values for the DNB critical heat flux.

Once the applicable CHF criteria is determined and post-CHF phenomena are found to exist, an appropriate heat transfer coefficient must be determined. Transition or film boiling occurs in region 1 when $\Delta T_{max} \le \Delta T \le \Delta T_{min}$, where ΔT_{max} and ΔT_{min} correspond to superheats at the maximum and minimum heat fluxes. For transition boiling, the boiling heat transfer coefficient is usually calculated using some type of proration method between coefficients at the minimum and maximum heat fluxes. Popular proration parameters include the fraction of wetted surface (Ragheb and Cheng [1979], Dhir and Liaw [1989]) and relationships involving the superheats ΔT_{max}, ΔT_{min}, and ΔT and/or heat fluxes (e.g., Tong and Young [1974]). A standard convective boiling coefficient can be used at the first critical point whereas the Bromley correlation [1950] can be used in the proration for the film boiling heat transfer coefficient. If transition boiling does not exist, proration between the critical heat fluxes is not necessary, and the film boiling coefficient can be determined using the Bromley correlation [1950].

A proration method can also be used to determine the heat transfer coefficient in region 3 (annular mist flow). Because the wall is either fully or partially wet, the heat transfer coefficient in this region lies somewhere between the convective boiling coefficient given by eqn. (2) and the sensible vapor coefficient. Possible proration parameters include the fraction of wetted surface and the vapor quality. Instead of applying a pure sensible vapor coefficient, the Groenveld [1975] expression for mist flow in tubes can be used by applying the same dimensionless groupings with shellside coefficients and exponents. This expression accounts for the difference in vapor-phase heat transfer due to the presence of liquid droplets in the bulk flow.

Fluid Mixtures

For fluid mixtures, evaporation at the vapor-liquid interface depletes the interface of the lighter component(s), causing an increase in the interfacial temperature and consequently a reduction in the interfacial superheat. The excess heavy molecules at the interface also produce an additional liquid-phase mass transfer resistance that inhibits the transport of the more volatile components to the interface. Although fluid mixtures sometimes improve the thermophysical properties of the liquid phase relative to the individual components, the combined effect of thermophysical properties, VLE, and transport phenomena usually causes a reduction in the boiling heat transfer coefficient.

A mixture correction factor for the nucleate boiling heat transfer coefficient was derived by Schlünder [1983] using film theory. The binary form of the coefficient is given by

$$F_c = \cfrac{1}{1 + \cfrac{h_{nbl}}{q_{nb}}(y_i - x_i)\cfrac{dT_{bub}}{dx_i}\left[1 - \exp\left(-\cfrac{q_{nb}}{\rho_l \lambda \beta_{nb}}\right)\right]}$$

(20)

where y_i is the mole fraction of the light component in the bulk vapor phase and x_i is the mole fraction of the light component in the bulk liquid phase. The mass transfer coefficient (β_{nb}) in eqn. (20) can be back-calculated from pool boiling mixture data and correlated in terms of the nucleate boiling heat flux, density ratio, and bulk composition. Schlünder also approximated the derivative in eqn. (20) with the difference between pure component saturation temperatures ($T_{sat,2} - T_{sat,1}$), producing a more practical expression. For multicomponent mixtures, Schlünder proposed a functional form for eqn. (20) that used standard thermodynamic mixture rules and multicomponent VLE relationships. Unfortunately, this approach is somewhat cumbersome for multicomponent mixtures and is not practical for most fluid mixtures.

To get around the difficulties associated with complex VLE phenomena, Thome and Shakir [1987] modified eqn. (20) for multicomponent mixtures by using the mixture boiling range as a VLE parameter. The mixture boiling range in this case is defined as the difference between dew point and bubble point temperatures in the bulk fluid. Thome and Shakir found this approach could be used successfully to describe the nucleate boiling correction factor for fluid mixtures of ethanol-water, methanol-water, n-propanol-water, and acetone-water.

A mixture correction factor (F_{cc}) for forced convection can be adapted from the functional form for the nucleate boiling mixture correction factor given by eqn (20). In this case, the heat flux, heat transfer coefficient, and mass transfer coefficient for two-phase forced convection boiling are used instead of those for nucleate boiling. The mass transfer coefficient can be back-calculated from two-phase evaporation experiments (e.g., for a falling film). Palen et al. [1994] successfully correlated the forced convection mass transfer coefficient for falling films in terms of the Sherwood number, which is a function of the two-phase Reynolds number and the Schmidt number.

EXAMPLES OF SHELLSIDE CONVECTIVE BOILING IN SHELL-AND-TUBE HEAT EXCHANGERS

An HTRI computer program applying the relationships discussed in this paper, as well as other proprietary methods, was used to simulate some of the local shellside flow regimes and heat transfer mechanisms that can occur in a typical industrial heat exchanger. Proprietary and open literature data were used to develop and confirm the heat transfer methods in the computer program. The exchanger used in the simulation was a TEMA E-shell exchanger with a shell diameter of 0.864 m and bundle length of 5 m. Table 1 gives the geometrical specifications of the exchanger. A variety of shellside process fluids were tested in the same heat exchanger to simulate various convective boiling conditions. The fluids used for the numerical simulation were p-xylene (vacuum, 34.3 kPa), n-pentane (high pressure, 1.37 MPa, P_r = 0.4), and a mixture of propane, butane, hexane, heptane, and octane.

Table 1. Heat Exchanger Geometrical Specifications.

TEMA shell type	AEL
Shell I. D.	0.864 m
Baffle type	Single segmental
Baffle cut	27.5%
Baffle spacing	0.540 m
Number of Crosspasses	9
Tube bundle length	5 m
Number of Tube Passes	2
Tube type	Plain
Tubecount	666
Tube O.D.	19 mm
Tube pitch	25.4 mm

During each run, the condensing steam temperature, used as the in-tube heat source, was raised to increase the heat exchanger's duty. As the duty increased, the mean temperature difference and the outlet vapor quality of the process fluid increased, forcing the exchanger to experience a variety of flow regimes and heat transfer mechanisms.

Vacuum Operation

Figure 2 shows the average heat transfer coefficient as a function of the outlet vapor quality for pure p-xylene under vacuum conditions (34.3 kPa). The tube pitch in this case was changed to 28.7 mm to prevent excessive pressure drop on the shell side of the exchanger. The p-xylene flow rate was fixed at 2 kg/s for each run. As the quality increases, the average heat transfer coefficient reaches a maximum and then begins to decrease in magnitude. The reason for this maximum in the heat transfer coefficient is depicted by Figure 3, which shows the local heat transfer coefficient as a function of the fraction of the total tube length for different outlet vapor qualities. For vapor qualities of approximately 0.6 and less, the heat transfer mechanism in the entire exchanger is solely convective boiling (CB) given by eqn. (2), and the heat transfer coefficient increases from the shell inlet to the outlet. The net effect of increasing the outlet vapor quality in the flow boiling regime is to increase the average heat transfer coefficient in the exchanger.

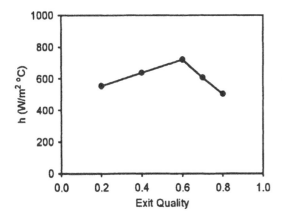

Figure 2. Average Heat Transfer Coefficient for p-xylene.

As the temperature driving force is increased enough to produce an outlet vapor quality of 0.7, the fluid enters the annular mist-flow regime (AM) at an axial location of 0.75, and the latter 25 percent of the exchanger exhibits a significantly lower heat transfer coefficient. In this case, the annular mist heat transfer coefficient is very low for two reasons: (1) the wall is only partially wet (on the front side of the tube) and (2) a relatively low forced convection coefficient dominates eqn. (2) because the liquid mass flow rate is small due to the low total mass flow rate and a high degree of vapor quality. Additional turbulence generated by two-phase vapor shear at the vapor-liquid interface is not large enough to compensate for these effects. The lower local heat transfer coefficients are responsible for the declining average coefficient with increasing quality

exhibited by Figure 2. For an outlet vapor quality of 0.8, nearly 50 percent of the exchanger is in the annular-mist flow regime further reducing the average heat transfer coefficient. It should be noted that the heat transfer coefficients used for the annular mist flow regime are based on limited data. The mechanisms associated with the annular mist-flow regime in tube bundles are very complicated and warrant further experimental and theoretical study.

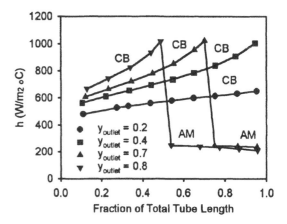

Figure 3. Local Heat Transfer Coefficient for p-xylene.

High-Pressure Operation

Figure 4 shows the average heat transfer coefficient as a function of outlet vapor quality for pure n-pentane at an inlet pressure of 1.37 MPa. For each run, the n-pentane flow rate was set to 100 kg/s and the inlet fluid was subcooled 8 °C below the inlet saturation temperature. In this case, the average heat transfer coefficients are at least one order of magnitude higher than those for the p-xylene under vacuum conditions because nucleate boiling and forced convection heat transfer mechanisms are important. In contrast, the predominant mechanism for p-xylene boiling under vacuum conditions is forced convection boiling. The pressure-drop limitations under vacuum conditions restrict the p-xylene mass flow rate, which produces lower convective boiling heat transfer coefficients. The average heat transfer coefficient for n-pentane increases to a maximum and then levels out to a relatively constant value. If the outlet vapor quality levels are pushed to higher values, the heat transfer coefficient eventually decreases to a value for film boiling and finally approaches a value equal to that for a pure sensible vapor.

Figure 5 shows the local heat transfer coefficient for n-pentane as a function of the fraction of the total tube length for outlet vapor qualities of 0.1 and 0.2. Only outlet vapor qualities of 0.1 and 0.2 are shown to illustrate the effect of subcooled boiling. For an outlet quality of 0.1, the first half

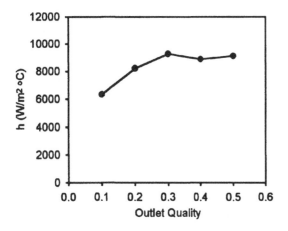

Figure 4. Average Heat Transfer Coefficient for
n-pentane.

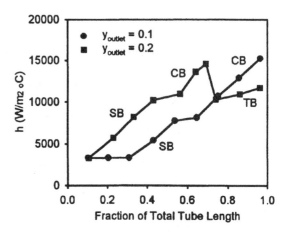

Figure 5. Local Heat Transfer Coefficient for
n-pentane.

of the exchanger (50 percent) experiences sensible heating and subcooled boiling (SB), while the second half of the exchanger is in the convective boiling regime. The early stages of the subcooled flow region (first three points) show a nearly zero slope, suggesting that nucleate boiling is completely suppressed and the liquid-phase sensible heat transfer coefficient is constant. Once nucleate boiling kicks in near 0.3, the slope for subcooled boiling increases to a value similar to that for convective boiling.

For an outlet vapor quality of 0.2, subcooled boiling, convective boiling, and transition boiling (TB) exist at various locations within the exchanger. Because the tubeside steam temperature is higher in this case, nucleate boiling is not suppressed near the inlet of the exchanger as indicated by the positive slope between approximately 0.1 and 0.4. Convective boiling begins earlier as well, generating higher local heat fluxes that eventually exceed the DNB CHF near a fractional tube length of 0.7. Beyond 0.7, the exchanger falls into transition boiling.

Convective Boiling of a Fluid Mixture

Figure 6 shows the average heat transfer coefficient as a function of the outlet vapor quality for a five-component mixture of 0.15 mole propane, 0.25 mole butane, 0.05 mole hexane, 0.30 mole heptane, and 0.25 mole octane. The mixture has an inlet pressure of 0.345 MPa and a flow rate of 24 kg/s. The boiling range of this mixture, defined as the difference between the dew point at the outlet and the bubble point at the inlet, is approximately 200 °C. The figure shows that the average heat transfer coefficient for the mixture remains approximately constant up to an outlet

quality of about 0.32 and then begins to decrease as the quality is increased further. The reasons for this functional behavior are illustrated in Figure, 7 which shows the local heat transfer coefficient. At an outlet quality of 0.165, convective boiling persists throughout the entire exchanger, and the heat transfer coefficient is relatively constant. As the temperature driving force is increased, resulting in an outlet vapor quality of 0.268, transition boiling develops at the inlet of the exchanger and extends over about 15 percent of it's surface. Convective boiling is present throughout the

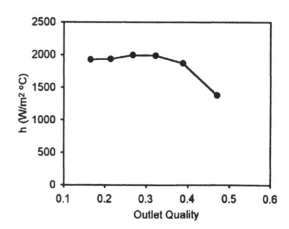

Figure 6. Average Heat Transfer Coefficient for the
Five-Component Mixture.

rest of the exchanger. The reason for this seemingly unusual behavior (i.e., transition boiling followed by convective boiling) is because of the change in the local liquid-phase superheat that results when a mixture boils off the more volatile component(s). Figure 8 depicts the local liquid-phase superheat for each outlet quality plotted in Figure 7. For a vapor quality of 0.268, the superheat is fairly large (38 °C) at the inlet because the bubble point at the vapor-liquid interface is low due to the presence of the more volatile component(s). The large superheat forces the mixture into transition boiling. As the fluid flows through the exchanger, the volatile component(s) boil off, and the vapor-liquid interface temperature increases, resulting in a lower effective liquid superheat. Eventually the superheat reaches a value low enough to prevent transition boiling. At this point, the fluid drops into a convective boiling flow regime as illustrated in Figure 7 by the abrupt increase in the local heat transfer coefficient near 0.2 for an outlet quality of 0.268. Thus, not only is the heat transfer coefficient reduced because of mixture effects during convective boiling, but the wall superheat (or MTD) decreases axially because the interfacial saturation temperature increases according to changes in interfacial composition. Therefore, the MTD available to a fluid mixture is significantly less than that available for a pure fluid with identical thermophysical properties. The VLE effect is compounded by the frictional pressure drop that changes the local pressure used in the VLE calculation. From a design perspective, not accounting for VLE effects on the MTD and the additional mass transfer resistance during convective boiling in heat exchangers can result in an undersized exchanger that cannot meet design specifications.

Understanding when and where these local phenomena (e.g., transition/film boiling) occur in a heat exchanger can have a significant effect on the exchanger's design. For example, if the process mixture given above develops a fouling layer when the fluid experiences transition/film boiling, fouling will occur near the entrance of the exchanger. If an overall calculational method is used, the local phenomena are not observed, and the average heat flux will probably not be large enough to exceed the critical heat flux because roughly 80 percent of the exchanger experiences convective boiling; the high coefficients associated with convective boiling dominate the overall heat transfer coefficient. Using the coefficient given by the overall method may yield the correct amount of heat exchanger surface but does not give any indication that fouling can occur near the inlet of the exchanger. The result is an undesirable fouling layer that can reduce the exchanger's performance.

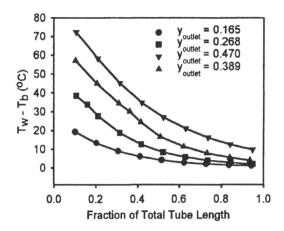

Figure 8. Local Wall Superheat for the Five-Component Mixture.

Figure 7 also shows that if the outlet quality is increased to 0.389, the transition boiling region expands to approximately 35 percent of the exchanger surface area, with the remaining area experiencing convective boiling. Figure 8 shows that this happens because the large superheat driving force extends further downstream in the exchanger due to the increased heat source temperature. In this case, an overall calculational procedure may produce an average heat flux that exceeds the critical heat flux and consequently apply transition or film boiling to the entire exchanger. The result is a significantly over-surfaced exchanger that may actually yield heat fluxes below critical, thereby producing much higher boiling heat transfer coefficients and outlet qualities significantly higher than process specifications. This problem may be alleviated during operation by reducing

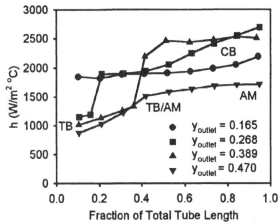

Figure 7. Local Heat Transfer Coefficient for the Five-Component Mixture.

the steam temperature or heat exchanger duty.

The remaining curve shown in Figure 7 is that for an outlet vapor quality of 0.470. For this outlet quality, the large temperature driving force and interfacial shearing by the vapor phase cause the exchanger to experience several different flow regimes. For reasons stated earlier, transition boiling occurs near the inlet of the exchanger (axial location < 0.4). As the vapor quality increases, causing an increase in the vapor interfacial shear stress, the transition boiling process is disrupted and a transition boiling/annular mist-flow regime (TB/AM) exists in the middle portion of the exchanger. The local heat transfer coefficient increases in this region because large pockets of vapor are broken into smaller bubbles resulting in a greater wetted wall area. The transition boiling/annular mist-flow regime exists over the range 0.4 ≤ axial location ≤ 0.8. As the vapor quality and interfacial shear continue to increase, the annular mist-flow regime forms and persists throughout the remainder of the exchanger (i.e., axial location > 0.8.)

CONCLUSIONS

The relationships and subsequent examples presented in this paper demonstrate that the local heat transfer mechanisms and flow regimes occurring in heat exchangers are strongly dependent on process conditions. Changes in process fluid pressure and heat source temperature can dramatically affect the heat exchanger's design. The design is complicated further for fluid mixtures because of (1) additional mass transfer resistance that reduces the heat transfer coefficient and (2) local changes in VLE due to variable composition and pressure.

RESEARCH NEEDS

The predictive model given in this report contains many assumptions; therefore, more fundamental studies of convective boiling phenomena in tube bundles are necessary to verify and improve current and future boiling models and correlations. A list of future studies for both crossflow and longitudinal (window) flow in tube bundles includes:

(1) Two-phase flow studies under vacuum conditions
(2) Functional expressions for the nucleate boiling and forced convection mass transfer coefficients for mixture correction factors
(3) Partial and fully-developed subcooled boiling
(4) Verification of the Jensen-Tang CHF criteria for vacuum and high pressure conditions
(5) CHF criteria for subcooled boiling in tube bundles
(6) Subcooled transition/film boiling heat transfer coefficients
(7) Transition/film flow boiling heat transfer coefficients
(8) Annular mist/mist flow boiling heat transfer coefficients

In addition, further attention should be given to the details and influence of pressure drop and VLE phenomena on heat transfer coefficients for fluid mixtures in bundle crossflow and windowflow.

NOMENCLATURE

A -   Tube bundle outer surface area

C_A -   Coefficient in the two-phase frictional multiplier for shellside heat transfer and pressure drop

C_{TD} -   Taitel-Dukler flow regime parameter

D_0 -   Tube outer diameter

F_c -   Mixture correction factor for nucleate boiling

F_{cc} -   Mixture correction factor for forced convection evaporation

F_e -   Fin efficiency correction factor for nucleate boiling

F_s -   Fin surface-area correction factor for nucleate boiling

F_{tp} -   Two-phase correction factor for the liquid-phase heat transfer coefficient

G -   Mass velocity

g -   Acceleration due to gravity

h_b -   Convective boiling heat transfer coefficient

h_{fc} -   Forced convection heat transfer coefficient

h_l -   Liquid-phase heat transfer coefficient

h_{nb} -   Nucleate boiling heat transfer coefficient

h_{nbl} -   Nucleate boiling heat transfer coefficient on a single tube

h_{nc} -   Natural convection heat transfer coefficient

k_l -   Liquid thermal conductivity

L -   Tube bundle length

m_l -   Exponent for the two-phase heat transfer correction factor

P_r -   Reduced pressure

q -   Heat flux

q_1 -   Critical heat flux for transition/film boiling (region 1)

q_2 -   Critical heat flux for the region between transition/film boiling and annular mist flow (region 2)

q_3 -   Critical heat flux for annular mist flow (region 3)

q_b -   Convective boiling heat flux

q_{fc} -   Forced convection heat flux

q_{max} -   Single tube maximum heat flux corresponding to transition/film boiling

q_{min} -   Minimum flow boiling heat flux

q_{nb} -   Nucleate boiling heat flux

q_{nc} -   Natural convection heat flux

Re -   Shellside Reynolds number

T_b -   Bulk temperature

T_{bub} -   Dew point temperature

$T_{sat,i}$ -   Saturation temperature for component i

T_w -   Wall temperature

x_i - Liquid-phase mole fraction for the more volatile component

X_{tt} - Martinelli parameter for turbulent-turbulent flow

y - Local vapor quality

y_{12} - Vapor quality separating CHF regions 1 and 2

y_{23} - Vapor quality separating CHF regions 2 and 3

y_i - Vapor-phase mole fraction for the more volatile component

y_{outlet} - Outlet vapor quality

Y_0 - Characteristic bubble diameter in the Bennet et al. [1980] nucleate boiling suppression factor

Greek

α - Nucleate boiling suppression factor

β_{nb} - Nucleate boiling mass transfer coefficient

ΔT - Wall superheat

ΔT_{max} - Maximum wall superheat corresponding to q_{max}

ΔT_{min} - Minimum wall superheat corresponding to q_{min}

λ - Latent heat of vaporization

μ_l - Liquid viscosity

μ_v - Vapor viscosity

ϕ_l - Two-phase frictional multiplier for the liquid phase

Ψ - Jensen-Tang [1994] Convective boiling bundle correction factor for q_{max}

ρ_l - Liquid density

ρ_v - Vapor density

ρ_{tph} - two-phase homogeneous density

σ - Surface tension

REFERENCES

Bennett, D. L., and Chen, J. C., 1980, "Forced Convective Boiling in Vertical Tubes for Saturated Pure Components and Binary Mixtures," *AIChE J.*, Vol. 26, No. 3, pp. 454-461.

Bennett, D. L., Davis, M. W., and Hertzler, B. L., 1980, "The Suppression of Saturated Nucleate Boiling by Forced Convective Flow," AIChE Symp. Series, Vol. 76, No. 199, pp. 91-103.

Bromley, J. A., 1950, "Heat Transfer in Stable Film Boiling," *Chem. Engr. Prog.*, Vol. 46, No. 5, pp. 221-227.

Chen, J. C., 1966, "Correlation for Boiling Heat Transfer to Saturated Fluids in Convective Flow," *I&EC Process Design and Development*, Vol. 5, No. 3, pp. 322-329.

Dhir, V. K., 1991, "Nucleate and Transition Boiling Heat Transfer under Pool and External Flow Conditions," *Int. J. Heat and Fluid Flow*, Vol. 12, No. 4, pp. 290-314.

Dhir, V. K., and Liaw, S. P., 1989, "Framework for a Unified Model for Nucleate and Transition Pool Boiling," *ASME J. Heat Transfer*, Vol. 111, pp. 739-746.

Dowlati, R., Chan, A. M. C., and Kawaji, M., 1993, "Measurement of Void Fraction and Pressure Drop in Flow Boiling of R-113 Across Horizontal Tube Bundles," ASME Paper No. 93-HT-17, New York.

Forster, H. K., and Zuber, N., 1955, "Dynamics of Vapor Bubbles and Boiling Heat Transfer, *AIChE J.*, Vol. 1, pg. 531.

Groenveld, D. C., 1975, "Post-Dryout Heat Transfer: Physical Mechanisms and a Survey of Prediction Methods," *Nuclear Engineering and Design*, Vol. 36, pp. 17-26.

Ishihara, K., Palen, J. W., and Taborek, J., 1977, "Critical Review of Correlations for Predicting Two-Phase Flow Pressure Drop Across Tube Banks," ASME Paper No. 77-WA/HT-23, New York.

Jensen, M. K., and Hsu, J. T., 1988, "A Parametric Study of Boiling Heat Transfer in a Horizontal Tube Bundle," *ASME J. Heat Transfer*, Vol. 110, pp. 976-981.

Jensen, M. K., and Tang, H., 1994, "Correlations for the CHF Condition in Two-Phase Crossflow Through Multitube Bundles," *ASME J. Heat Transfer*, Vol. 116, pp. 780-783.

Klausner, J. F., Mei, R., Bernhard, D. M., and Zeng, L.Z., 1993, "Vapor Bubble Departure in Forced Convection Boiling," *Int. J. Heat Mass Trans.*, Vol. 36, No. 3, pp. 651-662.

Kutateladze, S. S., 1961, "Boiling Heat Transfer," *Int. J. Heat Mass Transfer*, Vol. 4, pp. 31-45.

Leroux, K. M., and Jensen, M. K., 1992, "Critical Heat Flux in Horizontal Tube Bundles in Vertical Crossflow of R113, *ASME J. Heat Transfer*, Vol. 114, pp. 179-184.

Lienhard, J. H., 1988, "Burnout on Cylinders," *ASME J. Heat Transfer*, Vol. 110, pp. 1271-1286.

Lockhart, R. W., and Martinelli, R. C., 1949, "Proposed Correlation of Data for Isothermal Two-Phase Two-Component Flow in Pipes," *Chem. Engr. Prog.*, Vol. 45, No. 1, pp. 39-48.

McAdams, W. H., 1954, Heat Transmission, 3rd ed., McGraw-Hill, New York.

Palen, J. W., and Yang, C. C., 1983, "Circulation Boiling Model for Analysis of Kettle and Internal Reboiler Performance," Heat Exchangers for Two-Phase Flow Applications, ASME HTD-Vol. 27, pp. 55-61.

Palen, J. W., Wang, Q., and Chen, J .C., 1994, "Falling Film Evaporation of Binary Mixtures," *AIChE J.*, Vol. 40, No. 2, p. 207.

Ragheb, H. S., and Cheng, S. C., 1979, "Surface Wetted Area during Transition Boiling in Forced Convective Flow," *ASME J. Heat Transfer*, Vol. 101, pp. 381-383.

Rohsenow, W. M., 1962, "A Method of Correlating Heat Transfer Data for Surface Boiling of Liquids," *Trans. ASME*, Vol. 84, p. 969.

Schlünder, E. U., 1983, "Heat Transfer in Boiling of Mixtures," *Int. Chem. Eng.*, Vol. 23, No. 4, pp. 589-599.

Steiner, D., and Taborek, J., 1992, "Flow Boiling Heat Transfer in Vertical Tubes Correlated by an Asymptotic Model," *Heat Transfer Engineering*, Vol. 13, No. 2, p. 43.

Stephan, K., and Abdelsalam, M., "Heat Transfer Correlations for Natural Convection Boiling," *Int. J. Heat Mass Transfer*, Vol. 23, pp. 73-78.

Taitel, Y., and Dukler, A. E., 1976, "A Model for Predicting Flow Regime Transitions in Horizontal and Near Horizontal Gas-Liquid Flow," *AIChE J.*, Vol. 22, pp. 47-55.

Thome, J. R., and Shakir, S., 1987, "A New Correlation for Nucleate Pool Boiling of Aqueous Mixtures," AIChE Symp. Series, No. 83, pp. 46-51.

Tong, L. S., and Young, J. D., 1974, "A Phemomenological Transition and Film Boiling Heat Transfer Correlation," *Proc. 5th Int. Heat Trans. Conf.*, Tokyo, Paper B.3.9, Vol. IV, pp. 120-124.

Webb, R. L., and Gupte, N. S., 1992, "A Critical Review of Correlations for Convective Vaporization in Tubes and Tube Banks," *Heat Transfer Engineering*, Vol. 13, No. 2, pp. 58-81.

Zeng, L. Z., and Klausner, J. F., 1993, "Nucleation Site Density in Forced Convection Boiling," *ASME J. Heat Transfer*, Vol. 115, pp. 215-221.

BOILING IN COMPACT HEAT EXCHANGERS/
INDUSTRIAL PRACTICE AND PROBLEMS

L E Haseler and D Butterworth
Heat Transfer and Fluid Flow Service
AEA Technology, UK

ABSTRACT

A number of industrial applications of boiling in compact heat exchangers are examined, to identify features particular to these applications, and to the exchanger geometries, which need to be taken into account in boiling heat transfer calculations. Particular emphasis is given to recent developments of compact heat exchangers Some existing methods for calculating boiling heat transfer coefficients are then reviewed, to identify how they can be adapted for use with compact heat exchangers. A number of areas of uncertainty remain, in predicting the boiling performance of compact heat exchanger in industrial use.

INTRODUCTION

The term 'compact heat exchanger' can be used for a wide range of devices. The objective of this paper is to identify a number of applications where boiling occurs in compact heat exchangers, to point out some of the special features, which differentiate this boiling process from boiling in conventional tubular exchangers, and to highlight some of the problems in modelling boiling heat transfer in compact exchangers.

There is no universally agreed definition of what constitutes a compact heat exchanger. One broad definition might use a shell-and-tube heat exchanger with ¾ inch tubes as a reference, and define anything with a significantly smaller hydraulic diameter, or a significantly larger surface area density (area per unit volume) as being compact. It is probably more useful to identify a number of particular features, which are often found in compact heat exchangers, and which will affect the boiling process in these exchangers. The features are

High surface area density, and low hydraulic diameter (d_h): These two parameters are closely related, although the exact form of the relationship depends on the detail of the exchanger geometry, and the wall thickness. For the notional case of an exchanger comprising plain parallel plates of separation a, the surface area density is $2/d_h$ while for tubes with pitch $4d_h/3$, the surface area density is $9\pi/16d_h$, which is similar.

Low driving temperature differences: If these are required by a process, then a high exchanger surface area is needed. Exchangers with a high surface area density are usually appropriate.

Complex passage shapes: Compact heat exchangers are often formed from sheet metal, folded or formed to give the required structures, which sustain the fluid working pressure, and provide some heat transfer enhancement. At the simplest, the flow channels may be simple straight non-circular channels, but very often, not only is the cross-section non-circular, but the passages are not straight, and/or the cross-section varies along the flow path.

Fins: In some compact heat exchangers, the high surface area density requires that, in addition to the primary surface area, which gives direct heat transfer between fluids, some of the surface acts as a fin, with conduction along the fin metal playing a significant part in the heat transfer.

Special materials: When the fluids used impose a requirement for special heat exchanger material, there is clearly an advantage in going to compact exchangers which are relatively small, and if possible, are manufactured out of relatively thin plate, to minimise the cost of the material used.

This paper will concentrate on two main forms of compact heat exchangers, the plate heat exchanger (gasketed plate and frame and its derivatives) and the plate-fin heat exchanger. Mention will be made in passing of a range of other types of exchangers, in particular the printed circuit exchanger, which has a niche for applications at higher temperature and/or pressures. In each case, a brief description of the exchanger will be given, followed by a number of examples of applications involving boiling. A more general review will then be provided of the problems involved in modelling boiling heat transfer in compact heat exchangers.

PLATE HEAT EXCHANGERS

Exchanger Structure

Plate heat exchangers (PHE's) are made from stacks of plates onto which a pattern of corrugations has been pressed. An example is shown in Figure 1. Plates can be from 0.3m to over 2m long, while their width is typically 20 to 40 percent of their length. The two streams flow in alternate channels between plates, entering and leaving via ports in the corner of the plates. In each set of channel there is usually a gasket round the edge of the plate, and around the ports carrying the fluid in the other channels. The stack of plates is clamped together in a frame. The exchanger can be entirely dismantled for cleaning, which makes it very suitable for the food industry and other hygienic applications.

The plate thickness in plate heat exchangers is normally in the range 0.6 to 0.9 mm, while the plate spacing lies between 2.5 and 4 mm, except for special wide gap plates sometimes used for vapours or fibrous materials. The hydraulic diameter for flow between plates approximates to twice the plate spacing. It incorporates a factor to allow for the undulatory nature of the path, based on the ratio of the actual plate area to the projected area.

For duties where the use of gaskets is undesirable, plates pairs can be made with a laser weld round the edge instead of a gasket. Gaskets are used between the pairs, for the less hazardous fluid.

Operating pressures up to 25 bar are standard, while somewhat higher pressures can be achieved using heavy duty frames. Temperatures are usually limited by the gasket. Upper limits of 160 to 180 deg C apply to most gasket materials. Asbestos based gaskets can take operating temperatures up to 240 deg C, but do not have the elasticity of other materials used for gaskets. The APV company have recently announced a graphite based gasket, which is claimed to operate up to temperatures of 400 deg C.

A variant on the conventional PHE is the plate in shell heat exchanger. This is illustrated in Figure 2, (courtesy of APV). The plates are circular, and welded into a stack which fits into a cylindrical shell. Operating pressures up to 350 deg C, and pressures up to 100 bar can be achieved. Although apparently very different from the conventional rectangular plate, the underlying flow passage structure of the two exchanger types is the same.

Another variant is the brazed plate exchanger. These have a similar basic plate structure to a conventional PHE, but the stack of plates is brazed together. This eliminates the need for a either a frame or gaskets, and gives a very inexpensive exchanger. Plate lengths are usually one metre or less.

The common feature of all plate heat exchangers is the use of corrugations in the plates, giving both support against internal pressures and heat transfer enhancement. The simplest plate pattern is the 'washboard plate', with the corrugations perpendicular to the plate axis, and alternate plates arranged so that flow is along a sinusoidal path of constant cross section.

The more common type of plate uses crossed corrugations, that is, the corrugation patterns in adjacent plates are at an angle to each other, giving lattice of support points where they touch, and a complex flow channel shape between the plates. The corrugations are usually formed as chevrons. There may be a single chevron pattern, or, as in Figure 1, multiple chevrons across the plate width. Other variants have the chevron pattern running along the length of the plate. In all cases, however, the local flow geometry has the same crossed corrugation pattern. For single phase flows, the larger the angle between the corrugations and the plate axis, the larger the heat transfer coefficient and pressure gradient.

Boiling in Plate Heat Exchangers

Plate heat exchangers are used in a number of applications involving boiling. They are used in both large and small scale refrigeration systems, with either ammonia or organic refrigerants. In refrigeration applications the exchanger can be in either the 'flooded' or 'dry' evaporator mode. In the 'flooded' duty, the exchanger forms part of a thermosiphon loop, being fed with liquid from a vessel located above the exchanger, and returning the partially evaporated refrigerant back up to the vessel.

In the 'dry' duty, the exchanger is located directly after the expansion valve in the refrigeration circuit, so that a two phase mixture, with quality (vapour mass fraction) of around 15 percent is fed into the bottom of the exchanger. Obtaining a reasonably uniform distribution of the phases among the plate channels is clearly important, and sometimes a proprietary phase distribution device is included in the inlet port.

Plate heat exchangers are also used in shipboard de-salination units, as both evaporator and condenser. Specially formed titanium plates are used, giving a large flow area for the water vapour at 0.1 bar, from the sea water evaporating at 40 degrees C. Engine cooling water at 60-80 degrees C is used as the heat source.

In the food industry too, plate heat exchangers are used in concentrating products by evaporating off water. This is sometimes done in a multiple-effect system, with various stages operating at different, (sub-atmospheric) pressures. The vapour evaporated in one stage is condensed to supply heat to a lower pressure stage.

The facility for dismantling PHE's for cleaning is not the only reason for their use in the food industry. The products being concentrated in the exchanger are often viscous, and may contain suspended solids. Although locally the flow path in a PHE, with crossed chevrons, is complex, overall the flow approximates to plug flow through the exchanger. Unlike shell and tube exchangers, there are no recirculation zones, where suspended solids might deposit.

Correspondingly, the residence time for flow through a plate exchanger is relatively well defined. Because of the short path lengths and high velocities, the residence time is also relatively small. There is thus less risk that some parts of the flow will remain in the exchanger for significantly longer than others, which would clearly be undesirable for heat sensitive products.

Plate heat exchangers are also used for the partial evaporation of fluids which are extremely viscous, and possibly non-Newtonian as well. This can present a particularly complex situation. As the fluid evaporates, its viscosity changes both because its composition changes and because its temperature changes. In addition, the very high viscosity leads to very large pressure changes within the exchanger. As the fluid flows through the exchanger, significant evaporation occurs because of the pressure change, as well as because of the heat input.

The open technical literature contains very few data on two phase flows in PHE's. Manufacturers, however, undoubtedly have their own proprietary data, built up from experience. Unlike most other exchangers, the PHE can easily have its flow area increased, by adding extra plates, if the heat transfer achieved is too low, or the pressure drop is too high. If a design mistake did arise because of limited information it would be relatively inexpensive to rectify.

PLATE-FIN HEAT EXCHANGERS

Exchanger Structure

A typical multistream brazed aluminium plate-fin heat exchanger is illustrated in Figure 3. It consists of a block (core) of alternating layers (passages) of corrugated fins. The layers are separated from each other by parting sheets, and sealed along the edges by means of side bars. In most plate-fin exchangers, the central 'effective length' of the block consists of finning laid parallel to the block axis, to give true counterflow heat exchange among the streams. At the end of the exchanger, as illustrated in Figure 3, there are pads of finning laid at an angle, known as distributors. These direct the flow to gaps (ports) in the side bars so as to connect the heat transfer passages with the headers and nozzles, which are welded onto the outside of the block.

Large plate-fin heat exchangers are made of aluminium, and can be up to 1.2m wide, 1.2m deep (the stack height) and 6.2m long. They are most frequently used in low temperature applications, where they give the benefit of a multi-stream capability, ensuring that all the cold streams produced in a process can be used to cool the incoming warm streams. The six manufacturers of these exchangers have recently formed the Brazed Aluminium Plate-fin Heat Exchanger Manufacturers Association. The manufacturers are

Altec (USA)
IMI Marston (UK)
Kobe Steel (Japan)
Linde AG (Germany)
Nordon Cryogenie (France)
Sumitomo Precision (Japan).

The manufacturers association has recently produced a standards document for brazed aluminium plate-fin heat exchangers, from which some of the figures in this report are taken.

Brazed aluminium exchangers can be used for streams at pressures up to 100 bar, and although most commonly used for cryogenic applications, they can be used at temperatures up to 200 degrees C, if appropriate alloys are used for the headers and nozzles.

Within the plate-fin core, each stream flows in a number of layers, each of which is divided into a number parallel subchannels, by the finning. Fin heights are typically between 5 and 9 mm, while fin frequencies, in the core heat transfer region, are typically 15-20 fins/inch (590-787 fins/m). Subchannel hydraulic diameters are thus a few millimetres at most.

There are four basic finning geometries, as illustrated in Figure 4. All manufacturers make plain, perforated and serrated (offset strip) fins. Some make wavy fins, others prefer serrated fins with a long serration length. The presence of perforations gives a small enhancement over plain fins for single phase performance. Perforated finning is often used for boiling, because the perforations are seen as providing a facility for the equalisation of flows among the subchannels, mitigating against local blockage or pressure fluctuations arriving from the evaporation process.

Serrated fins give a significant increase in both heat transfer and pressure drop over plain fin values. The standard length of serrations is 3 mm (or 1/8 inch). Using a longer length (12 or 15 mm) gives a fin of performance intermediate between plain/perforated fins and standard serrated fins.

A number of developments are being undertaken to produce plate-fin heat exchangers capable of operating at higher temperatures and pressures, and in materials other than aluminium. Stainless steel exchangers have been used for some time in vehicle and aerospace applications, mainly for single phase duties. These are mainly relatively small exchangers - blocks with sides below 0.3m. Some manufacturers, however, can supply larger brazed stainless steel plate-fin units for process industry applications. IMI Marston produce exchangers up to about 0.6m by 0.6m by 1.5m long.

Stainless steel plate-fins are geometrically similar to brazed aluminium plate-fins, but normally have fin heights below 5 mm, because of the poor metal thermal conductivity. They also use plain fins, because other fin types are difficult to manufacture in stainless steel. The braze metal used for stainless steel exchangers is copper. The effect of the braze on process fluids has sometimes been of concern to potential users. IMI Marston are therefore developing diffusion bonding techniques for stainless steel plate-fin exchangers to avoid problems with the braze.

Another development of plate-fin heat exchangers capable of high pressure, high temperature applications is being undertaken by Rolls Laval (a joint venture between Rolls Royce and Alfa Laval) using a diffusion bonding technology developed for titanium turbine blades. The manufacturing technique, illustrated in Figure 5. Three sheets of titanium are diffusion bonded at selected positions, and then expanded by pressurising the unbonded regions between the plates. This gives what is basically a plate-fin geometry, though the subchannels are trapezoidal in shape, and rather larger than the subchannels in aluminium plate-fins. These exchangers can operate at temperatures above 550 degrees C and at pressures above 200 bar. Although to date all applications of the new technology are single phase, they could also be used for boiling applications.

Plate-fin Thermosiphon Reboilers

Aluminium plate-fin heat exchangers are most commonly used in cryogenic applications. Some of the simplest boiling applications are in air separation, where a nearly pure stream is evaporated in a plate-fin thermosiphon. For example, in the basic double-column separation unit, the reboiler of the upper, low pressure column is heated by condensation of a high pressure nitrogen stream at the top of the lower column. The thermosiphon is located inside the upper column, so that it is partially immersed in the column sump, and the boiling stream has direct entry at the bottom of the exchanger, and leaves directly at the top, without passing through any pipework, headers or distributors.

A schematic form of the temperature profiles of the boiling and condensing streams is shown in Figure 6. The condensing stream, in downflow, experiences little pressure or composition change, and is approximately isothermal. The liquid stream enters the plate-fin thermosiphon subcooled, and is first heated to its boiling point. Then, as it boils, its pressure, and hence saturation temperature, decrease. Very small driving temperature differences are used, because larger values would require higher condensing stream pressures, needing much higher energy input to the compressors which drive the process. The pinch temperature difference, in mid-exchanger, is around or below one degree Kelvin.

Plate-fin thermosiphons are designed to operate with exit qualities in the range 0.1 to 0.15. Predicting the performance of such exchangers is not easy, since not only must accurate heat transfer coefficients be known, but single and two phase pressure drops must be calculated precisely. The local boiling stream pressure in the two phase region has a significant effect on the driving temperature difference, in addition to the effect of the thermosiphon flowrate on heat transfer coefficients.

A further complication in oxygen re-boilers is the risk of build up of solid acetylene. This can in principle happen if there is boiling to dryout in any part of the exchanger. Ignition of the acetylene in the oxygen, with explosive consequences, may result. Good distribution of the boiling flow throughout the exchanger is thus of fundamental importance. The upper quality limit of 15% is intended to maintain slug/churn flow, as there is perceived to be a risk of dryout in the annular flow found at higher qualities. Perforated fins have traditionally been used to assist in flow distribution, but serrated fins are also now considered for use.

To avoid the pressure drop and power penalties associated with the liquid head for a thermosiphon reboiler, falling film vaporisation is being considered as an alternative. The pressure drop of the boiling stream thus becomes negligible, so the temperature difference across the exchanger becomes uniform. The extra complexity of a liquid pump to re-circulate the un-evaporated liquid back to the top of the exchanger is more than compensated for by the benefits of reduced power input, and/or exchangers with larger driving temperature differences.

Plate-fin Kettle Reboilers

A relatively new development is the plate-fin kettle reboiler. This is similar to a conventional kettle reboiler, but instead of a tube bundle, a plate-fin core is built into the kettle shell, as shown in Figure 7, courtesy of IMI Marston. The boiling is in vertical upflow, with no header or distributor region, while the condensing stream flows horizontally along the exchanger.

Unusually for a plate-fin exchanger, the two streams are in crossflow, but if the condensing stream is isothermal, this is of relatively little significance. If the condensing stream heat transfer coefficient is higher near the end of the exchanger where it enters, then more vigorous boiling, and higher thermosiphon flowrates might be expected there.

Other Plate-fin Boiling Applications

In a cryogenic process, boiling will often occur in the main exchanger, in which the major temperature reduction required by the process is achieved. The boiling stream can be either a product stream, a waste stream, or a refrigerant stream. There may be more than one coolant (cold) stream in an exchanger. Some may be boiling, others gas-only-streams. The boiling streams can be either pure components or mixtures, but more importantly, they will usually completely vaporise in the exchanger, and then the resulting vapour continue to act as a coolant, in the same set of layers in the exchanger.

A straightforward example of a two-stream exchanger, involving a boiling multicomponent refrigerant, can occur in natural gas liquefaction processes. As the refrigerant boils, its temperature increases, and the temperature of the vapour produced must also be raised. The composition of the refrigerant will have been selected so that its temperature enthalpy profile matches as closely as possible that of the gas being condensed. A perfect match is in general not possible, so the temperature difference between the two streams will vary significantly along the exchanger. There will be one (or more) pinch points within the exchanger, at which very small temperature differences, of only a degree or so will occur.

The stream temperature profiles used in the process design, and in the exchanger design, all assume equilibrium between the liquid and vapour phases of the stream. It is important that the boiling process achieves a close approximation to such equilibrium. There must be an effective mechanism for heating up any vapour generated, as it flows upwards with the liquid. Plate-fin heat exchanger passages, which maintain close contact between the phases will promote such mechanisms. Surfaces such as serrated fins, which lead to relatively high gas coefficients, are also likely to be helpful.

In two stream exchangers, the driving temperature difference between streams is at least well defined. In multistream exchangers the situation is more complicated. This applies particularly in main exchangers where one of the cold streams is a pure product, which boils and is superheated within the exchanger. While it is boiling, such a stream will have a high heat transfer coefficient, and be the main sink of heat for the hot streams. After it has boiled, it may well make a relatively small contribution to the cold streams' heat load.

This presents two problems in plate-fin heat exchanger design. The first relates to the number of layers for this stream. In the boiling region, relatively few layers would be required, because of the high heat transfer coefficient. In the vapour superheating region, many more layers may be needed. Changing the number of layers a stream occupies part way along an exchanger is possible in principle, but unlikely in practice. Changing layers is only usually done if there are otherwise partly empty layers which must be filled. Some compromise on the required number of layers must therefore be made.

The second problem relates to the location of the stream in question in the layer pattern. It is desirable to spread the cold stream load uniformly throughout the layer pattern. If there are layers which represent a large fraction of the heat load in a boiling region, but a small fraction in the vapour superheating region, the distribution of heat load across the layer pattern cannot be uniform throughout the exchanger. This may have an effect on thermal performance, but more seriously, it may lead to thermally induced stresses in the exchanger.

Prediction of the stream temperature profiles in such an exchanger is important if the possibility of thermal stress related damage is to be assessed. The calculation process must allow for heat conduction through the exchanger via the fin metal, in parallel with the single phase and/or boiling heat transfer processes which are occurring.

PRINTED CIRCUIT HEAT EXCHANGERS

Exchanger Structure

The printed circuit heat exchanger (PCHE), manufactured by the Heatric Company, uses diffusion bonding technology in its manufacture. The term 'printed circuit' is used because the flow channels are manufactured by chemically etching into a flat plate. The plates are then stacked and diffusion bonded together, to give an exchanger capable of operation at pressures up to 1000 bar and temperatures up to 900 degrees C. The exchangers can be manufactured in either stainless steel or a range of higher alloys.

Figure 8 shows a cross-section though the passages in a printed circuit heat exchanger. The size of the passages is normally between 0.5 and 2.0 mm deep, and the cross-section approximates to a semi-circle. Figure 9 illustrates a chemically milled plate. It can be seen that a zig-zag pattern is used. Various combinations of crossflow and counterflow of the streams are employed as required.

Boiling in PCHE's

Applications in which printed heat exchangers are particularly useful include those involving boiling with a large temperature difference between the streams. Examples might be vaporisation of liquid natural gas by sea water, or by process fluids at 100 or 200 deg C. As boiling occurs, there can be severe temperature fluctuations within the exchanger, with consequent thermal stressing of the exchanger metal. With aluminium plate-fin heat exchangers, which might otherwise have been used for LNG vaporisation, there may be a risk of failure due to fatigue. The much more robust PCHE avoids this problem, while still providing a high surface area density.

PCHE's are also used for more straightforward boiling duties involving hydrocarbons. Normally there would be liquid phase input, but two phase input is also possible. Sometimes it is possible to accept the risk of maldistribution of the phases for a two phase input, but in general it is preferable to separate the phases, before input. A key feature of PCHE's is that flow is along a set of channels in parallel, so that any flow distribution achieved at input is maintained throughout the exchanger.

When PCHE's are used for boiling with large temperature differences, a particular problem is calculating when and where the critical heat flux may occur. Most critical heat flux data are for flat plates or tubes, and the heat fluxes and temperature differences used are not easily related to those in a PCHE application.

BOILING THEORY FOR COMPACT HEAT EXCHANGERS

Basic Boiling Methods

Models of boiling in compact heat exchanger passages must start from models for in-tube boiling, where there is much more extensive data, and the geometry is simple. The literature contains many in-tube boiling correlations. These have recently been reviewed by Webb and Gupte (1992). For applications to compact heat exchangers, two particular methods will be considered here: Chen (1963) and Steiner and Taborek (1992).

In general, the methods for saturated flow boiling are based on the combination of correlations for two different physical phenomena, nucleate boiling and convective boiling. For application to compact heat exchangers, it is important to understand the assumptions behind the methods.

Nucleate Boiling

The nucleate coefficient h_n, is found from a pool boiling coefficient, h_p, which may be multiplied by a suppression factor S to account for convective effects.

$$h_n = S\,h_p \qquad (1)$$

Pool boiling coefficients increase with temperature difference (wall superheat), ΔT, but by their nature, do not depend on stream mass flux or quality. In pool boiling, the dominant heat transfer mechanism is associated with the release of bubbles of vapour, so that increasing ΔT increases the rate of bubble departure, which increases the disturbance to the liquid film through which heat is transferred to make vapour, and hence increases the heat transfer coefficient.

The physical basis of the suppression factor is that the steeper temperature gradient near the wall, which will apply when convective boiling is present, will reduce the amount of superheated liquid in contact with a growing bubble. Suppression factors, however, are based on correlations, rather than physical models of the bubble growth process.

Pool boiling coefficients are often correlated by a power law dependence on heat flux. Simple algebra can convert this to a (different) power law dependence on ΔT. When used as part of a flow boiling model, it is important to use the ΔT form. The driving force for the nucleation of bubbles is the local wall temperature difference. The heat flux associated with nucleate boiling is not known separately, only the total heat flux, including both nucleate and convective effects is known.

Convective Boiling

The convective boiling coefficient h_c is generally correlated by multiplying a liquid coefficient h_l, by some factor F, representing the two phase enhancement.

$$h_c = F\,h_l \qquad (2)$$

The convective coefficient is assumed to involve heat transfer through a liquid film, the structure of which is determined by the local flow conditions (mass flux and quality), but not the rate of heat transfer (no dependence on ΔT).

Combining Coefficients. There are a number of options for combining the convective and nucleate coefficients into an overall boiling coefficient h_b. One method, designed to give asymptotically correct results in the limit when one or the other mechanism dominates, uses the relation

$$h_b = (h_n{}^3 + h_c{}^3)^{1/3} \qquad (3)$$

This form was used by Steiner and Taborek (1992). It is illustrated in Figure 10, where boiling coefficient is plotted as a function of quality (vapour mass fraction). Steiner and Taborek assumed that the boiling suppression factor S is unity, so their nucleate coefficient is independent of mass flux and quality.

The addition of the nucleate and convective components using the mean cube approach is an ad hoc device. It acknowledges that the convective and nucleate mechanisms occur in parallel, and are broadly additive, but that they interfere with each other in some way, so that the total heat transfer is less than the simple sum of the two. The form of the mean cube approach is such that explicit suppression factors are not needed.

It may be noted that a higher exponent than three in the averaging process represents more interaction. In the limit where the exponent is infinite, this is equivalent to asserting that whichever is the more effective mechanism, convective or nucleate boiling, completely suppresses the other, so the overall coefficient is simply the larger of the two.

The older, well established, Chen (1963) correlation, by contrast, uses simple addition of the convective and nucleate terms. It has therefore to incorporates an effective nucleate boiling suppression factor, which depends on mass flux and quality, to account for the absence of heat flux dependence when convective heat transfer dominates. Chen's convective term, however, is unaffected by the nucleate term.

The form of the interaction between the nucleate and convective components of boiling is a real physical question, and not just an artifice of the combination method. One key issue is whether the nucleate boiling term is affected by mass flux and velocity, and an explicit suppression term is required.

Steiner and Taborek (1992) quote extensive nucleate boiling data showing no significant dependence on mass flux or quality. However, as Webb and Gupte (1992) point out, a small convective component which increases with both mass flux and quality could approximately balance out the suppression of the nucleate term.

Wadekar (1995) shows that a simple addition approach for nucleate plus convective terms, which also incorporates a reduction in the convective coefficient because of nucleate boiling, can predict coefficients which are effectively independent of wall superheat in the convective region.

HTFS has unpublished data for both tubes and plain plate-fins, which indicate small decreases in coefficient with both mass flux and quality. Since convective coefficients always increase with these parameters, suppression of nucleate boiling appears a possible explanation.

Size Dependence

An obvious question to ask of the various boiling methods is what dependence on hydraulic diameter they contain, and whether it is plausible.

For the convective component, both Chen (1963) and Steiner and Taborek (1992) contain no explicit dependence on hydraulic diameter, but both contain an implicit dependence in their use of a single phase convective coefficient. The effect of size on the convective boiling coefficient is thus significant.

For the nucleate component, there seems no clear reason why passage size should have an effect, unless the size is comparable with the bubble departure diameter, when any effect might be considerable. The nucleate term in the Chen correlation has no dependence on passage size.

The basic methods for nucleate boiling, on which Steiner and Taborek built, also have no size dependence. In their overall correlations, however, these authors introduce a significant size dependence in a multiplier $(d / 0.01)^{-0.4}$. They deduce this from data shown in Figure 11 (Figure 13 from their paper, based on Steiner, 1988). These data cover the range 1 to 20 mm, but are concentrated in the range 5-20 mm. Below 5 mm, the evidence for this multiplier which they present is less convincing.

Other data are provided by Galezha et al (1976) who undertook an interesting series of experiments, in which they measured boiling heat transfer coefficients on five different plate-fin surfaces, four serrated, one plain, using both R12 and R22 boiling in upflow. They developed correlating equations of the form

$$h = C A q^n \qquad (4)$$

where A was a physical property function, depending on reduced pressure, reduced temperature and molecular weight, and C was a number, which was different for each fin type. In general the parameter C was larger for the fins with smaller hydraulic diameters, and smaller serration lengths.

The apparatus used in these studies, however, was a thermosiphon loop, in which the flowrate was not measured, and may well have been different for all the fins. If all the data were in the nucleate boiling dominated region, as the authors believed, the flowrate would not matter. The exponent n on heat flux q was however rather small (1/3). For nucleate boiling an exponent between 0.5 and 0.75 would be more usual. The presence of either a subcooled boiling region, or a convective element in the boiling, would reduce the apparent exponent on heat flux. The conclusions about size dependence of nucleate boiling coefficients must thus be viewed with some caution.

Wambsganns et al (1993) measured boiling coefficients for horizontal flow in a 2.92 mm diameter tube, and the same authors (Tran et al, 1993) then reported parallel data for boiling in a small, horizontal rectangular channel, 4.06 mm high, and 1.70 mm wide, using a similar range of heat fluxes, mass fluxes and qualities. They found that a simplified Stephan and Abdelsalam (1980) correlation, in which the coefficient depends on the heat flux to the power 0.746, but not on mass flux or quality, fitted the tube data, but underpredicted the channel data. The channel data also showed evidence of a lower exponent at lower heat fluxes, suggesting there was an element of convective boiling.

Unfortunately, the tube experiments used R113, while the channel experiments used R12, so it is not easy to see whether the differences are due to size or shape effects, or to the use of different fluids. The fact that the test sections were horizontal also means that the lessons may not be directly applicable to industrial compact heat exchangers, where upflow boiling is most common.

There is thus a significant question requiring clarification. Size dependence at diameters of 5 mm and above is hard to understand, but is claimed to exist. At lower diameters, where size dependence might be plausible, the evidence is unclear.

Applying Boiling Methods to Compact Exchangers

A number of questions arise when applying boiling methods developed for tubes to compact heat exchanger surfaces. There is relatively little openly published data on boiling in compact heat exchangers, so it is inevitable that calculation methods will be used in the regions where they have little or no experimental backing. Ensuring that the calculation method, at the very least, extrapolates plausibly into such regions, is therefore important. This is not an easy task, and many areas of uncertainty remain.

A fundamental question is to what extent the convective enhancement to single phase heat transfer, produced by various compact heat exchanger surfaces, carries over into boiling heat transfer. Before looking at the overall picture, however, there are some important issues of detail.

The Convective Enhancement Factor.
One issue is how the convective enhancement is to be calculated. In both the Chen (1963) and Steiner and Taborek (1992) methods the enhancement factor is expressed as a function of physical properties and quality. While Steiner and Taborek use an ad hoc fitted relation, Chen's enhancement factor F is a function of the Martinelli parameter, χ. This can be expressed in three ways

$$\chi^{-1} = (x/(1-x))^{0.9} (\rho_l/\rho_g)^{0.5} (\eta_g/\eta_l)^{0.1} \qquad (5)$$

$$\chi^{-1} = (x/(1-x)) (f_g\rho_l/f_l\rho_g)^{0.5} \qquad (6)$$

$$\chi^{-1} = (x/(1-x))^{1-m/2} (\rho_l/\rho_g)^{0.5} (\eta_g/\eta_l)^{m/2} \qquad (7)$$

Equation 5 is the purely physical property based form, while Equation 6, with friction factors evaluated at liquid only and vapour only Reynolds numbers, is a more fundamental definition of the Martinelli parameter. Equation 7 uses the exponent m in the friction factor relation

$$f = a\,Re^{-m} \qquad (8)$$

The three equations give the same results for turbulent flow in a plain tube, where $m = 0.2$, but will not do so for compact surfaces. Sakaue (1994) reports that, using the more fundamental form based on friction factors, his plain fin boiling data agreed with the Chen correlation. This seems a sounder approach than using the physical property based form.

Evaluating Liquid-only Coefficients.
A closely related problem is the method of calculating the liquid coefficient in the Chen (1963) method, which uses the liquid only coefficient, based on the liquid fraction of the flow, and liquid phase properties.

The problem may be illustrated using Figure 12, which shows typical compact heat exchanger single phase performance data, in the form of plots of Colburn j factor against Reynolds number for plain and serrated fins. The single phase heat transfer coefficients, just before boiling begins, might be based on j factors at Reynolds number Re_0. These are the convective heat transfer coefficients at quality zero. Different Re_0 are shown for the two fin types, since plain fins often operate at higher Reynolds numbers than serrated fins.

The difficulty can arise because liquid-only coefficients are calculated at Reynolds numbers below Re_0. If Re_0 is sufficiently high that turbulent flow is well established, then the dependence of the coefficient on the Reynolds number, for plain fins is broadly similar to that for turbulent flow in a tube, and there is not a problem.

As shown in Figure 12, however, for serrated fins the Reynolds number dependence of the heat transfer coefficient is significantly different from that for turbulent flow in a tube. Simply using single phase performance data for the compact surface to evaluate the liquid only heat transfer coefficient is thus a dubious procedure. Some alternative method of evaluating liquid only coefficients is thus required. One option for extrapolation to lower Reynolds numbers is to use the tangent at the all-liquid point. Another is to use a slope parallel to the plain fin line, as illustrated in Figure 12.

Enhanced Coefficients - Plate-fin.
A fundamental question for flow boiling in compact heat exchangers is the extent to which enhancement of the single phase heat transfer coefficient, above the values for a plain channel, carry over to convective boiling coefficients.

It is certain that there can be an effect. Boiling coefficients on plate-fin surfaces with serrated fins have been published by Robertson (1979) and Carey and Mandrusiak (1986). Both gave significantly higher values than for plain fins. Robertson's data could be fitted quite well by just using the convective boiling part of the Chen correlation, and ignoring the nucleate part. By contrast, Sakaue (1994) published boiling data on plain plate-fin surfaces, indicating that the Chen correlation gives a reasonable fit, provided that the nucleate term and convective term are both included.

Enhanced Coefficients - Wire Inserts.
There is, however, also evidence that surfaces giving significant enhancement to single phase heat transfer can give no such enhancement in flow boiling. Measurements were made by HTFS, in vertical tubes containing inserts produced by the Cal-Gavin Company. Figure 13 shows the form of insert and the test section used.

The inserts have been very successful in increasing heat transfer coefficients in single phase applications. They are simply pushed into a tube, and have the added benefit, for fouling applications, that they can both reduce the wall temperature (which may mitigate fouling) and can be removed from the tube, bringing the fouling deposits with them, for cleaning.

The experimental set-up used by HTFS involved a long vertical tube divided into three sections. There were inserts in the top and bottom sections, but the central section was plain tube. Figure 14 shows an example of stream and wall temperature profiles obtained for boiling. In the corresponding single phase case, wall temperatures were very much lower in the regions with inserts, indicating a coefficient five or more times the plain tube value. For boiling, however, the wall temperatures are indistinguishable in the three regions, indicating no increase in boiling coefficient due to the insert.

Data such as that shown have been obtained both in the region where nucleate boiling would be expected to dominate, and in regions where both nucleate and convective mechanisms could be significant. In the nucleate dominated region, it is perhaps not unexpected that the insert, (which adds little to the surface area) has no significant effect. Nevertheless, it is of interest to note that nucleate boiling is apparently unaffected by a device which must severely affect flows adjacent to the wall. In cases where, for the plain tube surface, convective boiling plays a significant part, it is more surprising that adding an insert which dramatically affects the single phase coefficient apparently has no influence whatsoever on the boiling coefficient.

The data were obtained using a pure fluid, refrigerant 113. This is, of course, an unlikely application as far as the insert is concerned. It is, however, possible to envisage boiling situations where the presence of such wire loop inserts might be valuable. For example, in boiling a multi-component, they might enhance the vapour phase heat transfer, with benefits to the overall boiling coefficient.

It must also be conceded that tubes with wire loop inserts would not normally come under the definition of compact heat exchangers, but the lessons they provide have implications for these exchangers. Single phase heat transfer enhancement do not always carry over into convective boiling heat transfer.

Enhanced Coefficients - Plate Exchangers

Very significant enhancement to the single phase coefficients in plate heat exchangers is achieved by increasing the angle which the corrugation pattern makes with the flow direction. Provided that the angle is not too large, the single-phase flow is locally along the corrugation troughs in a plate. There is a change in direction at the edge of the plate or at the centre of the chevron, so that the flow then goes down the troughs of the plate forming the other wall of the flow channel. Interaction between the two criss-crossing flows leads to heat transfer enhancement.

With two phase flows, it seems unlikely that vapour will always follow the same rather complex path as would single phase liquid. Single phase enhancement would therefore not be expected to carry over fully into the boiling region. For boiling in plate heat exchangers, the few openly available publications include Panchal et al (1983), Marvillet (1992) and Osterberger and Slipcevic (1990). Data on plate geometries and on measurement conditions are however not always complete. Further data are thus an important requirement. HTFS is working in this area, and has some data which indicate that 'hard' plates,

with high single phase heat transfer, do not necessarily have correspondingly high boiling coefficients.

Boiling on Fins

In plate fin exchangers, the secondary (fin) surface area can be five or six times the primary surface area. It is common practice to assume a constant coefficient over both primary and secondary surfaces, and then to evaluate a fin efficiency using the standard hyperbolic tangent (tanh) formula, using the fin half-height, because the fins are double-ended. The temperature gradient along the fin, required for conduction, means that the metal to stream temperature difference varies along the fin. For a constant coefficient, the temperature difference is given by a hyperbolic cosine function of position along the fin. For boiling, the assumption of uniform heat transfer coefficient along the fin will not in general be true.

If the coefficient varies exactly as temperature difference to some power, Liaw and Yeh (1994) showed that there is an exact solution for the temperature profile, in terms of hypergeometric, rather than hyperbolic, functions. The fin efficiency, which depends on the temperature gradient at the fin root, can thus also be expressed in terms of hypergeometric functions. This solution does not apply, however, if there are other considerations such as a convective element to the heat transfer, or a minimum temperature difference for nucleation, or dryout at high temperature differences.

If it can be assumed that the heat transfer coefficient increases with temperature difference, which will be true unless dryout occurs, then some simple and general deductions can be made. Any local decrease in coefficient must make the overall heat transfer worse. Any increase must improve it.

Using the temperature difference on the primary surface, ΔT_0, it is possible to evaluate a coefficient h_0, and then calculate the fin efficiency which would apply if this coefficient applied everywhere. Using h_0 with this efficiency will give an overestimate of the actual heat transfer.

It is also possible to calculate iteratively the coefficient h_2, calculated at ΔT_2, which, if it applied everywhere along the fin would give a mid-fin temperature difference ΔT_2. Since the mid point is the minimum temperature difference, the true coefficient at elsewhere on the fin must be above h_2. Using h_2 with the fin efficiency calculated for uniform h_2 thus gives an underestimate of the heat transfer.

The uniform coefficient assumption can thus be used to determine upper and lower limits to the heat transfer, defined by h_0 and h_2. If the fin efficiency is not too low, and the temperature dependence of the coefficient is not too great the two limits may be sufficiently close that taking a mean value is a reasonable approximation to the true solution.

When using brazed aluminium plate-fin heat exchangers, it is normal practice to try to keep fin efficiencies to about 0.7 or above, by selecting an appropriate fin thickness.

For plate fin exchangers made of materials other than aluminium, however, much lower fin efficiencies will be inevitable. An understanding of the way heat transfer coefficients depend on temperature difference may be much more important.

CONCLUSIONS

A number of compact heat exchanger geometries have been described, including types recently developed. A range of boiling applications in these exchangers have been identified, together with special features that need to be taken into account in modelling the boiling heat transfer.

Some current boiling methods, based on flow in tubes, have been reviewed to identify how they should be applied to compact exchangers. It is important to understand the basis of the convective and nucleate boiling components in the methods, and the form of the interaction between them

One question is whether the convective enhancement factor should be calculated on a purely physical property basis, or whether it should be linked to the frictional loss behaviour of the exchanger. Another question is how exchanger performance data should be extrapolated, when liquid-only convective coefficients are needed.

The fundamental question for boiling in compact exchangers is however the extent to which enhancements to single phase heat transfer affect boiling. Examples have been cited both where there appears to be enhancement in both single and two phase heat transfer (serrated fins) and where the enhancement is single phase only (wire loops).

With finned surfaces, there can in principle be very significant variation of temperature difference, and hence coefficient along the fin. In practice, for aluminium exchangers the corresponding uncertainties in heat transfer are probably relatively small, but for other metals, and for duties involving higher temperature differences, the effects may be significant.

ACKNOWLEDGEMENTS

Thanks are due to Dr V V Wadekar for many helpful discussions, to Mr M. Gough of Cal Gavin for permission to use Figure 14, and to APV, Heatric, Rolls-Laval and The Brazed Aluminium Plate Fin Heat Exchanger Manufacturers Association, for the use of other figures showing their products.

NOMENCLATURE

F	2 phase enhancement fraction
f	friction factor
h	heat transfer coefficient
m,n	exponents
q	heat flux
Re	Reynolds number
S	boiling suppression factor

Greek

ρ	density
η	viscosity
χ	Martinelli parameter

Subscripts

c	convective
g	gas
l	liquid
n	nucleate
p	pool boiling

REFERENCES

Carey, V. P., and Mandrusiak, G. D., 1986, "Annular Film-flow Boiling of Liquids in a Partially Heated, Vertical Channel with Offset Strip Fins, "*International Journal of Heat and Mass Transfer*, Vol. 29, No. 6, pp. 927-939.

Chen, J. C., 1963, "A Correlation for Boiling Heat Transfer to Saturated Fluids in Convective Flow," *ASME-AIChE Heat Transfer Conference*, Boston, MA, USA, 11-14 August 1963, Paper 63-HT-34, 11 pp.

Galezha, V. B., Usyukin, I. P. and Kan, K. D. 1976, "Boiling Heat Transfer with Freons in Finned-plate Heat Exchangers" *Heat Transfer Sov. Res.* Vol. 8, No. 3, pp. 103-110.

Liaw, S. P., and Yeh, R. H., 1994, "Fins with Temperature Dependent Surface Heat Flux - II. Multi-boiling Heat Transfer," *International Journal of Heat Transfer*, Vol. 37, No. 10, pp. 1517-1524.

Mandrusiak, G. D., and Carey, V. P., 1989, "Convective Boiling in Vertical Channels with Different Offset Strip Fin Geometries," *Journal of Heat Transfer*, Vol. 111, No. 1, pp. 156-165.

Marvillet, Ch., 1992, "Welded Plate Heat Exchangers as Refrigerants Dry-Ex Evaporators," *EUROTHERM Seminar No. 18 (Design and Operation of Heat Exchangers)*, Germany, ISBN 0387537716, pp. 255-268.

Osterberger, R. and Slipcevic, B., 1990, "Wärmeübergang beim Blasensieden in Plattenverdampfern," *Ki Klima*, Kälte-Heizung 11.

Panchal, C. B., Hillis, D. L. and Thomas, A., 1983, "Convective Boiling of Ammonia and Freon 22 in Plate Heat Exchangers," *ASME/JSME Themal Eng. Joint Conference*, Hawaii, ASME Book I00158-B, Vol. 2, pp. 261-268.

Robertson. J. M., 1979, "Boiling Heat Transfer with Liquid Nitrogen in Brazed-aluminium Plate-fin Heat Exchangers," *AIChE Symposium Series*, Vol. 75, No. 189, *18th National Heat Transfer Conference*, San Diego, California, 6-8 August 1979, 14 pp.

Robertson, J. M., Lovegrove, P. C., 1983, "Boiling Heat Transfer with Freon 11 (R11) in Brazed Aluminium, Plate-fin Heat Exchangers," *Journal of Heat Transfer*, Vol. 105, No. 3, pp. 605-610.

Sakaue, S., 1994, "Boiling Heat Transfer Characteristics of Nitrogen in a Thermosyphon Condenser-reboiler", *Low Temp. Eng. and Cryogenics Conference*, London, July 1994.

Steiner, D., 1988, "Wärmeüberagung beim Sieden Gesättigter Flüssugkeiten", Sect. Hbb, in *VDI Wärmeatlas*. VDI Verlag, Düsseldorf.

Steiner, D. and Taborek, J., 1992, "Flow Boiling Heat Transfer in Vertical Tubes Correlated by an Asymptotic Model," *Heat Transfer Engineering*, Vol. 13, No. 2, pp. 43-69.

Stephan, K. and Abdelsalam, M, 1980, "Heat transfer correlations for natural convection boiling", *Int. J. Heat Mass Transfer*, Vol. 23, pp. 73-80.

Tran, T. N., Wambasganss, M. W., France, D. M. and Jendrzejczyk, J. A. 1993 "Boiling heat transfer in a small horizontal, rectangular channel", *AIChE Symp. Ser.*, Vol. 89, No. 295, pp. 253-261.

Wadekar, V. V., 1995, "An Alternative Model for Flow Boiling Heat Transfer", *Paper submitted to Convective Flow Boiling Conference, Banff, Canada, April 30-May 5, 1995.*

Wambasganss, M. W., France, D. M., Jendrzejczyk, J. A. and Tran, T. N., 1993 "Boiling heat transfer in a horizontal small diameter tube ", *J. Heat Transfer.*, Vol. 115, No. 4, pp. 963-972.

Webb, R. L. and Gupte, N. S., "A Critical Review of Correlations for Convective Vaporization in Tubes and Tube Banks," *Heat Transfer Engineering*, Vol. 13, No. 3, pp. 58-81.

Figure 1 A typical plate heat exchanger plate

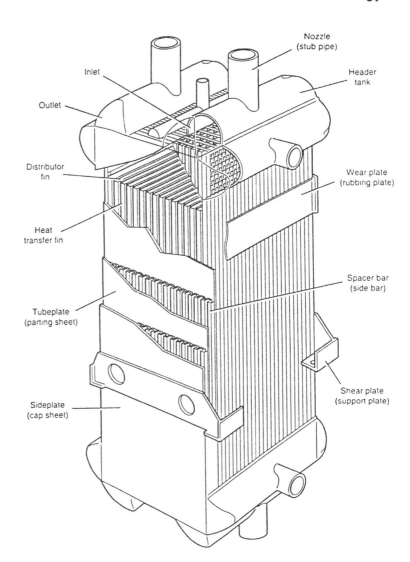

Figure 3 A plate-fin heat exchanger

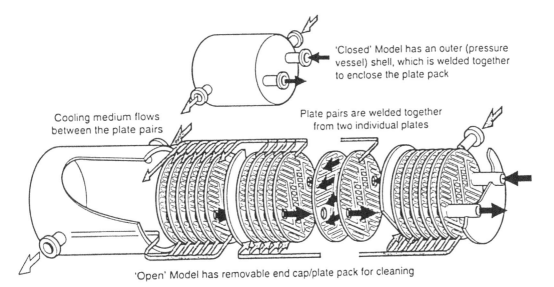

Figure 2 A plate and shell heat exchanger

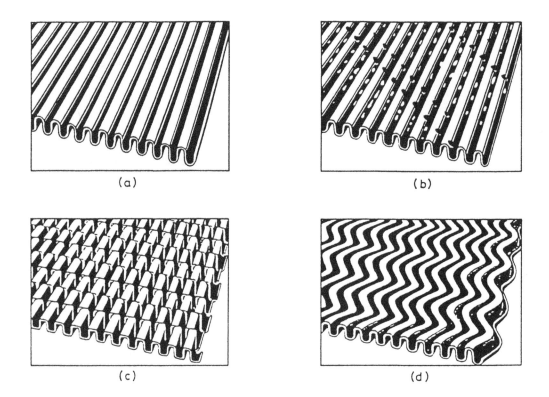

Figure 4 Fin types (a) plain; (b) perforated; (c) serrated (offset strip); (d) wavy (herringbone)

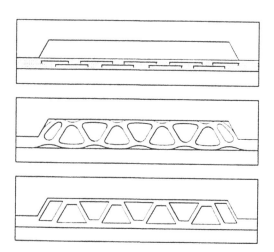

Figure 5 Manufacturing a Rolls-Laval diffusion bonded plate-fin exchanger
Top   Three sheets with local diffusion bonding in mould
Middle   Pressurised gas introduced to blow up fins
Bottom Final shape of one layer of the exchanger

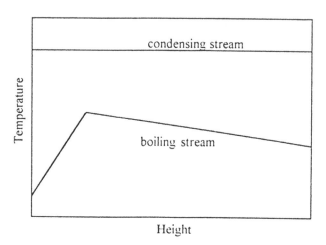

Figure 6 Typical temperature profile in a plate-fin thermosiphon reboiler

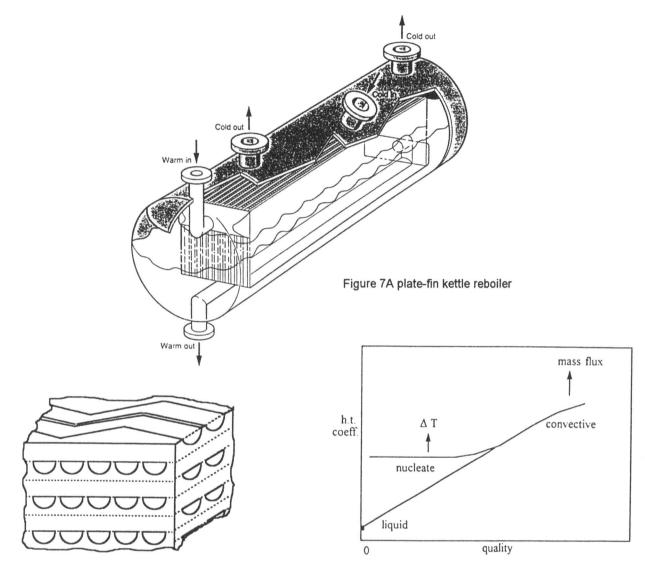

Figure 7A plate-fin kettle reboiler

Figure 8 A printed circuit heat exchanger (section)

Figure 10 Typical boiling heat transfer coefficient as a function of quality

Figure 9 Plan view of part of a printed circuit heat exchanger plate

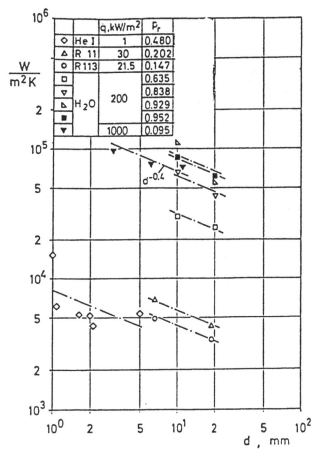

		q,kW/m²	P_r
◇	He I	1	0.480
△	R 11	30	0.202
○	R 113	21.5	0.147
□			0.635
▽		200	0.838
◁	H_2O		0.929
■			0.952
▼		1000	0.095

Figure 11 The effect of tube diameter on nucleate flow boiling (from Steiner and Taborek, 1992)

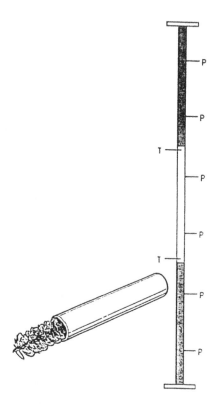

Figure 13 Cal Gavin wire loop insert and test section . (Inserts in shaded regions; P - pressure tappings; T - stream thermocouples; 50 wall temperature thermocouples not shown.)

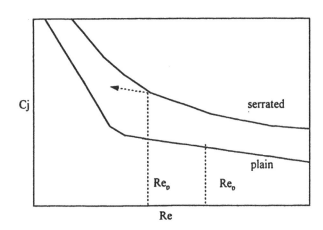

Figure 12 Colburn j factor for plain and serrated fins, as a function of Reynolds number (log scales). Arrow shows extrapolation direction for liquid only values

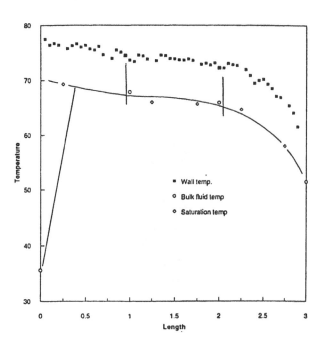

Figure 14 Wall and stream temperature profiles along the test section   (pressure = 3bar mass flux = 281.9 kg/m²s heat flux = 51.54 kW/m²)

BOILING INCEPTION
AND SUBCOOLING BOILING

A PHOTOGRAPHIC STUDY OF NUCLEATION CHARACTERISTICS OF CAVITIES IN FLOW BOILING

Satish G. Kandlikar, Michael D. Cartwright, and Viktor R. Mizo
Department of Mechanical Engineering
Rochester Institute of Technology
Rochester, New York

ABSTRACT

In the present investigation, the effects of wall superheat and flow velocity on the nucleation characteristics of cavities of different radii have been experimentally determined for flow boiling of subcooled water near atmospheric pressure in a narrow 3 mm x 50 mm rectangular channel. The study has been conducted with high speed video camera obtaining pictures at a maximum rate of 6000 frames per second. The natural cavities present on a polished aluminum surface serve as nucleation sites.

INTRODUCTION

Since the systematic study of nucleate boiling started with Nukiyama's (1934) experiment, the attention has been focused on the individual bubbles, their nucleation, growth, and departure from individual cavities, cavity sizes and their distribution on the surface, cavity shape, bubble frequency and the associated heat transfer rates and mechanisms. Extensive research has been done on the pool boiling topic. Early work of Hsu (1962) on nucleation in pool boiling and its extension by Bergles and Rohsenow (1964) provide the basis for analyzing nucleation characteristics in flow boiling. From their study the influence of flow on activation of a cavity could be predicted using the "tangency" criterion.

In the models proposed in literature to predict heat transfer coefficient in flow boiling, there are two distinct regions identified, one being the nucleate boiling dominant region and the other convective boiling dominant region. In the nucleate boiling region, heat transfer is predominantly by nucleate boiling with some contribution from convective component still present. In the convective boiling dominant region, however, the nucleate boiling contribution becomes smaller with increasing quality. Complete suppression of nucleate boiling is assumed to occur at higher qualities in many models, although this is not conclusively proven. Models utilizing the contribution from both components to varying degrees have been quite successful (Kandlikar, 1990b, 1991a, 1991b).

The conclusion of complete suppression has been based on two observations: (i) the heat transfer coefficient shows little dependence on the imposed heat flux, and (ii) there are no bubbles observed in the flow through normal visual observation without any significant magnification or high speed photography under flow conditions.

Kandlikar and Stumm (1993) presented experimental data on departure bubble diameter and contact angles under subcooled flow boiling conditions, and a control volume approach was presented to predict the departure bubble diameter as a function of flow parameters. The experimental work is now extended in the present investigation to include the surface temperature and heat flow rate measurement.

OBJECTIVES OF THE PRESENT WORK

The main objective of the present work is to obtain nucleation characteristics of a polished aluminum surface under subcooled flow conditions with water near atmospheric pressure. The effect of flow, surface temperature, and subcooling will be investigated to establish the cavity sizes at inception and to determine the range of active cavities under a given set of conditions. The effect of flow on the suppression of nucleation activity will also be investigated.

EXPERIMENTAL SETUP

The experimental setup is same as the one used by Kandlikar and Stumm (1993). A brief description of the setup is given below.

The experimental setup consisted of a horizontal, rectangular flow channel 3 mm x 50 mm cross-section with a heated aluminum rod of 10 mm diameter placed in the center of the lower (50 mm wide) wall. Water from the constant temperature bath near atmospheric pressure flows through the channel. The accuracy of flow and temperature measurements are ±3 percent and ±0.1 °C respectively. The bubble nucleation and its growth are observed through a microscope. Top-views of the bubbles were obtained by looking down directly on the bubble through a viewing window. Side views of the bubbles were obtained by looking at a front-surface silicon mirror placed at 45 degrees adjacent to the heated surface, parallel to the flow. A microscope and a video camera were employed to provide an effective magnification of up to 1350X on the video monitor. The video camera was able to capture images at a rate of up to 6000 frames per second using Kodak Ektapro camera with image intensifier. A schematic of the experimental setup is shown in Figure 1. Further details of the experimental apparatus may be found in Kandlikar (1992, 1994) and Kandlikar (1993).

The surface of the heated aluminum rod exposed to the flow was polished on a cloth covered metallographic polishing wheel using 1 micron particle size alumina in water suspension. The polished aluminum surface was placed flush with the bottom wall of the flow channel. Figure 2 shows a schematic view of the heater assembly. Four thermocouples are placed at locations 1-4 in the 10 mm section at equal distance apart along the aluminum rod. A circular heater is wrapped around the 25 mm diameter lower section of the aluminum rod. A special plastic bushing made of Torlon was used near the channel wall in which the aluminum rod was press-fitted to avoid leakages. Experimental measurement of the contact angle for the water-aluminum combination was made and was observed to be between 52 and 56 degrees for very slow growing bubbles under low velocity flow, and was seen to be unchanged throughout the experiments.

EXPERIMENTAL PROCEDURE

Experiments were conducted to observe bubble nucleation over cavities for different flow and heater conditions. The constant temperature bath was filled with distilled water. Prior to any experiments, it was heated to 90 °C and kept circulating through the test section to allow any trapped gases to be removed. The procedure was repeated over several days before taking any data.

Water temperature and flow rate were set at desired values. The heater was then powered with a DC source in small increments. After allowing sufficient time to attain steady-state (usually less than 10-15 minutes), bubble activity was observed through the microscope using appropriate magnification and frame rate for microscope and the high-speed camera respectively. Many times, the presence of any bubble activity could be observed only by recording at a high speed (1000 to 6000 fps), and playing it back at slow speed (30 fps). The heater input was increased in small steps and bubble activity observed over specific sites was recorded. Simultaneous measurements of water and heater temperatures were made and recorded. Each run was identified with specific run identifier recorded and displayed by the camera. A log was also maintained to clearly identify the flow and temperature data with the visual recordings.

HEATER SURFACE TEMPERATURE ESTIMATION

Heater surface temperature and heat flux were obtained from the experimental data by using a finite difference nodal network as shown in Figure 2. Thermocouple readings correspond to nodes 1-4. From the known physical properties, the resistance values at different locations were calculated. A commercial software package THERMONET developed by the first author was employed to analyze the system and iterate for the surface temperature at node 5 and the surface heat flow rate. From this data, heat transfer coefficient can be calculated. It is estimated that the error in the calculated surface temperature is within ±0.2 C and the heat transfer coefficient is within less than ±5%.

EXPERIMENTAL RESULTS AND DISCUSSION

There are two types of data reported in this paper. The first type consists of the radii of the cavities nucleating shown for different wall superheat values for a given set of conditions of bulk temperature and flow rate. Figures 3, 4 and 5 fall in this category. The other type of data presented shows the heat transfer coefficient as a function of the wall superheated plotted for different set of conditions. These data provide an overall picture of the heat transfer mechanism occurring at the heater surface.

The active cavities were identified from the bubble frame sequence during the video playback. The cavities were of non-uniform shapes, and their diameter was determined to be approximately the mean of the horizontal and vertical dimensions of the cavity. For long narrow cavities, the width was taken as the cavity diameter. Although it is possible that the actual vapor bubble formation was observed at the mouth of the cavity, a bubble may have nucleated inside the cavity and then grown to cover the cavity opening. Such a phenomenon was observed when an artificial cavity of 100 μm diameter was indented on the heater surface, and bubbles were seen nucleating from inside the cavity. For smaller diameter cavities, the bubbles were always seen to be nucleating from the entire cavity opening. For the measurements made in this study, the cavity dimensions were in the range of 5 to 40 micrometers, and the bubbles are believed to be nucleating from these cavities.

Figures 3 and 4 show the cavity radii for which active nucleation was observed. The bulk temperature of water was maintained constant at 60 °C and the two flow rates corresponding to Re=3801 and 5068 are represented. For each flow rate, the heater power was systematically increased in small steps. After allowing sufficient time for reaching steady state, the surface was scanned with high speed camera to record any bubble nucleation activity. The film was replayed at a slower rate and the diameter of cavities were measured on the screen. From the known magnification, the cavity radii were then calculated.

Referring to Figs. 3 and 4, at Re=3801, we see that the cavities are nucleating at lower wall superheat values as compared to Re=5068. An increase in flow rate requires a higher degree of wall superheat to activate the cavities of same sizes. For Re=3801, it can be seen that cavities in the range from 2.8 to 5.8 μm radii are nucleated. A shift toward smaller cavity radii can be seen with increasing superheat. For Re=5068, the range of active cavities clearly shifts toward smaller radii with increasing wall superheat. For a wall superheat of about 9 °C, the range of active cavity radii is only 2.0 to 3.8 μm. A larger number of smaller cavities are seen to become active at this superheat.

Figure 5 shows a similar plot showing active cavity radii at different wall superheat values for a bulk temperature of 80 °C. The flow rates are not too far apart, with Reynolds numbers of 1664 and 1997 for the two data sets. At 80 °C bulk temperature, nucleation was seen only for these low values of Re. At higher values of Re, the nucleation was apparently suppressed. However, later it was confirmed from the set of graphs shown next that nucleate boiling (or similar phenomenon) was going on which yielded a higher value of heat transfer coefficient than the single phase value.

Figure 5 shows quite a scatter of active cavity radii and no clear trend could be established, probably due to small change in the Reynolds numbers between the two sets.

Next five figures, Figs. 6-10 are obtained by conducting a thermal analysis of each individual data point obtained during the test as described earlier using THERMONET software package. Figure 6 shows a plot of the heat transfer coefficient on the aluminum test section surface plotted as a function of wall superheat for Re of 3801 and 5068. For low values of wall superheat, the heat transfer coefficient is almost constant at its single phase value. As the wall temperature increases, the heat transfer coefficient begins to gradually increase, indicating the onset of nucleate boiling. This onset occurs in the range of 3 to 8 °C range of wall superheat. As the wall superheat is increased further, the heat transfer coefficient continues to increase. The two curves are seen to run parallel to each other in the range of wall superheat plotted, although the contribution due to nucleate boiling is quite large at higher values of wall superheat. It may be noted that there was no bubble activity detected at larger values of wall superheat indicating that the bubbles are perhaps growing very rapidly and leaving at a higher frequency from the small cavities. This point is also discusses under growth rate curves reported by Kandlikar, Mizo and Cartwright(1995) in another paper presented in the same conference.

Figure 7 shows a similar plot as Fig. 6, but for a bulk temperature of 70 °C and Re values of 1467 and 2934. The trends are same as noted in Fig. 6.

The next three plots, Figs. 8, 9 and 10 show h versus wall superheat for the same flow rate (Re is different in the plots due to property variation with temperature), but for three different bulk temperatures of 80, 85 and 90 °C. The h values at low superheat in Figs. 8 and 9 correspond to the single phase values, and onset of nucleating boiling can be detected with an increase in h at higher wall superheat values.

Figure 10 however displays a completely different trend. Here, h is quite high even at low values of wall superheat, and it decreases with increase in wall superheat. This trend may seem quite surprising as the nucleate boiling is suppressed at higher values of wall superheat. It is however seen clearly in the bubble growth rates reported for this case by Kandlikar, Mizo and Cartwright (1995). The visual observations at high speed (4000 fps) indicate that the bubble growth rate increases at higher wall superheat, and the bubble departure radius becomes smaller. The bubble frequency becomes very rapid and with further increase in superheat, the activity becomes too fast to be captured even at 6000 fps. Whether the cavities are completely suppressed or are active under a different mode is a topic of ongoing theoretical and experimental study in the Thermal Analysis Laboratory at RIT.

CONCLUSIONS

An experimental study is presented on the nucleation characteristics of cavities under subcooled flow boiling of water on aluminum heater surface. The effects of three major parameters have been analyzed and the conclusions are summarized below.

Effect of Flow Rate

Increasing flow rate requires a larger wall superheat to activate the cavities. The range of active cavities also shifts toward smaller radii. Larger cavities start becoming inactive with increasing flow rates.

Effect of Subcooling

The effect of subcooling is quite complex on nucleating cavity radii. It is strongly coupled with flow rate. However, the bubble activity becomes very rapid at lower values of subcooling (higher bulk temperatures).

Effect of Wall Superheat

As the wall superheat increases, smaller cavities are activated and the heat transfer coefficient increases with increased nucleation activity. However at higher bulk temperatures, the nucleation seems to be too rapid but reducing the nucleate boiling contribution. This phenomenon needs further investigation.

Inception of Nucleation

Exact conditions of nucleation inception are very difficult to obtain. From the data presented here and additional data we plan to to obtain, the inception criterion could be represented by the curve enveloping the data on the lower and left bound of wall superheat versus cavity radius plots.

ACKNOWLEDGEMENTS

The authors gratefully acknowledge the support provided by the Mechanical Engineering Department at RIT. The help provided by David Hathaway and Tom Locke in the Machine Shop is sincerely appreciated.

NOMENCLATURE

T_s - wall temperature, °C
T_b - bulk temperature, °C
r_c - cavity radius, m
Re - Reynolds number

REFERENCES

Bergles, A.E, and Rohsenow, W.M., 1964, "The Determination of Forced-Convection Surface Boiling Heat Transfer," Journal of Heat Transfer, Vol. 86, pp. 365-372.

Hsu, Y.Y., 1962, "On the Size Range of Active Nucleation Cavities on a Heating Surface," Journal of Heat Transfer, Vol. 84, pp. 207-216.

Kandlikar, S.G., 1990a, "A Mechanistic Model for Flow Boiling Heat Transfer," Paper presented at the 1990 ASME Winter Annual Meeting, Dallas, Nov., Single and Multiphase Convective Heat Transfer, eds. M.A. Ebadian, K. Vafai, and A. Levine, eds., ASME HTD-Vol. 145, pp. 61-69.

Kandlikar, S.G., 1990b, "A General Correlation for Two-phase Flow Boiling Heat Transfer Coefficient Inside Horizontal and Vertical Tubes", Journal of Heat Transfer, Vol. 102, pp. 219-228.

Kandlikar, S.G., 1991a, "A Model for Predicting the Two-Phase Flow Boiling Heat Transfer Coefficient in Augmented Tube and Compact Heat Exchanger Geometries," Journal of Heat Transfer, Vol. 113, Nov., pp.966-972.

Kandlikar, S.G., 1991b, "Correlating Heat Transfer Data in Binary Systems," Paper presented at the National Heat Transfer Conference, Minneapolis, July 1991, Phase Change Heat Transfer, E. Hensel et al., eds., ASME HTD-Vol. 159.

Kandlikar, S.G., 1992,"Bubble Behavior and Departure Bubble Diameter of Bubbles Generated Over Nucleating Cavities in Flow Boiling," Pool and External Flow Boiling, Proceedings of The Engineering Foundation Conference on Pool and External Flow Boiling, March 22-27, ASME.

Kandlikar, S.G., 1994,"Measurement of Departure Bubble Diameter and Advancing and Receding Contact Angles in Subcooled Flow Boiling of Water," Paper presented at the joint ASME and ISHMT Conference, Jan. 5-8, BARC, Bombay, India.

Kandlikar, S.G., and Stumm B.J., 1993, "A Control Volume Approach for Investigating Forces on a Departing Bubble under Subcooled Flow Boiling," ASME Heat Transfer Conference, Colorado Springs. Also accepted for publication in Journal of Heat Transfer, to appear in Nov. 1995 issue.

Kandlikar, S.G., Mizo, V.R., and Cartwright, M.D., "Investigation of Bubble Departure Mechanism in Flow Boiling using High-speed Photography," Proceedings of International Conference on Convective Flow Boiling, ASME Foundation Conference, Banff, Canada, April 30-May 5.

Nukiyama, S., 1934, "The Maximum and Minimum Values of Heat q Transmitted from Metal Surface to Boiling Water under Atmospheric Pressure," Journal of Society of Mechanical Engineers (Japan), Vol. 37, pp. 367-374, 553-554.

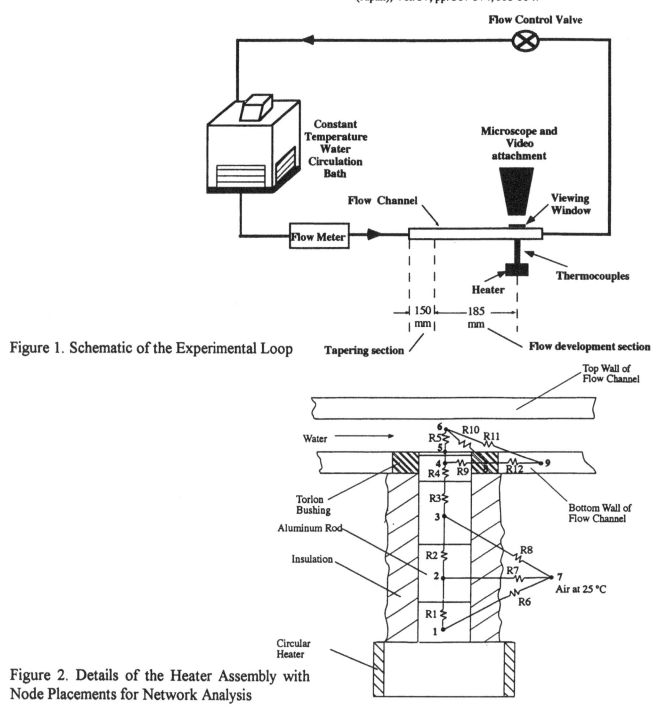

Figure 1. Schematic of the Experimental Loop

Figure 2. Details of the Heater Assembly with Node Placements for Network Analysis

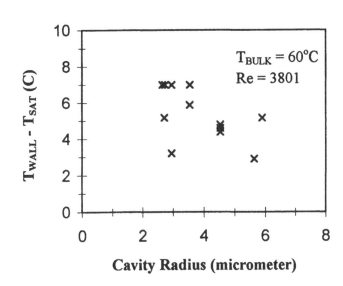

Figure 3. Effect of Wall Superheat on Nucleating Cavity Radii at Tb = 60 °C, Re = 3801.

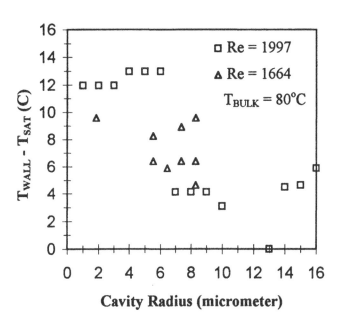

Figure 5. Effect of Wall Superheat and Flow Rate on Nucleating Cavity Radii at Tb = 80 °C.

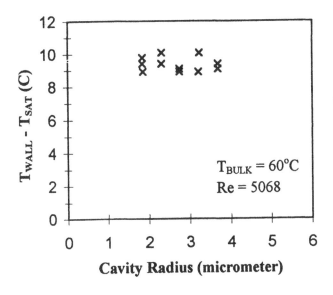

Figure 4. Effect of Wall Superheat on Nucleating Cavity Radii at Tb = 60 °C, Re = 5068.

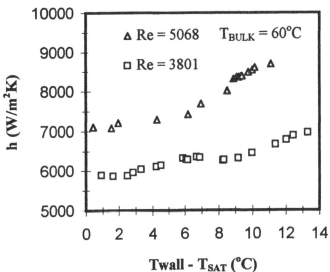

Figure 6. Effect of Wall Superheat and Flow Rate on Heat Transfer Coefficient at Tb = 60 °C.

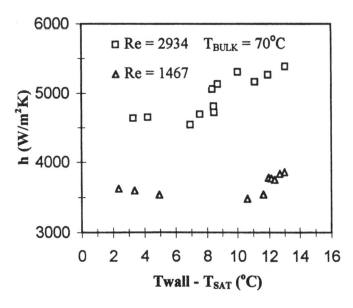

Figure 7. Effect of Wall Superheat and Flow Rate on Heat Transfer Coefficient at Tb = 70 °C.

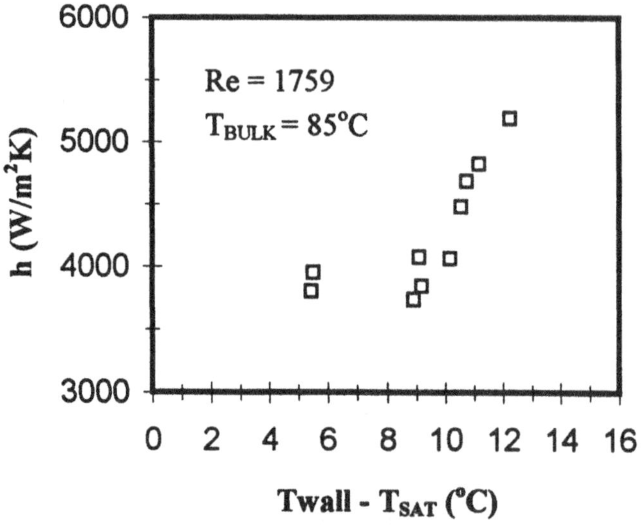

Figure 9. Effect of Wall Superheat on Heat Transfer Coefficient at Tb = 85 °C and Re = 1759.

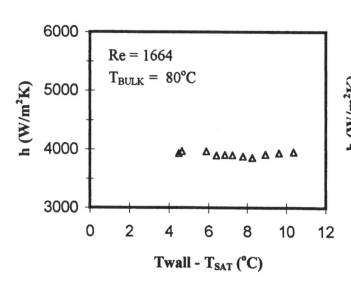

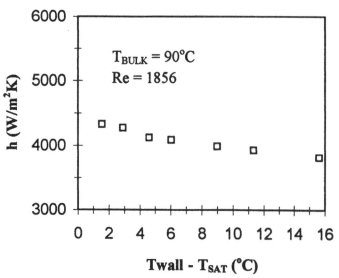

Figure 8. Effect of Wall Superheat on Heat Transfer Coefficient at Tb = 80 °C and Re = 1664.

Figure 10. Effect of Wall Superheat on Heat Transfer Coefficient at Tb = 90 °C and Re = 1856.

INFLUENCE OF OBSTRUCTIONS ON THE INITIATION OF SUBCOOLED BOILING IN AN ANNULAR CHANNEL

S.Toda, H. Hashizume, Y. Fujiwara, and J. Ogawa
Department of Nuclear Engineering
Tohoku University
Aoba-ku, Sendai, Japan

M. Mori
Nuclear Power R & D Center
Tokyo Electric Power Company
Tsurumi-ku, Yokohama, Japan

ABSTRACT

Fuel rods in the BWR core are held in position by tie plates at the bottom and by grid plates at the top, and in between by spacers. Such obstructions in a channel that are designed in order to improve the mixing between the bubble layer and the liquid near the fuel wall cause an additional pressure drop and influence the subcooled boiling behavior. However, the influence of these obstructions has not been evident especially on the inception of subcooled boiling. In the present work, the effect of the turbulent mixing caused by the obstructions was analyzed experimentally, and the mechanism to retreat the initiation front of subcooled boiling downstream was clarified. Experiments were performed using a small boiling loop with a vertical annular channel test section that had a heated SUS tube simulating the fuel rod. Video image data of incipient boiling were analyzed by the image processing technique, and compared with numerical analyses carried out to simulate the incipient subcooled boiling with low void fraction and its upstream region of single phase flow. From the work, it was cleared that, in the case of short distance from the tie plate as the upstream obstruction, the initiation of the subcooled boiling was suppressed, and the position of boiling front moved downstream, that depends on thermohydrodynamic conditions and the obstruction arrangements.

INTRODUCTION

Subcooled boiling heat transfer in an annular channel with forced convection has been investigated in many works, in relation to fuel bundle heat transfer in a BWR core, in which fuel rods are held in position by tie plates at the bottom, by grid plates at the top, and in between by spacers. In a nuclear reactor, such structural complexity might bring about thermohydraulic instability problems interacted with nuclear reactivity. Obstructions such as tie plates or spacers in a fluid channel are designed to improve the mixing between the thin superheated layer containing fine vapor bubbles and the subcooled bulk liquid, but bring an additional pressure drop

and influence the heat transfer characteristics due to turbulent agitation growing behind the obstructions. Since the subcooled boiling is caused by a thermodynamic nonequilibrium in liquid, the turbulent mixing of the superheated liquid near the heated wall with the fairly subcooled bulk might influence strongly the initiation of surface boiling. Generally one can see many works for a normal straight tube or an annuli done by Levy, Staub (1968), and Saha and Zuber (1974), and others, however, the subcooled boiling influenced by turbulent mixing caused by obstructions in the channel has been studied less.

In the present work, the effect of the turbulent mixing by the obstructions in an annular channel is analyzed experimentally and the mechanism to shift the initiation front of subcooled boiling downstream is clarified by using computer codes developed by authors based on the turbulent models.

EXPERIMENTAL APPARATUS AND PROCEDURE

Experiments were performed using a small boiling loop as shown in Fig. 1, in which distilled water was circulated by a circulating pump through a pre-heating tank, boiled in a test section, and then cooled in a primary cooling tank, and the steam was completely condensed in a secondary cooling tank (condenser). The flow channel of the test section was vertical and annular, and composed of a Pyrex glass pipe with 3 cm in inner diameter and a simulated fuel rod of SUS tube with 1 cm in outer diameter located at the center axis of the Pyrex glass pipe, and heated directly by electricity. Figure 2 shows the fuel rod in the channel fabricated with obstructions simulating a spacer and tie plate, which were held in location at a distance of 460 mm between them. The test channel was covered by a acrylic resin square pipe, and the gap space between them was filled with distilled water, through which we could take a picture without any optical distortion caused by refraction.

The photographic image pictures of inception of subcooled boiling were recorded at time intervals of 1/60 second with 1/1000 second electrical shutter on video tape by a video camera. The pictures were sufficiently clear without shading.

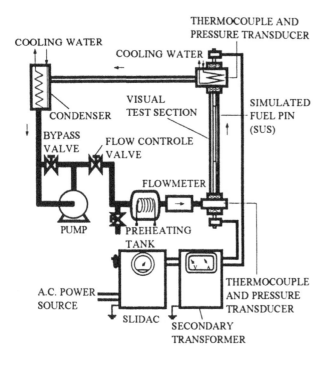

Fig. 1   Outline of experimental apparatus.

Table 1   Experimental conditions.

Flow Velocity, U, [m/s]	0.10 - 0.24
Heat Flux, q, [W/cm^2]	14 - 28
Inlet Fluid Temperature, T_{in}, [°C]	89 - 96.5
Inlet Pressure, [MPa]	0.111 - 0.120
Outlet Pressure, [MPa]	0.105 - 0.106
Reynolds Number, Re	6000 - 13000
Peclet Number, Pe	12000 - 29000

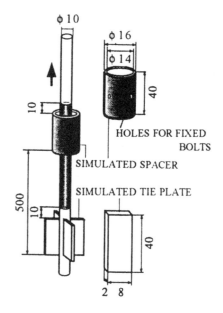

Fig. 2   Obstruction arrangement in the test channel.

Lighted backward from the opposite side of transparent test section, the vapor-liquid interface was taken as a black line due to the different refractive indexes between both phases. An image processing technique developed by Toda and Hori (1992) was applied for analyzing the time-varying image data of incipient subcooled boiling layer on the fuel wall in a subcooled water flow, of which details are explained later.

Three cases of experiments were performed as follows:
(1) Under the normal flow condition by using a normal straight fuel rod
(2) Under the obstructed flow condition by using a fuel rod with only a lower tie plate
(3) Under the obstructed flow condition by using a fuel rod with both a spacer and tie plate

These experiments were carried out under the experimental conditions listed in Table 1, and the experimental procedure was as follows; Firstly the heating power and flow rate were adjusted by controlling both an electric voltage controller and a flow control valve. Next, regulating the electric power of the pre-heating tank, the temperature of water at the inlet was set up at a steady value. After several minutes, the flow rate, pressure and temperature of water in the test section were measured, and then photographic image data of the incipient boiling layer on the fuel wall were taken by the video camera and analyzed by the image processing technique. The observed area of the test channel ranged from the entrance of the heated section to 80 mm downstream, where single picture element was equivalent to the length of 0.48 mm. The heat flux of the fuel rod was estimated from the calibrated data obtained by the preliminary experiment. The wall temperature of fuel rod was measured directly by thrmocouples of which data were compared with the numerical analyses.

IMAGE PROCESSING TECHNIQUE

It was generally difficult to distinguish the initiation point of subcooled boiling by visual inspection, so that the image processing technique as mentioned before was applied for discriminating the boiling inception point from the photographic image data taken by the video camera. The video image processor had a function to convert the two-dimensional analog data of video image picture into digital values of dots matrix with different gray levels that include the information of concentration from 0 to 1 at each position. A proper threshold level of concentration could differentiate the liquid phase and vapor phase, that is called as the "two-valuing" process. In order to decrease the error, the threshold value was assumed to be the level of back ground estimated from the image data of non-boiling test channel at the same experimental condition

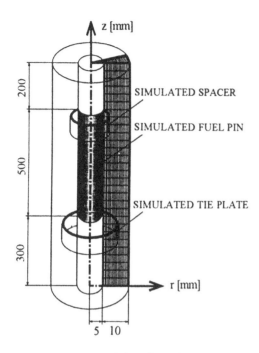

Fig. 3 Computational domain in the test channel.

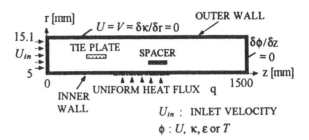

Fig. 4 Boundary conditions in numerical analyses (I)
and (II).

and Fig. 4 the boundary conditions. The cylindrical coordinate system, where the z coordinate axis was denoted in the direction of flow and the r coordinate in the radial direction, was considered with uniform grid spacing, together with 150 and 19 grid points respectively in the flow and the radial directions for the analysis (I) or 102 and 22 grid points for the analysis (II). The fluid was assumed to be Newtonian and incompressible, and to have constant fluid properties. In the analysis, the tie plate and spacer were substituted for their equivalent cylinders as shown in Figs. 3 and 4 of which both cross-sectional and lateral surface areas were equal to their original ones. The inflow boundary conditions were as follows: the flow was steady, and in the analysis (I), the inlet velocity, k and ε were uniform, and the temperature was constant to be 80℃, or in the analysis (II), the inflow temperature was the same with the experiments that was from 94 to 96.5℃. The outflow boundary was not restricted. The total thermal power loaded in the heated section was 2.2 kW, and the other walls were assumed to be thermally insulated.

In the analysis (II), the governing equations were discretized by the finite volume method based on the control volume, and the QUICK method (Leonard (1979)) was applied to the convection terms in the momentum equations. The power-law scheme was applied to the scalar values of k and ε, and temperature. The SIMPLEC algorithm (van Doormaal and Raithby (1984)) was employed for calculation of flow and temperature fields, and the wall functions were used as the boundary conditions near the wall.

On the other hand, in the analysis (I), the calculations were performed by using the SIMPLE algorithm (Patankar(1980)) and applying the pressure correction only to lines in the flow direction. The power law scheme was used to the convection and diffusion terms.   In both analyses, the staggered grid method was applied for calculating the flow fields.

For calculation of the temperature field with subcooled boiling when the wall temperature exceeded the saturation temperature of liquid, the modified enthalpy method was introduced together with the following equivalent thermal capacity,

$$\rho C = \frac{\rho H_{fg}}{T_b - T_a} \frac{T_w - T_s}{2\,(T_w - T_l)} \qquad (1)$$

where $(T_b - T_a)$ is the temperature difference during phase change, ρ the density of fluid, C the specific heat of fluid, H_{fg} the

just before the boiling was initiated. From this image processing analysis, we could know what phase exist at the concerned position, liquid (1) or vapor (0). The void fraction was calculated statistically from the data of about 100 pictures, and corrected by introducing the correction factor for converting the projected two-dimensional data to three-dimensional values in the actual channe. This factor was calibrated by the preliminary experiments that bubbles in the channel flow was replaced by small nylon balls with several sizes.

NUMERICAL ANALYSIS

Two dimensional analyses in the cylindrical coordinate were performed numerically. In the upstream region of the channel where the subcooled boiling was not incepted, the flow was assumed to be the single phase and not to be disturbed by subcooled boiling developed in the downstream region. Therefore, in the region from the tie plate to the upstream front of incipient boiling point, the turbulent convective flow and temperature distribution can be analyzed by the following time-averaged governing equations: the r, z continuity equation, momentum equations and energy equation, and (I) the transport equations based on the k-ε model where k and ε are the turbulence kinetic energy and its dissipation respectively, or (II) the turbulent equation based on the mixing length models modified by van Driest (1956). The analyses based on (I) and (II) were applied for the detailed analysis of hydrodynamic structure influenced with the obstructions and the global analysis for heat transfer characteristics resppectively.

Figure 3   shows the computational domain of the test channel, that is annular and symmetrical with respect to the axis,

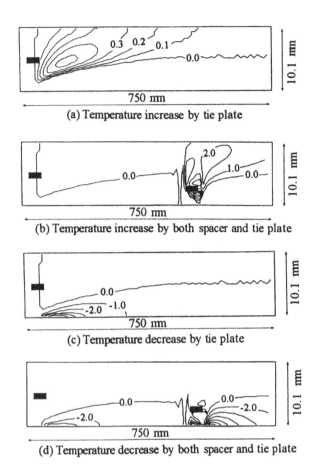

(a) Temperature increase by tie plate

(b) Temperature increase by both spacer and tie plate

(c) Temperature decrease by tie plate

(d) Temperature decrease by both spacer and tie plate

Fig. 5   Temperature distribution simulated by the numerical
analysis (I) based on the k-ε model, q= 14 W/cm^2 and
T_{in} =80 ℃.

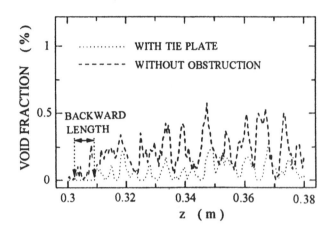

Fig. 6   Onset of subcooled boiling, q= 14 W/cm^2,
T_{in}=74.0 - 86.0 ℃.

ound that obstructions enhance the turbulent heat transfer in
their downstream flows, and the cooled liquid near the wall
might suppress the boiling to make the boiling initiation point
move downstream.

INFLUENCE OF OBSTRUCTIONS ON BOILING INITIATION
Onset point of Boiling

Experiments were made under   the inlet temperature T_{in} from
74 to 86℃ and another parameters set at the lowest conditions
shown in Table 1. Figure 6 shows the growth of void fraction
from the onset point of boiling on the fuel rod with or without
the tie plate. The void fraction is fluctuated due to both
generation and collapse of bubbles in the subcooled thin layer
near the wall. From the figure, it can be found that, due to
turbulent effect caused by the tie plate, the onset point of
boiling moves downstream, that   is,   about 6 mm in this
experiment.

Point of Bubble Detachment   (Point of Net Vapor Generation)

Figure 7 shows the example of comparisons between void
fraction distributions developing along the fuel rod without
obstruction, with only a tie plate, or with both a tie plate and
spacer, of which original image data were analyzed by the image
processing technique. In the figure, Re denotes the Reynolds
number.

Figure 8 shows the change of cross section-averaged void
fraction distributed along the fuel rod, and due to turbulent
agitation caused by obstructions, the point of bubble
detachment (or the point of net vapor generation) moves
downstream. It can be seen that, in the case with both the tie
plate and spacer, the tie plate effect is distinguished, while the
spacer has no effect. This indicates that the downstream
obstruction is ineffective and also does not   interact each other.
This figure shows the case that the boiling initiation occurs
near behind the obstruction (tie plate)   and is influenced   by it.

latent heat of evaporation, T_w the heated wall temperature, T_s
the saturation temperature of fluid and T_l is the liquid
temperature in the closest grid point to the wall. This thermal
capacity absorbs the latent heat transferred with subcooled
boiling in the thin layer near the heated wall. From comparisons
with temperature distributions measured on the fuel wall, this
calculation method was confirmed to be applicable to the
analysis of the subcooled surface boiling.

TEMPERATURE IN THE OBSTRUCTED CHANNEL

The numerical analysis (I) based on the k-ε model cleared
the influence of the tie plate and spacer as shown in Fig. 5,
where q is the heat flux and T_{in} the inlet temperature of fluid.
The figure shows clearly that a pair of opposite zones with
temperature-increase and temperature-decrease are generated in
the down streams just after these obstructions, that is, the fluid
near the wall is cooled down and the fluid in the main flow
heated up. Enhanced by this turbulent mixing, the heat
absorbed in the thin thermal boundary layer near the wall is
transferred efficiently to the main flow.   From these results, it is

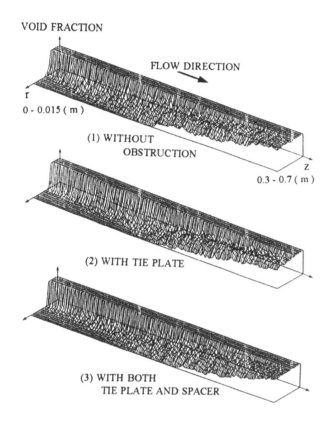

Fig. 7 Distribution of void fraction along the channel measured by experiments, q= 28 W/cm^2, Re=12000 and T_{in} =95.5 ℃.

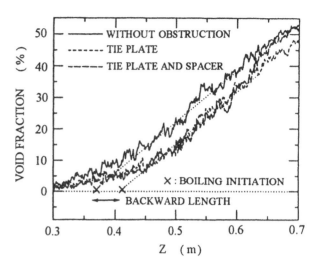

Fig. 8 Influence of obstructions upon the averaged void fraction distribution, q= 28 W/cm^2, Re=12000 and T_{in} =95.5 ℃.

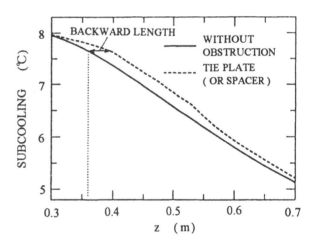

Fig. 9 Subcooling temperature distribution along the channel estimated by the numerical analysis (II), q= 28 W/cm^2, Re=12000 and T_{in} =95.5 ℃.

As shown in Fig. 9, the subcooling temperature of fluid distributed along the fuel rod and the backward length of the movement of incipient boiling point due to influence of obstructions are well predicted by the present numerical analysis (II). From the analysis, the increase of subcooling temperature of fluid caused by the upstream obstruction was correlated by the following equation;

$$\Delta(\Delta T_{sub}) = 0.018\ S_c^{-0.1}\ S_s^{-0.3} \qquad (2)$$

where ΔT_{sub} is the subcooling temperature, S_c and S_s the cross-sectional area and the total lateral surface area of the upstream obstruction (tie plate) respectively. Figure 10 shows not only the similar case that the boiling is suppressed by the upstream obstruction (tie plate), but also the interesting phenomenon that the void fraction increases in the downstream due to turbulent mixing effect of the tie plate compared with the case with no obstruction effect. The bulk fluid near the concerned zone is heated locally by the enhanced turbulent heat transfer, where condensation of bubbles is suppressed and the void fraction in the fluid is increased.

However, under the condition that the boiling is initiated far from the tie plate, the cooled-down zone caused behind the obstruction as has been shown in Fig. 5 is apart from the boiling initiation zone and does not interact each other. We can see this case in Fig. 11 that any influence of obstruction upon the boiling initiation is not recognized. Therefore, it is concluded that the influence of obstruction depends on the relative backward length of initiated boiling region against the upstream obstruction, and it is correlated by the following criteria within the present study;

Re > 19000 : influenced (Boiling initiation
point moves downstream.) (3)

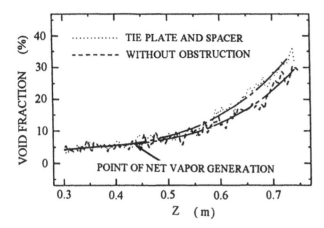

Fig. 10   Influence of obstructions upon the averaged void
fraction distribution, $q=14$ W/cm^2,   Re=6500 and
T_{in}=95.0 ℃.

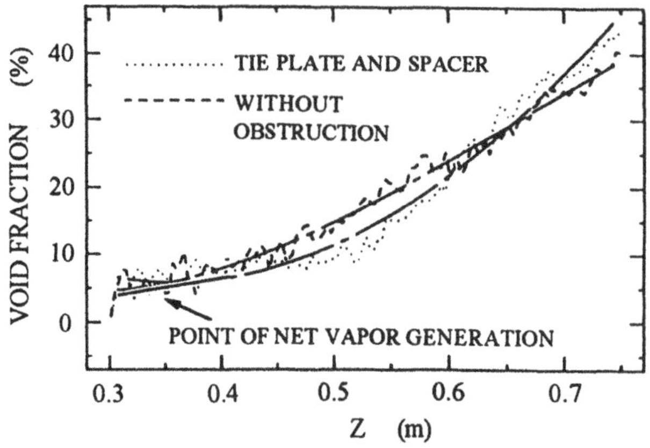

Fig. 11   Averaged void fraction distribution not influenced
by obstructions, $q=14$ W/cm^2,   Re=6500 and T_{in}=96.5 ℃.

$$Re < 53000 \quad : \quad \text{not influenced} \qquad (4)$$

where

$$Re = \frac{L\,U}{v} \qquad (5)$$

L is the length from the tie plate to the point of bubble
detachment measured in the channel without obstruction, U the
averaged fluid velocity and v the kinematic viscosity of fluid.

CONCLUDING REMARKS

To investigate the effect of the turbulent mixing by the
obstructions, tie plate and spacer of the BWR fuel rod,
fundamental experiments and numerical analyses were performed.
In the experiments, a small boiling loop was used with the test
section composed of a simulated fuel rod of SUS tube and the
visible transparent annular channel. Subcooled boiling
phenomena were   observed by the video camera, and
photographic image data of incipient boiling layer near the
heated fuel wall were analyzed by the image processing
technique.

As for the numerical analyses, two dimensional conditions
were carried out based on the turbulent model, the k-ε model
or the mixing length model, together with the modified enthalpy
method applicable to the low void fraction layer in the initiated
subcooled boiling region. The detailed analyses show clearly
that obstructions enhance the turbulent heat transfer in their
downstream flows, and the cooled liquid near the wall might
suppress boiling and the boiling initiation point move
downstream.

From experimental results of distribution of void fraction
along the channel, it is cleared that the tie plate influence
strongly upon the retreat of the incipient boiling front toward
the downstream region, but the spacer has no evident influence.
This fact indicates that the downstream obstruction has no
influence on the initiation of subcooled boiling. Within the

present experiments, the increase of subcooling temperature of
fluid caused by the upstream obstruction was correlated well by
equation (2), that leads to the backward movement of the
boiling initiation. Detailed discussions on the interacting
influence of obstructions with the initiated subcooled boiling
region were also performed. The influence of obstruction was
varied depending upon relative arrangements of obstructions in
the channel and its thermohydrodynamic conditions, that is, the
backward length of the initiated boiling region against the
upstream obstruction.

REFERENCES

Leonard, B.P., 1979, "A Stable and Accurate Convective
Modeling Procedure Based on Quadratic Upstream
Interpolation," *Computer Method in Applied Mechanics and
Engineering*, Vol. 19, pp. 59-98.

Levy, S., 1967, "Forced Convection Subcooled Boiling -
Prediction of Vapor Volumetric Fraction," *Int. J. Heat Mass
Transfer*, Vol. 10, pp. 951-965.

Patankar, S.V., 1980, *Numerical Heat Transfer and Fluid
Flow*, Hemisphere Pub. Co., pp. 126-133.

Saha, P. and Zuber, N., 1974, "Point of Net Vapor Generation
and Vapor Void Fraction in Subcooled Boiling," *Proc. 5th Int.
Heat Transfer Conf.*, Vol. 4, pp. 175-179.

Staub, F.W., 1968, "The Void Fraction in Subcooled Boiling
- Predication of the Initial Point of Net Vapor Generation,"
Trans. ASME, Ser. C, Vol.. 90, pp. 151-157.

Toda, S. and Hori, Y., 1992, "Void fraction distribution in
two-phase single component condensing flow," *Dynamics of
Two-Phase Flows*, CRC Press, pp. 275-293.

Van Doormaal, J.P. and Raithby, G.D., 1984, "Enhancement
of Simple Method for Predicting Incompressible Fluid Flow,"
Numerical Heat Transfer, Vol. 7, pp. 147-163.

Van Driest, E.R., 1956, "On Turbulent Flow near a Wall,"
Aero. Sci., Vol. 23, pp. 1007.

ON THE NET VAPOUR GENERATION PHENOMENON
IN LOW PRESSURE AND LOW MASS FLUX
SUBCOOLED FLOW BOILING

O. Zeitoun and M. Shoukri
Mechanical Engineering Department
McMaster University
Hamilton, Ontario, L8S 4L7, CANADA

ABSTRACT

An experimental investigation into low pressure subcooled flow boiling in a vertical annular test section was undertaken. A traversing single-beam gamma densitometer was used for measuring the axial distribution of the area-averaged void fraction and a high speed video system was used to visualize the subcooled flow boiling phenomena at various axial locations along the test section. The high speed photography results confirmed the fact that bubble detachment was not the cause of the NVG phenomenon. A new NVG model was developed based on the balance between vapour generation and condensation rates at the NVG point. The model is consistent with the Saha and Zuber's model (1974) for thermally controlled net vapour generation. The predictions of the proposed model are in good agreement with available low pressure NVG data for subcooled flow boiling in vertical conduits.

1. INTRODUCTION

The importance of two-phase flow and boiling studies has increased significantly in the last few decades with the fast development of nuclear power reactors. This led to an enormous number of investigations especially in high pressure flow; typical of power reactors. Meanwhile, the development of small research reactors, requires detailed information on low pressure flow boiling. Under some abnormal operating conditions, subcooled boiling may be encountered in the reactor core. Knowledge of the detailed void fraction distribution is important for many reasons. The void fraction distribution affects the pressure drop, heat transfer rates and flow stability in the reactor core. It also affects the characteristics of neutron moderation and, consequently, the reactor power.

When subcooled liquid enters a heated channel, boiling starts when the wall temperature exceeds the saturation temperature enough to cause vapour nucleation. This point is commonly called the onset of nucleate boiling (ONB). Bubbles are generated on the heating surface and condensed in the sub-cooled bulk. The net amount of vapour generation is determined by the difference between the generation rate and the condensation rate. The subcooled boiling continues downstream from the ONB point but the void fraction does not grow significantly because of the high subcooling. This trend continues until the void fraction starts to increase significantly at a point called the net vapour generation (NVG), or the onset of significant voids (OSV). The subcooled flow boiling along a heated channel can, accordingly, be divided into two regions; the highly subcooled region upstream of the NVG point and the slightly subcooled region downstream of this point.

Available NVG models can be classified into three groups: (i) thermal-hydrodynamically based models, which are based on correlating the heat transfer rate at the NVG point, (ii) models in which the NVG point is considered the point of the first bubble departure from the heating surface, and (iii) models which considers the ejection of bubbles from a wall bubbly layer into the subcooled bulk to be the point of NVG. The models of Griffith et al. (1958), Ahmad (1970), Hancox and Nicoll (1971), Saha and Zuber (1974) and Ünal (1975) fall into the first category. Bowring (1962) was the first to introduce the idea of predicting the NVG point by modelling the point of bubble departure from the heating surface. Many models were later developed based on the same idea such as the models of Levy (1967), Rogers et al. (1987), Rogers and Li (1992) and Lee and Bankoff (1992). Based on visual observations and high speed photography, Dix (1970) introduced the idea of modelling the NVG as the point of bubble ejection from the bubbly layer adjacent to the heating surface. When the bubbly layer thickness reaches a certain critical size the bubbles are no longer stable in the bubbly layer and are ejected into the subcooled core. Serizawa (1979) formulated an analytical model by assuming that the point of bubble ejection was the inflection point of the axial void fraction profile.

A critical review of existing NVG models, carried out by Zietoun (1994), demonstrated the lack of agreement about the mechanisms causing the NVG phenomenon, particularly at low pressure. Recent data obtained for low pressure subcooled flow boiling by Rogers et al. (1987), Donevski and Shoukri (1989), Dimmick and Selander (1990), and Bibeau and Salcudean (1990) have shown that the low void fraction region, preceding the NVG point, in low pressure flow tends to be longer than that in high pressure and is characterized by almost uniform void fraction. Zietoun (1994) compared available data on low pressure NVG with all existing models. He showed that Saha and Zuber's model (1974) was able to predict the data trends but the accuracy of the predictions was poor. He also showed that the model of Rogers et al. (1987), which is based on the bubble detachment criterion, predicted most of the data within ±40%.

The objectives of the present work are to obtain accurate data on subcooled flow boiling under low pressure conditions, investigate the nature of the NVG phenomenon in low pressure flow boiling and examine the applicability of available NVG models.

2. EXPERIMENTAL ARRANGEMENTS
2.1 Test Facility
A schematic of the test loop is presented in Figure 1.

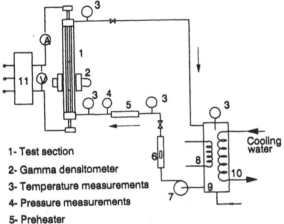

1- Test section

2- Gamma densitometer

3- Temperature measurements

4- Pressure measurements

5- Preheater

6- Flowmeter   9- Holding tank

7- Pump   10- Tank cooler

8- Tank heater   11- DC power supply

Figure 1 Test loop

The low pressure circulating loop consisted mainly of a holding tank in which the distilled-degassed water temperature was controlled by an immersed electric heater and a cooling coil, a circulating pump, a preheater and the annular test section. The test section was a 1.15 m vertical concentric annular test section. The inner tube was a 12.7 mm O.D. and the outer tube was a 25.4 mm O.D. plexiglass tube to allow visual observation. The inner tube was made of three sections; the middle section was thin-walled stainless-steel tube (0.25 mm thickness), preceeded and followed by 34 cm and 50 cm long, thick-walled copper tubes respectively. The entire inner tube was connected to a 55 kW DC power supply. Accordingly, subcooled flow boiling took place in the middle section.

The experiments carried out in the present investigation covered the following range of test conditions:

Mass flux: $100 < G < 500$ kg/m^2 s

Heat flux: $200 < q < 750$ kW/m2

Inlet subcooling: $10 < \theta_{in} < 33°C$

Pressure: $1.07 < P < 1.8$ bar

2.2 Measurements
The experiments were carried out by adjusting the inlet mass flux, wall heat flux and inlet subcooling. The axial void fraction was measured using a single-beam gamma densitometer. It consisted of a 75 mCi Cobalt-57 sealed line source and a NaI (Tl) scintillator. The gamma beam was collimated as a thin beam wide enough to cover the cross section for accurate area-averaged measurements. The densitometer was mounted on a traversing table to obtain the void fraction profile at two centimeters intervals along the test section. Zietoun (1994) showed that for the present densitometer design, the sensitivity to water content was 20%, the statistical error was less than 2% and the uncertainty in the void fraction measurements was in the range of ±4% within the range of 2% $< \alpha < 30\%$.

The test section inlet and exit temperature were measured using calibrated platinum resistance temperature detectors while the axial liquid subcooling was measured at ten centimeters intervals along the test section using 32 gauge calibrated J type thermocouples with uncertainties better than ±0.2°C. Preliminary experiments showed that measuring the local subcooling at the centre of the annular gap was consistent with the area-averaged liquid temperature along the channel. A photographic study of the boiling phenomenon was also undertaken using a high speed video camera which was mounted on the traversing table such that the phenomenon may be examined at various axial locations along the test section.

Additional information about the test facility, experimental measurements and uncertainties were reported by Zietoun (1994).

3. EXPERIMENTAL RESULT
3.1 Void Fraction Profiles and the NVG Point
Void fraction profiles were measured along the subcooled boiling at different levels of mass flux, heat flux and inlet subcooling. Typical measured void fraction profiles along the subcooled boiling regions are shown in Figure 2. The results showed that the void fraction profile was almost flat at the beginning of the heating section, i.e in the high subcooled region. The void fraction started to increase significantly at a certain point defining the location of the NVG point. As shown in Figure 2, the NVG point is not sharply defined and it rather falls in a range not at a point. The location of the NVG point was defined by fitting the void fraction data in the highly subcooled boiling region with a straight line, the point at which the profile deviates from the straightline fit was considered the NVG point as shown schematically in Figure 3. The local subcooling at the NVG point was calculated from the heat balance between the heating section inlet and the NVG point,

$$\theta_d = \theta_{in} - \frac{q P_h z_d}{G A C_p} \qquad (1)$$

Equation (1) was used in the present data reduction procedure to be consistent with other investigators with whose results comparisons will be made. More importantly, the measured local liquid subcooling values were in good agreement with the results obtained from the above equation, as shown by Zietoun (1994), confirming that the latent heat used for evaporation is much less than the sensible heat gained by the subcooled liquid.

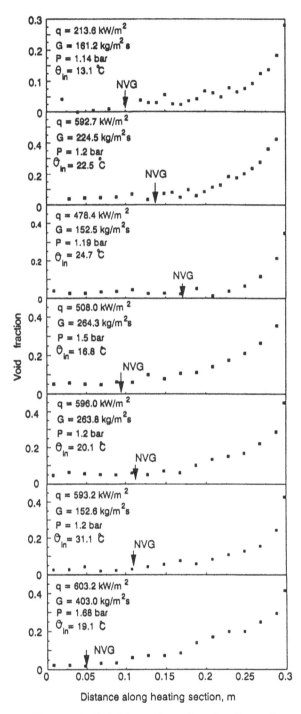

Figure 2 Void fraction distribution along the heated section

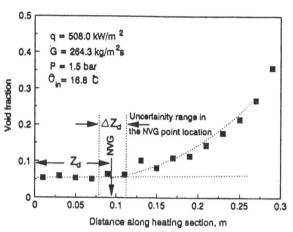

Figure 3 Uncertainty in NVG point location

3.2 Visual Observations

The high speed video system was used to visualize the flow at two centimetres increments along the subcooled flow boiling region. Extensive visual observations were reported by Zeitoun (1994). However, in the present paper, we shall focus on the region upstream of, and near the NVG point. Figure 4 shows typical bubble behaviour upstream of, and near the NVG point. In these photographs, the heater edge appears at the right side of the image and the wall of the plexiglass tube and a one-millimetre reference tube appear at the left side. Some individual bubbles are identified to show the bubble cycle as it nucleates, grows and slides along the heating surface, detaches and collapses in the subcooled liquid bulk. Upstream of the NVG point, the photographs show that after nucleation, the bubbles grow while attached to, or slowly sliding along, the heating surface until they lift off. After detachment, the bubbles condense very rapidly in the subcooled bulk. The photographs confirm that bubbles consistently detach from the heating surface upstream of the NVG point and accordingly, at least under the present test conditions, bubble detachment is not the cause of the NVG phenomenon. Moreover, the concept of a bubbly layer from which a bubble ejected to mark the occurrence of NVG, Dix (1970) and Serizawa (1979), cannot be supported by the present observations. The above observations are in agreement with the recent results of Bibeau (1993). Recent data by Bibeau and Salcudean (1990) also provide indirect support of the above observation. They measured void fraction profiles in subcooled flow boiling for upward and downward flows. The interesting feature of their results was that the NVG point occurred at higher subcooling in the case of downward flow. This is contrary to the bubble detachment models because bubbles would be expected to detach earlier, i.e. at higher subcooling due to the buoyancy force in upward flow.

4. A SIMPLE NET VAPOUR GENERATION MODEL

Examining the measured void fraction profiles in the highly subcooled region in subcooled flow boiling as obtained in the present study and by Bibeau (1993), Dimmick and Selander (1990), Stangl and Mayinger (1990) and Donevski and Shoukri (1989), showed that the void fraction starts from a zero value at the ONB point and increases, initially with a high gradient, to form a flat plateau with an almost constant void

Uncertainty analysis for the liquid subcooling at the NVG point was carried out. It was found that for all the void fraction profiles obtained in the present work, the maximum uncertainty $\Delta z_d/2$ in the NVG point location is ± 2 cm. The range of uncertainty in the liquid subcooling at the NVG point, was consequently found to be in the range $\pm 5\%$ to $\pm 15\%$.

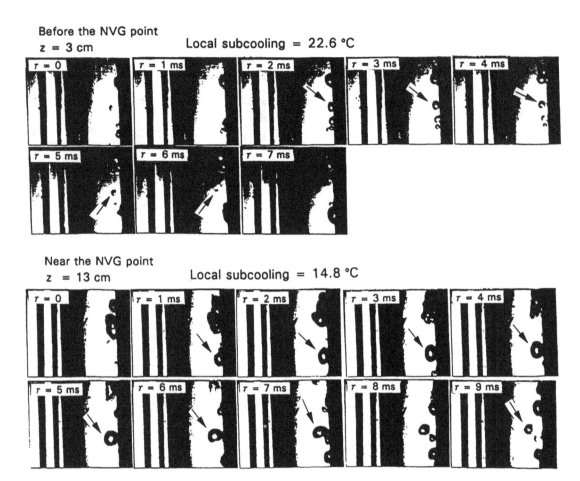

Before the NVG point
z = 3 cm   Local subcooling = 22.6 °C

Near the NVG point
z = 13 cm   Local subcooling = 14.8 °C

Figure 4 Bubble growth-collapse cycle before, near and after the NVG point for run no. B4
(q = 478.4 kW/m² , G = 152.5 kg/m² s, inlet subcooling = 24.7 °C and Z_d = 17.0 cm)

fraction. This flat plateau continues until the NVG point is reached where the void fraction starts to increase significantly, as shown schematically in Figure 5. From the above description, it is clear that the NVG point is the inflection point of the void fraction profile as suggested by Serizawa (1979). The high speed photographic results showed that the NVG phenomenon was not initiated by any dramatic change in bubble behaviour or the flow field. It also confirmed that the NVG phenomenon is not characterized by the first bubble detachment, from the wall, or ejection from a wall bubbly layer.

The void fraction profile described above is the result of the competition between the vapour generation at the heating surface and condensation at the subcooled liquid-bubble interface. By examining this profile, one can assume the profiles of the vapour generation and condensation rates, as shown schematically in Figure 6. At the ONB, the void fraction is zero and consequently the interfacial area concentration is zero. Accordingly, the condensation term will be zero and the absence of the condensation component results in the initial high gradient of the void fraction at the ONB point. This high gradient cannot continue because of the high subcooling which suppresses the net void generation due to condensation. This trend will continue until the NVG point, after which the vapour generation rate will exceed the vapour condensation rate causing significant net vapour formation. As observed from the high speed photographic results, there are two types of bubbles, attached and detached. The attached bubbles grow due to vaporization caused

by the heat transfer through the microlayer at the wall and the heat transfer from the thin superheated liquid layer adjacent to the wall. Condensation occurs at the subcooled liquid-bubble interface if the bubble penetrates the subcooled core while still attached to the heating surface. The detached bubbles are mostly influenced by condensation. The analysis of this problem is complicated due to the large number of parameters controlling the evaporation and condensation processes. However, a simplified model is given below.

The vapour generation term, responsible for evaporation, includes that portion of the applied heat flux responsible for evaporation through the microlayer beneath the bubbles as well as the component which causes evaporation at the superheated liquid-bubble interface within the superheated liquid layer, in the vicinity of the heating surface. The last part can be neglected by assuming that the superheated layer is very thin. Based on this assumption, the vapour generation term is only proportional to the applied heat flux and the apparent contact area between the attached bubbles and the heated surface. The vapour generation term, energy rate per channel length, can be estimated as follows:

$$G_v = C_1 \, q \, N_{ba} \, a_c \qquad (2)$$

where a_c is the contact area between an attached bubble and the heated surface and N_{ba} is the number of attached bubbles per unit length of the heating section.

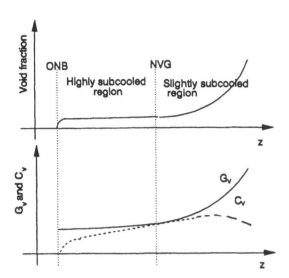

Figure 5 Speculated vapour generation and condensation functions

For attached and detached bubbles, the condensation occurs at the subcooled liquid-bubble interface. The condensation rate can be represented by:

$$C_v = C_2 \, a_b \, N_{bt} \, h_c \, (T_s - T_d) \quad (3)$$

where a_b is the bubble surface area, N_{bt} number of bubbles per unit length of the channel and h_c is the interfacial condensation heat transfer coefficient between the bubbles and the subcooled liquid. C_1 and C_2 are proportionality constants which include the uncertainty due to approximation in the above two equations. At the NVG point, the vapour generation term will be equal to the vapour condensation term, i.e.

$$q/(T_s - T_d) = [C_2 \, N_{bt} \, a_b / C_1 \, N_{ba} \, a_c] \, h_c \quad (4)$$

Akiyama (1973) proposed a correlation for the interfacial condensation coefficient in subcooled boiling in the form,

$$N_c = \frac{h_c \, D_s}{k} = 0.37 \ Re_b^{0.6} \ Pr^{1/3} \quad (5)$$

where the bubble Reynolds number was estimated based on the mean bubble diameter and the bubble relative velocity, i.e. $Re_b = \rho_l \, U_b D_s / \mu$. Substituting equation (5) into (4) we get,

$$\frac{q \, D_s}{(T_s - T_d)k} = \left[\frac{0.37 \, C_2 \, N_{bt} \, a_b}{C_1 \, N_{ba} \, a_c} \right] Re_b^{0.6} \ Pr^{1/3} \quad (6)$$

where D_s is the mean bubble diameter. The term inside the square brackets in the above equation is a complex function of active nucleation site density, bubble frequency and bubble growth-collapse time. However, these parameters will affect both the numerator and the denumerator. Accordingly, the net effect of these parameters on the square bracket term is assumed negligible. Based on this assumption, the above equation can be approximated to:

$$\frac{q \, D_s}{(T_s - T_d) \, k} = C_{\alpha} \ Re_b^{0.6} \ Pr^{1/3} \quad (7)$$

where C_{α} is an empirical constant. In the above equation the mean bubble diameter can be calculated from the correlation proposed by Zeitoun and Shoukri (1995),

$$\frac{D_s}{\sqrt{\sigma / g \Delta \rho}} = \frac{0.0683 \, (\rho_l / \rho_g)^{1.326}}{Re^{0.324} \left[Ja + \dfrac{149.2 \, (\rho_l / \rho_g)^{1.326}}{Bo^{0.487} \, Re^{1.6}} \right]} \quad (8)$$

and the bubble relative velocity can be calculated based on the drift flux model of Zuber and Findlay (1965),

$$U_b = 1.53 \left[\frac{g \, \sigma \, (\rho_l - \rho_g)}{\rho_l^2} \right]^{1/4} \quad (9)$$

Based on the above, the present data was used to calculate the value of the empirical constant C_{α}. It was found to equal unity. The left hand side of Equation (7) represents the Nusselt number based on the bubble diameter as the characteristic length. Saha and Zuber (1974) defined the Nusselt number based on the hydraulic diameter of the channel and showed that it is a constant for the thermally controlled region (Pe < 70,000), which is the case for the data presented in this work. Saha and Zuber's model for the thermally controlled region was formulated as:

$$\frac{q D_h}{(T_s - T_d)k} = 455 \quad (10)$$

which can be written in terms of Equation (7) as

$$\frac{q \, D_s}{(T_s - T_d) \, k} = 455 \left[\frac{D_s}{D_h} \right] \quad (11)$$

Comparing Equations (7) and (10) one can see that the proposed equation includes bubble dynamics effects through the bubble Reynolds number.

The predictions of Equation (7) were compared with available net NVG data. In this regard, Equation (7) was solved iteratively using Equations (8) and (9). The comparisons are shown in Figure 6. It is shown that the current model can predict the data within ±30%. One reason for this wide margin is the uncertainty in determining the exact location of the NVG point.

5. CONCLUSIONS

Experiments were carried out on subcooled flow boiling in an annular vertical channel under low pressure and low mass flux conditions. The boiling phenomenon was visualized using a high speed video camera while the axial void fraction profiles were measured using a traversing gamma densitometer. The high speed photographic study demonstrated that the NVG phenomenon is not initiated by bubble detachment from the heating surface.

A simple model was developed in the present work based on the balance between vapour generation and condensation at the NVG point. The model was shown to be capable of predicting existing low pressure data well.

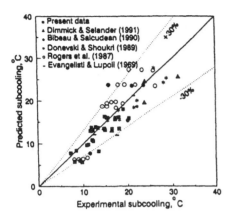

Figure 6 Comparison between the proposed
NVG model and experimental data

NOMENCLATURE

A	Channel cross sectional area	m^2
a_b	Bubble surface area	m^2
a_c	Contact area between a bubble and the wall	m^2
Bo	Boiling number, $q/G\, h_{fg}$	
C_p	Liquid specific heat	J/kg K
D_h	Channel hydraulic diameter	m
D_s	Mean bubble diameter	m
G	Mass flux	$kg/m^2\,s$
h_c	Condensation coefficient	W/m^2 K
h_{fg}	Latent heat	J/kg
Ja	Jakob number, $\rho_l\, C_p\, (T_s - T_l)/\rho_g\, h_{fg}$	
k	Liquid thermal conductivity	W/m K
N_{ba}	Number of attached bubbles per unit length	m^{-1}
N_{bt}	Number of bubbles per unit length	m^{-1}
P	Pressure	Pa
Pe	Liquid Peclet number	
P_h	Heated perimeter	m
Pr	Liquid Prandtl number	
q	Applied heat flux	W/m^2
Re	Flow Reynolds number, $G\, D_h/\mu$	
Re_b	Bubble Reynolds number, $\rho_l\, U_b\, D_s/\mu$	
T_d	Liquid temperature at the NVG point	°C
T_l	Liquid temperature	°C
T_s	Saturation temperature	°C
U_b	Bubble relative velocity	m/s
z_d	NVG axial location	m

Greek Symbols:

α	Average void fraction.	
μ	Liquid viscosity	Pa s
ρ_l	Liquid density	kg/m^3
ρ_g	Vapour density	kg/m^3
σ	Surface tension	N/m
θ_{in}	Inlet subcooling = $T_s - T_{in}$	°C
θ_l	Liquid subcooling = $T_s - T_l$	°C

REFERENCES

Ahmad, S. Y. 1970 "Axial Distribution of Bulk Temperature and Void Fraction in a Heated Channel with Inlet Subcooling", J. Heat Transfer, Vol. 92, pp. 595-609.

Akiyama, M. 1973 "Bubble Collapse in Subcooled Boiling", Bulletin of JSME, Vol. 16, pp. 570-575.

Bibeau, E. L. 1993 "Void Growth in Subcooled Flow Boiling for Circular and Finned Geometries for Low Values of Pressure and Velocity", Ph.D. Thesis, The University of British Columbia.

Bibeau, E. L. and Salcudean, M. 1990 "The Effect of Flow Direction on Void Growth at Low Velocities and Low Pressure", Int. Comm. Heat Mass Transfer, Vol. 17, pp. 19-25.

Bowring, R. W. 1962 "Physical Model Based on Bubble Detachment and Calculation of Steam Voidage in the Subcooled Region of a Heated Channel", HPR-10, Institute for Atomenergi, Halden.

Dimmick, G. R. and Selander, W. N. 1990 "A Dynamic Model for Predicting Subcooled Void: Experimental Results and Model Development" EUROTHERM Seminar # 16.

Dix, G. E. 1970 "Vapour Void Fraction for Forced Convection with Subcooled Boiling at Low Flow Rates", General Electric Report No. NEDO-10491.

Donevski, B. and Shoukri, M. 1989 "Experimental Study of Subcooled Flow Boiling and Condensation in Annular Channel", McMaster University, Dept. Mech. Eng., Thermofluids Report No. ME/89/TF/R1.

Griffith, P., Clark, J. A. and Rohsenow, W. M. 1958 "Void Volumes in Subcooled Boiling", ASME Paper 58-HT-19, U.S. National Heat Transfer Conf., Chicago.

Hancox, W. T. and Nicoll, W. B. 1971 "A General Technique for the Prediction of Void Distributions in Non-Steady Two-Phase Forced Convection", Int. J. Heat Mass Transfer, Vol. 14, pp. 1377-1394.

Lee, S. C. and Bankoff, S. G. 1992 "Prediction of the Onset of Significant Void in Downflow Subcooled Nucleate Boiling" HTD-Vol 197, Two-Phase Flow and Heat Transfer, ASME, pp. 93-100.

Levy, S. 1967 "Forced Convection Subcooled Boiling: Prediction of Vapour Volumetric Fraction", Int. J. Heat Mass Transfer Vol. 10, pp 951-965.

Rogers, J. T. and Li, J. 1992 "Prediction of the Onset of Significant Void in Flow Boiling of Water", HTD-Vol. 217, Fundamentals of Subcooled Flow Boiling, ASME, pp. 41-52.

Rogers, J. T., Salcudean, M., Abdullah, Z., McLeod, D. and Poirier, D. 1987 "The Onset of Significant Void in up-Flow Boiling of Water at Low Pressure and Low Velocities", Int. J. Heat Mass Transfer, Vol. 30, No. 11, pp. 2247-2260.

Saha, P. and Zuber, N. 1974 "Point of Net Vapour Generation and Vapour Void Fraction in Subcooled Boiling", Proceeding of Fifth Int. Heat Transfer Conf., Tokyo, Vol. IV, pp. 175-179.

Serizawa, A. 1979 "A Study of Forced Convection Subcooled Flow Boiling", Two-Phase Momentum, Heat and Mass Transfer in Chemical Processes and Energy Engineering Systems, ed. Drust, F., Tsiklauri, G. V. and Afgan, N. H., Vol. I, pp. 231-242.

Stangl, G. and Mayinger, F. 1990 "Void Fraction Measurement in Subcooled Forced Convection Boiling with Refrigerant 12", Experimental Heat Transfer, Vol. 3, pp. 323-340.

Ünal, H. C. 1975 "Determination of the initial point of Net Vapour Generation in Flow Boiling System" Int. J. Heat Mass Transfer, Vol. 18, pp.1095-1099.

Zeitoun, O. 1994 "Subcooled Flow Boiling and Condensation", Ph. D. Thesis, McMaster University, Hamilton, Ontario, Canada.

Zeitoun, O. and Shoukri, M. 1995 "Bubble Behaviour and Mean Diameter in Subcooled Flow Boiling", Accepted for publication, J. of Heat Transfer.

Zuber, N. and Findlay, J. 1965 "Average Volumetric Concentration in Two-Phase Flow System", J. Heat Transfer, Vol. 87, pp. 453-462.

VAPOR BUBBLE GROWTH IN HIGHLY SUBCOOLED
HETEROGENEOUS BOILING

W. Chen[1], R. Mei[2], and J.F. Klausner[1]
[1]Department of Mechanical Engineering
[2]Department of Aerospace Engineering
Mechanics, and Engineering Science
University of Florida
Gainesville, Florida USA

ABSTRACT

Detailed theoretical and computational analyses are carried out to investigate vapor bubble growth in highly subcooled heterogeneous boiling. The vapor bubble growth process is considered to be controlled by the unsteady energy transfer process among four regions: a wedge shaped liquid microlayer, a vapor bubble whose shape is assumed to be a truncated sphere, a subcooled bulk liquid, and a solid heater. Finite difference solutions for the solid thermal field is obtained on expanding coordinates as the bubble grows. A linear temperature profile across the liquid microlayer is assumed based on our previous investigation of saturated boiling. The unsteady energy equation in the subcooled liquid is strongly coupled with the momentum equation due to the rapid bubble growth and collapse; the momentum equation is approximated by the Laplace equation by assuming an inviscid flow. The bubble radius R(t) is determined by simultaneously solving the temperature fields in the solid and the subcooled liquid. The computational study for the solid and the subcooled bulk liquid thermal fields elucidates the energy transfer process. It is revealed that the subcooled bulk liquid flow has a significant effect on the vapor bubble growth and collapse. Close agreement between the predicted and measured bubble radius requires specification of the ratio of the microlayer base radius to the bubble radius.

INTRODUCTION

In previous investigations by Mei et al. (1995a,b), detailed theoretical and numerical analyses of vapor bubble growth in saturated heterogeneous pool boiling have been studied over a wide range of boiling conditions. In the formulation, the simultaneous coupling of the thermal fields in a thin liquid microlayer and in the solid heater takes into account the continuously decreasing heating surface temperature beneath the growing bubble. In the parametric investigation, four dimensionless parameters were identified to characterize the vapor bubble growth rate and associated temperature fields in the solid and liquid microlayer. The numerical analysis has elucidated the detailed energy transfer process for the solid thermal field. In saturated pool boiling, the bulk liquid surrounding the bubble cap is nearly saturated, and thus the heat and mass transfer in this region is small. In this paper, the bubble growth in a subcooled boiling condition is examined. It is usually quite difficult to maintain constant subcooling in a pool boiling experiment due to the transient nature of energy transfer. However, since the time scale of the transient is typically much larger than the time scale of the bubble, it is appropriate to focus on the rapid growth and collapse of the bubble in which a quasi-steady surrounding thermal field is assumed. In subcooled boiling, the unsteady heat transfer between the vapor bubble dome and the bulk liquid is no longer negligible due to the effect of subcooling. During the early stages of growth, the vapor bubble dome obtains energy from the surroundings because a large portion of the bubble surface area is within the thermal boundary layer. Thus, the overall energy for the bubble growth comes from the liquid microlayer and the vapor bubble dome. As the bubble continues to grow, a large portion of the bubble dome protrudes the thermal boundary layer. At the same instant, the energy into the bubble from the microlayer is smaller than that lost from the dome to the surrounding subcooled liquid, and the bubble starts to collapse. In this study, the vapor bubble growth process is considered to be controlled by the unsteady energy transfer among four regions: a wedge shaped liquid microlayer, a vapor bubble whose shape is assumed to be a truncated sphere, a subcooled bulk liquid, and a solid heater. A linear temperature profile across the liquid microlayer is assumed based on the investigation by Mei. et al. (1995b). The unsteady energy equation in the subcooled liquid which is assumed to be axisymmetric with respect to the axis normal to the heating surface, is strongly coupled with the momentum equation. The momentum equation is approximated by the Laplace equation by assuming an inviscid flow. Good agreement is achieved for the predicted and measured bubble growth rate.

The present model assists in developing a basic understanding of the bubble growth mechanism during subcooled boiling.

THEORETICAL FORMULATION

1. Vapor Bubble and Liquid Microlayer

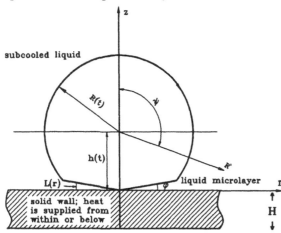

Figure 1 Sketch for the growing bubble, liquid microlayer, subcooled bulk liquid, and heating solid.

Consideration is given to an isolated vapor bubble growing from a solid heating surface into a subcooled liquid pool as shown in Fig. 1. The bubble shape is assumed to be a truncated sphere on a wedge-shaped microlayer. This assumption alleviates the requirement to specify the movement of the three-phase contact point and thus greatly simplifies the analysis. In the parametric investigation for saturated boiling (Mei et al., 1995b), it was established that the temperature profile in the liquid microlayer is practically linear and the liquid temperature in the microlayer is given by

$$T_\ell(r,z,t) = T_{sat} + \Delta T_{sat}(r,t)\left(1 - \frac{z}{L(r)}\right), \qquad (1)$$

where $\Delta T_{sat}(r,t) = T_s(r,z=0,t) - T_{sat}$, $L(r) = r\phi$ is the local microlayer thickness, and ϕ is the microlayer wedge angle which is generally very small. During the bubble growth and collapse, the change in the bubble volume V_b can be obtained through an energy balance on the microlayer liquid-vapor interface and the subcooled bulk liquid-vapor interface,

$$\rho_v h_{fg} \frac{dV_b}{dt} = \int_{A_m} -k_\ell \frac{\partial T_\ell}{\partial n}\bigg|_{z=L(t)} dA_m$$

$$+ \int_{A_b} -k_\ell \frac{\partial T_{\ell\ell}}{\partial n}\bigg|_{R'=R(t)} dA_b, \quad for\ \phi \ll 1, \qquad (2)$$

where $T_{\ell b}$ is the temperature of the subcooled liquid, A_m is the surface area of wedge on the microlayer liquid-vapor interface, A_b is the surface area of the vapor bubble dome on the subcooled bulk liquid-vapor interface, and R' is the spherical coordinate in the radial direction attached to the moving bubble. Denoting $R_b(t)$ as the microlayer base radius and $R(t)$ as the bubble radius, the bubble volume can be expressed as

$$V_b = \frac{4}{3}\pi R^3 f(C(t))$$

$$= \frac{4}{3}\pi R^3 \left(1 - \frac{3}{4}\left(1 - \sqrt{1 - C^2(t)}\right)^2 + \frac{1}{4}\left(1 - \sqrt{1 - C^2(t)}\right)^3\right) \qquad (3)$$

where $C(t) = R_b(t)/R(t) \leq 1$ is the bubble shape factor. In the limit as the bubble shape approaches a hemisphere, $C(t) \to 1$, and as the bubble shape approaches a sphere, $C(t) \to 0$. Based on experimental evidence by Cochran and Aydelott (1966) and Ellion (1954), the bubble shape in subcooled pool boiling changes with time from a hemispherical dome in the early stage to a spherical dome. Thus, C varies with time. The observation that $C(t) \to 0$ at the later stage of bubble growth-collapse is due to the high surface tension of water, so that the microlayer size decreases with time.

2. Subcooled Bulk Liquid

By assuming axisymmetry for the temperature and velocity fields, the unsteady energy equation for the subcooled bulk liquid in cylindrical coordinates (r,z) is

$$\frac{\partial T_{\ell\ell}}{\partial t} + u_r \frac{\partial T_{\ell\ell}}{\partial r} + u_z \frac{\partial T_{\ell\ell}}{\partial z} = \alpha_\ell \left(\frac{1}{r}\frac{\partial}{\partial r}\left(r\frac{\partial T_{\ell\ell}}{\partial r}\right) + \frac{\partial^2 T_{\ell\ell}}{\partial z^2}\right). \quad (4)$$

where α_ℓ is the thermal diffusivity, (u_r, u_z) are the liquid velocity components in the (r,z) directions. The initial condition specified for the liquid thermal field is based on Marcus and Dropkin's (1965) investigation of the thermal boundary layer above a copper surface in saturated nucleate pool boiling of water:

$$\frac{T_{\ell\ell} - T_{sub0}}{T_w - T_{sub0}} = \begin{cases} 1 - w\frac{z}{\delta}, & \frac{z}{\delta} \leq 0.57 \\ D\left(\frac{z}{\delta}\right)^{-\lambda}, & \frac{z}{\delta} \geq 0.57 \end{cases} \qquad (5)$$

where T_w is the average temperature of boiling surface, w and λ are experimentally determined constants based on the measurements of the subcooled liquid temperature, $D = (1 - 0.57\ w)0.57^\lambda$, and δ is the extrapolated superheated-layer thickness. It is noted that $w = 1$ in saturated nucleate pool boiling. In subcooled pool boiling, equation (5) is adopted by noting that the value of w, which can be obtained from the data of Wiebe and Judd (1971) may be a function of heat flux and degree of subcooling and varies from 1 to 1.75 for the data considered here. λ is set to be 1.2 in the numerical computation. The extrapolated superheated-layer thickness, δ, is estimated by Hsu (1962) using the expression given below which was obtained assuming the thermal layer is governed by one-dimensional unsteady heat conduction,

$$T_w - T_{sub0} = T_{sat} - T_{sub0} + \frac{2AC_3}{\delta}$$

$$+ \sqrt{\left(2(T_{sat} - T_{sub0}) + \frac{2AC_3}{\delta}\right)\left(\frac{2AC_3}{\delta}\right)}, \qquad (6)$$

where C_3 is equal to 1.6, A is equal to $2\sigma T_{sat}/\rho_v h_{fg}$, and σ is the surface tension. Hsu has demonstrated the usefulness of equation (6) for the prediction of the incipience point over a range of subcooling and pressure. As suggested by Wiebe and Judd (1971), equation (6) is used in this study to estimate δ.

Since there is no strong mean flow over the bubble, the flow induced by the growth of the bubble is mainly of inviscid nature. Thus the liquid velocity field may be determined by solving the Laplace equation $\nabla^2\Phi = 0$ for the velocity potential Φ and the velocities are given by,

$$u_r(r,z) = \frac{\partial\Phi}{\partial r} \quad and \quad u_z(r,z) = \frac{\partial\Phi}{\partial z}. \qquad (7)$$

Since the bubble is a growing truncated sphere, the velocity field needs to be solved at each time step. If the vapor bubble is exactly hemispherical, which is the case during the early growth stage, the liquid velocities are

$$u_r(r,\psi) = \dot{R}(t)\,\frac{R^2(t)}{r^2}\sin^2(\psi) \qquad (8)$$

$$and \quad u_z(r,\psi) = \dot{R}(t)\,\frac{R^2(t)}{r^2}\cos^2(\psi),$$

where $\dot{R}(t) = dR(t)/dt$ and ψ is measured from the axis of symmetry, i.e., z-axis. The general boundary conditions for Φ will be given in the transformed coordinates when a non-hemispherical bubble is considered.

3. Solid heater

The energy equation for the solid is

$$\frac{\partial T_s}{\partial t} = \alpha_s\left[\frac{1}{r}\frac{\partial}{\partial r}\left(r\frac{\partial T_s}{\partial r}\right) + \frac{\partial^2 T_s}{\partial z^2}\right] + \frac{\alpha_s}{k_s}q''', \quad for\ 0<r<\infty,\ -H<z<0 \quad (9)$$

subject to the initial condition

$$T_s = T_{sat} + \Delta T_{sat0} \qquad\qquad at\ t=t_0<t_c \qquad (10)$$

and the boundary conditions

$$\frac{\partial T_s}{\partial r} = 0 \qquad\qquad at\ r=0\ and\ r\to\infty, \qquad (11)$$

$$\frac{\partial T_s}{\partial z} = -q''/k_s \qquad\qquad at\ z=-H, \qquad (12)$$

$$k_\ell\frac{\partial T_\ell}{\partial z} = k_s\frac{\partial T_s}{\partial z} \qquad at\ z=0,\ \ 0<r\le R_b(t), \qquad (13)$$

$$-k_s\frac{\partial T_s}{\partial z} = \bar{h}(T_s - T_{sub0}) \qquad at\ z=0,\ \ r>R_b(t) \qquad (14)$$

where α_s and k_s are the thermal diffusivity and conductivity of the solid, t_c is the bubble collapse time scale, T_{sub0} is the subcooled bulk liquid temperature at infinity, $\bar{h}$ is the average convective heat transfer coefficient in the subcooled bulk liquid, q'' is the heat flux (on the average), and q''' is the volummetric heating (on the average).

4. Non-dimensionalization and grid generation

To obtain a better understanding of the physics of the problem, the following dimensionless variables are introduced,

$$\tau = \frac{t}{t_c},\quad \bar{r}_s = \frac{r}{R_b(t)},\quad \bar{r}_{b\ell} = \frac{r}{R(t)},\quad \bar{z}_s = \frac{z}{H},\quad \bar{z}_{b\ell} = \frac{z}{R(t)},$$

$$\bar{R}(\tau) = \frac{R(t)}{R_c(t_c)},\quad \theta_s = \frac{T_s - T_{sat}}{\Delta T_{sat0}},\quad \theta_{b\ell} = \frac{T_{b\ell} - T_{sub0}}{T_{sat} - T_{sub0}}, \qquad (15)$$

where $R_c(t_c)$ is defined by Mei et al. (1995a) (see nomenclature). Following Mei et al. (1995a), by using equation (15) and applying coordinate transformations,

$$\bar{r}_s = \bar{r}_\infty\{1 - S_r\tan^{-1}[(1-\zeta)\tan(1/S_r)]\} \qquad for\ 0\le\zeta\le1, \qquad (16)$$

and $\bar{z}_s = -1 + S_z\tan^{-1}[(1-\varepsilon)\tan(1/S_z)] \qquad for\ 0\le\varepsilon\le1, \qquad (17)$

equation (9) becomes

$$\frac{\partial\theta_s}{\partial\tau} - \frac{\dot{R}_b t_c}{R_b}\bar{r}_s\frac{\partial\theta_s}{\partial\zeta}\frac{\partial\zeta}{\partial\bar{r}_s}$$

$$= \frac{\phi\,f(C(\tau))}{\alpha\,Ja\,C^2(\tau)}\frac{1}{\bar{R}^2(\tau)}\frac{1}{\bar{r}_s}\frac{\partial}{\partial\zeta}\left(\bar{r}\frac{\partial\theta_s}{\partial\zeta}\frac{\partial\zeta}{\partial\bar{r}_s}\right)\frac{\partial\zeta}{\partial\bar{r}_s} \qquad (18)$$

$$+ Fo\frac{\partial}{\partial\varepsilon}\left(\frac{\partial\theta_s}{\partial\varepsilon}\frac{\partial\varepsilon}{\partial\bar{z}_s}\right)\frac{\partial\varepsilon}{\partial\bar{z}_s} + \frac{\alpha_s t_c q'''}{k_s\Delta T_{sat0}},$$

where α and κ are defined in the nomenclature, $\bar{r}_\infty$ is the radial position within which equation (9) is applied, the constant S_r determines the percentage of the grid number allocated to the region of $\bar{r}_s\le1$, and S_z is a constant characterizing the stretching in the z-direction.

For the subcooled bulk liquid, the energy equation in dimensionless form is

$$\frac{\partial\theta_{b\ell}}{\partial\tau} + \left(u_r - \dot{R}\bar{r}_{b\ell}\right)\frac{t_c}{R}\frac{\partial\theta_{b\ell}}{\partial\bar{r}_{b\ell}} + \left(u_z - \dot{R}\bar{z}_{b\ell}\right)\frac{t_c}{R}\frac{\partial\theta_{b\ell}}{\partial\bar{z}_{b\ell}}$$

$$= \frac{t_c\alpha_\ell}{R^2}\left(\frac{\partial^2\theta_{b\ell}}{\partial\bar{r}_{b\ell}^2} + \frac{1}{\bar{r}_{b\ell}}\frac{\partial\theta_{b\ell}}{\partial\bar{r}_{b\ell}} + \frac{\partial^2\theta_{b\ell}}{\partial\bar{z}_{b\ell}^2}\right). \qquad (19)$$

To solve the Laplace equation for Φ and energy equation for $\theta_{b\ell}$, an algebraic grid generation technique is used for the bulk liquid outside the growing bubble. Figure 2 shows the computational grids for the numerical simulation. The clustering of grids near the bubble surface is necessary to resolve large temperature gradients. The $\bar{r}_{b\ell}$ and $\bar{z}_{b\ell}$ within the physical domain are calculated as

$$\bar{r}_{b\ell}(i,j) = D(j)\sin(\gamma(i,j)), \qquad (20)$$

$$\bar{z}_{b\ell}(i,j) = D(j)\cos(\gamma(i,j)) + \bar{z}_b,$$

where

$$\omega(j) = \frac{\pi}{2} + \sin^{-1}\left(\frac{\bar{z}_b}{D(j)}\right), \qquad (21)$$

$$\gamma(i,j) = \omega(j)\frac{i-1}{M},$$

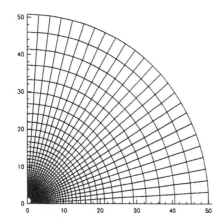

Figure 2 The computational grids for the numerical simulation

$$D(j) = 1 + (\bar{R}_\infty - 1)\left\{1 - S_{R'}\tan^{-1}\left[(1-\eta)\tan(1/S_{R'})\right]\right\},$$

and $\bar{z}_h = \sqrt{1-C^2}$. $\bar{R}_\infty$ is the radial position at infinity and the constant $S_{R'}$ determines the percentage of the grid number near the bubble surface. Figure 3 illustrates the grid parameters.

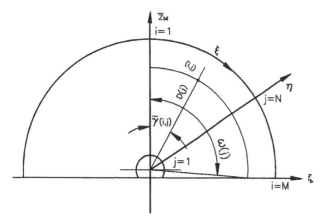

Figure 3 Grid parameters

The transformed energy and Laplace equations for the subcooled bulk liquid are

$$\frac{\partial \theta_{bl}}{\partial \tau} + A_r J\left(\bar{z}_{bl_\eta}\frac{\partial \theta_{bl}}{\partial \xi} - \bar{z}_{bl_\xi}\frac{\partial \theta_{bl}}{\partial \eta}\right) + A_z J\left(\bar{r}_{bl_\xi}\frac{\partial \theta_{bl}}{\partial \eta} - \bar{r}_{bl_\eta}\frac{\partial \theta_{bl}}{\partial \xi}\right)$$

$$= \frac{t_c \alpha_l}{R^2}J^2\left(a\frac{\partial^2\theta_{bl}}{\partial\xi^2} + 2b\frac{\partial^2\theta_{bl}}{\partial\eta\partial\xi} + c\frac{\partial^2\theta_{bl}}{\partial\eta^2} + d\frac{\partial\theta_{bl}}{\partial\eta} + e\frac{\partial\theta_{bl}}{\partial\xi}\right) \quad (22)$$

$$+ \frac{t_c\alpha_l}{R^2}\frac{J}{r_{bl}}\left(\bar{z}_{bl_\eta}\frac{\partial\theta_{bl}}{\partial\xi} - \bar{z}_{bl_\xi}\frac{\partial\theta_{bl}}{\partial\eta}\right)$$

and

$$J^2\left(a\frac{\partial^2\Phi}{\partial\xi^2} + 2b\frac{\partial^2\Phi}{\partial\eta\partial\xi} + c\frac{\partial^2\Phi}{\partial\eta^2} + d\frac{\partial\Phi}{\partial\eta} + e\frac{\partial\Phi}{\partial\xi}\right)$$

$$+ \frac{J}{r_{bl}}\left(\bar{z}_{bl_\eta}\frac{\partial\Phi}{\partial\xi} - \bar{z}_{bl_\xi}\frac{\partial\Phi}{\partial\eta}\right) = 0 \quad (23)$$

where $J = \dfrac{1}{\bar{r}_{bl_\xi}\bar{z}_{bl_\eta} - \bar{z}_{bl_\xi}\bar{r}_{bl_\eta}}$ is the Jacobian, and

$$a = \bar{r}_{bl_\eta}^2 + \bar{z}_{bl_\eta}^2, \quad b = \bar{r}_{bl_\eta}\bar{r}_{bl_\xi} + \bar{z}_{bl_\eta}\bar{z}_{bl_\xi}, \quad c = \bar{r}_{bl_\xi}^2 + \bar{z}_{bl_\xi}^2,$$

$$d = J(\beta_1\bar{z}_{bl_\xi} + \beta_2\bar{r}_{bl_\xi}), \quad e = J(\beta_2\bar{r}_{bl_\eta} + \beta_1\bar{z}_{bl_\eta}),$$

$$\beta_1 = a\bar{r}_{bl_{\xi\xi}} - 2b\bar{r}_{bl_{\xi\eta}} + c\bar{r}_{bl_{\eta\eta}}, \quad \beta_2 = a\bar{z}_{bl_{\xi\xi}} - 2b\bar{z}_{bl_{\xi\eta}} + c\bar{z}_{bl_{\eta\eta}},$$

$$A_r = (u_r - \dot{r})t_c/R, \quad and\ A_z = (u_z - \dot{z})t_c/R.$$

The temperature boundary conditions are $\theta_{bl} = 1$ along the vapor-liquid interface, and the temperature gradient is zero along the line of symmetry. The heat flux at the solid heater is equal to the heat flux to the subcooled bulk liquid along the wall. The far-field boundary condition for θ_{bl} is approximated by zero heat flux. The boundary conditions for velocity potential are zero velocity potential gradient along the wall and the line of symmetry. The far-field boundary condition is approximated by a zero flux condition. The boundary condition at the vapor-liquid interface is $\partial\Phi/\partial n = \dot{R} + \dot{h}\cos\psi$, where $\dot{h} = dh(t)/dt$ is the bubble vertical velocity due to its moving center, and h(t) is the distance from the wall to the bubble center. Equation (22) is solved using the ADI method (Anderson et al., 1984) and equation (23) is solved using SOR by line (Anderson et al., 1984).

RESULTS AND DISCUSSION

The bubble shape factor is assumed to vary as

$$C(t) = \begin{cases} 1 & , \ 0 \le t \le t_s \\ \sin\left(\dfrac{\pi}{2}\left(2 - \dfrac{1}{1+(t/t_s-1)^2}\right)\right) & , \ t_s < t \end{cases} \quad (24)$$

where t_s is the time when the bubble shape begins to change from a hemisphere. Due to a lack of experimental information, the determination for t_s employs a dimensionless parameter $\mu = \dot{R}(t)\,t/R(t)$ which characterizes the relative growth rate. It is noted that μ_s (i.e. $t=t_s$) is determined by matching the computed growth rates with the experimental data in order to achieve the best fit. In this study μ_s ranges from 0.23 to 0.48.

Two parameters, ϕ and μ_s, need to be specified for the present computation. Through a comparison of predicted growth rates with experimental results, ϕ and μ_s can be determined via the best fit. There exists only one combination of ϕ and μ_s for the best agreement with experimental data.

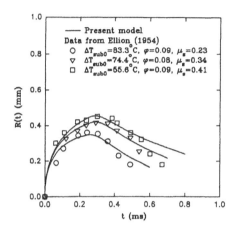

Figure 4a   Comparison of the bubble growth rate between the present model and the experimental data of Ellion (1954).

It is found that the solution for $\theta_{b\ell}$ is much more sensitive to the grid resolution in the r-direction than that for u or Φ due to the large convection term on the right hand side of Eq. (19). The final computational grid is chosen based on: i) the comparison of $\theta_{b\ell}$ computed at a large distance from the bubble and an asymptotic solution outside a hemispherical bubble which can be obtained using the method of characteristics (Chen, 1995); and ii) the comparison of $\theta_{b\ell}$ for the successive doubling of grids in the r-direction.

To validate the present model and to understand the effect of subcooling on the growth rate, comparisons are made between the predicted and the measured R(t) by Ellion (1954) and Cochran and Aydelott (1966). Figure 4a shows comparisons in the high subcooling range, $\Delta T_{sub0} = 55.6°C \sim 83.3°C$, and Ja=83. The data are taken from Ellion (1954) who used distilled degassed water as the boiling liquid. The heating surface consists of a 304 stainless-steel strip with a thickness of 0.11 mm. For each subcooling there are 5 sets of data at the same boiling conditions. The data shown in Figure 4a consist of a set which falls within the average value of the experimental results. The values of w are 1.65, 1.70, and 1.75 for $\Delta T_{sub0} = 55.6$, 74.4, and 83.3°C. Close agreement between the predicted and the measured growth rates is observed for most of the time. Figure 4a also shows that the maximum R(t) and the bubble lifetime increase with an increase in bulk liquid temperature. In the later stage of collapse, the bubble radius is over-predicted. It may be due to an improper specification of the shape factor, C(t) for later time. In Figure 4b the computed bubble volume is compared with the measured data of Cochran and Aydelott (1966). The subcooling ΔT_{sub0} ranges from 7.4°C ~ 21.4°C, Ja=25, and the boiling liquid is also water. The heater material is a 0.13 mm thick chromel strip with w=1.05 and 1.15 for $\Delta T_{sub0} = 7.4$, and 21.4°C.   Good agreement for the bubble volume is observed for most of the time. A little discrepancy is observed at the later stage of collapse. It is also evident that the vapor bubble has a larger volume when the bulk liquid temperature is higher. In Figure 5 the numerical results for the total heat transfer rate between the liquid microlayer and the vapor bubble dome are shown. The boiling condition

corresponds to that of Ellion (1954) with $T_{sub0}=55.6°C$. It is evident that the microlayer continuously supplies energy to the bubble. At the early stage, the bubble grows rapidly within the thermal boundary layer, in which only a small portion of the top of the dome is exposed to the subcooled bulk liquid. Because of the rapid increase in the bubble surface area, the total heat transfer rate increases during the early stage. When the bubble is collapsing at the later stage, a large portion of the surface area is exposed to the thermal boundary layer again. This results in a decreasing rate of bubble collapse. The bubble collapse time scale is 0.67 ms. It is interesting to note that the computational results are not sensitive to the values of w. A 10% variation in w results in little change in the growth rate when all other parameters are held fixed.

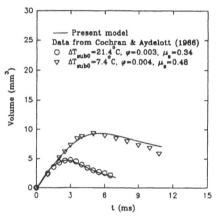

Figure 4b   Comparison of the vapor bubble volume between the present model and the experimental data of Cochran and Aydelott (1966).

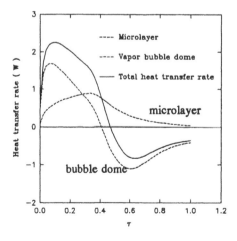

Figure 5   The total heat transfer rate between the liquid microlayer and the vapor bubble dome.

Figure 6 shows the development of the thermal layer at the surface of the solid heater at Fo=0.1 and Ja=83 in which $T_{sub0}=55.6°C$. The circular markers are the locations of the microlayer edge. During bubble growth, the temperature behaves similarly to the saturated boiling case (Mei et al.,1995b). It is clearly seen that the thermal field moves to the right following the bubble growth. During the bubble collapse period, the surface temperature quickly recovers and the thermal field shifts back to the left. Figure 7 shows the four computed temperature contours at four different times, τ=0.1, 0.25, 0.5, 1. For Fo=0.1, the heater is considered to be thick and the asymptotic behavior (Mei et al., 1995b) is valid for most of the time. During the bubble growth period, it is very easy to see that the temperature contours in the solid penetrate down and move to the right in a similar way as the bubble grows. When the bubble collapses, the temperature contours move to the left following the movement of the microlayer edge.

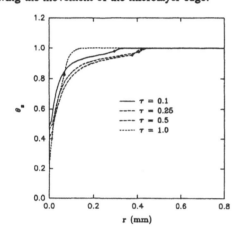

Figure 6   Development of the thermal layer at the surface of the solid heater

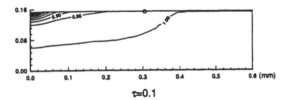

τ=0.1

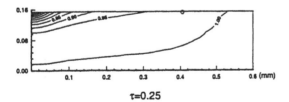

τ=0.25

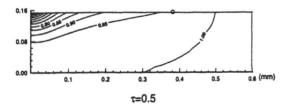

τ=0.5

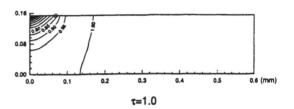

τ=1.0

Figure 7   Temperature contours for the solid heater at four different times.

Figure 8 shows the computed isotherms in the subcooled liquid surrounding a growing and collapsing vapor bubble. It is seen that there exists an unsteady thermal boundary layer around the vapor bubble. During the early stage of growth, we indeed see that most of the bubble surface is exposed to the thermal boundary layer, while most of the bubble surface is outside the thermal boundary layer when the bubble is collapsing.

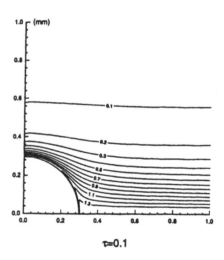

τ=0.1

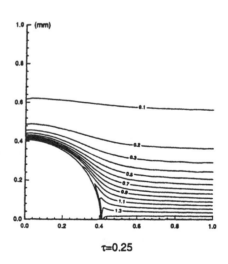

τ=0.25

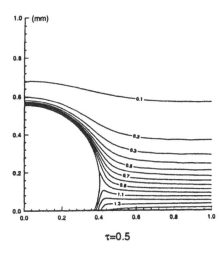

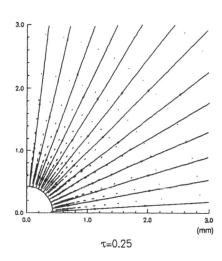

τ=0.25

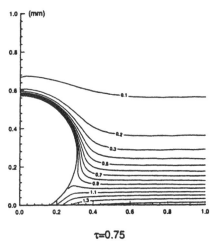

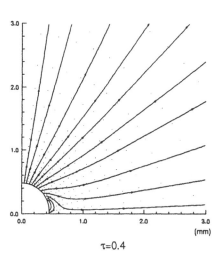

τ=0.4

Figure 8 The computed isotherms in the subcooled liquid surrounding a growing and collapsing vapor bubble at four different times.

Figure 9 shows the computed velocity vectors and the streamlines. At $\tau=0.25$, the velocity field is a source flow. For $\tau<0.4$, the velocity field is due to a source and a moving doublet. The source results from the growth of the bubble while the moving doublet is caused by the motion of the bubble center in the z-direction. At $\tau=0.4$, the velocity vectors are almost zero. At this point, the bubble growth rate is close to zero, and the bubble begins to collapse. The velocity field can be represented by sink and doublet flows moving downwards at $\tau=0.5$. The bubble continues to collapse. At $\tau=0.75$, the flow field is mainly a sink flow and the bubble continues to collapse

CONCLUSIONS
Detailed theoretical and numerical analyses are presented to elucidate the effect of the energy transfer and hydrodynamics on the growth rate of a vapor bubble in subcooled pool boiling. The bubble growth or collapse rates are obtained by combining the energy transfer from the liquid microlayer and from the bulk liquid. It is shown that the velocity field in the subcooled bulk liquid has a significant effect on the vapor bubble growth and collapse. The present study for vapor bubble growth is limited to pool boiling in the isolated bubble regime.

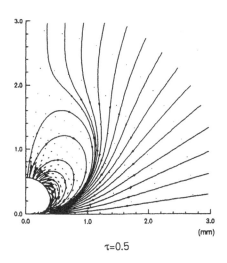

τ=0.5

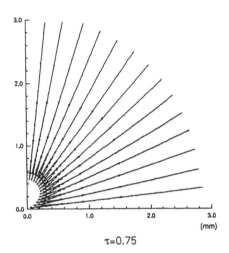

$\tau = 0.75$

Figure 9 The computed velocity vectors and the streamlines
at four different times.

NOMENCLATURE

A_m: area of microlayer wedge
A_b: surface area of vapor bubble dome
C: bubble shape parameter
$C_{p\ell}$: liquid specific heat
C_{ps}:   solid specific heat
$f(C)$:   bubble volume factor
Fo: Fourier number ($= \alpha_s t_c / H^2$)
h: vertical distance form the center of bubble to the wall
$\dot{h}$ bubble vertical velocity
h_{fg}:   liquid latent heat of vaporization
H: solid heater thickness
k_ℓ & k_s: liquid and solid thermal conductivity
Ja: Jacob number
$L(r)$:   the liquid microlayer thickness
Pr_ℓ: liquid Prandtl number
$\dot{q}_{evap}$: the rate of heat removal from the heating surface due to microlayer evaporation
q'':   heat supply from the bottom of the heater
q''':   the internal heat generation per unit volume
r: radial coordinate
$\bar{r}_s$: dimensionless coordinate in the solid heater
$\bar{r}_{b\ell}$: dimensionless coordinate in the bulk liquid
R':   radial direction in the spherical moving coordinate
$R(t)$:   bubble radius
$R_b(t)$:   radius of the liquid microlayer underneath the bubble
$R_c(t_c)$:   bubble radius for a constant wall superheat

$$\left(= \left[\frac{C \, Ja \, \alpha_\ell t_c}{\phi f(C)} \right]^{1/2} \right).$$

$\bar{R}(t)$: dimensionless bubble growth rate
t: time
t_c: vapor bubble growth and collapse time scale
$T_{b\ell}$:   bulk liquid temperature
T_ℓ & T_s: liquid and solid temperature
T_{sat}:   saturated temperature
T_{sub0}:   initial bulk liquid temperature

T_w: average temperature of boiling surface
u_r :   bulk liquid velocity in the r direction
u_z :   bulk liquid velocity in the z direction
V_b: the bubble volume
z: coordinate in the direction normal to the heating surface
$\bar{z}_s$:   dimensionless coordinate in the solid heater
$\bar{z}_{b\ell}$: dimensionless coordinate in the bulk liquid
$\bar{z}_h$:   grid parameter ($= h/R$)

Greek symbols

α: liquid-to-solid thermal diffusivity ratio (α_ℓ / α_s)
α_ℓ & α_s: liquid and solid thermal diffusivity
δ: extrapolated superheat-layer thickness
T_{sat0}:   initial superheat at incipience ($= T_w - T_{sat0}$)
ϕ: microlayer wedge angle
Φ: velocity potential
κ: liquid-to-solid thermal conductivity ratio (k_ℓ / k_s)
ν_ℓ: liquid kinematic viscosity
θ_s: dimensionless temperature of wall superheat
$\theta_{b\ell}$: dimensionless temperature of bulk liquid
ρ_ℓ & ρ_v: liquid and vapor density
τ: dimensionless time ($= t/t_c$)
ξ and η: computational coordinates in the bulk liquid
ζ and ϵ: computational coordinates in the solid

REFERENCES

Anderson, A.A., Tennehill, J.C., and Pletcher, R.H., 1984, *Computational Fluid Mechanics and Heat Transfer*, McGraw Hill.

Chen, W., 1995, Vapor Bubble Growth in Heterogeneous Boiling, Ph.D. Thesis, University of Florida, Gainesville.

Cochran, T.H. and Aydelott, J.C., 1966, "Effects of subcooling and gravity level on boiling in the discrete bubble region," *NASA Tech. Note* TN D-3449.

Ellion, M.E., 1954, " A study of the mechanism of boiling heat transfer", Ph.D. Diss., Calif. Inst. Technology, Jet Prop. Lab. Rept., Memo. 20-88.

Hsu, Y.Y., 1962, " On the size range of active nucleation cavities on a heating surface," *J. Heat transfer*, Vol. 84, No. 3, pp. 207-216.

Marcus, B.D. and Dropkin, D., 1965, " Measured temperature profiles within the superheated boundary layer above a horizontal surface in saturated nucleate pool boiling of water," *J. Heat Transfer*, Vol. 87, No. 3, pp. 333-341.

Mei, R., Chen, W.C., and Klausner, J.F., 1995, "Vapor bubble growth in heterogeneous boiling –Part I. Formulation," *Int. J. Heat Mass Transfer*, Vol. 38, No. 5, pp. 909-919.

Mei, R., Chen, W.C., and Klausner, J.F., 1995, "Vapor bubble growth in heterogeneous boiling –Part II. Growth rate and thermal fields," *Int. J. Heat Mass Transfer*, Vol. 38, No. 5, pp. 921-934.

Wiebe, J.R. and Judd, R.L., 1971, " Superheat layer thickness measurements in saturated and subcooled nucleate boiling," *J. Heat Transfer*, Vol. 93 , pp. 455-461.

VOID FRACTION AND RECONDENSATION RATE IN SUBCOOLED FORCED CONVECTIVE BOILING WITH FREON R 12

G. Stängl and F. Mayinger
T. U. Munich, Germany
Arcisstr. 21

ABSTRACT

This paper presents an investigation and results of volumetric void fraction and recondensation of bubbles in forced convective subcooled boiling. Dichlordifluormethane (CCl_2F_2) served as test fluid. The data were taken at different heat fluxes in a 12- to 25- bar pressure range, the mass fluxes have been varied from 500 to 3000 kg/m²s with an inlet subcooling in a range of 10 to 50K. The experiments have been conducted in an annular test channel with an 0.016m inner diameter and an 0.03m outer diameter. The inner tube of the annulus was heated by direct current. The void fraction data were gauged with a γ - densitometer and a specially designed impedance void probe. The experimental results reveal that the void fraction is nearly constant from the onset of nucleation boiling to a subcooling of about $\Delta T = 10K$. A method for predicting the void fraction based on the drift flux model was calculated. With the help of this equation, a model for the recondensation rate could be derived and will also be presented in this paper.

INTRODUCTION

The knowledge of the onset of nucleate boiling, void fraction, and pressure drop in subcooled boiling is important for the practical layout and design of liquid-cooled heat-generating systems operating with high heat fluxes. The phenomena occurring during subcooled boiling are explained in Figure 1. This figure, however, is intended to give an overview of the physical phenomena occurring in subcooled boiling and does therefore not consider the specific behaviour of refrigerants. A fluid, which is subcooled at the channel inlet, flows through an uniformly heated tube. Because of the single-phase convective heat transfer, a superheated thermal boundary layer is established near the heated wall. Here, in small cavities, the first nucleation sites are activated and the first bubbles are generated (z_{sb}). This phenomenon is called "onset of nucleate boiling" (ONB). The average bulk temperature at ONB is still considerably below the saturation temperature, and the generated bubbles

are small and therefore condense as soon as they come in contact with the subcooled bulk flow. Consequently, the void fraction is very low in this region. The average bulk temperature increases with increasing channel length. Therefore, the bubble condensation rate decreases and the bubbles grow, tend to retreat from the heated wall, and drift into the bulk flow (z_{ndp}). Beyond the latter point the temperature near the center

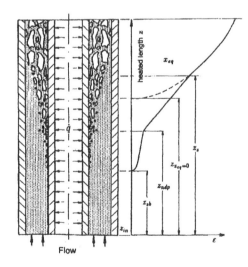

FIG. 1 VOID FRACTION DEVELOPMENT

of the tube is still below the saturation temperature, and thermodynamic non-equilibrium continues to exist until this temperature reaches saturation (z_b). Beyond that point, the released heat contributes only to liquid evaporation. Point $z_{xeq=0}$ reveals the saturation temperature of the flow when the first law of thermodynamics is applied neglecting the thermodynamic non-equilibrium. In subcooled boiling, not only the heat transfer coefficient increases in comparison to single-phase flow, but also the void fraction, which causes a change of the fluid dynamic characteristics, such as the pressure drop. Subcooled boiling phenomena have been investigated by several authors. Dix and Jain et al. have experimented in the region $z_{sb} - z_{ndp}$ using Freon 113 as test fluid, and Rouhani using water; all of them, however, have worked in low-pressure ranges.

EXPERIMENTAL SETUP

The experiments are conducted in a closed loop made of stainless steel. The loop is designed to operate with refrigerant 12 up to the critical pressure (41.6 bar). A vertically arranged test section having an annular cross section with a hydraulic diameter of 0.014 m is mounted in the loop. The inner tube consists of three sections. The unheated sections at the inlet and outlet are made of copper; the heated section is a stainless steel tube with a 16-mm outer diameter and a 1.5-mm wall thickness. This tube is supplied by direct current to guarantee uniform heat flux. The void fraction was gauged by means of a γ - densitometer and specially designed impedance probes. More detailed information of the test loop and the respective measuring devices have been published in earlier publications of the authors.

VOID FRACTION IN SUBCOOLED BOILING

As an example of the experimental results concerning the void fraction Fig. (2) shows the volumetric void fraction plotted over the local subcooling, which in this case is more revealing than the equilibrium quality defined below. This drawing illustrates the volumetric void fraction formation under different pressures. It is obvious that the void

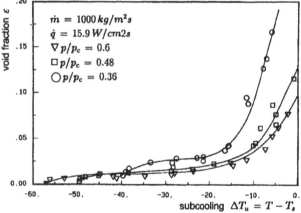

FIG. 2 VOID FRACTION FORMATION AT DIFFERENT PRESSURES

fraction decreases under increasing pressure. The overall behaviour of the void fraction can be explained by a combination of physical phenomena. The recondensation rate decreases with reduced subcooling of the liquid. Therefore the void fraction increases. On the other hand, the bubbles are able to detach from the heated wall and move inward to the bulk flow. This will be supported by even very little pressure differences due to the radial velocity differences in the flow. The differences in density will make the detached bubbles faster than the liquid flow surrounding them. This phenomenon is effected by decreasing the local void fraction gauged with the impedance probes. This behaviour of the void fraction at various mass fluxes can be explained by the decreasing

radius of the bubbles and decreasing thickness of the boundary layer with increasing pressure. Therefore the bubbles detach later from the heated wall and condense more rapidly in the subcooled bulk flow, due to the smaller diameter and the smaller density difference between the phases.

The influence of the mass flux density on the void fraction formation is shown in Fig. (3). Within these test runs, pressure and heat flux were kept as constant as possible, the tests were however, performed with different mass fluxes. In this case, the void fraction is plotted over the equilibrium quality x_{eq} defined by eq. 1 :

$$x_{eq} = \frac{h_f - h_{sl}}{h_{lg}} = \frac{c_p \, [T(z) - T_s]}{h_{lg}} \tag{1}$$

The expected behaviour of the void formation, decreasing void fraction

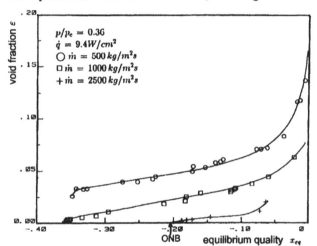

FIG. 3 VOID FRACTION FORMATION AT DIFFERENT MASS FLUX DENSITIES

by increasing mass flux can be physically explained with the decreasing thickness of the superheated or saturated boundary layer. This leads to an earlier condensation of the bubbles attached at the wall. Moreover, the single phase heat transfer between the wall and the liquid increases with increasing mass flux and consequently, the walls super heating necessary for the bubble growth will be reached in lower subcooling ranges. The arrow marked with ONB indicates the onset of nucleate boiling calculated with the equation published by Bräuer. The calculated value of the equilibrium quality for the ONB and the measured value are in a very good agreement. (i. e. $x_{eq,cal} = 0.209$, $x_{eq,meas} = 0.19$).

The changes in the void fraction formation by various heat flux densities is illustrated in Fig. 4. In this figure, only subcoolings up to $\Delta T = -40 \, [K]$ are shown, due to limitation of the test loop. The pressure and mass flux are kept constant. An interesting detail in this figure is the slight decrease of the void fraction measured with a heat flux of $\dot{q} = 3.8 \, [W/cm^2]$. As set out above, this can be explained with the detaching and floating of the bubbles in the bulk flow. Moreover, due to the detaching of the bubbles, the turbulence in the boundary layer is increased which supports the heat transfer between the heated wall and the single phase liquid flow. Rouhani proved this physical phenomenon experimentally.

Modelling of the results

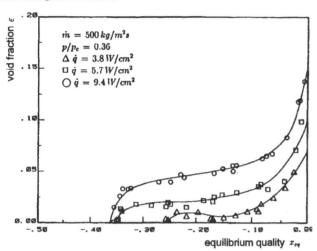

FIG. 4 VOID FRACTION FORMATION AT DIFFERENT HEAT FLUX DENSITIES

To calculate the true quality $\dot{x}$, defined by eq. 2

$$\dot{x} = \frac{\dot{M}_g}{\dot{M}} \qquad (2)$$

with the total mass flow ($\dot{M}$) and the vapor mass flow ($\dot{M}_g$) can be calculated by using an energy balance. In this case, the measured radial temperature profile must be corrected. To fit the data, the well-known drift flux model, proposed by Zuber and Findlay, is used. In this model, the relationship between the volumetric void fraction ε and the true quality $\dot{x}$ is given by eq. 3

$$\varepsilon = \frac{\dot{x}/\rho_g}{C_0 \left[\dot{x}/\rho_g + (1 - \dot{x})/\rho_l\right] + V_{jg}/\dot{m}} \qquad (3)$$

where V_{jg} is the vapor drift velocity and C_0 is the distribution parameter. The drift velocity was set to

$$V_{jg} = 1.18 \left[\frac{\sigma g (\rho_l - \rho_g)}{\rho_l^2}\right]^{0.25} \qquad (4)$$

with the surface tension (σ) and the acceleration due to gravity (g). Using a stepwise regression analysis of about 500 data, measured in the above-described test section and the saturation data published by Friedel, the distribution parameter C_0 was fit to

$$C_0 = \dot{\varepsilon}\left[1 + 1.049 Fr^{-0.05} (1 - Ja)^{0.164} \frac{\rho_g}{\rho_l}\left(\frac{1 - \dot{x}}{\dot{x}}\right)^{0.864}\left(1 - \frac{p}{p_c}\right)^{0.124}\right] \qquad (5)$$

with the average volumetric flow concentration $\dot{\varepsilon}$ defined by

$$\dot{\varepsilon} = \frac{1}{1 + [(1 - \dot{x})/\dot{x}] (\rho_g/\rho_l)} \qquad (6)$$

In consideration of the geometry and the mass flux, the Froud number

$$Fr = \frac{\dot{m}^2}{g d_{hy} \rho_l^2} \qquad (7)$$

was included. To indicate the degree of subcooling, i.e., the inlet subcooling of the fluid, a modified Jacob number was inserted, which is defined by

$$Ja = \frac{h_{sl} - h_e}{h_{lg}} \qquad (8)$$

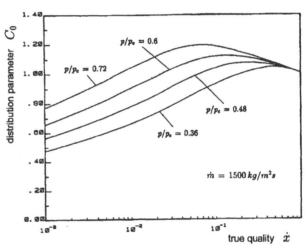

FIG. 5 CALCULATED DISTRIBUTION PARAMETER C_0

Figure 5 shows the calculated distribution parameter C_0 for a mass flux density $\dot{m} = 1500\ [kg/m^2 s]$ at various pressures. Figure 7 illustrates the measured quality $\dot{x}$, gauged at the mass flow rate

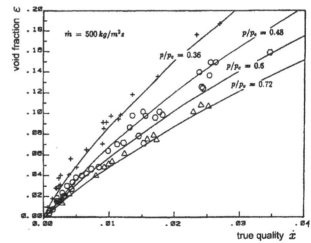

FIG. 6 TRUE QUALITY AND VOID FRACTION AT DIFFERENT PRESSURES

$\dot{m} = 500\ [kg/m^2 s]$ at several heat flux densities and pressures, which is plotted versus the volumetric void fraction. The lines indicate the calculated void fraction with eq. (3) using the distribution parameter calculated with eq. (5). The data of refrigerant 12 under subcooled conditions ($T \leq T_s$) and a pressure range $12 \leq p \leq 40$ [bar] are regarded in eq. (4). The subcooling is limited by $50 \leq \Delta T \leq 0K$. The mean deviation between the measured values and the prediction with the presented calculation method is 12%.

Model of Recondensation

Now, the true quality at hand, it is possible to determine the mass flow of the void which has recondensated due to the thermodynamic non-equilibrium. Models for the recondensation rate had been published by Ulrych, Dix, Huges and Rouhani and Axelsson and other. They all used slightly different approaches to calculate the recondensation rate. The model which is to be demonstrated here is based on a macroscopic view of the void fraction of subcooled boiling processes. Based on the conception that the present true quality can be obtained from the difference of the theoretically formed bubbles in the boundary layer of the heated wall and the recondensated bubbles, the eq. 10

$$\dot{M}_g = \dot{M}_{g,theo} - \dot{M}_{rc} \qquad (9)$$

can be used. If the theoretical void mass flow can be determined, the fraction of recondensation can be obtained by the present void mass flow received by the measured values. This theoretically formed steam amount can be determined by the eq. 10

$$\dot{Q}_{tot} = \dot{Q}_V + \dot{Q}_{sp} + \dot{Q}_{mc} + \dot{Q}_{co} \qquad (10)$$

using the energy balance of the boundary layer (see Fig. 6). In this equation $\dot{Q}_V$ stands for the heat flow which is used for the evaporation of the saturated or overheated fluid, which can be found at the heated wall or in the boundary layer. $\dot{Q}_{sp}$ is the heat flow transferred by single-phase forced convection and $\dot{Q}_{mc}$ is the heat flow transferred by micro-convection. $\dot{Q}_{co}$ is the heat flow which is transferred by condensation at the bubble top/head to the subcooled fluid. This fraction of the total transfer of heat flow is small compared to the heat flow which is transferred by micro-convection and can therefore be neglected according to Forster and Greif. This means that equation 10 can be simplified to

$$\dot{Q}_{tot} = \dot{Q}_V + \dot{Q}_{sp} + \dot{Q}_{mc} \qquad (11)$$

The term micro-convection designates the disturbance of the boundary layer by detaching bubbles letting subcooled fluid enter the boundary layer. The detaching bubble pulls some fluid from the boundary layer along in addition to its volume. A schematic representation of this process can be seen in fig. 7. This method of balancing the various heat transfer mechanisms is analogous to that of Rouhani. To simplify the calculation of the individual heat flows, the following assumptions are made:

   - the boundary layer has saturation temperature (compare Larsen and Tong)
   - the subcooled fluid has the same temperature level. This simplification was also used by Larson and Tong.

With these assumptions the various heat flows in equation 12 can be determined as follows. The heat flow which contributes to the evaporation of the fluid is described by eq. 12

$$\dot{Q}_V = \dot{M}_{g,theo} \, h_{fg} \qquad (12)$$

and contains the theoretically formed void mass flow $\dot{M}_{g,theo}$, which is required later to determine the recondensating void mass flow.

The heat flow $\dot{Q}_{mc}$ transferred by micro-convection can be described by the equation 13

$$\dot{Q}_{mc} = B \dot{V}_{Bu} \rho_f (h_{f,s} - h_f) \qquad (13)$$

according to Bowring. The empirical factor B takes account of the mass of the fluid which follows at the immediate condensation of the bubble at the heated wall, respectively after the bubbles have detached from the subcooled area. Hun and Griffith have determined this factor to the value two by photographic measurements and theoretical considerations of this process, i.e. when a bubble detaches from the heated wall, not only the volume of the void bubble is displaced by the subcooled fluid, but saturated, respectively overheated, fluid is drawn along with the bubble from the boundary layer. This process in subcooled boiling activities is also significant for the heat transfer. Using the before mentioned assumptions the growth of a bubble leads to a displacement of saturated fluid from the boundary layer into the subcooled fluid. With these assumptions the factor B was determined to three.

The heat flow $\dot{Q}_{sp}$ transferred by single phase forced convection was determined using the well-known equation by Colburn. It was applied by Rouhani in this rather unconventional form and presented more precisely by Mayinger.

According to this relation the single phase heat transfer coefficient α_{sp},

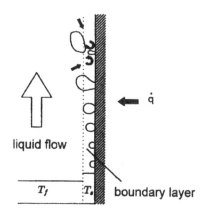

FIG.7 SCHEMATIC ILLUSTRATION OF THE MICRO CONVECTION

in this case between the saturated boundary layer and the subcooled fluid flow in the boiling channel, can be determined by eq. 14.

$$\alpha_{sp} = 0.023 \dot{m} c_p Pr^{(-2/3)} Re^{-0.2} \qquad (14)$$

The transferred heat flux density by the single phase force convection can thus be obtained by eq. 15

$$\dot{q}_{sp} = \alpha_{sp}(T_s - T_f) \qquad (15)$$

After having filled in the various heat fluxes and the algebraic transformation one receives an equation (which is equal to eq. 16 except the limits of the integral) for the recondensated void mass flow in a control element of the length Δz. Due to the differences in physical properties, surface tension and evaporation enthalpy, between

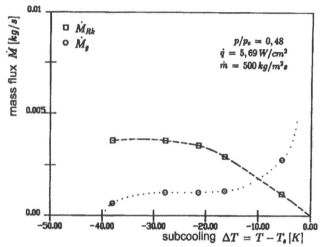

Fig. 8 EXISTING AND RECONDENSATED VAPOR

water and R12 used in this experiment, a strict distinction between the range where the bubbles adhere to the wall and the area of detaching bubbles cannot be made. Pictures with a high speed camera, as shot by Bräuer, reveal that the bubbles detach very soon after boiling begin. Responsible for this early detachment is the considerably lower surface tension of R12 as compared to water. Therefore, this equation can be written integrally from the beginning of the bubble formation (ONB) z_{sb} to the point z, where the fluid temperature equals the saturation temperature. Using this equation the proportion of the recondensated void can be determined within the limits of predetermined assumptions. The results where the fluid temperature equals the saturation temperature, are as follows: the calculation of the recondensation rate are shown as an example in fig. 8. Here the void fraction and the recondensating void mass flux acc. to eq. 16 are drawn above the subcooling range ΔT. The mass flux of the void $\dot{M}_g$ present in the balance volume with the length Δz at a given subcooling temperature describes an almost horizontal line after its steep rise following the onset of boiling, as does the volumetric void fraction with which it is connected. At a subcooling rate of approx. $\Delta T_u = -15\ K$ it begins to increase exponentially. The recondensated void flux (acc. to eq. 16) shows a similar behaviour but decreases to the value 0 when reaching saturation temperature. This behaviour can be explained physically by the bubbles' early detachment, due to the relatively low

surface tension. Recondensation is highly determined by the fluid temperature, i.e. the bubble condensate faster at a high subcooling than at a low one letting a great proportion of the developed void condensate again after the onset of boiling. When subcooling decreases the bubbles in the fluid flow can exist longer and the present void flux increases proportionally, or resp. the recondensating void flux decreases until this process stops completely when reaching saturation temperature.

Fig. 9. shows the course of the recondensating mass flux at a higher heating performance and a higher mass flux density. This behaviour lets us conclude that the bubbles already condensate at the beginning of the heated course and not only at the boiling surface, but can detach from the surface of the inner tube and condensate in the subcooled fluid. These figures show that eq. 16 represents for the physical processes of recondensation under subcooled boiling to a high degree and allows the calculation of the recondensation with a known existing true quality.

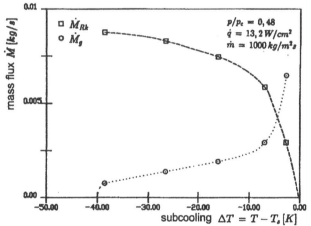

FIG. 9 EXISTING AND RECONDENSED VAPOR

CONCLUSION

By applying two different measurement techniques, i. e. a γ-densitometer and a specially designed differential impedance probe, measurements of void fraction in subcooled boiling in steady state conditions at very high subcooling rates and very low void fractions had beeen performed. Using these data and including data from other authors, an equation of the true quality by means of the Zuber - Findly drift flux model has been developed. Knowing the true quality, the recondensation rate can be obtained by calculating the theoretically formed void fraction
according to an energy balance on the boundary layer of the liquid. Accordingly, the equations presented in this paper are also applicable for other fluids, for example water.

$$\dot{M}_{rc} = \int_{z_{\lambda}}^{z_s} \frac{[\dot{q}U_{heat} - \alpha_{sp}(U_{heat} + 2\delta/(T_s - T_f(z)))]z + \dot{m}A\dot{x}(z)h_{fg}}{h_{fg} + B\rho_f(z)c_p(z)(T_s - T_f(z))}\ dz \qquad (16)$$

NOMENCLATURE

A	area $[m^2]$
c_p	specific heat capacity of the liquid $[kJ/kg\,K]$
C_0	distribution parameter
d_{hy}	hydraulic diameter $[m]$
Fr	Froud number
g	gravity acceleration $[m/s^2]$
h	specific heat $[kJ/kg]$
h_{fg}	specific heat of vaporization $[kJ/kg]$
Ja	Jacob number
L	length $[m]$
$\dot{m}$	mass flux density $[kg/m^2s]$
$\dot{M}$	mass flux $[kg/s]$
p	pressure $[bar]$
$\dot{q}$	heat flux density $[kJ/m^2s]$
$\dot{Q}$	heat flux $[kJ/s]T$ temperature $[^0C]$
ΔT	subcooling $[\Delta T = (T - T_s); K]$
U	circumference $[m]$
V_{jg}	vapor drift velocity
$\dot{x}$	true quality
z	axial coordinate $[m]$

Greek Symbols

α	heat transfer coefficent
ε	volumetric void fraction
$\dot{\varepsilon}$	volumetric flow concentration
ρ	density $[kg/m^3]$
σ	surface tension $[N/m]$

Subscripts

c	critical
co	condensation
e, in	inlet
g	gas phase
$heat$	heated
l	liquid phase
ndp	net vapor production
mc	micro-convection
rc	recondensation
s	saturation
sb	onset of nucleate boiling
sp	single phase
$theo$	theoretical
tot	total

REFERENCES

Bowring, R. W. "Physical Model based on Bubble Detachment and Calculation of Steam Voidage in the Subcooled Region of a Heated Channel", *Institut for Atomenergie, Halden, Norway, HPR 10*,1962

Bräuer, H. "Wärmeübergang und Siedebeginn bei unterkühltem Sieden unter Zwangskonvektion". *Diss. T.U. München*, 1988

Dix, G. E."Vapor Void Fractions for Forced Convection with Subcooled Boiling at Low Flow Rates", *Ph.D thesis, Univ. of California*, 1970.

Friedel, L. "Modellgesetze für den Reibungsdruckverlust in der Zweiphasenströmung". *Diss. Univ. Hannover*, 1974.

Han, C., Griffith, P., "The Mechanism of Heat Transfer in Nucleate Pool Boiling", PartI and II. *Int. j. Heat Mass Transfer, Vol. 8, pp887 - 913*,1965

Huges, E. D., Paulsen, M. ; Agee, L. P. "A Drift Flux Model of Two Phase Flow". *A.S.M.E., Nuclear Technology, Vol 54*, 1981.

Jain, P. K.; Nourmohammadi,R. and Roy, R. "A Study of Forced Convective Subcooled Boiling in Heated Annular Channels", *NucEear Eng. Design, vol. 60*, 1980

Larson, P. S., Tong, L. S. "Void Fraction in Subcooled Boiling". *Heat Transfer Vol. 91, pp. 472*, 1969

Mayinger, F. "Strömung und Wärmeübergang in Gas - Flüssigkeitsgemischen", *Springer Verlag, Berlin*, 1982.

Rouhani, S. Z."Calculation of Steam Volume Fraction in Subcooled Boiling", *Trans. ASME*, 1968.

Rouhani, S. Z; Axelsson, E., "Calculation of Void Volume Fraction in the Subcooled and Quality Region". *Int. J. Heat and Mass Transfer Vol. 13, pp. 383 - 393*,1970

Stängl, G., Mayinger, F. "Void Fraction Measurement in Subcooled Forced Convective Boiling". *Exp. Heat Transfer Vol. 3 No. 3*. 1990

Ulrych, G.," Strömungsvorgänge mit unterkühltem Sieden in Brennstabbündeln wassergekühlter Reaktoren". *Diss. T.U. Braunschweig* 1976

Zuber, N. and Findlay, J. A."Average Volumetric Concentration in Two Phase Flow Systems", *J. Heat Transfer, vol 87*, 1975.

Subcooled Convective Boiling of Aqueous Surfactant Solutions

Yu-Min Yang and Jer-Ru Maa
Chemical Engineering Department
National Cheng Kung University
Tainan, Taiwan 70101
R.O.C.

ABSTRACT

Subcooled convective boiling of aqueous surfactant solutions of sodium dodecyl surfate over relatively wide ranges of concentration and heat flux was carried out in a vertical annulus. The experimental results show that a small amount of surfactant additive makes the nucleate boiling heat transfer coefficient, h, considerably higher, and that there is an optimum additive concentration for higher heat fluxes. Beyond this optimum point, further increase in additive concentration makes h lower. The dependence of h on surfactant concentration could not be correlated with solution surface tension. Nevertheless, the effect of surfactant additive on boiling is attributed to the departure from equilibrium surface tension which is produced by extension of the vapor-liquid interface during growth and coalescence of vapor bubbles in the vicinity of boiling surface. The surface tension of a streching surface in surfactant solution is higher than the static value because the surfactant molecules cannot diffuse to the adsorbed layer promptly. This dynamic surface effect can be expressed by the $Y(= C(d\sigma/dC)^2)$ value which is also shown to be a representation of the elasticity of an adsorbed film. The similarity between the increment of h due to the addition of surfactant and Y curves indicates that the dynamic surface effect may play an important role in the boiling process of aqueous surfactant solutions.

INTRODUCTION

Surfactant Effect

Relatively small amounts of certain surfactants in solution have been known to significantly alter the rate of boiling heat transfer(Wu et al., 1995). Few studies have been carried out on the effect of surface active materials on the convective boiling of water. Strobe et al.(1939) found that the lowering of water surface tension by a surfactant additive enhanced drastically the boiling-film heat transfer coefficient in a long-tube vertical evaporator. The derived empirical correlation revealed that coefficient is proportional to the minus 2 power of the surface tension of the liquid. Frost and Kippenhan(1967) studied forced convection boiling of water and aqueous surfactant solutions in a vertical annulus. Different energy transfer processes which control the bubble growth were proposed for boiling of water with and without surfactant additive. It was concluded that latent heat transport contributes approximately half of the measured heat flux in water and significantly more in water having reduced surface tension by surfactant additive. Shah and Darby(1973) found that the evaporative heat transfer coefficients for vertical downward water film flow over a flat plate were strongly dependent on surfactant concentration. The dependence on surfactant concentration could not be correlated with corresponding changes in surface tension, but was due to foaming and could be explained in light of a theory of foam stability. Shibayama et al.(1980) studied the effect of surface tension on boiling heat transfer when a horizontal haeting surface is covered with a flowing liquid film of water with and without surfactant. Heat transfer coefficient was found to be proportional to the minus 0.4 power of the surface tension of the liquid. These investigations are listed in Table 1.

Role of Surface Tension

Because the concentration is usually very low, the addition of surfactant to water causes no significant change in physical properties except that the surface tension is depressed cosiderably. Theoretically, surface tension is an important variable for boiling. The rate of nuclei formation is proportional to $e^{-\sigma^3}$. Thus, samall decreases in σ should cause large increases in the number of nuclei. The cavitation theory predicts that the force required to rupture a liquid in tension is proportional to $\sigma^{3/2}$. Liquid with large surface tension, therefore, should be difficult to fracture. On the other hand, considering the equation of state only, no reason is apparent why a surfactant should affect the matastable state. Therefore, on nucleation theoretical grounds only, a decrease in σ by surfactant additive may have no effect or it may increase the rate of heat transfer. It can-

Table 1 Surfactants that enhance the forced convection boiling heat transfer rate of water

References	Flow Type	Heating Surfaces	Surfactants
Strobe et al.(1939)	pipe	Tube	Duponol
Frost and Kippenhan(1967)	annulus	Tube	Ultra Wet 60L
Shah and Darby(1973)	film	Plate	Joy
Shibayama et al.(1980)	film	Plate	Sodium oleate; Rapisool B80; Puluronic type F98, F88, F208
Yang and Maa(This work)	annulus	Tube	Sodium lauryl sulfate

not decrease the heat flow. Surface tension also affects different fundamental processes in nucleate boiling differently, such as bubble growth, bubble departure, and bubble shape etc. However, the way through which nucleate boiling heat transfer may be varied by the changes in these processes is not so obvious. A number of studies have been made in the past to relate the nucleate boiling heat transfer coefficient to the surface tension of a boiling surfactant solution by the equation

$$h = \sigma^n \qquad (1)$$

However, there is very large discrepancy among the n values based on different correlations reported in the literature. This is because the surfactant additives may influence the heat and mass transfer mechanism in the vicinity of the vapor/liquid interface in a rather complicated manner, their overall effect on boiling heat transfer is neither simple nor direct, and cannot be expressed by the value of surface tension of the boiling liquid alone. Thus one cannot expect that a solution will boil in the same way as a "pure" liquid with the same surface value. Past boiling heat transfer correlations for surfactant solutions in the literature, as expressed in the form of equation (1) cannot be very meaningful.

The objectives of this work are to report the experimental data obtained by using a small loop of convective boiling which is closer to the operating conditions of industrial operations and to explain the experimental results by vapor/liquid interfacial effects.

THEORETICAL

Dynamic Surface Tension

The repeated ebullition cycle in nucleate boiling demonstrates a "fast" periodic process of vapor/liquid interface formation. According to the experimental results of bubble growth dynamics in aqueous surfactant solutions at a relatively low heat flux of 23 kw/m^2, the average bubble ebullition periods were found to range from 70-20 ms for various surfactant concentrations(Wu et al., 1993). It is known that non-equilibrium surface tensions are of paramount importance in the practical application of surfactants. The value of the dynamic surface tension and the rate at which it approaches equilibrium are thus critical for phenomena or process with time scale comparable or small than the adsorption equilibrium time. Because the time required for equilibrium to be reached covering from milliseconds to hours, various techniques have been developed for measuring the

dynamic surface tensions (Miller et al., 1994).

When the surface area of a solution is suddenly created by the formation of a bubble with fixed size, a part of the bulk liquid is forced to enter the surface layer. Just after the creation of the surface area, the surface-layer composition is very similar to that of the bulk and can become identical if the rate of increase of the area is sufficiently fast. The layer will keep this composition for only a very short time after the end of creation of the surface area, because diffusion tends to restore the equilibrium composition, which for solutions is usually quite different from that of the bulk. The same but vastly more complicated situation occurs after the creation of an ever changing fresh surface as when a vapor bubble grows at aa active nucleation site.

Dynamic Surface Effect and Film Elasticity

The coalescence of vapor bubbles near the heating surface is intensely involved in the processes of boiling heat transfer. As shown in Fig. 1(A), when two vapor bobbles approach one another during their growth period, a liquid film may form between them and the liquid in the film flows outward and causes the vapor/liquid interfaces on both both sides to stretch. In a pure liquid, the interfaces are free to move along with the liquid in the film and the rate of thinning is therefore controlled by the inertial of the liquid pushed away from the film. Small amounts of surfactant may cause additional resistances to the outward motion of the liquid film between the approaching vapor bubbles. As the interface is stretched, due to the increase of interfacial area, the surfactant molecules to be adsorbed at the interface have to be supplied from the bulk liquid to the interface. Because the rate of convective diffusion is finite, the concentration of the surfactant in the vicinity of the stretching interface is lower than that for the case without stretching as shown in Fig. 1(B) and 1(C). The adsorption on the interface is the equilibrium value corresponding to this lower concentration and is thus lower than that of the static interface. Hence, the local dynamic rise in surface tension caused by the stretching of the interface has the effect of retarding further stretching. Vapor bubble coalescence in surfactant solutions will thus be drastically different from that in a pure liquid. Consequently, a different nucleate boiling heat transfer rate is expected.

Andrew(1960) considered the case of a steady uniform expansion at the surface of a liquid with infinite depth by solving the corresponding convective diffusion equation, and found that the local rise in surface tension due to the dynamic surface effect described above can

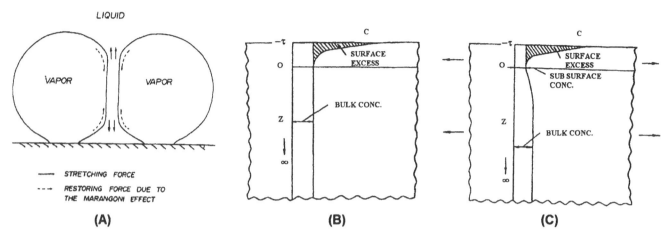

Fig.1. (A) Coalescence of two vapor bubbles near the heating surface. (B) Concentration gradient of surfactant near the interface of a static film. (C) Concentration gradient of surfactant near the interface of a expanding film.

be expressed, for a dilute solution, by the equation:

$$\Delta\sigma = \frac{1}{RT}(\frac{\pi S}{2D})^{1/2}C(\frac{d\sigma}{dC})^2 \qquad (2)$$

where S is the stretching rate of the interface:

$$S = \frac{1}{A}(\frac{dA}{dt}) = -\frac{1}{h}\frac{dh}{dt} \qquad (3)$$

The additional resistance to the stretching of vapor/liquid interface in a surfactant solution can thus be quantitatively represented by Eq.(2). It is postulated that the resistance to the coalescence of the vapor bubbles and the enhancement of the nucleate boiling heat transfer rate in a surfactant solution relative to that in a pure liquid are due to the same effect. If we define

$$Y \equiv C(\frac{d\sigma}{dC})^2 \qquad (4)$$

then Y is proportional to $\Delta\sigma$, and the value of Y is a measurement of the dynamic surface effect caused by a certain surfactant additive under a given condition. The presence of a solute capable of being adsorbed at the vapor/liquid interface is required for a system to resist excessive localized thinning of the laminae surrounding the vapor bubbles. This important property of adsorbed film is known to be the film elasticity. The effectiveness of solute in generating surface restoring forces when a surface is subject to a sudden deformation may be represented by the dimensionless elasticity number(Palmer and Berg, 1972):

$$E_l \equiv -\frac{\Gamma}{\mu D}(\frac{\partial\sigma}{\partial C}) \qquad (5)$$

With the assumption that the Gibbs adsorption equation applies, and neglecting activity coefficient effects, the elasticity number becomes

$$E_l = \frac{C}{RT\mu D}(\frac{\partial\sigma}{\partial C})^2 \qquad (6)$$

It is interesting to note that the film elasticity has the same dependence on $C(\frac{d\sigma}{dC})^2$ as the dynamic rise in surface tension by comparing Eqs.(4) and (6).

EXPERIMENTAL

Figure 2 is a schematic diagram of the convective boiling loop of this work. The capacity of receiver (1) is 15 liters, a reflux condenser (8b) is installed on top of this receiver in order to avoid concentration variation of the solution due to evaporation loss. During its sojourn in (1), the liquid is preheated by the first electric preheater (9), but one has to be careful not to bring the liquid temperature up too high so that no cavitation will occur as the liquid entering the centrifugal pump (3). The power of this pump is 1/4 HP. The liquid flow rate is adjusted by valves (4a) and (4b) and is measured by an orifice meter (2). The liquid is preheated again to the desired temperature in a tubular second electric preheater (5) before entering the boiling assembly. The liquid and vapor from the boiling assembly return to the receiver after passing through a cooler-condenser (8a). The material of construction of all components in this boiling loop is stainless steel except the cooler-condenser (8a) and the Pyrex glass outer tube of the boiling assembly.

The boiling assembly as shown in Fig.4 is the heart of the apparatus. The heating element is a seamless stainless steel tube of 3.41 mm O.D. and 2.81 mm I.D. It is housed in a Pyrex glass tube of 32 mm O.D. and 28 mm I.D. As the liquid flows upward through the annulus space between the glass tube and stainless steel heating elements, it is heated up and boils. The surface of the heating element is polished by No. 1200 sand paper and its both ends are welded to 6 mm O.D. thick wall copper tubes. The electrical resistance of these two copper tubes is negligibly small. They are insulated from the

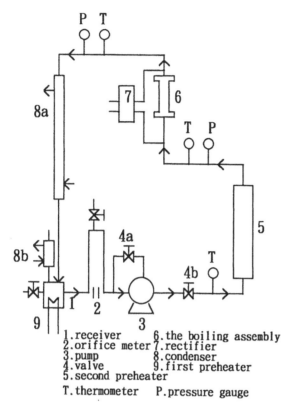

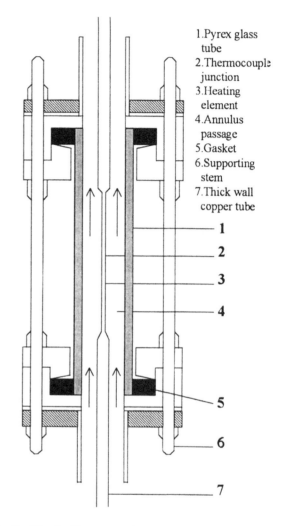

1. Pyrex glass
   tube
2. Thermocouple
   junction
3. Heating
   element
4. Annulus
   passage
5. Gasket
6. Supporting
   stem
7. Thick wall
   copper tube

1. receiver   6. the boiling assembly
2. orifice meter   7. rectifier
3. pump   8. condenser
4. valve   9. first preheater
5. second preheater

T. thermometer   P. pressure gauge

Fig.2. Schematic diagram of the experimental apparatus.

Fig.3. The boiling assembly.

rest of the apparatus and serve as parts of the wiring system for the heating current. A Cole Parmer T type thermocouple of 0.076 mm diameter is inserted in the heating element to measure its inner wall temperature. The location of the thermocouple joint is at the center of the heating element.

The experimental setup of this work is a heat flux controlled system. The current for the heating element in the boiling assembly is supplied by a rectifier (7) with a capacity of 20 volt and 500 amp. The current and voltage across the heating element are recorded by a recorder. The electric current for each of the first and second preheaters is supplied by a variable transformer. The fluid temperature in boil assembly is taken to be the average value of the inlet and the outlet temperatures, and the subcooling, ΔT_{sub}, is equal to the saturation temperature of water corresponding to the system pressure minus this average value. The temperature of the outer surface of the heating element, T_o, can be computed from its inner wall temperature, T_i, by the following equation:

$$T_i - T_o = \frac{S_e}{4k}(R_o^2 - R_i^2) - \frac{S_e}{2k}R_i^2 ln\frac{R_o}{R_i} \qquad (7)$$

where S_e is the power density, k is the thermal conductivity, R_i and R_o are the inner and outer radii of the heating element, respectively.

The additive used in this work is sodium laury sulfate (SLS, $C_{12}H_{25}NaSO_4$). It is water soluble, anionic

surfactant. Because the concentration is always very low, the addition of SLS to water causes no significant change in physical properties except that the surface tension is reduced considerably. The surface tension data of aqueous solution of SLS of various concentrations at different temperatures were determined by the capillary rise medthod and documented (Chang, 1985).

During each run, one has to wait until the flow rate, the current, the voltages, and the temperature of various points reach steady state before data are taken. Because the boiling heat transfer rate is very sensitive to the state of the heating surface, boiling of pure water is carried out until the reproducibility of the boiling curve becomes very good before the beginning of each set of experiments with the addition of various amount of additive. After the last experiment of each set, another run using pure water is also made to check if there is any significant change of the condition of the heating surface during this set of experiments.

RESULTS AND DISCUSSION

We can compute the values of Y from the surface tension data determined under various additive concentration. One of the curves in Fig. 5 shows Y as a function of concentration for the dilute solutions of SLS in water at $80°$ C ($\Delta T_{sub}=20$). Y has a maximum at a concentration of about 150 ppm, this means that the dynamic surface effect is the greatest at this concentration, it decreases gradually after this concentration is exceeded and vanishes when the concentration is sufficiently high.

The appearance of the boiling of water containing surface active additive is considerably different from that of pure water. Under the same heat flux, the surface active additive makes the number of vapor bubbles on the heating surface much larger, the size of the bubbles smaller, and the coalescence of the bubbles more difficult. Moreover, vapor bubbles formed is unison rather than at individual sites.

Figure 4 shows a set of boiling curves for water containing SLS of various concentrations. Among them there are two for zero SLS concentration. The open circles are data taken before the beginning of this series of experiments and the solid circles are data taken afterward. The agreement between these two curves showe that the reproducibility of the boiling curves of thiswork is satisfactory and that no serioue change of surface condition of the heating element occurs when this series of experiment is carried out.

Results in Fig. 4 show that when the heat flux is relatively low, the boiling curve shiftl leftward as the concentration of the surfacf active additive is increased; when the heat flux becomes higher, the boiling curves for higher additive concentrations incline rightward and gradually reach to the curve for pure water The relationship between Y and the additive concentration in Fig. 5 shows that the dynamic surface effect vanishes as the additive foncentrationexceeds certain value,1,000 ppm in this case. From the pointof view of the dynamic surface effect the boiling behavior of such a solution should be the same as that of pure water. As the boiling heat flux reaches about $100w/cm^2$, vapor bubbles are generated violently from the heating surface also increases with the increase of the heat flux. The number of vapor bubbles per unit area is higher, the chances for the coalescence of the vapor bubbles are even larger, and the dynamic surface effect is consequently more significant at higher heat flux. This explains the phenomenon in Fig.4 that the slopes of the boiling curves for solutions of high surface active additive concentration become less steep and approaches to the position of the boiling curve for pure water as the heat flux is getting higher.

Nucleate convective boiling heat transfer coefficient, h, is defined as the heat flux divided by the saturation superheat of the heating surface. Figure 5 shows the Y value and enhancement of the heat transfer coefficient, (h-h'), as functions of additive concentration, where h' is the nucleate boiling heat transfer coefficient for pure water. This figure indicates that (h-h') has a maximum and that when the heat flux is increased the maximum becomes more prominent and its position shifts to lower additive concentration. Both of the Y and (h-h') curves in this figure have a maximum, and the distance between these two maxima becomes closer as the heat flux is higher. This is apparently because that as the heat flux is increased more vapor bubbles are generated, the coalescence between bubbles becomes more frequent and the dynamic surface effect plays a more important role in this heat transfer process. Hence, the dynamic surface

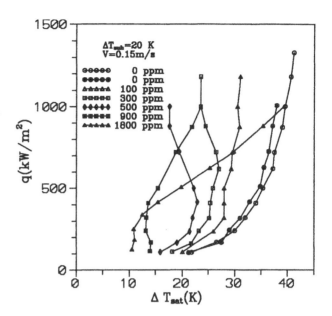

Fig.4. Boiling curves of aqueous solutions of SLS.

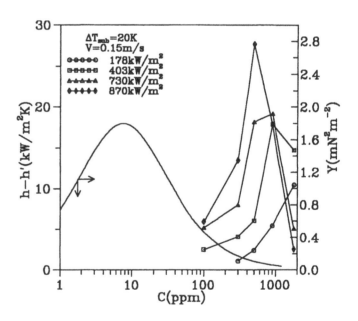

Fig.5. Y and (h-h') as function of concentration for aqueous solution of SLS.

effect is a very influential factor in causing the nucleate boiling heat transfer coefficient of water containing surfactant higher than that of pure water.

The experimental evidence and analysis for convective boiling heat transfer in this work did find their counter parts in the work of saturated pool boiling heat transfer (Tzan and Yang, 1990; Yang, 1990). It is reasonable to doubt if an exceedingly complex phenomenon like boiling of surfactant solution can be explained simply by a function such as Y. For example, subcooled boiling is based on the cycle of generation and condensation of bubbles. Thus evaporation on the bubble surface causes concentrated surfactant solution and condensation causes diluted solution. This change of surfactant concentration around bubbles might influence subcooled boiling process. As it is a general interfacial property of surfactant solution, however, the dynamic surface effect or film elasticity is deemed important in boiling heat transfer of surfactant solution.

CONCLUSIONS

Flow boiling experiments of this work verify again that small amount of surface active additive makes nucleate boiling heat transfer coefficient considerably higher. There is an optimum additive concentration for each heat flux. Beyond this optimum point, further increase in additive concentration makes the boiling heat transfer coefficient lower. The dynamic surface effect is closely related to the growth and coalescence of the vapor bubbles and is a very influential factor to cause the nucleate boiling heat transfer coefficiennt to increase.

NOMENCLATURE

A = interfacial area, m^2

C = concentration, ppm

D = diffusion coefficient, $m^2 s^{-1}$

E_l = film elasticity, Nm^{-1}

h = heat transfer coefficient, $kWm^{-2}K^{-1}$; film thickness, m

k = thermal conductivity, $Wm^{-1}K^{-1}$

n = constant,

q = heat flux, kWm^{-2}

R = gas constant, $Jmol^{-1}K^{-1}$

S = surface expansion rate per unit area, s^{-1}

Se = power density, Wm^{-3}

t = time, s

T = temperature, K

ΔT = temperature difference, K

V = velocity, ms^{-1}

Y = function defined by Eq.(4), $N^2 m^{-2}$

Greek Symbols

σ = surface tension, Nm^{-1}

$\Delta\sigma$ = dynamic elevation in surface tensionn due to expansion, Nm^{-1}

μ = viscosity, $kgm^{-1}s^{-2}$

Γ = surface excess, $molm^{-2}$

Subscripts

i = inside,

o = outside,

sub = subcooling,

sat = saturation,

REFERENCES

Andrew, S.P.S., 1960, "Frothing in Two-Component Liquid Mixtures," International symposium on Distillation, P.A. Rottenturg, ed., Institution of Chemical Engineer, London, pp. 73-78.

Chang, C.S., 1985, "Surfactant and Boiling Heat Transfer," Master Thesis, National Cheng Kung University, Tainan, Taiwan.

Frost, W., and Kippenhan, C. J., 1967, "Bubble Growth and Heat Transfer Mechanisms in the Forced Convection Boiling of Water Containing a Surface Active Agent," International Journal of Heat and Mass Transfer, Vol.10, pp. 931-949.

Miller, R., Joos, P., and Fainerman, V. B., 1994, "Dynamic Surface and Interfacial Tensions of Surfactant and Polymer Solutions," Advances in Colloid and Interface Science, Vol. 49, pp. 249-302.

Palmer, H. J., and Berg, J. C., 1972, "Hydrodynamic Stability of Surfactant Solutions Heated from Below," Journal of Fluid Mechanics, Vol. 51, pp. 385-402.

Shah, B. H., and Darby, R., 1973, "The Effect of Surfactant on Evaporative Heat Transfer in Vertical Film Flow," International Journal of Heat and Mass Transfer, Vol.16, pp. 1889-1903.

Shibayama, S., Katsuta, M., Suzuki, K., Kurose, T., and Hatano, Y., 1980, "A Study on Boiling Heat Transfer in a Thin Liquid Film," Heat Transfer (Japan Research), Vol. 9, No. 4, pp. 12-40.

Strobe, G. W., Baker, E. M., and Badger, W. L., 1939, "Boiling-Film Heat Transfer Coefficients in a Long-Tube Vertical Evaporator," Industrial and Engineering Chemistry, Vol. 31, No. 2, pp. 200-206.

Tzan, Y.L., and Yang, Y.M., 1990, "Experimental Study of Surfactant Effects on Pool Boiling Heat Transfer," ASME Journal of Heat Transfer, Vol. 112, pp. 207-212.

Wu, W.T., Hu, C.L., and Yang,Y.M., 1993, "Surfactant Effect on Boiling Incipience and Bubble Growth Dynamics of Surface Boiling in Water", Journal of Chinese Institute of Chenical Engineers, Vol. 24,No. 2, pp. 111-118.

Wu, W. T., Yang, Y. M., Maa, J. R., 1995, "Enhancement of Nucleate Boiling Heat Transfer and Depression of Surface Tension by Surfactant Additives," ASME Joural of Heat Transfer, Vol. 117, in press.

Yang, Y.M., 1990, "Dynamic surface Effect on boiling of Aqueous Surfactant Solutions," International Communications in Heat and Mass Transfer, Vol.17, No.6, pp.711-727.

FORCED-CONVECTION SUBCOOLED NUCLEATE BOILING
AND ITS APPLICATION IN MICROGRAVITY

Tien-Chen Wang, Trevor J. Snyder and Jacob N. Chung
School of Mechanical and Materials Engineering
Washington State University, Pullman Washington USA

ABSTRACT

Forced-convection boiling heat transfer experiments with Freon-113 were performed on thin gold-film heaters under earth gravity and microgravity. The microgravity environment, which was used to unmask the buoyancy effect, was achieved in a 0.6-second airbag decelerated drop tower at Washington State University. The main objective was to experimentally investigate the latent heat transport mechanism and the "sliding bubble" model proposed by Tsung-Chang and Bankoff (1990) for subcooled forced-convection nucleate boiling. In general, during microgravity the heat transfer is enhanced the greatest if individual spherical bubbles rather than vapor chunks are maintained by the forced flow. For this case, we found individual spherical bubbles which slid on the heater surface at 25 to 90% of the free-stream velocity. Forced-convection subcooled nucleate boiling was found to have minimum reliance on gravity and therefore is a feasible and efficient heat transfer mechanism for microgravity applications.

INTRODUCTION

Forced-convection subcooled nucleate boiling has been recognized as one of the most efficient heat transfer mechanisms (Tsung-Chang and Bankoff, 1990). Many fundamental investigations were found in the literature concerning the basic mechanisms of forced-convection boiling that are responsible for the enhancements in heat transfer coefficient over the single-phase forced-convection heat transfer. The main question is, "why does the nucleation and formation of bubbles improve the heat transfer so drastically?" However, the theories have been divided. Gunther (1951) suggested that turbulent mixing or microconvection between growing and collapsing bubbles is the mechanism responsible for the enhancement. The second theory (Bankoff et al., 1956, Bankoff, 1959, 1962, Snyder and Robin, 1969) is based on the latent heat transport where condensation at the cap of the bubble and

evaporation from the surrounding microlayer are suggested as the key transport mechanisms of latent heat.

Recently Tsung-Chang and Bankoff (1990) re-examined the latent heat transport theory. They proposed that "sliding" of bubbles on the heater surface would result in the augmentation of microlayer evaporation under the bubble by a factor of two or possibly more over a stationary bubble. The first objective of the current research is to verify the Tsung-Chang and Bankoff "sliding bubble" model in a carefully designed experiment. In highly subcooled and high-velocity forced-convection nucleate boiling, bubbles grow, slide and collapse on the heater surface, therefore buoyancy is of minimum reliance. In order to further obviate the masking effect of buoyancy, our experiment was performed under low-velocity and microgravity conditions such that the physical phenomenon was more clearly recorded by our experimental system. The second objective of this paper is to study the feasibility of this highly effective heat transfer mechanism in a microgravity environment.

Cochran (1970) reported the only microgravity forced-convection boiling results in the open literature. He investigated the bubble sizes and two-phase bubbly flow patterns for low heat flux and slightly subcooled boiling in water. Major findings include the existence of a bubble boundary layer and a linear relationship between the size of the bubbles and the saturation layer thickness. The current study deals with boiling heat fluxes that are one to two orders of magnitude higher than those in Cochran's (1970) experiment.   Cooper et al. (1983) experimentally demonstrated a sliding single bubble due to forced flow under microgravity. Only a single bubble was studied, therefore it did not offer any information for forced-convective microgravity boiling heat transfer.

DESIGN OF EXPERIMENT

Since the main objective was to examine the forced-convective boiling mechanism, the experiment design was based on the principle that the momentum of the external flow must play a major role in the nucleate boiling process. Therefore, the maximum pump capacity was employed in our experiment which maintained a free-stream velocity of 7.7±0.1 cm/s over the heater surface.

Due to the physical limitations, the experimental system was open to the atmosphere. The bulk fluid temperature in the experiment was also conveniently set to approximately that of the room temperature. This resulted in a relatively constant subcooling of 17±2 °C for our experiment.

The flow boiling apparatus used for this study consisted of a flow control device, preheater, boiling test section, condenser, lighting, video and data acquisition system. Spectrophotometric grade Freon-113 (1,1,2 trichlorotrifluoroethane 99+%) was chosen as the test fluid. Because of the high solubility of dissolved gases in Freon-113 a typical degassing procedure was followed which included boiling at reduced pressure. A 0.6 second drop tower was utilized in this study which provided a microgravity level of approximately $a/g = 3 \times 10^{-3}$. Further details can be found in Wang (1993).

Visualization of nucleation, growth, and departure of vapor bubbles was desired along with the mean surface temperature and mean heat flux. A semi-transparent thin gold-film heater (2.54cm x 2.54cm) was utilized in this study. The heater surface consists of a 400 Angstroms thick gold-film while the current and voltage leads consist of a 5000 Angstroms gold-film. This design provided a very flat heating surface with no protrusion into the flow. Heater construction and calibration was performed based on the procedure reported by Oker and Merte (1981). The uncertainty of the gold-film heater resistance, surface temperature and heat flux are 0.00148 ohms, 1.16 °C, and 380 W/m² respectively. This resulted in an uncertainty in heat transfer coefficient of 96 W/m²K

RESULTS AND DISCUSSION
The Heat Transfer Coefficient

The heat transfer coefficient was calculated according to the following definition:

$$q'' = h\left[T_s - T_{sat}(nominal)\right] \tag{1}$$

where q'' is the heat flux from the heater surface, h is the heat transfer coefficient, T_s is the mean heater surface temperature, and $T_{sat}(nominal)$ is the nominal saturation temperature. Variations in the system pressure due to the transient gravity field resulted in about 1°C difference in the calculated superheat and were neglected. Therefore, in our case, $T_{sat}(nominal)$ was defined as 45 °C - the saturation temperature for the ambient pressure of 93.1 kPa. The heat transfer to the boiling fluid is assumed to be equal to the electrical power supplied to the heater because for a

polycarbonate substrate the heat loss from the non-wetted surface of the substrate is 3% or less (Wang, 1993). The maximum change in heat flux was found to be approximately 0.35% and therefore the heat flux on the heater surface was assumed constant. The resistance calibration did not allow measurement of the spatial temperature variation and, therefore, only provided the determination of a mean heat transfer coefficient. The heat transfer values should only be interpreted as an indication of the condition of the heater surface. For example, enhancements or reductions in heat transfer efficiency, believed to be caused from increased microlayer evaporation or the formation of dry spots on the heater surface, could be determined by changes in the mean surface temperature (or heat transfer coefficient) as compared to the average values measured in normal gravity, and the initial values encountered when transitioning to microgravity. A 1-D transient model was formulated and it was shown that the thermal capacitance of the polycarbonate substrate could be neglected when calculating the temperature (Wang, 1993).

Figure 1 shows the heat transfer coefficient plot for all four heat fluxes under pool and forced-convection boiling.

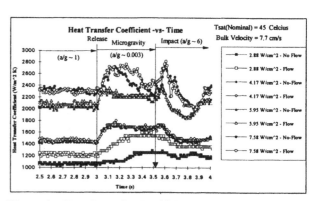

Figure 1 - Heat Transfer Coefficient for Pool and Forced-Convection Boiling of R-113

The first arrow in the graphs indicating "release" represents the moment at which the experiment was released and microgravity started. The second arrow indicates when the experimental rig hit the airbag and microgravity ended. There are three important results from the heat transfer coefficient plot. First, the heat transfer coefficients are invariably higher during microgravity for both pool and flow boiling except at the highest heat flux. The present findings for the case of pool boiling are in good agreement (both quantitative and qualitative) with some previous published results (Merte, Lee and Ervin, 1993) which showed that nucleate pool boiling heat transfer is enhanced as the buoyancy normal to the heater surface is reduced and that it is degraded as the buoyancy normal to the heater surface is increased. The second important result is the sharp decline in the heat transfer coefficient during microgravity for the highest heat flux case (7.58×10^4 W/m^2). The forced flow only delayed the drop in heat transfer coefficient. The drop in the heat transfer coefficient is thought to have been caused by the drying-up of some portion of the heater surface due to the formation of vapor agglomerates which stuck to the surface. The forced flow lost its effectiveness when the vapor grew to a certain size. We will further verify this point later in the flow visualization study. The last important finding is on the effectiveness of the forced flow. Owing to the fact that the forced velocity is constant in all the cases, it is apparent that the forced-convection is only effective for the lowest heat flux case (2.88×10^4 W/m^2), whereas for the other cases, the differences in heat transfer coefficient are negligible between pool and forced convection boiling. For the lowest heat flux, the differences in heat transfer coefficient grew even larger during microgravity, which is believed to be due to the ability of the flow to prevent the formation of vapor chunks by maintaining single and sliding bubbles. For the other cases, the heating is so high that the bubbles merge before the flow could move them. The flow visualization, presented later, will also confirm the above suggested mechanism for the forced-convection boiling.

In an effort to further examine the forced-convection effects in a microgravity environment, the percent increase in heat transfer coefficients under forced-convection over those of 1g pool boiling are shown in Figure 2.

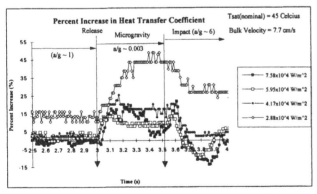

Figure 2 - Percent Change of Heat Transfer Coefficient under Forced-Convection over that of 1g Pool Boiling

Under 1g, the forced-flow made a difference only for the lowest heat flux case. Approximately 15% increase in the heat transfer coefficient was measured for the heat flux of 2.88×10^4 W/m^2. Some of this was certainly due to

increased convection in addition to single and sliding bubbles and, therefore, not a result of any boiling phenomena. During microgravity, both the forced-convection and pool boiling heat transfer coefficients are 10 to 20% higher than those of 1g pool boiling for the three higher heat fluxes. For the lowest heat flux case, the forced-convection heat transfer coefficient enjoyed a 45% increase under microgravity. We attribute this sizable increase in heat transfer coefficient to the sliding individual bubbles on the heater surface according to the numerical model of Tsung-Chang and Bankoff (1990), and further verification of this is given later in the visualization section.

For the heat transfer results in Figure 1, it is reassuring to note that they are well represented by the Rohsenow correlation for nucleate boiling given in the following form by Mills (1992),

$$Nu = \frac{Ja^2}{C_{nb}{}^3 Pr_l{}^m} \qquad (2)$$

Where Nu is the Nussel number, Ja is the Jakob number, and Pr_l is the Prandtl number of the boiling fluid. The empirical constants m and C_{nb} for our data are 4.1 and 0.0075, respectively. Rohsenow suggested for all organic fluids m = 4.1, while Mills (1992) only gave C_{nb}=0.0068 for scored copper and C_{nb}=0.008 for polished stainless steel. Therefore, our gold-film surface acts approximately between a copper surface and a stainless steel surface in the nucleate boiling regime.

Visualization Study

To compliment the measured data, a set of video recordings are presented. Under normal gravity the addition of the forced flow simply swept the bubbles along the flow direction. For the case of microgravity, Figure 3 provides a comparison between pool and forced-convection boiling flow patterns in the vicinity of the heater surface for heat fluxes of 2.88×10^4 and 5.95×10^4 W/m^2. The time for each picture is 0.5 seconds after the release and represents a general picture of the vapor bubble dynamics for each heat flux. We found the following general trends after examining the visualization figures. During microgravity pool boiling, the bubbles did not rise off the heater surface, and as a result, they coalesced to form what might be called 'vapor agglomerates.' These large bubbles of various shapes have very rough surfaces and tend to cover up the heater surface, which appeared to cause some portions of the surface to become unwetted during the trailing portion of microgravity.

Pool Boiling (2.88×10^4 W/m^2)

Forced-Convection Boiling (2.88×10^4 W/m^2)

Pool Boiling (5.95×10^4 W/m^2)

Forced-Convection Boiling (5.95×10^4 W/m^2)

Figure 3 - Comparisons of Pool and Forced-Convection Boiling Bubble Dynamics

The drying up of the surface due to vapor agglomerates contributed to the drop in heat transfer coefficient for the high heat flux cases as discussed above. Therefore, under microgravity the bubble swarm of terrestrial boiling was, in general, replaced by bubble slugs or agglomerates. The average size of the bubbles was seen to increase with the heat fluxes. For all the cases except the lowest heat flux case we found that immediately around the large bubbles there were swarms of smaller bubbles which were nucleating and detaching from the heater surface. The smaller bubbles appeared to coalesce with the larger bubbles. When a smaller bubble coalesced with a larger bubble the surface where the smaller bubble originated was left wet and this transport mechanism sustained the nucleate boiling process locally. If a small bubble nucleated away from a large bubble, and could not coalesce, then it could form a large bubble itself as other smaller bubbles coalesced with it. This appeared to be a strong mechanism controlling the bubble dynamics for pool boiling in microgravity for the conditions of this study. While these effects were dominant during pool boiling, they could also be seen during the forced flow cases for the higher heat flux runs.

The second finding is concerned with the differences between pool and forced convection boiling under microgravity. With forced flow, the bubbles experienced a shear force along the flow direction. In the absence of gravity, the bubble dynamics depends on the balance between the drag due to the forced flow and the surface tension. The surface tension force prevails if the contact area between the bubble and the heater surface is larger than the bubble frontal surface area experiencing the forced flow, otherwise the drag force dominates. For the former, a hemispherical bubble would tend to stick to surface while a spherical bubble, which belongs to the latter case, would most likely slide on the heater surface. Based on the visualization photographs, we found that for medium to high heat fluxes, the coalesced vapor chunks tend to form hemispherical or dome shape with large base areas as a result of rapid vapor generation rates. Therefore at medium and high heat fluxes ($4-7 \times 10^4$ W/m^2) in the vicinity of the heater surface, the surface tension dominated which left the forced flow with negligible effects on the bubble dynamics and heat transfer. The heat transfer coefficients presented previously also support this conclusion. The only noticeable difference between pool and forced-convection boiling for medium to high heat fluxes was a reduced vapor doom height for the forced convection runs as compared to pool boiling.

For the lowest heat flux case under microgravity as shown in Figure 3, the two-phase pattern of the forced-convection boiling was fundamentally different from that of the pool boiling. Owing to the low heating rate, the bubbles did not have an opportunity to coalesce, which resulted in the dominance of the flow drag over the surface tension. Basically, it was found that the bubbles are spherical and most of them are isolated from each other. Based on our image analysis, the average individual bubble size was calculated at 1 to 2 mm while those of vapor agglomerates were on the order of 1 cm and up. Figure 4 is specially marked to demonstrate the "sliding" phenomenon.

0.133 sec

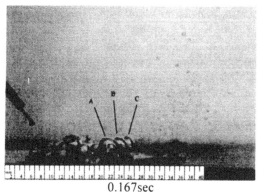

0.167sec

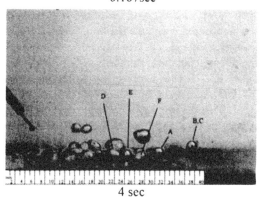

4 sec

0.433 sec

Figure 4 - Tracking of Individual Sliding Bubbles, 2.88×10^4 W/m^2

These Figures provide strong evidence that the isolated spherical bubbles were "sliding" on the heater surface. In order to obtain quantitative results, we marked the individual bubbles and tracked their movements on the heater surface. The sliding velocities of bubbles A-C were measured by digitizing the images and tracking the centroid of the bubble.

As shown in Figure 5, the patterns of bubble velocities are quite consistent between the three bubbles tracked in the analysis. The velocities of the individual bubbles fluctuated at relatively large amplitudes between 25 to 90% of the free-stream velocity during the first half of the transient and was possibly due to some surface tension effects. After detachment from the heater surface their velocities appeared to converge to between 80 to 90% of the free-stream velocity and was even measured larger than the free-stream. This was probably due to the reduction in flow area due to the vapor itself.

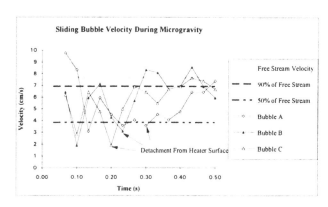

Figure 5 - Sliding Bubble Velocity

This result appears to be in agreement with that of Gunther (1951). For forced-convection nucleate boiling in water, Gunther (1951) indicated that the bubbles were sliding at

80% of the free-stream velocity of 1.7 m/s. As shown in Figure 9, the bubbles grew in size from the beginning of microgravity until the tops of the bubbles reached the subcooled liquid (approximately t = 0.167 second for this case). After that, the sizes of the bubbles were found nearly constant due to the balance between condensation on the bubble cap and vaporization from the microlayer. The finding of this nearly constant bubble size also contributes to the verification of the latent heat transport mechanism (Bankoff et al., 1956, Bankoff, 1962, Snyder and Robin, 1969).

It is our conclusion that the significantly higher heat transfer coefficient for forced-convection boiling than that for pool boiling in microgravity for the lowest heat flux is apparently due to the combination of the prevention of vapor chunk formation by the flow and the sliding of individual spherical bubbles. The current experimental results also help verify the mathematical model of Tsung-Chang and Bankoff (1990) that the sliding of bubbles in highly subcooled forced-convection nucleate boiling strongly augments the heat transfer through microlayer evaporation.

Based on the measured heat transfer coefficient data and the flow visualization, it is reasonable to suggest that for a given heat flux, if the forced-flow field is strong enough to prevent bubble agglomeration and to maintain single and sliding bubbles, subcooled forced-convection nucleate boiling would be a feasible and efficient heat transfer mechanism in microgravity.

CONCLUSION

A microgravity experiment was performed to further examine the momentum and energy transport mechanisms during forced-convection subcooled nucleate boiling. Under microgravity, we were able to study exclusively the interaction between the forced flow field and the vapor bubbles without the masking effect of buoyancy. Our flow visualization results clearly support the latent heat transport mechanism and the "sliding bubble" model proposed by Tsung-Chang and Bankoff (1990) as the key transport mechanism in subcooled forced-convection nucleate boiling. For a given heat flux level, it is important that the velocity field is adequate to move the single bubbles and prevent the formation of vapor chunks which tend to stick to the heater surface and create drying. With high enough flows, a steady and stable nucleate boiling condition could be maintained under microgravity.

ACKNOWLEDGMENT

This work was supported by NASA Grant NAG3-1387. The authors gratefully acknowledge the interest and support of Dr. Fran Chiaramonte of the NASA Lewis Research Center.

NOMENCLATURE

a	acceleration
g	gravitational field
h	mean heat transfer coefficient
q''	mean heat flux
T_s	mean surface temperature
$T_{sat(nominal)}$	nominal saturation temp (45 °C)
V	velocity

REFERENCES

Bankoff, S.G., Colahan, W.J., Jr., and Bartz, D.R.,1956, "Summary of Conference of Bubble Dynamics and Boiling Heat Transfer held at the Jet Proplsion Laboratory," Memo No. 20-137, Jet Propulsion Laboratory, Pasedena, CA.

Bankoff, S.G., 1962, " A Note on Latent Heat Transport in Nucleate Boiling," AIChE J., Vol.8, pp. 63-65.

Cochran, T.H., 1970, "Forced-Convection Boiling Near Inception in Zero Gravity," NASA TN D-5612.

Gunther, F.C., 1951, "Photographic Study of Surface-Boiling Heat Transfer to Water With Forced Convection," Trans. ASME, Vol. 73, pp. 115-124.

Merte, H. JR., Lee, H. S., Ervin, J. S., 1993, "Transient Nucleate Pool Boiling in Microgravity - Some Initial Results," *Int. Symp. Microgravity Science and Applications*, Paper J-5.

Mills, A.F., 1992, Heat Transfer, Irwin, Homeward, Illionois.

Oker, E., and Merte, H. JR., 1981, "Semi-Transparent Gold Film as Simultaneous Surface Heater and Resistance Thermometer for Nucleate Boiling Studies," *Journal of Heat Transfer*, Vol. 103, pp. 65-68.

Snyder, N.W., and Robin, T.T., 1969, "Mass-Transfer Model in Subcooled Nucleate Boiling," ASME *Journal of Heat Transfer*, Vol. 91, pp. 404-412.

Tsung-Chang, G., and Bankoff, S.G., 1990, "On the Mechanism of Forced-Convection Subcooled Nucleate Boiling," *Journal of Heat Transfer*, Vol. 112, pp. 213-218.

Wang, Tien-Chen, 1993, <u>Transient Effects of Microgravity on Pool and Low Velocity Flow Boiling Heat Transfer</u>, Master Thesis, Department of Mechanical and Materials Engineering, Washington State University.

PARAMETRIC STUDY OF ULTRA-HIGH CHF
IN HIGHLY SUBCOOLED WATER FLOW
INSIDE SMALL DIAMETER TUBES

Issam Mudawar[†] and Morris B. Bowers[††]
Boiling and Two-Phase Flow Laboratory
School of Mechanical Engineering
Purdue University
West Lafayette, Indiana, U.S.A.

ABSTRACT

A data base was obtained for ultra-high CHF, with values mostly exceeding 10^4 W cm^{-2}, using high mass velocity water flow through small diameter tubes. The parametric trends of ultra-high CHF were ascertained with respect to key flow and geometrical parameters over a wide range of exit pressures. Pressure drop for most conditions was fairly constant over the entire range of heat fluxes, from the single-phase flow condition corresponding to zero heat flux up to CHF, proving CHF was triggered even with negligible net vapor production. CHF increased with increasing mass velocity, increasing subcooling, decreasing tube diameter, and decreasing heated length. CHF increased with increasing pressure up to about 30 bars, became fairly constant up to about 100 bars, and then decreased for higher pressures. A new CHF correlation was developed for pressures ranging from 10.3 to 172.4 bars corresponding to CHF values as high as 2.76×10^4 W cm^{-2}.

1. INTRODUCTION

Practical cooling technologies are presently needed for dissipating ultra-high heat fluxes in applications such as x-ray medical devices, high-power lasers and fusion reactors. According to Boyd (1985), heat loads for fusion reactor components require cooling schemes that could dissipate heat fluxes on the order of 10^4 W cm^{-2}. Since these heat fluxes exceed the critical heat flux (CHF) attainable with common flow boiling systems, the ability to both greatly increase and predict the magnitude of CHF is of paramount importance to these ultra-high heat flux applications.

The key parameters affecting flow boiling CHF for any cooling configuration are mass velocity, subcooling, pressure, hydraulic diameter, heated length, and, of course, the coolant itself. As shall be shown below, previous studies provide ample evidence ultra-high CHF is possible with flow boiling of water at high mass velocities in small diameter tubes.

In perhaps the most extensive study of ultra-high CHF, Ornatskii and Vinyarskii (1965) reported an increase in CHF with increasing exit subcooling for water flowing through 14 mm long tubes ranging in diameter from 0.4 to 2 mm. Recent high heat flux studies have confirmed the strong influence of subcooling on CHF. These include studies by Celata *et al.* (1993) and Vandervort *et al.* (1994).

Another parameter that greatly influences CHF is the mass velocity. Boyd (1988, 1989) reported a linear increase in CHF with increasing mass velocity. Similar results have been reported by others including Ornatskii and Vinyarskii, who covered mass velocities up to $G = 9 \times 10^4$ kg m^{-2} s^{-1}.

With regards to the tube length and diameter effects on CHF, Bergles (1963) observed an increase in CHF with decreasing heated length and decreasing tube diameter. Vandervort *et al.* also measured an increase in CHF with decreasing L/D, the length-to-diameter ratio, especially for L/D below 10. They concluded the high CHF values achieved with short tubes are related to the state of flow development along the tube. They also reported an increase in CHF with decreasing diameter for diameters smaller than 2.0 mm; the diameter effect was most significant at high mass velocities. This contradicts the earlier findings of Ornatskii and Vinyarskii who reported a diminution in the diameter effect for large mass velocities.

The effect of pressure on CHF is far more complicated than those of the parameters just mentioned. Boyd (1989) reported a weak effect due to pressure for CHF values around 500 W cm^{-2}; however, at heat fluxes above 1500 W cm^{-2}, an increase in outlet pressure yielded a significant rise in CHF. Bergles also reported an increase in CHF with increasing pressure for pressures between 1.38 and 5.86 bars. On the other hand, Ornatskii and Vinyarskii measured an increase in CHF with decreasing outlet pressure, and Vandervort *et al.* found no significant outlet pressure effect on CHF for pressures below 25 bars.

In order to dissipate ultra-high heat fluxes from flat surfaces such as supercomputer chips and fusion blankets, small diameter grooves can also be formed into solid metal slabs. Bowers and Mudawar (1993) proposed thermal design correlations for such slabs. Complementing their experimental investigation, they later

---

[†] Professor and Director of the Purdue University Boiling and Two-Phase Flow Laboratory; author to whom correspondence should be addressed
[††] Graduate Student

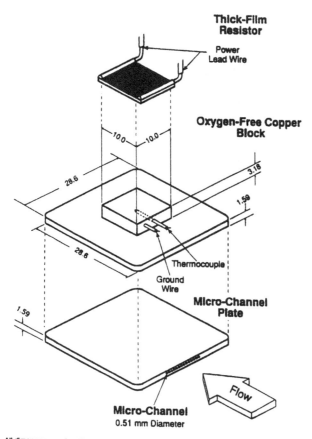

Thick-Film
Resistor

Power
Lead Wire

Oxygen-Free Copper
Block

10.0   10.0

28.6   3.18

1.59

28.6

Thermocouple

Ground
Wire

Micro-Channel
Plate

1.59

Micro-Channel
0.51 mm Diameter

Flow

All dimensions are in millimeters

(a)

presented methods for optimizing the design of these slabs for specific electronic cooling applications as illustrated in Fig. 1a (Bowers and Mudawar, 1994a,b). The groove diameter and spacing between grooves were optimized analytical for such concerns as temperature uniformity along the slab surface, flow rate, pressure drop, and CHF. Similar work was also performed by Jimenez and Mudawar (1994) in relation to immersion cooling of avionic hardware, Fig. 1b, using small hydraulic diameter cooling channels.

It is the ultimate objective of the present study to extend the methodology of designing grooved slabs to applications involving ultra-high heat fluxes. To accomplish this objective, experiments were performed with water flowing through small diameter tubes and a large data base was obtained to ascertain the effects of mass velocity, subcooling, heated length, diameter, and pressure on CHF. A new dimensionless correlation will be presented for ultra-high CHF.

2. EXPERIMENTAL APPARATUS
Test Section

Each of the test sections used in the present study consisted of a capillary tube made from either 304 stainless steel or Cu-Ni 30%. Power cables were connected to two stainless steel blocks mounted to the tube ends. Heat was generated in the capillary tube by passing an electrical current (up to 700 amps and 20 volts) from one stainless steel block through the tube itself and returned to the power supply from the other stainless steel block. The electrical resistance of the capillary tube wall was much greater than the remainder of the electrical path; over 97% of the supplied electrical power was dissipated uniformly along the inner surface of the tube.

CHF tests were conducted for different combinations of flow parameters such as flow rate, inlet temperature, and outlet pressure. While performing a given test, these parameters were continuously monitored, and the heat flux, calculated from power measurements and heated area, was increased in very small increments. Following each power increment, steady state was achieved and additional flow adjustments were made in order to maintain the

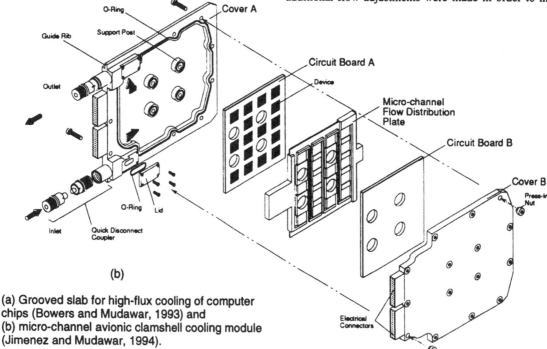

(b)

Fig. 1   (a) Grooved slab for high-flux cooling of computer
chips (Bowers and Mudawar, 1993) and
(b) micro-channel avionic clamshell cooling module
(Jimenez and Mudawar, 1994).

appropriate test conditions. CHF values were very high resulting in physical "burnout" of the tube. Thus, the criterion used for determining CHF was the tube burnout.

Flow Loop

The present study demanded a flow loop capable of withstanding pressures up to the critical point. As shown in Fig. 2, water was fed from a large reservoir to a high pressure piston pump which both increased the pressure and circulated the fluid through the loop. Submerged inside the tank were both a cooling coil and an immersion heater. These were used to heat or cool the reservoir water, thereby providing a means for controlling the test section's inlet temperature. Located immediately at the pump exit was an unloader valve, a device that protected the pump by forcing the flow to a recirculating loop in the event of an excessive increase in flow resistance downstream. Leaving the unloader valve, the fluid entered a pulsation dampener that reduced fluctuations in both flow rate and pressure. A portion of the flow then entered the test section while the balance returned to the reservoir through a by-pass.

The flow rate entering the test section as well as the outlet pressure were controlled by valves located in the test section line and the by-pass. A turbine flow meter used to measure the flow rate was located upstream of the test section. Both the inlet and outlet pressures were measured using pressure gauges; however, for pressure drops below 3.5 bars, a differential pressure transducer was used to acquire more accurate readings of the pressure drop.

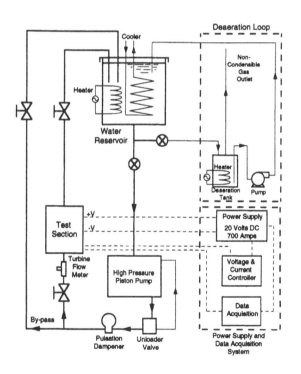

Fig. 2   Flow loop.

3. EXPERIMENTAL INVESTIGATION
Data Matrix

The bulk of the data was obtained for tube diameters of D = 0.902 and 0.406 mm and a nominal heated length of L = 5.6 mm.

Additional tests were performed with tube diameters of 0.635, 0.500, 1.08, and 2.54 mm and heated lengths up to 30.8 mm. The acquired data base covers broad ranges of mass velocity, G = 5.0 × 10^3 to 1.3 × 10^5 kg m^{-2} s^{-1}, and exit pressure, P_o = 3.4 to 172 bars. The corresponding CHF values ranged from q_m = 0.94 × 10^3 to 2.76 × 10^4 W cm^{-2}, the upper limit being the highest uniform CHF ever reported in the literature. This value surpasses the prior CHF record of q_m = 2.28 × 10^4 W cm^{-2} achieved by Ornatskii and Vinyarskii in 1965 for uniformly heated tubes.

Prevention of Premature Burnout

The tubes used throughout this study were thin-walled to provide sufficient electrical resistance for generating the required high heat fluxes. However, associated with the high rate of heat generation was a large temperature gradient even for the small wall thicknesses used. Figure 3 shows the temperature distribution across the wall for tube diameters of 0.902 and 0.406 mm. These diameters and the wall thicknesses presented are representative of the bulk of the data base. Figure 3 compares temperature distributions for an ultra-high heat flux of q = 3.0 × 10^4 W cm^{-2} and a moderately high flux of q = 3.0 × 10^3 W cm^{-2}. The temperature distribution was calculated using an analytical one-dimensional heat diffusion model, assuming uniform heat generation in a cylindrical cross section with an adiabatic outer boundary and a known inner surface temperature. The inner temperature was set equal to the saturation temperature of water at 31.0 bars which corresponds to a large fraction of the present data base. As illustrated, the higher heat flux creates a temperature difference between the inner and outer surfaces of approximately 500 and 400 °C for the 0.902 and 0.406 mm tubes, respectively. Even though the gradient is very large, the outer surface temperature remains below the melting point for 304 SS by more than 600 °C. This shows determining CHF by tube burnout is a valid technique, since the tube would suddenly burst only as a result of a sudden large increase in the inner surface temperature. This criterion was equally valid for the tubes made from Cu-Ni 30%.

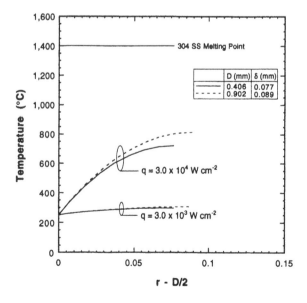

Fig. 3   Temperature distribution across stainless steel tube wall for different tube diameters and different heat fluxes.

Pressure Drop Characteristics

Pressure drop results for ultra-high heat flux conditions are presented in Fig. 4. Shown is the pressure drop versus heat flux at mass velocities of 4.0×10^4 and 1.0×10^5 kg m^{-2} s^{-1} for three tube diameters. Also indicated is the length to diameter ratio for each based on both the heated length, L, and total length, L_{tot}, which includes the unheated inlet and outlet regions of the tube. All of these tubes had a nominal heated length of 5.5 mm. Referring to the three cases corresponding to the lower mass velocity, the pressure drop increased with decreasing tube diameter with the smallest diameter of 0.406 mm yielding a total pressure drop of approximately 20 bars. An increase in mass velocity to 1.0×10^5 kg m^{-2} s^{-1} resulted in large increases in pressure drop for both the 0.902 and 0.406 mm tubes with the smaller diameter yielding a very high pressure drop of approximately 100 bars. Ultra-high heat fluxes above 2.0×10^4 W cm^{-2} were achieved only with very high mass velocities similar to those indicated in Fig. 4 at the expense of very large pressure drops. However, for each case, there was little variation in pressure drop over the entire range of heat flux up to CHF. Beginning at low heat fluxes, where the flow was that of pure liquid, an increase in heat flux caused a slight decrease in pressure drop due to a reduction in liquid viscosity with increasing temperature. This trend was generally sustained until the heat flux neared CHF where a slight rise in the pressure drop was detected. The coolant entered the test section subcooled, and as a result of the large mass velocities required to achieve ultra-high CHF, remained subcooled at the exit even for the largest heat flux, resulting in a predominately single-phase pressure drop.

These cases are representative of the bulk of the data obtained in the present study and clearly show that pressure drop should be an issue of major practical significance when designing a system for ultra-high heat flux cooling.

heated length of $L = 5.5$ mm. A broad range of mass velocity is shown, from a moderate value of 5.0×10^3 to a large value of 1.3×10^5 kg m^{-2} s^{-1} which surpasses the mass velocities of all previous CHF studies. Figure 5 shows an increase in CHF with increasing mass velocity for both tube diameters; the smaller diameter exhibited a general trend of a greater CHF than the larger diameter for equal mass velocities.

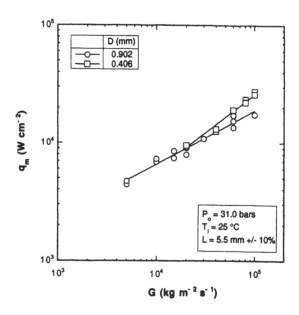

Fig. 5   Effects of mass velocity and diameter on CHF.

Effect of Heated Length

Figure 6 shows the effect of heated length on CHF for mass velocities between 5.0×10^3 and 1.5×10^4 kg m^{-2} s^{-1}, where the heated length was varied from 5.5 to 31.0 mm. For both mass velocities, CHF decreased monotonically with increasing heated length; however, the effect of heated length was more pronounced for the shorter tubes.

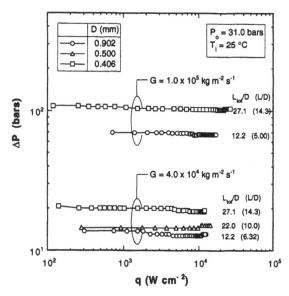

Fig. 4   Pressure drop versus heat flux for high mass velocities.

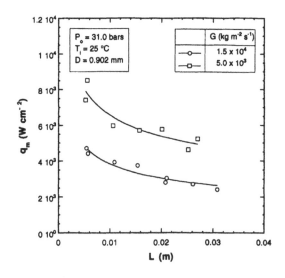

Fig. 6   Effect of heated length on CHF.

4. PARAMETRIC TRENDS
Effects of Mass Velocity and Tube Diameter

The effect of mass velocity on CHF at high pressure is presented in Fig. 5 for tube diameters of 0.902 and 0.406 mm and a nominal

Effect of Pressure

Figure 7 shows the variation of CHF with pressure at the exit of the heated length. CHF data are presented for a single nominal heated length of 5.5 mm and three mass velocities over a broad pressure range of 10 to 172 bars. For the two lower mass velocities, CHF increased appreciably with increasing pressure up to 30 bars, which was followed by a region from 30 to 100 bars where CHF showed little change with pressure. As the pressure was increased further, CHF began decreasing up to the highest pressure of 172 bars. Figure 7 also shows a reduced pressure effect for the highest mass velocity over the entire pressure range.

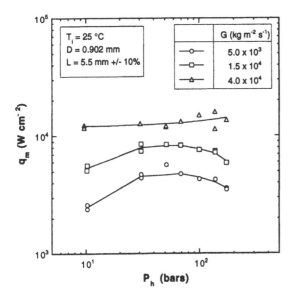

Fig. 7   Effect of pressure at exit of heated length on CHF.

5. CHF CORRELATION

Close examination of the CHF data base revealed certain parametric trends could not be reproduced using a single correlation for the entire data base. In particular, a natural division in the data base was noted between high and low pressures, the dividing line being around 10.3 bars. It was further noted this distinction was insignificant for mass velocities of 4.0×10^4 kg m^{-2} s^{-1} or greater.

One distinctive difference that was more significant at low mass velocities and low pressures was the effect of subcooling. The entire data base could be characterized as subcooled at the tube exit; however, the exit equilibrium quality varied significantly, from $x_o = -0.074$ to -2.37. Over such a broad range of quality, the physical mechanisms governing CHF are expected to be different. At high pressures, bubbles produced along the inner wall of the tube are small. Combined with a large negative exit quality, these bubbles will rapidly condense upon formation, and boiling heat transfer up to CHF should, therefore, be very localized near the wall. The only effect on the bulk flow by the large heat transfer rates is an increase in the sensible energy. In contrast, low pressures and small values of negative equality are conducive to net vapor generation, resulting in a more appreciable effect of subcooling on the bulk flow and a reduced CHF.

Based upon the observed trends, the data base was divided into regions of high and low pressure around 10.3 bars; the present correlation was developed for the high pressure range. Figure 8 shows the following dimensionless equation

$$\frac{q_m}{G\,h_{fg}} = 0.104\,We_L^{-0.24}\left(\frac{\rho_f}{\rho_g}\right)^{-0.29} \times$$
$$\left[1 - 1.2\left(\frac{L}{D}\right)^{-1.35}\right]\left[1 + 0.4\,\frac{c_{p,f}\,\Delta T_{sub,o}}{h_{fg}}\right].$$
$$(1)$$

correlates the CHF data for the indicated ranges of parameters, a total of 124 data points, with a mean absolute error of 12.9%. All the properties in this correlation are based on their saturation values corresponding the pressure at the exit from the heated section, P_h. Equation (1) shows all the CHF parametric trends can be represented by only four dimensionless parameters. They are the Weber number based upon the heated length, $We_L (= G^2 L/\rho_f \sigma)$, density ratio, ρ_f/ρ_g, heated length to diameter ratio, L/D, and a subcooling parameter, $c_{p,f}\Delta T_{sub,o}\,h_{fg}^{-1}$.

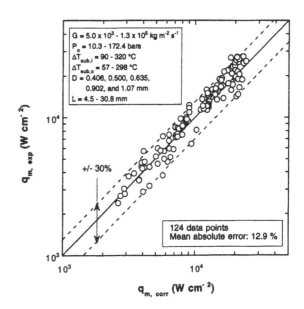

Fig. 8   CHF correlation.

A model by Lee and Mudawar (1988), which was later adopted by Katto (1990a,b) and Celata et al. (1994), suggests dryout of a liquid sublayer beneath discrete vapor blankets is the trigger mechanism for CHF in subcooled vertical upflow. CHF in the present study was always detected as burnout near the downstream end of the heated section. This is evidence, as was suggested by Lee and Mudawar, there exists a boiling development length along the heated wall, starting with subcooled liquid at the inlet, followed by a region of discrete bubble nucleation, and ending with very thin discrete vapor blankets towards the downstream end of the heated section, where CHF occurs.

6. CONCLUSIONS

A data base of ultra-high CHF was obtained to ascertain the effects of mass velocity, subcooling, heated length, tube diameter, and pressure on CHF. Key findings from the study are as follows:

(1) Ultra-high CHF values in excess of 10^4 W cm^{-2} can be achieved with subcooled water flow at high mass velocities in small diameter tubes ($D < 1$ mm) of short heated lengths ($L < 6$ mm). The present study yielded the highest recorded CHF value of 2.76×10^4 W cm^{-2} which surpasses the prior record of 2.28×10^4 W cm^{-2} for a uniformly heated tube achieved by Ornatskii and Vinyarskii in 1965.

(2) Pressure drop for flow boiling at high mass velocity in a short, small diameter tube varies little from the single-phase pressure drop corresponding to zero heat flux.

(3) CHF increases with increasing mass velocity, increasing subcooling, decreasing diameter, and decreasing heated length. However, the effect of pressure is not monotonic. CHF increases with increasing pressure below 30 bars, becomes fairly constant between 30 and 100 bars, and decreases above 100 bars.

(4) Two parametric regions of ultra-high CHF, low pressure and high pressure, were identified which define operating conditions with different flow structures and resulting CHF mechanisms. A new CHF correlation was developed which accurately predicts the high pressure CHF data corresponding to exit pressures exceeding 10 bars.

ACKNOWLEDGMENT

The authors are grateful for the support of the Office of Basic Energy Sciences of the U.S. Department of Energy (Grant No. DE-FE02-93ER14394).

NOMENCLATURE

c_p	specific heat
D	tube i.d.
G	mass velocity
h_{fg}	latent heat of vaporization
L	heated length
L_{tot}	total length
ΔP	pressure drop
P_h	pressure at the end of heated length
P_o	tube exit pressure
q	heat flux based upon tube inside area
q_m	CHF based upon tube inside area
r	coordinate in radial direction
T	temperature
ΔT_{sub}	liquid subcooling, $T_{sat} - T$
T_{sat}	saturation temperature
We	Weber number, $G^2 L (\sigma \rho_f)^{-1}$
x	thermodynamic equilibrium quality

Greek Symbols

ρ	density
σ	surface tension

Subscripts

corr	CHF correlation
exp	experimental data
i	inlet
m	maximum (critical heat flux)
o	outlet
sub	subcooled
tot	total.

REFERENCES

Bergles, A. E., 1963, "Subcooled Burnout in Tubes of Small Diameter," ASME Paper 63-WA-182.

Boyd, R. D., 1985, "Subcooled Flow Boiling Critical Heat Flux (CHF) and its Application to Fusion Energy Components - Part I. A Review of Fundamentals of CHF and related Data Base," *Fusion Technology*, Vol. 7, pp. 7-29.

Boyd, R. D., 1988, "Subcooled Water Flow Boiling Experiments under High Heat Flux Conditions," *Fusion Technology*, Vol. 13, pp. 131-142.

Boyd, R. D., 1989, "Subcooled Water Flow Boiling at 1.66 MPa under Uniform High Heat Flux Conditions," *Fusion Technology*, Vol. 16, pp. 324-330.

Bowers, M. B. and Mudawar, I., 1993, "High Flux Boiling in Low Flow Rate, Low Pressure Drop Mini-Channel and Micro-Channel Heat Sinks," *Int. J. Heat Mass Transfer*, Vol. 37, pp. 321-332.

Bowers, M. B. and Mudawar, I., 1994a, "Two-Phase Electronic Cooling using Mini-Channel and Micro-Channel Heat Sinks - Part 1. Design Criteria and Heat Diffusion Constraints," *ASME J. Electronic Packaging*, Vol. 116, pp. 290-297.

Bowers, M. B. and Mudawar, I., 1994b, "Two-Phase Electronic Cooling using Mini-Channel and Micro-Channel Heat Sinks - Part 2. Flow Rate and Pressure Drop Constraints," *ASME J. Electronic Packaging*, Vol. 116, pp. 298-305.

Celata, G. P., Cumo, M. and Mariani, A., 1993, "Burnout in Highly Subcooled Water Flow Boiling in Small Diameter Tubes," *Int. J. Heat Mass Transfer*, Vol. 36, pp. 1269-1285.

Jimenez, P. E. and Mudawar, I., 1994, "A Multi-Kilowatt Immersion-Cooled Standard Electronic Clamshell Module for Future Aircraft Avionics," *ASME J. Electronic Packaging*, Vol. 116, pp. 220-229.

Ornatskii, A. P. and L. S. Vinyarskii, L. S., 1965, "Heat Transfer Crisis in a Forced Flow of Underheated Water in Small Bore Tubes," *Teplofizika Vysokikh Temperatur*, Vol. 3, pp. 441-451.

Vandervort, C. L., Bergles, A. L. and Jensen, M. K., 1994, "An Experimental Study of Critical Heat Flux in Very High Heat Flux Subcooled Boiling," *Int. J. Heat Mass Transfer*, Vol. 37, pp. 161-173.

Lee, C. H. and Mudawar, I., 1988, "A Mechanistic Critical Heat Flux Model for Subcooled Flow Boiling based on Local Bulk Flow Conditions," *Int. J. Multiphase Flow*, Vol. 14, pp. 711-728.

Katto, Y., 1990a, "A Physical Approach to Critical Heat Flux of Subcooled Flow Boiling in Round Tubes," *Int. J. Heat Mass Transfer*, Vol. 33, pp. 611-620.

Katto, Y., 1990b, "Prediction of Critical Heat Flux of Subcooled Flow Boiling in Round Tubes," *Int. J. Heat Mass Transfer*, Vol. 33, pp. 1921-1928.

Celata, G. P., Cumo, M., Mariani, A., Simoncini, M., and Zummo, G., 1994, "Rationalization of Existing Mechanistic Models for the Prediction of Water Subcooled Flow Boiling Critical heat Flux," *Int. J. Heat Mass Transfer*, Vol. 37, pp. 347-360.

SATURATED CONVECTIVE
BOILING

FLOW BOILING IN HORIZONTAL AND VERTICAL TUBES:THE EFFECT OF TUBE ORIENTATION ON HEAT TRANSFER

N. Kattan, J.R. Thome and D. Favrat
Industrial Energetics Laboratory
Swiss Federal Institute of Technology - Lausanne
CH-1015 Lausanne, Switzerland

ABSTRACT

As part of a comprehensive study on factors affecting flow boiling in horizontal tubes, local boiling heat transfer coefficients were measured for R-134a evaporating inside the same 12.00 mm bore, plain tube for horizontal flow, vertical upflow and vertical downflow over a wide range of mass velocities, vapor qualities and heat flux for a direct comparison of the effects on tube orientation on heat transfer. The experiments show a significant effect of flow direction on local heat transfer coefficients with upflow giving the highest values, horizontal flow giving significantly lower values and downflow yielding even lower values. A flashing/subcooling effect on the heat transfer process caused by the negative/positive pressure gradient for upflow/downflow, respectively, is proposed as an important new factor affecting flow boiling.

INTRODUCTION

A comprehensive, stepwise study on factors affecting flow boiling is being undertaken in the LENI laboratory. Presently, local boiling heat transfer coefficients are reported for R-134a evaporating inside the same 12.00 mm bore, plain tube for horizontal, vertical upward and vertical downward flow orientations at nearly identical pressures, mass velocites, vapor qualities and heat fluxes. The objective it to make a direct experimental comparison of local boiling heat transfer coefficients to determine the effects of flow direction on the intube evaporation process.

In industrial practice, flow boiling inside horizontal tubes nearly always refers to refrigerants evaporating in direct-expansion evaporators with an inlet vapor quality of 0.1-0.25 and an exit condition of a few degrees of superheat. Instead, the most widely quoted flow boiling correlations have been developed from large databanks for vertical upflow with the majority the data in the vapor quality range from 0-0.5. Some of these correlations have then been extended to evaporation inside horizontal tubes using a horizontal tube databank.

Several weak points exist in this approach. First of all, above the stratified flow threshold criterion, it is assumed that there is no tube orientation effect on heat transfer; below the threshold, the reduction in the heat transfer coefficient (because the tube circumference is only partially wetted with liquid and dry at the top) is predicted by adding an empirical correction term to the vertical tube correlation. However, these empirical corrections have been developed by statistical regression to improve the fit of the vertical tube correlation to the horizontal tube boiling databank rather than by direct comparison of experimental test data for vertical and horizontal flows at the same local test conditions. Consequently, effects other than stratification may be involved and the empirical correction to heat transfer is not based on a fundamental analysis of the flow. Hence, these are potentially important weak points in existing design correlations for horizontal tubes.

Another potential weakness is related to flow patterns. Two-phase flow patterns are known to be influenced by tube orientation since the vapor is buoyant and tends to migrate towards the top half of a horizontal or inclined tube. Hence flow boiling heat transfer coefficients can be expected to be affected by this modification of the flow pattern, even for unstratified flow regimes.

Yet another potential weakness is faced at high vapor qualities. In vertical upflow boiling, dryout or the critical heat flux is reached in the vapor quality range from $0.5 < x < 0.75$, thought to occur simultaneously around the tube circumference at that height. Hence vertical tube databanks, dominated by electrically-heated tube data, tend to contain few high vapor quality data points in the range $0.5 < x < 1.0$. In direct-expansion evaporators (i.e. with horizontal tubes) the heat fluxes are not large enough to reach the critical heat flux and dryout begins at the top of the tube because of a deficiency of liquid. Dryout then proceeds axially along the tube from the top towards the bottom until the liquid in the lower portion of the tube either dries out completely or becomes entrained as mist in the vapor flow [refer to Figure 1 taken from Collier and Thome (1994)]. Hence, complete dryout may not occur until the vapor quality approaches 1.0 in a horizontal tube and a significant length of the tube has a local heat transfer coefficient controlled by wet wall convective heat transfer over the bottom portion of the tube and mist flow heat transfer over the upper portion. The ability of vertical tube correlations to model heat transfer in horizontal tubes in this vapor quality region is therefore questionable.

As an earlier part of the same study, experimental data on flow patterns and the threshold between stratified and unstratified flow (or better denoted as transition from all wet wall to partially wet wall flows) for horizontal flows have been obtained and are reported elsewhere by Kattan, Thome and Favrat (1993, 1995b, 1995c). In general, they showed that the liquid Froude number criterion (Fr_L = 0.04, 0.05 or 0.25) used by many flow boiling correlations is incapable of delineating the transition between stratified and unstratified flow data obtained for the following five refrigerants: R-123, R-134a, R-502, R-402A and R-404A. The newer criterion proposed by Klimenko and Fyodorov (1990) was also shown not to be very reliable. Thus, more investigation is desirable in order to increase our knowledge of the phenomenological differences between evaporation in vertical and horizontal tubes, resulting in the present tests being made with R-134a.

TEST FACILITY AND EXPERIMENTAL PROCEDURES

The experimental work was conducted using the test facility at the Laboratoire d'Energétique Industrielle at the Ecole Polytechnique Fédérale de Lausanne. Two-phase experiments to measure heat transfer coefficients and pressure drops and record flow patterns can be run for horizontal, vertical or inclined tube orientations by rotating two test sections connected in series around a central pivot that supports the test sections. Connections to the rest of the flow circuit are by insulated, flexible stainless steel hoses. Two double-pipe test sections are mounted in series with two 90° bends to form a "U", where the refrigerant flows inside the central tube. To eliminate nonuniform circumferential effects of electrical heating on horizontal stratified flows, counter-current flow of hot water is utilized in the annulus.

A data acquisition system together with a PC computer acquires, analyzes and stores all data. The thermocouples are calibrated using a double precision method available in the acquisition system to obtain the highest accuracies possible, that is 0.03°C. For adiabatic conditions after leaving the rig off all night, all temperatures are within a maximum deviation of 0.07°C and most are within 0.03°C of the average. The refrigerant and water flow rates are measured with an estimated accuracy of 0.2% by two Coriolis mass flowmeters. The

energy balances between the water and refrigerant flows in the two test sections agree to within 2% maximum error (1% average error). Repeatability of the data is very good. It is estimated that heat transfer coefficients have a maximum error of about 5% or less, depending on the test conditions. No lubricating oil enters the refrigerant charge since a magnetic rotor driven-type pump is used to circulate the R-134a.

Figure 2 shows a diagram of one test section and its measurement locations. At the refrigerant inlet, the temperature is measured by thermocouple #718 located in the refrigerant flow stream (the tube is well insulated from this point up to the location of the four water-side thermocouples #614, #613, #501 and #615 with a tightly-fitted teflon sleeve). Similar arrangements exist for the other measured refrigerant temperatures. At each of these locations the absolute pressures (P402, P403, etc.) of the refrigerant are measured (to calculate the local saturation temperature and determine the heat transfer coefficients) and the differential pressure drops (Pdif406, etc.) across the two test sections are measured. In addition, tubular sight glass sections of 12.00 mm bore and 100 mm length are installed inline with the tube to see the flow patterns.

The two test sections are plain, drawn copper tubing of 12.00 mm bore with 1.00 mm tube walls. Each section is 3.013 m long for heat transfer measurements and is divided into three zones. On the water-side in the annuli, four thermocouples are installed at eight measurement locations at 0°, 60°, 120° and 180° from the top. Thus, there are six test zones to obtain heat transfer coefficients over narrow changes of vapor quality (essentially local values) during boiling tests.

The boiling heat transfer coefficients are determined using a modified Wilson plot method, using a new approach based on the Gnielinski correlation, applicable to transition and turbulent flows. The local saturation temperatures of R134a were calculated from its vapor pressure curve using absolute pressure measurements at the inlet and outlet of each test section (these were prorated to the center of each test zone). For a more complete description of the test facility and test procedures, refer to Kattan, Thome and Favrat (1995a).

CORRECTIONS TO HEAT TRANSFER IN STRATIFIED FLOWS

The most widely used stratified threshold criterion for applying vertical flow boiling correlations to horizontal flows is the liquid Froude number, defined as

$$Fr_L = G^2/[\rho_L^2 \, g \, D] \qquad (1)$$

In the Shah (1982) correlation the threshold was set at Fr_L < 0.04. Below this value both the convective boiling and nucleate boiling heat transfer coefficients determined by separate correlations are reduced when the value of his dimensionless parameter N is less than 1.0. However, when N > 1.0 only the convective boiling coefficient is reduced if $Fr_L < 0.04$ such that there is no effect if the nucleate boiling coefficient is the larger of the two coefficients. In contrast, Gungor and Winterton (1986) determined the best threshold value for Fr_L to be 0.05 by statistical analysis of their databank and their convection enhancement factor E was reduced by the multiplying factor

$$E_2 = Fr_L^{(0.1-2Fr_L)} \qquad (2)$$

while the boiling suppression factor is substantially reduced by another multiplying factor, $(Fr_L)^{1/2}$. In their later forced convection dominated correlation in Gungor and Winterton (1987), the vertical tube coefficient was reduced only by the multiplying factor E_2, again for $Fr_L < 0.05$. In a Shah-type of correlation developed by Kandlikar (1990), he evaluated his data bank for various values of Fr_L and retained Shah's value of 0.04. Wattelet et al. (1994) have also recently used this criterion in a new correlation, setting the threshold value at 0.25. Notably, none of the above mention papers incorporated the local vapor quality into Fr.

EXPERIMENTAL RESULTS AND COMPARISONS

Some representative local heat transfer coefficient data for upward, downward and horizontal flows are shown in Figures 3, 4 and 5 for refrigerant R-134a [note: with hot water heating, heat fluxes depend on the overall U_o and LMTD of each test zone and hence cannot be preset by the experimenter...thus data are presented in a narrow range of heat flux for comparison].

In Figure 3 for a mass velocity of 101 kg/m²s ($Fr_L = 0.055$), the horizontal data points are seen to be much lower than those for upward flow boiling. The horizontal tube heat transfer coefficients range from 0.56 to 0.72 those of vertical upflow, i.e. a reduction from 44 to 28%, depending on heat flux and vapor quality. The Gungor-Winterton (1987) correlation applicable to vertical and horizontal flows only predicts the horizontal data well while significantly underpredicting the vertical upflow data. The Steiner-Taborek (1992) vertical upflow correlation instead predicts the upflow data quite well.

In Figure 4 at a mass velocity of 197 kg/m²s ($Fr_L = 0.209$), heat transfer data are shown over a quality range from 7 to 91%. The vertical downflow data are significantly below the vertical upflow data and are also below the horizontal flow data for most test conditions. For instance, the downflow coefficients tend to be 1/3 to 1/2 below the upflow values where direct comparisons can be made. The horizontal coefficients are also below those for upflow at low heat fluxes but approach the same level as the heat flux and vapor quality rise. The Gungor-Winterton (1987) correlation predicts some of the data well, but becomes progressively conservative as the heat flux rises (apparently because the correlation assumes that convective effects dominate). The Steiner-Taborek correlation again works reasonably well for the upflow data to its

recommended limit of application, i.e. $x < 50-60\%$. In Figure 4(d) the coefficients fall off considerably at high vapor qualities and indicate that dryout must have occurred in both the upflow and horizontal flow orientations.

In Figure 5 at a mass velocity of 300 kg/m²s ($Fr_L = 0.484$), the tube orientation effect on heat transfer is less important compared to the two lower mass velocities but is still evident. Curiously, the horizontal data are below those of upflow at low heat fluxes but become equal or larger at the higher heat fluxes. Thus as flow velocity increases and heat flux increases, the difference between upflow and horizontal flow coefficients diminish while downflow coefficients remain below these values.

Figures 3-5 conclusively demonstrate that there are important differences in upflow, downflow and horizontal flow boiling. The lower coefficients for downflow compared to upflow were not expected...it is hypothesized that the buoyancy effect of the vapor, which opposes downward flow, may reduce its accelerating effect on the liquid and hence diminish the convective contribution to heat transfer. The lower coefficients for horizontal flow compared to vertical flow were expected at low flow rates as a result of flow stratification in the horizontal tube (such that the tube circumference is only partially wet) with similar values at high mass velocites with both flows in the annular flow pattern, which was born out by the experiments. However, the strong effect of heat flux might be explained by increased intensity of nucleate boiling in the stratified flow regime causing the whole tube circumference to become wet, which Shah apparently included empirically into his correlation using his factor N.

One other fundamental difference in the evaporation process as a function of tube orientation is hypothesized to be important. In upflow the local pressure falls with increasing height as the local static pressure drops, such that the liquid flow always tends to be superheated with respect to its local saturation temperature as it moves incrementally up the tube. This will promote flashing of liquid at the vapor-liquid interface of bubbly and annular flows. For the extreme case with all liquid R134a in a vertical 3.0 m high tube, this represents a change of 2.0 °C in T_{sat} available for flashing in the annular film, similar to the values of the wall superheat. In contrast, in downflow the liquid requires sensible heating to remain at the rising saturation temperature as the pressure increases along the flow path with larger static head. Hence there is an adverse subcooling effect on the heat transfer process in both bubbly and annular flows. In horizontal flow no static head effect is present and the saturation temperature gradient will only be due to the frictional and accelerational pressure gradients...thus little flashing occurs if the pressure drop is small as in the present tests. This flashing/subcooling effect of tube orientation on intube flow boiling heat transfer is apparently a new effect not reported previously. Up until now, it was assumed that it was sufficient to utilize the local saturation temperature only and neglect the temperature gradient effect on the heat transfer process.

CONCLUSIONS

The following conclusions have been drawn from the present experimental tests:

1. Heat transfer coefficients for vertical downflow are considerably lower than those for vertical upflow for the present test conditions.

2. Heat transfer coefficients for horizontal flow are significantly below those for vertical upward flow for mass velocities from 100-300 kg/m²s and liquid Froude numbers from 0.055 to 0.483), except for high heat fluxes.

3. At vapor qualities above 0.50, the local heat transfer coefficients for horizontal flow are similar to those for vertical upflow.

4. For evaporation in horizontal tubes, the existing flow boiling correlations need to be improved to account for the effects of flow stratification at low to medium mass velocities and at high vapor qualities.

5. One fundamental difference between boiling in upflow, downflow and horizontal flow is the flashing/subcooling effect of the pressure gradient on the process.

ACKNOWLEDGEMENTS

The present research was funded by the Swiss Federal Office of Energy (OFEN) and the R-134a was donated by DuPont de Nemours International S.A. of Geneva.

NOMENCLATURE

D Internal tube diameter (m)
Fr_L Liquid Froude number

G Total mass velocity of liquid plus vapor (kg/m² s)

g Gravitational acceleration (9.81 m/s²)
x Vapor quality

ρ_L Liquid density (kg/m³)

T_{sat} Saturation temperature (°C)

U Overall heat transfer (W/m² K)
E Multiplying factor in Gungor-Winterton correlation
N Dimensionless parameter in Shah correlation

Subscripts

L Liquid
o outside

REFERENCES

Collier, J.G. and Thome, J.R. 1994. Convective Boiling and Condensation, 3rd Edition, Oxford University Press, Oxford.

Kattan, N., Thome, J.R. and Favrat, D., 1993. Two-Phase Flow Patterns during Evaporation of the New Refrigerants in Tubes and Annuli, European Two-Phase Flow Group Meeting, Hannover, FRG.

Gungor, K.E. and Winterton, R.H.S. 1986. A General Correlation for Flow Boiling in Tubes and Annuli, Int. J. Heat Mass Transfer, Vol. 29, pp. 351-358.

Gungor, K.E. and Winterton, R.H.S. 1987. Simplified General Correlation for Saturated Flow Boiling and Comparison of Correlations with Data, Chem. Eng. Res. Des., Vol. 65, pp. 148-156.

Kandlikar, S.G. 1990. A General Correlation of Saturated Two-Phase Flow Boiling Heat Transfer Inside Horizontal and Vertical Tubes, J. Heat Transfer, Vol. 112, pp. 219-228.

Kattan, N., Thome, J.R. and Favrat, D. 1995a. R-502 and Two Near-Azeotropic Alternatives - Part I: Intube Flow Boiling Experiments, ASHRAE Trans., Vol. 101(1), Paper CH-95-12-3 (3878).

Kattan, N., Thome, J.R. and Favrat, D. 1995b. R-502 and Two Near-Azeotropic Alternatives - Part II: Two-Phase Flow Patterns, ASHRAE Trans., Vol. 101(1), Paper CH-95-14-3 (3879).

Kattan, N., Thome, J.R. and Favrat, D. 1995c. Two-Phase Flow Patterns for New Refrigerants inside Horizontal Tubes, Int. Symp. on Two-Phase Flow Modelling and Experimentation, Rome, Italy (Oct. 9-11).

Klimenko, V.V. and Fyodorov, M. 1990. Prediction of Heat Transfer for Two-Phase Forced Flow in Channels of Different Orientation, Proc. 9th Int. Heat Transfer Conf., Vol. 5, pp. 65-70.

Shah, M.M. 1982. Chart Correlation for Saturated Boiling Heat Transfer: Equations and Further Study, ASHRAE Trans., Vol. 88(1), pp. 185-196.

Steiner, D. and Taborek, J. 1992. Flow Boiling Heat Transfer in Vertical Tubes Correlated by an Asymptotic Model, Heat Transfer Engng., Vol. 13(2), pp. 43-69.

Wattelet, J.P., Chato, J.C., Souza, A.L. and Christoffersen, B.R. 1994. Evaporative Characteristics of R-12, R-134a, and MP-39 at Low Mass Fluxes, ASHRAE Trans., Vol. 100(1), Paper NO-94-2-1.

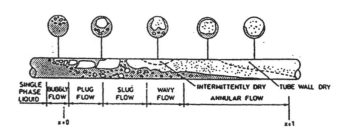

Figure 1. Flow patterns for evaporation in a horizontal tube [from Collier and Thome (1994)].

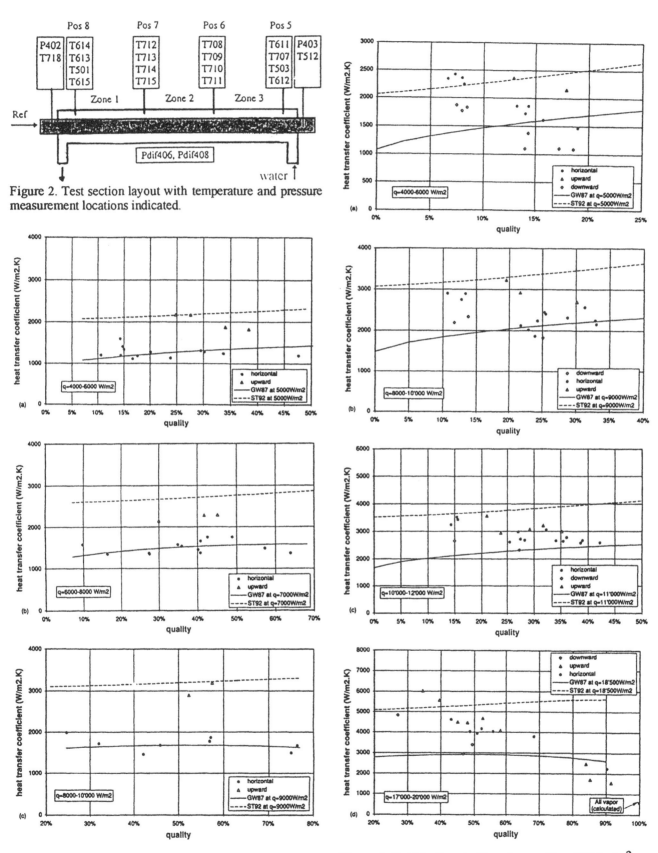

Figure 2. Test section layout with temperature and pressure measurement locations indicated.

Figure 3. R-134a flow boiling data at G = 101 kg/m2s at 10.5°C (4.22 bar).

Figure 4. R-134a flow boiling data at G = 197 kg/m²s at 10.5°C (4.22 bar).

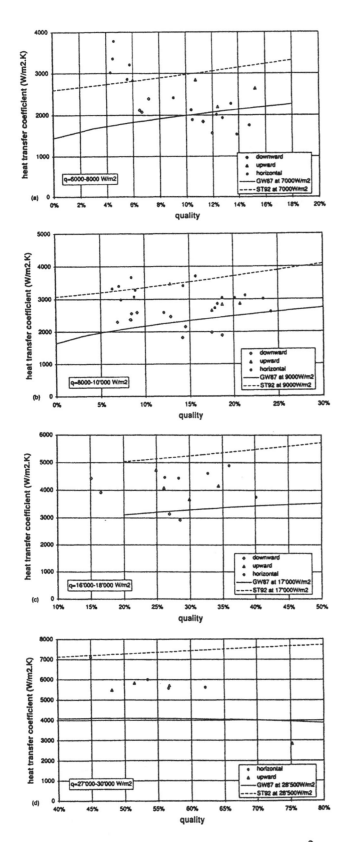

Figure 5. R-134a flow boiling data at G = 300 kg/m^{2}s at 10.5°C (4.22 bar).

CONVECTIVE FLOW BOILING IN COILED TUBES

Roderick E. Athey & Lawrence M. Saupe
Graham Manufacturing Company
Batavia, New York

ABSTRACT

Convective flow boiling of liquid nitrogen has been investigated using coiled-tube heat exchangers in order to determine the heat transfer characteristics. This project focused on the identification of the primary variables involved in establishing the flow boiling coefficient for nitrogen using a coiled-tube flow configuration.

The need for an accurate predictive correlation for this coefficient was recognized after an extensive testing program yielded results that did not correspond well with published information. A test method was developed which allowed for the accurate measurement of the boiling coefficient over a wide range of nitrogen flow rates, excess temperatures (T_{wall} - T_{sat}), and tube coil geometries.

In evaluating nitrogen boiling rate data using a large number of flow conditions and coil geometries, it was found that vaporization rates are a function of three main variables. These variables are:

(1) Mass velocity of the liquid nitrogen inside the tubes;

(2) Excess temperature involved in the boiling process;

(3) Coiled-tube geometry, specifically the ratio of the inside tube diameter to the average coil diameter.

A correlation has been developed which accounts for these variables. This correlation has been shown to be accurate in predicting the flow boiling coefficient within 10% for all cases tested.

INTRODUCTION

Graham Manufacturing makes coiled-tube heat transfer equipment (Heliflow heat exchangers) which is frequently utilized as cryogenic vaporizers, with 80% of this service dedicated to liquid nitrogen. This coiled-tube design is based on spiral coils held together between two flat surfaces. One of these surfaces is the base plate; the other is the end of the shell (or casing). Bolted together, the plate and shell confine a closed spiral-shaped fluid circuit outside the coil. Figure 1 illustrates a Heliflow configuration.

Previously, sizing of these Heliflows for liquid nitrogen use has been accomplished by utilizing nitrogen boiling data established using tube side boiling in a test Heliflow, with steam introduced on the shell side as the heating medium. It has generally been accepted that the vaporization rates were solely a function of the mass velocity of the nitrogen. The validity of this approach has been questioned for some time, but the procedure has remained in use due to lack of further information on the subject.

A literature search was conducted in an attempt to (1) verify the accuracy of the available heat transfer data, and (2) establish a reliable analytical approach for predicting boiling heat transfer coefficients. This search has yielded a number of interesting facts.

First, the heat transfer rates used to evaluate the Heliflow coiled-tube heat exchangers compared favorably with empirical data presented by others for forced convection film boiling in similar applications. Depending on the nitrogen flow rate, values of 110 to 850 W/m²-K have been reported in the published literature (References 1, 5, 6, and 7).

Second, in almost every case reported in the literature, an attempt was made to establish a suitable analytical correlation for the prediction of the boiling coefficients. In all of the literature examined, none of these correlations have been shown to have much reliability for convective flow boiling in coiled tubes. Often the author would try to establish an empirical approach by modifying the classic Dittus-Boelter relationship, or variations thereof. This approach has been reported to be in error by factors of as much as 4 to 26 (Reference 7). Initially our testing and evaluation for boiling rates followed the same general approach, and, like others, our attempts at predicting heat transfer coefficients using the Dittus-Boelter equation yielded inaccurate results. This approach was abandoned, and a new approach was established based upon the wide range of boiling data that has been collected during extensive testing.

Finally, indications have been reported in the literature that nitrogen vaporization rates might be a function of several variables other than the mass velocity of the liquid nitrogen. These variables could include such factors as the temperature of the system and the geometry of the coiled-tube heat exchanger.

CONTROLLING FLOW INSTABILITIES

The phenomenon of unstable operation in cryogenic vaporizers is well documented. Numerous accounts of flow instabilities during vaporization are presented in reports published by cryogenic researchers (References 2, 3, and 4). The problem of surging when using the Heliflow coiled-tube heat exchanger as a cryogenic vaporizer has been prevalent since its inception. The surging experienced is generally a 70 to 350 kPa pressure fluctuation originating near the Heliflow liquid inlet point, and usually of a periodic nature with frequencies in the range of 0.2 to 1.0 cycles per second. It was essential to eliminate (or at least significantly reduce) this flow instability to maintain the high rate of heat transfer available with the Heliflow heat exchanger.

An extensive testing program ultimately resulted in solving the flow instability problem. Based on the results of this testing, it became apparent that the most effective method of controlling the flow instability in the Heliflow was to utilize an accumulator prior to entering the coiled tubes. Figure 3 shows the effects that using an accumulator had on a typical flow instability situation.

TEST APPARATUS

The equipment used for measuring the nitrogen vaporization rates is shown in Figure 2. For all cases run during this testing program, liquid nitrogen (LN₂) vaporization was accomplished using tube side boiling.

Depending on the tube wall temperature desired, heating was achieved by using steam, water, ethlyene glycol, or methanol on the shell side.

LN_2 is stored in, and delivered from, a 6.06 m³ tank. The quality of the LN_2 entering the vaporizer is controlled through the use of a subcooler. This subcooler can deliver 3 K of subcooling at .0126 kg/sec, and up to 8 K of subcooling at 0.1 kg/sec. The subcooler is used in this system to ensure that the LN_2 is not being introduced in a mixed-phase condition. The intent is to assure that all of the surface area of the tubes is utilized in "boiling" the LN_2, not in increasing the sensible heat of vapor already present. An accumulator is introduced in the liquid nitrogen supply line just prior to the Heliflow inlet to act as a surge suppressor, allowing the vaporizer to operate under steady state (non-oscillatory) conditions.

Cryogenic temperature measurements are made with "T" type thermocouples, with all remaining temperature measurements taken using conventional RTDs. All pressure measurements are taken using absolute pressure transducers, backed up by direct reading pressure gauges. Pressure transducers in the cryogenic regime are insulated from the cold fluids by the use of stainless steel pig-tails. The data collected is sent to an electronic data acquisition system (DAS) and is used to calculate heat duties, flow rates, and other performance parameters.

Steam, ethylene glycol, and methanol shell side flow rates are measured using differential pressure across flow orifices. This pressure information is used by the DAS to calculate flow rates. Shell side water flow rates are indicated by high accuracy rotameters and manually input into the DAS. Nitrogen flow rates are determined at a location downstream from the primary vaporizer. After vaporization inside the Heliflow tubes, the nitrogen enters a second heat exchanger to superheat the vapor prior to flow measurement across a conventional critical orifice. Pressure and temperature before the orifice, and pressure after the orifice, are input to the DAS to calculate a flow rate.

An array of Heliflows permitted a systematic evaluation of boiling rates based upon the combined geometrical effects of tube diameter, tube length, and average tube bundle (helix) diameter. Each Heliflow could be fitted with a clear-ended casing when the test being conducted used water or ethylene glycol, which allowed for a visual evaluation of the shell side flow. Of prime importance in the shell side flow is the flow velocity distribution, and the subsequent formation of ice in low velocity areas. For all cases, if icing is present, it can be kept to a negligible amount by varying flow velocity and/or temperature to allow for the effective rating of each exchanger.

TEST PROCEDURE

Measurement of the boiling rates was achieved using nitrogen mass flow rates ranging from 75 to 600 kg/m²-sec For all cases, shell side flow was established prior to introducing liquid nitrogen into the vaporizer. Nitrogen flow was gradually increased to allow for the system piping to cool down, and also to establish an insulating ice barrier on the outside of the piping and accumulator.

The flow rate of the nitrogen was limited to the region where boiling was guaranteed over the entire tube length of the specific model of Heliflow being tested. The lower limit of nitrogen flow was assumed to be at the point where the vapor exiting the exchanger was 11 K to 17 K higher than the saturation temperature for the given nitrogen exit pressure. The upper limit for the nitrogen flow rate was assumed to occur at the point where the shell side duty began to indicate less heat transfer than that which was required to vaporize the nitrogen.

In addition to the array of Heliflows tested, 14 different tube wall temperatures were examined during this investigation. The three coldest tube wall conditions were established by using methanol on the shell side at temperatures of 191 K, 205 K, and 222 K. Three intermediate tube wall conditions were established using ethlyene glycol on the shell side at temperatures of 239 K, 255 K, and 272 K, and four more using water on the shell side at temperatures of 289 K, 305 K, 333 K, and 361 K. Finally, four elevated tube wall conditions were established by condensing steam at atmospheric pressure (373 K), 239 kPa (398 K), 446 kPa (420 K), and 929 kPa (450 K). In general, due to the relatively low boiling rate of liquid nitrogen, which is in the range of 110 to 567 W/m²-K compared to the shell side rate of 5,670 to 11,340 W/m²-K, the inside tube wall temperatures were always approximately 5 K colder than the average shell side fluid temperature.

Several mass and energy balance checks were employed during the testing process to assure that the data taken was reasonable. The primary check was the energy balance between the tube side (liquid nitrogen) and the shell side (methanol, ethylene glycol, etc.). Whenever the imbalance reached 5%, the test was rejected, and then repeated. A secondary check was done to affirm that boiling occurred over the entire length of the Heliflow tubes. Again, in the event that this did not happen, the test was rejected, and then repeated. Finally, the clear-ended casing allowed visual inspection to assure that excessive ice was not forming on the outside tube surfaces.

TEST RESULTS

The results of the extensive testing program that was carried out involved working with many Heliflow models operating under various conditions. Without a clear idea of the form that the data was going to take, our initial effort was to attempt to relate the measured heat transfer rates to the mass flow of the nitrogen passing through the Heliflow tubes. This endeavor followed the generally accepted form of the Dittus-Boelter equation, where the heat transfer coefficient "h" is a function of the flow characteristics through the tube, as well as the properties of the fluid. As has already been described, the actual data collected during this testing program did not correspond to the above approach. Figure 4 is a typical data curve which would be measured when testing any of the Heliflows. This particular Heliflow was fabricated with copper tubes, and had a tube wall temperature of 358 K.

The most obvious feature of this data curve is that it is a straight line, which means that the preferred form of the correlation should be:

$$N_{NU} = C1 \times N_{RE} + C2$$

Further testing with other Heliflow models at varying operating conditions revealed that the correlation was not quite so simple. Ultimately it was discovered that the liquid nitrogen boiling rates were a function of three main variables. These variables are:

(1) Mass velocity (Reynolds number) of the liquid nitrogen inside the tubes;

(2) Tube wall temperature;

(3) Heliflow geometry - specifically the ratio of the inside tube diameter to the average helix diameter.

The variables described above effect the heat transfer characteristics of the Heliflow in the following manner:

Mass Velocity

Boiling heat transfer coefficients increase as the mass velocity of the liquid nitrogen through the tubes increases. This increase in the boiling coefficient is essentially a linear function of the mass velocity. Figure 4 illustrates this linear relationship, where "h" is calculated from an energy balance across the exchanger based on measured values for mass flow and temperatures of each stream.

Tube Wall Temperature

Boiling heat transfer coefficients decrease as the tube wall temperature increases. This decrease in the boiling coefficient is an exponentially decreasing function of the tube wall temperature. This is shown in Figure 5 for a

Heliflow having a constant mass velocity of 390 kg/m²-sec (Note: The data presented in Figure 5 represents a composite of many tests, not a single test.) One possible explanation for this phenomenon is that an insulating vapor blanket forms at the tube wall around a liquid core at the center. This vapor blanket would become thicker as the tube wall temperature increases. The result would be an increase in thermal resistance to heat transfer through the fluid. (Due to the fact that the majority of the tube surface experiences vapor blanketing, this factor is significant.)

Heliflow Geometry

Boiling heat transfer coefficients exponentially increase as the ratio of the inside tube diameter to the average helix diameter (A/R) increases. This phenomenon is shown in Figure 6 for several different coiled-tube configurations. One reason for this enhancement of heat transfer may be due to centrifuging of the liquid core through the vapor blanket toward the tube wall as the fluid continuously changes direction in the spiral bundle. When the liquid core approaches the tube wall, this increases the tendency for nucleate boiling to occur, which would greatly enhance the heat transfer rates. We would expect that increasing the A/R ratio would increase the centrifugal force experienced by the fluid for a given mass velocity, and correspondingly improve the heat transfer capabilities.

In order to account for the fact that the tube wall temperature and the Heliflow geometry affect the overall heat transfer characteristics, it was necessary to introduce modifications to our basic approach. Basically, the slope of the linear ratio between the Nusselt Number and the Reynolds Number changes as the temperature and geometry change.

The correlation described in this paper takes the form:

$$N_{NU} = "M" \times N_{RE}$$

where the Reynolds Number is evaluated using vapor properties associated with the film temperature {(Tsat + Twall) / 2} and average pressure over the boiling region. The constant "M" can be obtained by using the graph in Figure 7. This graph requires knowledge of the coiled-tube A/R ratio and the inside tube wall temperature (assume 5K below the shell side average temperature). This correlation has been shown to be accurate to within 10% for all cases tested at 414 kPa. (Limited testing has been performed at an elevated pressure of 1100 kPa. At this higher pressure there is a decrease in the heat transfer coefficient of approximately 8%, but the correlation remains usable.)

Limitations in this study include a relatively fixed tube side boiling pressure (414 kPa), and also tube wall boiling temperatures ranging between 183 K and 433 K. Rating information for a tube wall temperature of 144 K is presented, but only as an extrapolation of known performance.

Example Calculation

Given: A Heliflow is to be used to vaporize 180 kg/hr of liquid nitrogen on the tube side. The heating medium is water at an average temperature of 335 K.

Step 1: Determine the A/R factor based on coil geometry. For this example, use a Heliflow having 4 tubes with an inside diameter of 0.0102 m (18 BWG) and an A/R of 0.068.

Step 2: Assume that the inside tube wall temperature is 5 K below the shell side average temperature. For this example, use an inside tube wall temperature of (335 K - 5 K) = 330 K.

Step 3: Determine "M" from Figure 7. M = 9.1E-4.

Step 4: Use nitrogen vapor properties evaluated at Tfilm = (Tsat + Twall) / 2. Tfilm = (92 + 330)/2 = 211 K. This gives a viscosity of 1.325E-5 kg/m-sec, and a thermal conductivity of 0.019 W/m-K.

Step 5: N_{RE} = Mass velocity * Tube ID / viscosity
= 153 kg/m²-sec * 0.0102 m / 1.325E-5 kg/m-sec
= 1.18E5

Step 6: N_{NU} = 9.1E-4 * 1.18E5 = 107.4

Step 7: h = N_{NU} * thermal conductivity / Tube ID
= 107.4 * 0.019 W/m-K / 0.0102 m
= 199.4 W/m²-K

CONCLUSIONS

An extensive testing program has been carried out with the purpose of establishing a method for rating coiled-tube heat exchangers for service as nitrogen vaporizers. Liquid nitrogen boiling coefficients have been measured for a number of Heliflow coiled-tube heat exchangers. As a result of this testing, an analytical formulation for predicting boiling heat transfer rates in coiled-tube heat exchangers has been developed. This technique accounts for the mass velocity of the nitrogen flowing in the tubes, the geometrical considerations of each style of coiled-tube heat exchanger, and the effect of the tube wall temperature. This correlation has been shown to be accurate in predicting the flow boiling coefficient within 10% for all cases tested.

NOMENCLATURE

C1, C2, M	constants determined by measurement
N_{RE}	Reynolds Number
N_{NU}	Nusselt Number
Tsat	saturation temperature, K

REFERENCES

1. Flanigan, V.J., and Park, E.L., 1971, "A Study of Film Boiling of Corresponding-States Fluids Over a Wide Pressure Range With Cylindrical Heaters," *Advances in Cryogenic Engineering*, Vol. 16, pp 402-411.

2. Friedly, J.C., Manganaro, J.L., and Kroeger, P.G., 1969, "Approximate Criterion for Prediction of Flow Oscillations in Supercritical Fluid Heat Exchangers," *Advances in Cryogenic Engineering*, Vol. 14, pp 258-270.

3. Park, E.L., Colver, C.P., and Sliepcevich, C.M., 1966, "Nucleate and Film Boiling Heat Transfer to Nitrogen and Methane at Elevated Pressures and Large Temperature Differences," *Advances in Cryogenic Engineering*, Vol. 11, pp 516-529.

4. Rogers, J.D., 1968, "Oscillations in Flowing and Subcritical Hydrogen," *Advances in Cryogenic Engineering*, Vol. 13, pp 223-231.

5. Thurston, R.S., and Rogers, J.D., 1967, "Pressure Oscillations Induced by Forced Convection Heating of Dense Hydrogen," *Advances in Cryogenic Engineering*, Vol. 12, pp 438-451.

6. Timmerhaus, K.D., and Schoenhals, R.J., 1974, "Design and Selection of Cryogenic Heat Exchangers," *Advances in Cryogenic Engineering*, Vol. 19, pp 445-462.

7. Williamson, K.D., and Bartlit, J.R., 1965, "Experimental Study of H$_2$O-LH$_2$ and H$_2$O-LN$_2$ Heat Exchangers," *Advances in Cryogenic Engineering*, Vol. 10, pp 375-381.

Fig. 1 Heliflow heat exchanger

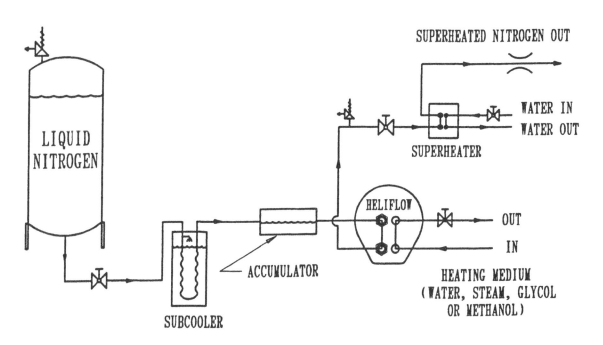

Fig. 2 Flow schematic of LN$_2$ vaporization test apparatus

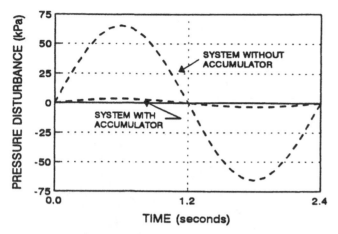

Fig. 3 Effect of accumulator on flow boiling instability

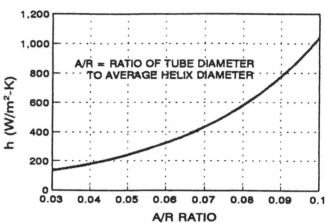

Fig. 6 Effect of coiled-tube geometry on the boiling coefficient

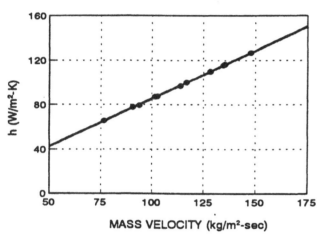

Fig. 4 Effect of mass velocity on the boiling coefficient

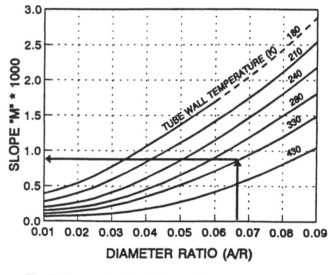

Fig. 7 Determination of Reynolds coefficient "M"

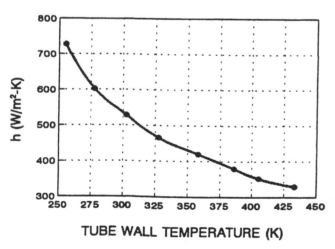

Fig. 5 Effect of tube wall temperature on the boiling coefficient

FORCED CONVECTIVE FLOW BOILING HEAT TRANSFER OF PURE REFRIGERANTS INSIDE A HORIZONTAL MICROFIN TUBE

Sh. KOYAMA*, J. YU*, S. MOMOKI**,T. FUJII*** and H. HONDA*
* Institute of Advanced Material Study, Kyushu University, Kasuga, Japan
** Faculty of Engineering, Nagasaki University, Nagasaki, Japan
*** Graduate School, University of East Asia, Shimonoseki, Japan

ABSTRACT

An experimental study is reported on the forced convective flow boiling of pure refrigerants HCFC22, HFC134a and HCFC123 inside a horizontal microfin tube. The local heat transfer coefficient defined based on the actual inside surface area is measured in the ranges of mass velocity of 200 to 400 kg/(m²s), heat flux of 5 to 64 kW/m² and reduced pressure of 0.07 to 0.24. Using the Chen-type model, a new correlation equation is proposed considering the enhancement effect of microfins on both the convective heat transfer and the nucleate boiling components. In the convective heat transfer component, the correlation to predict the heat transfer coefficient of liquid-only flow is determined from preliminary experiments on single-phase flow in microfin tubes, and the two-phase flow enhancement factor is determined from the present experimental data. For the nucleate boiling component, the correlation of Takamatsu et al. for smooth tubes is modified. The proposed equation correlates well the present experimental data, and is available for several microfin tubes which were tested by other researchers.

INTRODUCTION

Recently, different kinds of microfin tubes with high heat transfer performance and relatively low flow resistance are widely used in compact heat exchangers. The flow boiling in microfin tubes has been experimentally studied by many investigators. Among of them, Ito-Kimura (1979), Khanpara et al. (1987) and Kido et al. (1994) tested many kinds of microfin tubes and obtained a lot of first-hand heat transfer data; Yoshida et al. (1988) and Fujii et al. (1993) observed directly the flow pattern in microfin tubes; Miyara et al. (1988), Kandlikar et al. (1991), Murata et al. (1993) and Kido et al. (1994) proposed correlation equations for the flow boiling heat transfer coefficient in microfin tubes, respectively, but these proposed equations are only valid for their test ranges.

In the present study, three kinds of pure refrigerants HFC134a, HCFC123 and HCFC22 are used as working fluid to investigate the characteristics of flow boiling heat transfer in a microfin tube. Based on the present experimental data, a new correlation equation is proposed for heat transfer coefficient. The comparison between this equation and previous experimental data for other kinds of microfin tubes is also carried out.

EXPERIMENTAL METHOD

The experimental apparatus is shown in Fig.1. It consists of three main loops: a refrigerant loop, a water loop and a brine loop. In the refrigerant loop, the subcooled liquid refrigerant is delivered with a positive-displacement pump (1) through a desiccant filter (2), a mass flow meter (3), an electrical preheater (4), a mixing chamber (5) and a heat exchanger (6) to the test section (10) or (11). The refrigerant vapor generated in the test section condenses in two heat exchangers (14) and (15) to complete the cycle.

The test section shown in Fig.2 is a horizontal double-tube heat exchanger in which the refrigerant flows inside an inner tube and the heating water flows in an annulus. Depending on the requirement of the experiments, the flow condition in the double-tube heat exchanger can be set to a counter flow or parallel flow by means of valves in the refrigerant loop. In the test section, the inner tube is a 6 m long straight microfin copper tube. The outer tube is formed by two polycarbonate resin blocks each of which has a half-round ditch with a radius of 8 mm; so the dimension of the annular gap for water flow is 3 mm. To measure the local heat transfer rate, the annulus is divided into 14 subsections along the tube axis by brass blocks. The heat transfer rate in each subsection is treated as a local value.

Table 1 lists the geometrical parameters of the microfin tube. The microfin tube is 10.00 mm in outer diameter and 8.37 mm in mean inner diameter with which the cross-sectional area of a smooth tube equals to that of the microfin tube.

The heat flux q based on the real inside area of the tube that includes the base and the fin areas is defined as

$$q = Q/(\xi \pi d_i \Delta L) \qquad (1)$$

where d_i is the mean inner diameter of the tube, and ξ is the ratio of the real inside surface area of the microfin tube to the inside surface area of a smooth tube with diameter d_i. ΔL is the effective heating length of each subsection which is 210 mm for No.1, 2 ,13 and 14 subsections and 460 mm for other subsections. The heat transfer coefficient α is defined as

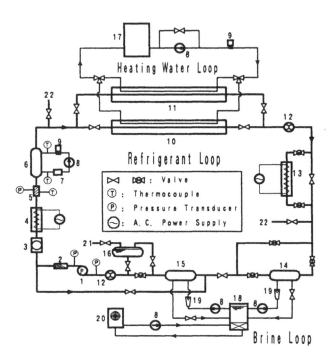

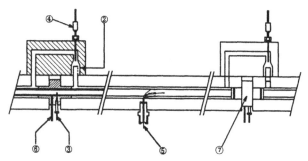

1	Positive displacement pump	12	Sight glass
2	Desiccant filter	13	After-heater
3	Mass flow meter	14	Condenser 1
4	Preheater	15	Condenser 2
5	Mixing chamber	16	Liquid reservoir
6	Heat exchanger	17	Heat source tank
7	Heat sink tank	18	Brine tank
8	Centrifugal pump	19	Float-type flow meter
9	Gear-type flow meter	20	Chilling unit
10	Test section(Smooth tube)	21	Charging port
11	Test section(Microfin tube)	22	Sampling port

Figure 1 Experimental apparatus

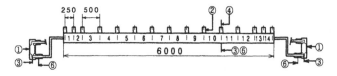

(a) Schema of test evaporator

(b) Detail of subsection

1	Mixing chamber (Refrigerant)	4	Resistant thermometer (Heating water)
2	Mixing chamber (Heating water)	5	Thermocouple (Tube surface)
3	Thermocouple (Refrigerant)	6	Pressure measuring port

Figure 2 Test evaporator

Table 1 Dimensions of test tube

		Unit	Microfin
Outside diameter	d_0	[mm]	10.00
Mean i.d.	d_i	[mm]	8.37
Root i.d.	d_r	[mm]	8.475
Tip i.d.	d_t	[mm]	8.139
Fin height	h_t	[mm]	0.168
Fin pitch	p	[mm]	0.44
No. of fins	n	[-]	60
Helix angle	γ	[deg]	18
Tip width	s	[mm]	0.073
Valley width	w	[mm]	0.245
Tip geometry			flat
Valley geometry			flat
Area ratio	ξ		1.52

$$\alpha = q/(T_{Wi} - T_{sat}) \tag{2}$$

where T_{Wi} and T_{sat} are the wall and the saturated refrigerant temperatures, respectively.

The variable ranges in the present experiments under the counter flow and parallel flow conditions are as follows; mass velocity: 200 - 400 kg/(m² s), heat flux: 5 - 64 kW/m² and reduced pressure: 0.07-0.24.

EXPERIMENTAL RESULTS
Single-phase Flow Heat Transfer

Figure 3 shows the present experimental results on heat transfer in the microfin tube together with previous experiments of Koyama et al. (1992) and Khanpara et al. (1987). The correlations of Koyama et al., Dittus-Boelter and Gnielinski with Filonenko's friction factor equation are also drawn in this figure. The correlation of Koyama et al., in which the Nu number is based on the nominal area, $\pi d_i \Delta L$, is transformed with the real heat transfer area as:

$$Nu = 0.028 Re^{0.8} Pr^{0.4} \tag{3}$$

In the range of Re < 10⁴, most of experimental results of the present study, Koyama et al. and Khanpara et al. are lower than the Gnielinski equation, which is caused by the definition of Nu number. It was confirmed by Koyama et al. that these data agreed with the Gnielinski equation when the nominal area was used in the definition of Nu number. For Re ≥ 10⁴, most of the experimental data are about 20% higher than the Dittus-Boelter equation and agree well with Eq.(3). This equation will be applied in the correlation of flow boiling heat transfer coefficients.

Two-phase Flow Heat Transfer

Figure 4(a) shows an example of experimental results of HFC134a under counterflow condition, where Z is the axial distance measured from the inlet of refrigerant, T_{RM} is the measured temperature of refrigerant, T_{RC} is the calculated bulk temperature of refrigerant, T_{wi} is the inner wall temperature, T_S is the temperature of heating water, q is the heat flux, and x is the vapor quality. In the refrigerant flow direction, T_{RM} increases until it researches saturated temperature, then it decreases gradually due to the pressure drop. In the saturated region, T_{RM} is in good agreement with T_{RC}. This indicates that the measured refrigerant pressure is very reliable. As the quality x exceeds 0.90, T_{Wi} increases sharply, which indicates that the dryout occurs in the tube. The temperature difference between T_{RM} and T_S increases with the progression of boiling, and so does the heat flux q before

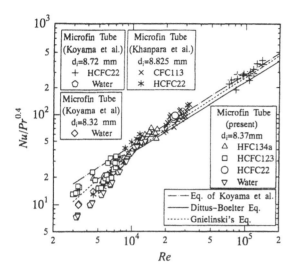

Figure 3   Single-phase heat transfer

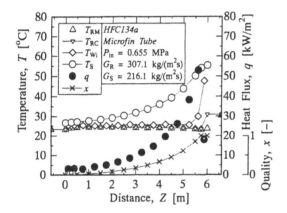

(a) Axial distributions of T, x and q

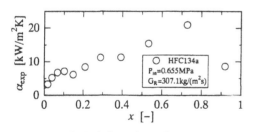

(b) Relation of x and α_{exp}

Figure 4   An example of experimental results

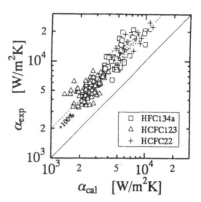

Figure 5 Comparison of present data with a correlation for smooth tubes

transfer coefficient is defined based on the real inside surface area and the maximum inner diameter of microfin tube. Kido et al. used the similar definition to Kandlikar's except the mean inner diameter d_i. Miyara et al. and Murata-Hashizume correlated their experimental data based on a mean diameter and a nominal inside surface area. It is found that for the data of HFC134a and HCFC22 with high heat flux Kandlikar's correlation gives a relatively good agreement, while for the HCFC123 data and some data of HFC134a and HCFC22 with low heat flux his correlation overpredicts. The reason is that in his correlation the convective heat transfer component, which is dominant at low heat flux condition, is overestimated. On the other hand the values predicted by the correlation of Kido et al. are lower than most of the present data. This reason is that in their correlation the nucleate boiling component may be underpredicted. Comparison between the correlation of Miyara et al. for HCFC22 and the present data shows that the prediction is in good agreement with the data of HFC134a and HCFC123, but is lower in some degree for the data of HCFC22 in high heat flux region. The Murata-Hashizume correlation agrees with only some HCFC123 data, but underpredicts for the data of HFC134a and HCFC22. This reason is that their correlation was developed only using the experimental data of HCFC123 at a pressure of 0.2 MPa; at this pressure the nucleate boiling contribution of HCFC123 is very weak.

DEVELOPMENT OF A CORRELATION FOR FLOW BOILING HEAT TRANSFER IN A MICROFIN TUBE

In the present study, Chen-type equation(1966) is adopted to correlate the present experimental data. The total heat flux q is contributed from the convective component q_{cv} and the nucleate boiling component q_{nb} as follows:

$$q = q_{cv} + q_{nb} \qquad (4)$$

The heat transfer coefficient α is given by

$$\alpha = \alpha_{cv} + \alpha_{nb} \qquad (5)$$

where $\alpha_{cv} = q_{cv}/(T_{Wi} - T_{sat})$ and $\alpha_{nb} = q_{nb}/(T_{Wi} - T_{sat})$.

Determination of Forced Convective Heat Transfer Coefficient

According to the results in single-phase flow heat transfer experiments, the heat transfer coefficient for liquid-only flow is calculated with the following equation.

$$\alpha_{lo} = 0.028 \, Re_{lo}^{0.8} \, Pr^{0.4} \lambda_l / d_i \qquad (6)$$

Figure 7 shows the relation of α / α_{lo} and $1/\chi_{tt}$. In general,

the dryout point. As reference, the relation between α_{exp} and x for this experiment is shown in Fig. 4(b).

Figure 5 shows the comparison between the present experimental data and the correlation for smooth tubes proposed by Yu (1995). In this figure the heat transfer coefficients of microfin tube are about 30 - 150% higher than those of smooth tubes. The averaged enhancement ratio is about 2.0.

Figures 6 (a), (b), (c) and (d) show the comparison between the present data and the correlations of Kandlikar (1991), Kido et al. (1994), Miyara et al. (1988) and Murata-Hashizume (1993) for microfin tubes, respectively. In Kandlikar's correlation the heat

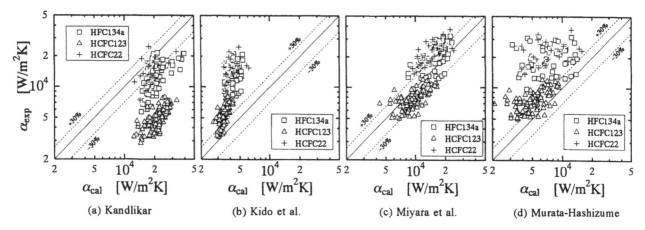

(a) Kandlikar   (b) Kido et al.   (c) Miyara et al.   (d) Murata-Hashizume

Figure 6 Comparison of present data with several previous correlations

the convective component is determined with this relation. It is found that the values of α/α_{lo} in the microfin tube are higher than the line for smooth tube obtained by Yoshida et al. (1990). It is also found that for one value of $1/\chi_{tt}$ the value of α/α_{lo} scatters in a range, that is related to the difference of nucleate boiling contribution. As the nucleate boiling component is fully suppressed, the total heat transfer coefficient α approaches to convective heat transfer component α_{cv}. Therefore, the data of which the nucleate boiling component is considered to be fully suppressed are used to determine the two-phase convection multiplier factor F in the microfin tube. The factor F is expressed as:

$$F \doteq \alpha/\alpha_{lo} = 1 + 2/\chi_{tt}^{0.88} + 0.8/\chi_{tt}^{1.03} \qquad (7)$$

Determination of Nucleate Boiling Heat Transfer Coefficient

In the microfin tube the fin height is usually lower than 0.3 mm, so the fin efficiency is near unit. Therefore, all of the inside surface area is effective for flow boiling heat transfer, that is, the fin surface acts as the same as the base surface in nucleate boiling condition. As a result, the microfin surface can be treated as a smooth surface with the area which equals to that of the microfin tube. In the present study, the nucleate boiling heat transfer coefficient α_{nb} is correlated based on the model of Takamatsu et al. for smooth tubes (1993). So the present proposed correlation is as follows:

$$\alpha = \alpha_{cv} + \alpha_{nb} \qquad (8)$$

$$\alpha_{cv} = 0.028\, Re_{tp}^{0.8}\, Pr_l^{0.4}\, \lambda_l/d_i \qquad (9)$$

$$Re_{tp} = F^{1/0.8}\, Re_{lo} \qquad (10)$$

$$Re_{lo} = G\,(1-x)\,d_i/\mu_l \qquad (11)$$

$$F = 1 + 2/\chi_{tt}^{0.88} + 0.8/\chi_{tt}^{1.03} \qquad (12)$$

$$\alpha_{nb} = K^{0.745}\, S\, \alpha_{pb} \qquad (13)$$

$$K^{0.745} = \frac{1}{1 + 0.875\eta + 0.158\eta^2 - 0.159\eta^3 + 0.7907\eta^4} \qquad (14)$$

$$\eta = \alpha_{cv}/(S\,\alpha_{pb}) \qquad (15)$$

$$S = (1 - e^{-\xi})/\xi \qquad (16)$$

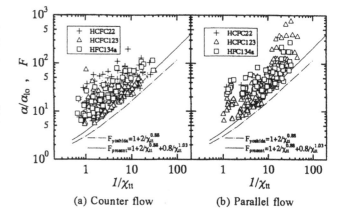

(a) Counter flow   (b) Parallel flow

Figure 7 Relation of α/α_{lo} and χ_{tt}

$$\xi = D_b\, \alpha_{cv}/\lambda_l \qquad (17)$$

$$D_b = C_1 \left(\frac{\rho_l\, Cp_l\, T_{sat}}{\rho_v\, h_{fg}} \right)^{1.25} \left\{ \frac{2\sigma}{g\,(\rho_l - \rho_v)} \right\}^{0.5} \qquad (18)$$

$$\alpha_{pb} = C_2 \times 207\, \frac{\lambda_l}{D_{be}} \left(\frac{q\, D_{be}}{\lambda_l\, T_{sat}} \right)^{0.745} \left(\frac{\rho_v}{\rho_l} \right)^{0.581} Pr_l^{0.533} \qquad (19)$$

$$D_{be} = 0.51\,\{2\sigma/[g\,(\rho_l - \rho_v)]\}^{0.5} \qquad (20)$$

As no correlation is available for the pool boiling heat transfer on the present microfin surface, the nucleate boiling component of smooth tube is modified for microfin tube based on the following consideration. Many pool boiling experimental results proved that the bulk superheat of initial boiling point on a finned surface is lower than that on a smooth surface. It is inferred from this fact that the local superheat near the base surface of microfin tube would be higher due to the effect of microfins, even though the bulk superheat is lower than that of smooth surface. For the same bulk superheat the local superheat adjacent to the wall surface in the microfin tube is higher than that in the smooth tube, so that the departure bubble size in the microfin tube is smaller than that in the smooth tube according to the statistic equilibrium theory and the Clausius-Clapeyron equation. On the other hand, in experimental results of the pool boiling on a low-fin surface conducted by Zhao et al. (1984), it was found that the heat transfer

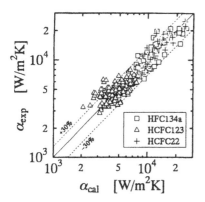

Figure 8 Comparison of the present experimental data with
the present correlation

coefficients based on the total surface area were higher than those of the smooth surface in the developing boiling region. This implies that the fin effect may cause additional enhancement to the pool boiling heat transfer besides the increase of surface area.

Based on the above consideration, the effects of microfin on the nucleate boiling are focused on two points. One is the size of active nucleate cavity decreased, and the other is the enhancement of nucleate boiling. These will lead the decrease of constant C_1 in Eq.(18) and the increase of constant C_2 in Eq.(19) compared with the values for smooth tube. Through a series of trial calculation the optimum values of C_1 and C_2 were determined as 1.0×10^{-5} and 2.8, respectively.

Comparison with Experimental Results

The comparison between the present experimental data and the present correlation is shown in Fig. 8. The mean and average deviation are 11.9% and -0.8%, respectively.

The experimental data of Kido et al. (1994), Khanpara et al. (1987), Miyara et al. (1988) and Fujii et al. (1993) are compared with the present correlation, as shown in Fig. 9. The mean and average deviations of the present comparisons for total experimental data are 18.0% and -3.8%, respectively.

CONCLUSIONS

1. The flow boiling heat transfer coefficients in the microfin tube reach about 2 times of those in a smooth tube at the same conditions. This enhancement of heat transfer is due to the increase of both convective component and nucleate boiling component.

2. It is found that the present experimental data can not be correlated well by previous correlations. Considering the effects of the differences between microfin surface and smooth surface, and modifying the correlation for smooth tube, a new correlation is proposed for the flow boiling heat transfer in a microfin tube based on the present experimental results in annular flow region. The mean deviation and mean absolute deviation between the present correlation and the present data are -0.8% and 11.9%, respectively.

3. It is confirmed that the present correlation is available for the different types of microfin tubes with different kinds of refrigerants.

ACKNOWLEDGEMENTS

The authors would like to thank Associate Professor H. Takamatsu of Kyushu University for his valuable advice in the present study. The authors also wish to appreciate Professor H. Uehara of Saga University and Dr. O. Kido of Matsushita Refrigeration Co. for their offering the unpublished data.

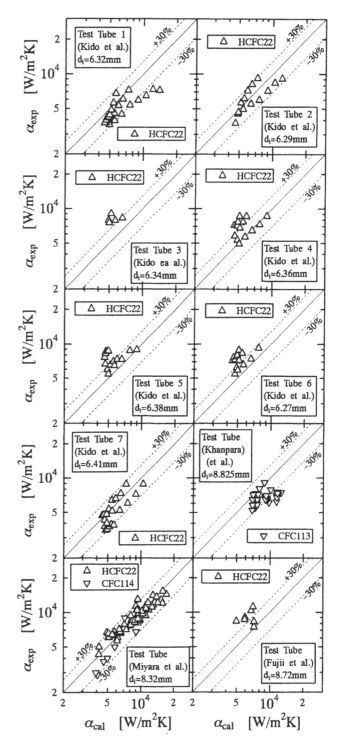

Figure 9 Comparison of the present correlation with several
previous experimental data

NOMENCLATURE

C_1, C_2	constants
Cp	isobaric specific heat [J/(kg K)]
D_b	bubble departure diameter [m]
D_{be}	equilibrium break-off diameter [m]
d_i	mean inner diameter [m]
F	two-phase convection multiplier factor
G	mass velocity [kg/(m^2 s)]
g	gravitational acceleration [m/s^2]
h_{fg}	latent heat of evaporation [J/kg]
K	heat-flux-fraction factor, q_{nb}/q
Nu	Nusselt number, $\alpha d_i / \lambda$
P	pressure [MPa]
Pr	Prandtl number, $\mu Cp/ \lambda$
Q	total heat transfer rate in a subsection [W]
q	heat flux [W/m^2]
r	departure bubble radius [m]
Re	Reynolds number, $G d_i / \mu$
Re_{lo}	Reynolds number of liquid-only flow, $G(1-x) d_i / \mu_l$
S	suppression factor
T	temperature [K]
v	specific volume, [m^3/kg]
x	quality

Greek Symbols

α	heat transfer coefficient [W/(m^2 K)]
ΔL	length of a subsection [m]
ΔT	local superheat [K]
λ	thermal conductivity [W/(m K)]
μ	dynamic viscosity [Pa s]
ξ	ratio of real inside surface area of a microfin tube to inside surface area of a smooth tube with diameter d_i
ρ	density [kg/m^3]
σ	surface tension [N/m]
χ_{tt}	Martinelli parameter, $\{(1-x)/x\}^{0.9}(\rho_v / \rho_l)^{0.5}(\mu_l / \mu_v)^{0.1}$

Subscripts

cal	calculated
cv	convective
exp	experimental
l	liquid
lo	liquid only
nb	nucleate boiling
pb	pool boiling
sat	saturated
tp	two-phase
v	vapor
Wi	inside surface of tube wall

REFERENCE

Chen, J.C., 1966, "Correlation for Boiling Heat Transfer to Saturated Fluids in Convective Flow," I&EC Process Design and Development, Vol.5, No.3, pp.322-329.

Ito, M., and Kimura, H., 1979, "Boiling Heat Transfer and Pressure Drop in Internal Spiral-Grooved Tubes,"(in Japanese) Trans. Jpn. Soc. Mech. Eng., Vol.45, No.389, pp.118-126.

Fujii, T., Koyama, S., Inoue, N., Kuwahara, K., and Hirakuni, S., 1993, "An Experimental Study of Evaporation Heat Transfer of Refrigerant HCFC22 inside an Internally Grooved Horizontal Tube,"(in Japanese) Trans. Jpn. Soc. Mech. Eng., Vol.59, No.562, pp.261-268.

Kandlikar, S.G., 1991, "A Model for Correlating Flow Boiling Heat Transfer in Augmented Tubes and Compact Evaporators," ASME Journal of Heat Transfer, Vol.113, pp.966-972.

Khanpara, J.C., Pate, M.B., and Bergles, A.E., 1987, "Local Evaporation Heat Transfer in a Smooth Tube and a Micro-fin Tube Using Refrigerants 22 and 113," ASME Trans. HTD., No.85, pp.31-39.

Kido, O., and Uehara, H., 1994, "Correlation for Evaporation Heat Transfer of Pure Refrigerant Inside an Internally Grooved Horizontal Tube,"(in Japanese) Trans of the JAR, Vol.11, No.2, pp.143-153.

Koyama, S., Inoue, N., Hirakuni, S., Kuwahara, K., and Fujii, T., 1992, "Heat Transfer and Pressure Drop of Single Phase Flow Inside Internally Grooved Tubes,"(in Japanese) Report of Institute of Advanced Material Study, Kyushu University, Vol.6, No.2, pp.79-86.

Miyara, A., Takamatsu, H., Koyama, S., Yonemoto, K., and Fujii,T., 1988, "Forced Convective Boiling of Nonazeotropic Refrigerant Mixtures of R22 and R114 inside a Horizontal Tube,"(in Japanese) Trans. Jpn. Soc. Mech. Eng., Vol.54, No.5505, pp.2523-2528.

Murata, K., and Hashizume, K., 1993, "An Investigation on Forced Convective Boiling of Nonazeotropic Refrigerant Mixtures," ASME Journal of Heat Transfer, Vol.115, pp.680-689.

Takamatsu, H, Momoki, S. and Fujii, T., 1993, "A Correlation for Forced Convective Boiling Heat Transfer of Pure Refrigerants in a Horizontal Smooth Tube ," Int. J . Heat Mass Transfer, Vol.36, No.14, pp.3351-3360.

Yoshida, S., Matsunaga, T., Hong, H.P., and Nishikawa, K., 1988, "Heat Transfer Enhancement in Horizontal Spirally Grooved Evaporator Tubes," Bulletin of JSME, Vol.31, No.3, pp.505-512.

Yoshida, S., Mori, H., Hong, H., Matsunaga, T. and Mataki, T., 1990, "Correlation for Heat Transfer of Refrigerants Evaporating in a Horizontal Smooth Tube," (in Japanese) Proc. of 27th National Heat Transfer Symp. of Japan, Vol.2, pp.607-609.

Yu, J., 1995, "Experimental Study on Flow Boiling of Pure and Mixed Refrigerants in Smooth and Microfin Tubes," Ph.D. Thesis, Kyushu University, Japan

Zhao, L.Q., Hirano, Y., Shimada, R., Kumagai, S. and Takeyama, T., 1984, "On the Pool Boiling of Round Surface with Fin Array," (in Japanese) Proc. of 21st National Heat Transfer Symp. of Japan, pp.397-399.

HEAT TRANSFER AND PRESSURE DROP CHARACTERISTICS OF FORCED CONVECTIVE EVAPORATION IN DEEP SPIRALLY FLUTED TUBING

Scott M. MacBain and Arthur E. Bergles
Department of Mechanical Engineering,
Aeronautical Engineering, and Mechanics
Rensselaer Polytechnic Institute
Troy, New York

ABSTRACT

The local heat transfer coefficient and pressure gradient have been studied for forced convective evaporation of R-12 in a horizontal, deep spirally fluted tube. Effects of local quality, mass flux, and heat flux were investigated. The reduced data clearly show nucleate boiling and bulk convective regimes at low and high qualities, respectively. For a performance evaluation, the data were compared to the Kandlikar (1989) correlation for a smooth tube with a diameter equivalent to the envelope diameter of the Turbotec tube. The results showed that the Turbotec tube has a heat transfer enhancement ranging from 1.8 to 2.7 in the nucleate regime, and a more substantial increase in the convective regime, with enhancements there ranging from 3.3 to 7.8. For a given mass flow rate and heat input, the pressure drop was found to be from 6 to 20 times as great as a smooth tube of equivalent envelope diameter and heated length.

INTRODUCTION

Research related to heat transfer enhancement continues to be an expanding field due to the ongoing interest in increasing the efficiency of heat exchanger systems. Various enhancement methods have been studied for single-phase, condensation, and boiling heat transfer. In particular, some of the enhancement techniques that have been studied for flow boiling include tubes with integral fins, twisted-tape inserts, corrugated tubes, and fluted tubes. The book by Thome (1990) discusses, in great detail, select data from the literature on several enhancements for flow boiling in tubes. In the present study, the technique to augment heat transfer is through the use of deep spiral flutes that introduce swirl into the flow, provide additional area, and present a roughness to the flow.

The results of this experiment are important for the design of coaxial water chillers that utilize this heat transfer augmentation technique. In these water chillers, with refrigerant evaporating on the tubeside and chilled water flowing in the annulus, the tubeside boiling heat transfer coefficient can be the controlling resistance for heat transfer. For this reason, data on the refrigerant side heat

transfer coefficient and pressure drop are necessary for accurate design of these heat-exchange systems.

A survey of the literature reveals that there have been several studies that investigate the single-phase enhancement of these tubes. Marto et al. (1979) performed experiments to determine the effects of tube geometry on heat transfer and pressure-drop. Bergles (1980) presented a comprehensive overview of single-phase heat transfer and pressure drop data available at the time, and more recently, Srinivasan et al. (1994), and Srinivasan and Christensen (1992) developed models for the single-phase behavior of these tubes. However, there has been only limited research into the two-phase characteristics of this tubing. Sami and Schnotale (1992), and Sami et al. (1994) have obtained data on the average heat transfer coefficient for pure R-22, pure R-134a, as well as non-azeotropic binary and ternary refrigerant blends, using a single horizontal spirally fluted tube. This research was performed using a coaxial evaporator with refrigerant flowing in the inner tube and water flowing in the annulus. These investigators studied the effect of mass flow rate, but did not determine the effect of the heat flux or local quality.

The scarcity of data on two-phase flow behavior of these tubes indicates a need for further study. The present investigation is a preliminary study that looks into the effects of local quality, mass flux, and heat flux on the evaporation heat transfer and pressure drop performance of this tube. Pure R-12 was used as the working fluid, due to the wealth of data available for comparison, and because the behavior of all pure refrigerants should be similar.

APPARATUS AND PROCEDURE

In this study, heat transfer coefficient and pressure drop were determined for a deep spirally fluted tube with R-12 as the operating fluid. The copper test section was 0.92 m long, had a bore diameter of 7.11 mm, an envelope diameter of 12.5 mm, a flute pitch of 7.94 mm and a wall thickness of 0.64 mm. (See Figure 1 for definition of the reference diameters.) The tube was heated by a 100 kW, 0-2500 A, direct-current power supply. The use of electric heating permits the determination of local heat transfer coefficients, in contrast to fluid heating, where average coefficients are typically obtained. The flow loop (Figure 2) consisted of a 5-HP reciprocating compressor (1), water cooled condenser (4), oil separators (3), manual and automatic bypass valves (24-26), subcooler (7), preheater (15), and test-section assembly (16). The test section thus simulated the evaporator in an air-conditioning system.

The basic spirally fluted tube data presented in this study are based on the bore diameter, because it more closely represents the flow area than does the envelope. To allow comparison with smooth tubes, a smooth tube was chosen with a diameter equal to the envelope diameter of the Turbotec tube. This contrasts the Turbotec tube and a smooth tube that it would most likely replace in an existing heat exchanger. This is the case because they have the same maximum outside dimension and could be directly substituted in coiled tubes or tube sheets.

The Kandlikar correlation (1989) was chosen to predict the performance of the smooth tube, because it included R-12 data in the correlation and it is applicable to horizontal tubes. Data presented for the smooth tube using this correlation have the same mass flow rate, heat input, and heated length as data presented for the spirally fluted tube. The heat transfer coefficient data for the smooth tube model are based on the envelope diameter of the tube.

The mass fluxes tested in the experiment were 100, 250, and 400 kg/m^2 s, and the heat fluxes tested were 10, 30, and 50 kW/m^2. The qualities tested ranged from a low of approximately 0.15, limited by the capacity of the subcooler, to a high of 0.9, which was a self-imposed limit to avoid burnout of the test section. The mass flux of 100 kg/m^{2}s was tested only with a heat flux of 10 kW/m^2, due to the fact that the tube experienced dryout at the higher heat fluxes.

The inlet pressure was measured using a strain-gage-type pressure transducer, and checked with a bourdon-type pressure gage. The pressure drop was measured with the same transducer and valves that allowed switching from inlet to outlet pressure. The outlet pressure transducer was located 0.3m from the outlet of the test section. To correct the effect of this short length of smooth tube on the overall pressure drop, it was modeled as adiabatic two-phase flow at the outlet quality. The prediction of this small correction was subtracted from the measured overall pressure drop, thus giving the expected actual value across the test section. The tube wall temperature was measured at 8 axial locations 100 mm apart by means of #30 BWG copper-constantan thermocouples attached to the root of the flutes at a position 90 degrees from the top of the tube. The thermocouples were referenced to an ice temperature bath maintained at 0 deg C. To avoid the imposition of the voltage from the direct heating on the temperature measurement, the thermocouples were attached by first adhering a small piece of 0.07 mm thick polymide tape to the tube and then epoxying the thermocouple to the tape. The flow rate was measured using a turbine-type flowmeter and one of three rotameters. The heat flux was determined by measuring the voltage drop and current in the test section. The voltage drop across the test section was measured directly, and the current was measured by means of a shunt at the power supply.

For each data point, the inlet pressure was maintained at a pressure of 320 kPa by manipulation of the bypass valves, the heat flux was set through use of a constant current adjustment on the DC power supply, and the mass flux was set through manipulation

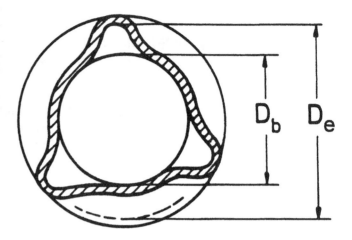

Figure 1. Turbotec Reference Diameters

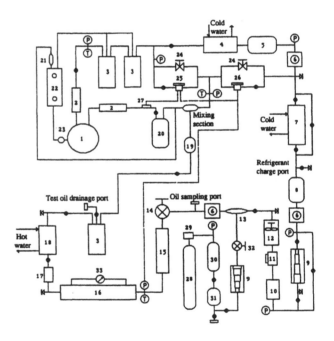

Figure 2. Test-Section Schematic

of the expansion valve and the bypass valves. Along with establishing the flow, heat, and pressure levels, the desired inlet quality was fixed by setting the electric preheater. These input conditions were affected until a steady-state was achieved at the desired levels, upon which pressure drop was measured. A linear interpolation from inlet to outlet pressure was assumed to determine the pressure at each axial location, therefore setting the local saturation temperature. With the local wall temperature, saturation temperature and heat flux known, the heat transfer coefficient was calculated.

Table 1 Heat Transfer Enhancement

	W=57.2 kg/h ($G_{b,t}$=400 kg/m² s)	
	x=0.2	x=0.8
Q (W)	$h_t A_b / h_s A_e$	$h_t A_b / h_s A_e$
206	1.9	4.3
617	1.8	4.2
1030	1.9	4.1
	W=35.8 kg/h ($G_{b,t}$=250 kg/m² s)	
206	2.7	4.9
617	2.1	3.7
1030	2.1	3.3
	W=14.2 kg/h ($G_{b,t}$=100 kg/m² s)	
206	2.1	7.8

RESULTS AND DISCUSSION

Local Heat Transfer Coefficient

The results of the experiment demonstrated that this tube offers an enhancement in the heat transfer performance over the entire range of test conditions. In particular, at low qualities, nucleate-boiling effects are clearly demonstrated. In this region, the enhancement ratio in comparison with a smooth tube of equivalent envelope diameter, at a fixed mass flow rate and heat input, as predicted by the Kandlikar correlation, ranges from approximately 1.8 to 2.7. The improvement in this regime, for a given length of tube, remains nearly constant over the range of mass flow rates and heat input tested (Table 1). More substantial improvement exists in the heat transfer performance for the range of high qualities where bulk convection dominates. The enhancement in this region for all test conditions was greater than 300%, with most data over 400% (Table 1). This enhancement was expected, because the swirl generated in the flow with this tubing should enhance this regime.

Heat Flux Effect. Figures 3 and 4 show the effect of the imposed heat flux on the local heat transfer coefficient. Both demonstrate that increasing the heat flux increases the heat transfer coefficient in the low-quality nucleate-boiling-controlled region, as would be expected. The enhancement in this region, shown in Table 1, is likely dominated by the increase in surface area. The "hypervapotron" effect, as discussed by Cattadori et al. (1993), may also play a role in the enhancement this tubing has in the nucleate boiling regime. This effect occurs in subcooled flow boiling with small side channels, such as flow over fins with a narrow spacing, or in this case, flow over the narrow flutes. Even in bulk boiling, flow can stagnate in these channels, especially at low mass fluxes, and the fluid in the channel then becomes superheated. The channel periodically purges itself of fluid by injecting vapor into the bulk flow, therefore enhancing the heat transfer through this purge/rewetting action. Due to difficulty with obtaining consistent temperature measurements at the tips of the spiral flutes, the present study does not explicitly investigate this effect; however, future work is planned.

This tubing demonstrates its most significant improvement in heat transfer in the high-quality, convection-dominated regime. Comparing Figures 4 and 5 shows how, in this region, the spirally

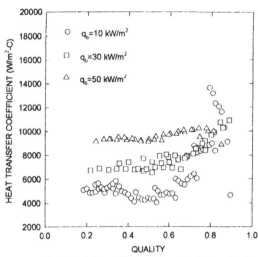

Figure 3. Heat-Flux Effect For Turbotec Tube (G_b=250 kg/m² s)

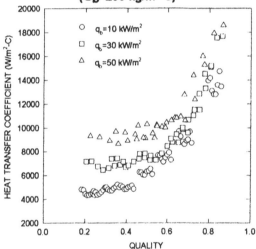

Figure 4. Heat-Flux Effect For Turbotec Tube (G_b=400 kg/m² s)

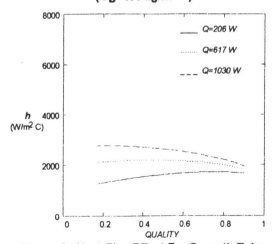

Figure 5. Heat-Flux Effect For Smooth Tube (Represented by the Kandlikar Correlation) (W=57.2 kg/h)

fluted tubing deviates from typical smooth tube behavior. The heat transfer coefficient of the Turbotec tube increases more substantially than the smooth tube, as predicted by the Kandlikar correlation, therefore increasing its enhancement in this regime. In this convection-dominated regime, the data demonstrate that the heat flux has little effect on the local heat transfer coefficient. Also, increasing heat flux does increase the quality at the departure from nucleate boiling (DNB).

Mass Flux Effect. Figures 6, 7, and 8 show the effect of mass flux on the heat transfer coefficient. They demonstrate that increasing mass flux has an inverse relationship with the transition quality for DNB. They also illustrate that increasing the mass flux increases the heat transfer coefficient in the convection-dominated regime. For mass fluxes of 250 and 400 kg/m^{2}s, the data show that increasing mass flux has no appreciable affect on the heat transfer coefficient in the nucleate-boiling regime. This disagrees with the increase in heat transfer coefficient with increasing mass flux for smooth tubes over this range of quality predicted by the Kandlikar correlation (Figure 9). This seems to run counter to the notion that mass flux has little effect on established nucleate boiling; however, the correlation predicts that for this range of quality, established nucleate boiling has been suppressed, and a convection dominated regime exists. The data for the mass flux of 100 kg/m^{2}s show a distinct decrease in the heat transfer from the 250 and 400 kg/m^{2}s mass fluxes (Figure 8). A comparison with single-phase liquid data calculated from a similar tube studied by Marto et al. (1979), showed that the heat transfer coefficient for 100 kg/m^{2}s was appreciably higher than that for single-phase liquid, indicating that it is still in the nucleate boiling regime. The explanation for the reduction in heat transfer coefficient at the lowest mass flux may lie with the location of these data on the boiling curve.

Pressure Drop

The pressure-drop data presented in this study were determined by finding the pressure drop across the test section for a given mass flow, heat input, and inlet quality. In Figures 10 through 12, the pressure drop measurements are plotted against the average quality of the test section for the given test conditions. The limited data collected reveal that the pressure-drop increases nearly linearly with increasing quality, thus validating the assumption of a linear profile in the calculation of the saturation temperature distribution.

The data were compared to single-phase liquid and vapor pressure drop data calculated from f(Re) data from a nearly identical tube tested by Marto et al. (1979), and it was determined that the data in the present study, at the smallest heat flux tested, closely follow the single-phase data. The two-phase pressure drop data extrapolated out to x=1 predict a pressure drop higher than single-phase vapor for all mass fluxes tested, likely due to momentum change effects in the boiling data.

The data were also compared to a homogeneous model for the evaporative pressure drop in a smooth tube. This model is an overly simple representation of the two-phase flow in the tube; however, Idsinga et al. (1976) compared it with the Baroczy (1968) and Thom (1964) correlations, along with several other representations of the flow, and found it to be the most accurate

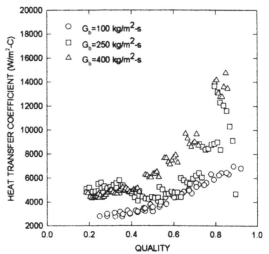

Figure 6. Mass-Flux Effect On Turbotec Tube
(q$_b$=10 kW/m^2)

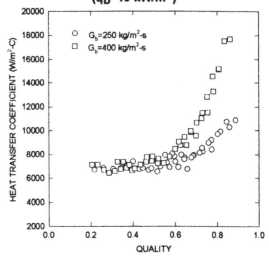

Figure 7. Mass-Flux Effect On Turbotec Tube
(q$_b$=30 kW/m^2)

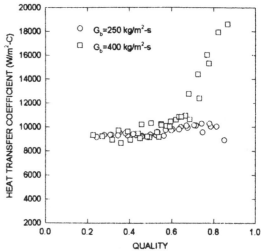

Figure 8. Mass-Flux Effect On Turbotec Tube
(q$_b$=50 kW/m^2)

Table 2 Pressure Drop Performance

W=14.3 kg/h ($G_{b,t}$=100 kg/m^2 s)

	x=0.4	x=0.8
Q (W)	delP_t/delP_s	delP_t/delP_s
206	6.3	14

W=35.8 kg/h ($G_{b,t}$=250 kg/m^2 s)

206	20	20
617	20	16

W=57.2 kg/h ($G_{b,t}$=400 kg/m^2 s)

206	15	15
617	12	13

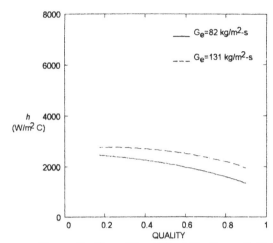

Figure 9. Mass-Flux Effect On Smooth Tube (Represented by the Kandlikar Correlation) (Q=1030 W)

predictor of pressure drop (for steam-water flow). The diameter of the smooth tube in the model was equivalent to the envelope diameter of the Turbotec tube, as used in the heat transfer performance comparison. The results of this comparison indicated that the pressure drop across the Turbotec tube ranged from approximately 6 to 20 times that of the smooth tube predicted by the model (Table 2).

Mass Flux Effect. Figures 10 and 11 show the effects of mass flux on the evaporative pressure drop. These Figures demonstrate that the pressure drop increased with increasing mass flux, and this effect increases with increasing quality. The smooth tube comparison found no relation between mass flux and the pressure drop increase over the smooth tube. (Table 2)

Heat Flux Effect. Figure 12 demonstrates the effect of heat flux on the two-phase pressure drop. Increasing heat flux increases the pressure drop across the Turbotec tube. This effect is much weaker than the effect of mass flux, and it diminishes nearly completely as the quality approaches saturated vapor. As with mass flux, heat flux was found to have no discernible effect on pressure drop increase over the smooth tube.

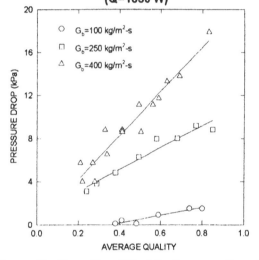

Figure 10. Mass-Flux Effect On Pressure Drop For Turbotec Tube (q_b=10 kW/m^2)

CONCLUSIONS

This investigation sought to lay a foundation for further research into the two-phase behavior of deep spirally fluted tubing. The data collected have been used to establish trends in the local heat transfer and pressure-drop characteristics for flow boiling in this tubing. From this investigation, the following conclusions can be drawn:

1. This tubing enhances heat transfer for the entire two-phase region. The most significant enhancement comes in the high-quality, bulk-convection-dominated regime.

2. Mass flux was found to have little effect on the local heat transfer coefficient at low qualities, counter to the smooth tube characteristics predicted by the Kandlikar correlation.

3. The pressure drop follows a linear profile with increasing quality and closely follows the single-phase pressure-drop data at saturated liquid and vapor conditions. The increase in pressure drop over a smooth tube ranged from 6 to 20.

RECOMMENDATIONS

Since this is a preliminary study, further work is planned with R-134a. Beyond alternate fluids, the authors believe that there is a need for more research into the evaporative heat transfer and

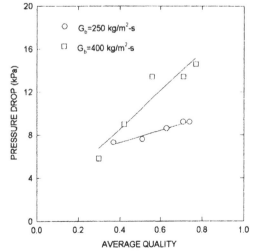

Figure 11. Mass-Flux Effect On Pressure Drop For Turbotec Tube (q_b=30 kW/m^2)

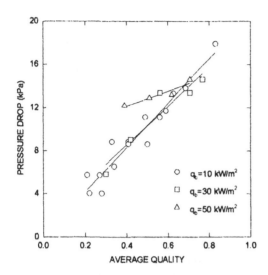

Figure 12. Heat-Flux Effect On Pressure Drop For Turbotec Tube (G_b=400 kg/m^2 s)

pressure-drop characteristics for this class of tubing. In particular, research into the effect of flute depth and pitch would be valuable. Also, the effect of oil in the fluid would aid in prediction of real-world applications. In addition, while there are data in the literature for the single-phase heat transfer and pressure-drop characteristics for this tubing, more detailed research would aid in developing correlations for not only single-phase, but two-phase flow as well.

NOMENCLATURE

D	tube diameter (m)
G	mass flux (kg/m^{2}s)
W	mass flow rate (kg/h)
q	heat flux (W/m^2)
h	heat transfer coefficient (W/m^2 C)
f	friction factor
x	quality
P	pressure (kPa)
Q	heat input to the test section (W)

Subscripts

b	based on bore diameter
e	based on envelope diameter
t	Turbotec tube
s	smooth tube

ACKNOWLEDGMENTS

This study was sponsored by Turbotec Products, Inc., Windsor, Connecticut. Dr. Sunil Raina, Engineering Manager of Turbotec, was especially helpful throughout the investigation. Mr. Fred Tesman, of the Rensselaer Staff, assisted with the power supply extension.

REFERENCES

Baroczy, C.J., 1968, "A Systematic Correlation for Two-Phase Pressure Drop", *Chemical Engineering Progress Symposium Series*, vol. 64, pp. 12-25.

Bergles, A.E., 1980, "Heat Transfer Characteristics of Turbotec Tubing", Iowa State University Heat Transfer Laboratory Report HTL-24, ISU-ERI-Ames-81018.

Cattadori, G., Gaspari, G.P., Celata, G.P., Cumo, M., Mariani, A., and Zummo, G., 1993, "Hypervapotron Technique in Subcooled Flow Boiling CHF", *Experimental Thermal and Fluid Science*, vol. 7, pp. 230-240.

Idsinga, W., Todreas, N., and Bowring, R., 1976, "An Assessment of Two-Phase Pressure Drop Correlations For Steam-Water Systems", *International Journal of Multiphase Flow*, vol. 3, pp. 401-413.

Kandlikar, S.G., 1990, "A General Correlation For Saturated 2-Phase Flow Boiling Heat Transfer Inside Horizontal and Vertical Tubes", *Journal of Heat Transfer*, vol. 112, pp. 219-228.

Marto, P.J., Reilly, D.J., and Fenner, J.H., 1979, "An Experimental Comparison of Enhanced Heat Transfer Condenser Tubing", *Advances in Enhanced Heat Transfer*, ASME, New York, pp. 1-9.

Sami, S.M., and Schnotale, J., 1992, "Comparative Study of Two Phase Flow Boiling of Refrigerant Mixtures and Pure Refrigerants Inside Enhanced Surface Tubing", *International Communications in Heat and Mass Transfer*, vol. 19, pp. 137-148.

Sami, S.M., Tulej, P.J., and Song, B., 1994, "Heat Transfer Characteristics of Ternary Blends Inside Enhanced Surface Tubing", *International Communications in Heat and Mass Transfer*, vol. 21, pp. 489-498.

Srinivasan, V., Vafai, K., and Christensen, R.N., 1994, "Analysis of Heat Transfer and Fluid Flow Through a Spirally Fluted Tube Using a Porous Substrate Approach", *Journal of Heat Transfer*, vol. 116, pp. 543-551.

Srinivasan, V., and Christensen, R.N., 1992, "Experimental Investigation of Heat Transfer and Pressure Drop Characteristics of Flow Through Spirally Fluted Tubes", *Experimental Thermal and Fluid Science*, vol. 5, pp. 820-827.

Thom, J.R.S., 1964, "Prediction of Pressure Drop During Forced Circulation Boiling of Water", *International Journal of Heat and Mass Transfer*, vol. 7, pp. 709-724.

Thome, J.R., 1990, *Enhanced Boiling Heat Transfer*. Hemisphere, New York.

ON THE EFFECT OF THE TEMPERATURE BOUNDARY CONDITION ON SINGLE BUBBLE DETACHMENT IN FLOW BOILING

Cees W. van der Geld   Wim G. van Helden   Piet G. Boot

Faculty of Mechanical Engineering
Eindhoven University of Technology
Eindhoven, The Netherlands.

ABSTRACT

A bubble generator has been developped that has a constant wall temperature, T, during bubble growth. Bubbles originate from an artificial cavity with a known mouth radius and T is directly measured.

The angle of detachment of bubbles is found to depend on the liquid velocity and on the temperature boundary condition that also affects the growth rate in a manner that has been explained.

In the force balance governing bubble motion in the direction normal to the wall the so-called expansion force and one of the lift forces are probably much less important than the other forces. New fitting coefficients are derived for all forces, although no definitive values have been obtained yet.

INTRODUCTION

The present study is a follow-up of an earlier study with direct injection of nitrogen (generator types 4,5 and 6) and with bubbles generated at an artificial cavity indirectly heated via a heating coil in the wall behind the cavity (generator type 1) [van Helden et al., 1995a]. In the new measurements described in this paper artificial cavities are heated via a strip of stainless steel (type 2) and via a very thin layer of platinum (type 3). The effect of different types of temperature boundary conditions on bubble detachment is examined.

It is attempted to explain the differences in detachment angle of various bubbles. The force balance governing bubble detachment as reported in the earlier study is re-examined.

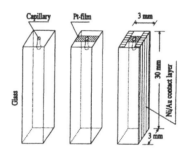

Figure 1: The thin film bubble generator and the manufacturing process.

EXPERIMENTAL

The test section has a square cross-section of 2x2 cm^2 and is positioned vertically with demineralised water flowing upward. More details are given in the earlier study [van Helden et al., 1995a] that also describes the type 1 generator as a capillary like the one on the left of Fig. 1. The entire glass is heated by a little coil at the back of the cavity. Two new bubble injectors have been applied that are described below.

Thin Film Generator

The thin film bubble generator is a small glass rod ($3x3x30 \ mm^3$) with a capillary of the reservoir type placed in the middle, see Fig. 1. A Platinum layer of 0.1 μm is deposited onto part of the top side by means of microwave sputtering. Two sputtered layers of Nickle/Gold connect the top ends with the back of the glass rod. The Platinum layer is heated electrically by a Constant Temperature Anemometer (CTA). This device

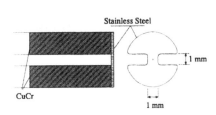

Figure 2: Cross-section and top view of the stainless steel generator.

controls the current through the metal film, keeping the electrical resistance constant at a preset value. Because pure metals have an increasing resistance with increasing temperature [Meaden, 1965] a certain resistance of the Platinum layer corresponds to a given temperature. By means of the CTA the temperature of the thin film is controlled and measured.

After testing and gauging the electrical resistance the bubble generator is cast and fixed in a stainless steel plug by means of a resin (Marocol 701 AB, a two-component epoxy). Since the heat capacity of the thin film is negligible the surrounding plug is not heated during operation.

Stainless Steel Generator

Local heating of type 2 generators is achieved with a direct current through a small strip of stainless steel, see fig.2. In the center a capillary has been made by means of spark erosion. The capillary leads to a reservoir that was made by drilling a 0.35 mm hole into the back of the stainless steel. At the backside a miniature (27 μm diameter) Chromel-Alumel thermocouple has been fixed. The electric current is fed to the heater plate by two CuCr wires. The whole system is cast into the same type of plug as described above. The surrounding plug is heated by the steel strip. This, together with the thermal inertia of the strip causes a different type of temperature boundary condition during bubble growth.

DETACHMENT AND HEATING MODE

From high speed cinerecordings trajectories and shapes of bubbles after detachment are measured. Figures 3 and 4 show typical bubble shapes at detachment to be different for glass and platinum, which might partly be due to different static contact angles, θ. The cavity mouth of generators 1,4,5,6 is made of glass with θ ca. 2° if the glass is very clean and ca. 48° otherwise, that of generator 2 is stainless steel having $\cos(\theta)$

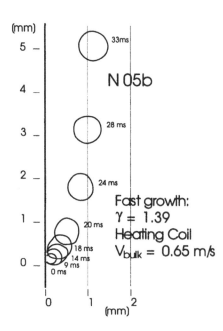

Figure 3: Bubble shape and trajectory after detachment from a heating coil generator (no. 1).

= 1.001 and the one of generator type 3 is platinum having $\cos(\theta) = 0.998$. Although the roughness of the surface and the edge of the cavity mouth allows the bubble foot to accommodate to the fluid stresses exerted θ is considered to affect the bubble shape. The advancing and receding contact angles have been measured for all bubbles individually.

Let γ denote the growth constant in:

$$R = R_0 + C_\gamma \cdot t^\gamma \qquad (1)$$

with R_0 a constant of the order of the capillary radius, t time and C_γ another coefficient. The three coefficients are determined by a least-squares fit to bubble radii measured. The fit is made for times in the range $[0.5t_d, t_d]$, t_d being the time of bubble growth to exclude the initial stage of bubble growth that usually differs from later stages. Figure 5 gathers some typical trajectories of the bubble midpoint after detachment. Let Φ denote the detachment angle, the angle between the wall and the tangent to the bubble trajectory at detachment. In the earlier study [van Helden et al., 1995a] different detachment angles for nitrogen and steam bubbles generated with the heating coil generator (type 1) were noted without a proper explanation. Figure 5 offers the following interpretation

Nitrogen bubbles detaching without liquid flow have about the same Φ as type 1 bubbles. The N_2 bubbles have a detachment radius of about 1.5 mm which is about five times the radius of type 1 bubbles. Fluid stresses are much more effective on larger bubbles that

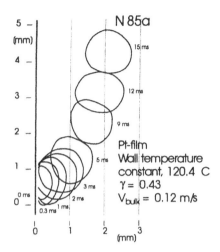

Figure 4: Bubble shape and trajectory after detachment from a Pt generator (no. 3).

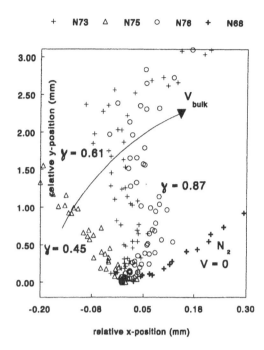

Figure 5: Trajectories of bubble midpoints after detachment, relative to the bubble midpoint at detachment. The data are also typical for type 1 bubbles with γ-values exceeding 1.

are more readily deformed. Nitrogen bubbles in liquid flow are indeed observed to have quite different advancing and receding contact angles. Without flow this is not the case and about the same Φ as for the smaller type 1 bubbles is therefore observed. The mean liquid velocity

Table 1: Results of the experiments performed with bubble generator number 2, the stainless steel generator; Θ_{bulk} is the liquid bulk superheat, $\overline{R_d}$ the mean detachment radius, $\overline{\gamma}$ the mean growth exponent.

Exp. nr.	V_{bulk} mm/s	T_{sat} $(^\circ C)$	Θ_{bulk} (K)	$\overline{R_d}$ (mm)	$\overline{\gamma}$
N75	0	101.8	0.7	0.46	0.45
N72	140	101.8	1.7	0.38	0.73
N73	175	101.8	1.5	0.35	0.61
N74	250	101.8	1.4	0.34	0.75
N78	190	101.7	1.7	0.36	0.74
N76	220	101.7	1.5	0.35	0.87
N77	190	101.7	1.5	0.29	1.08
N57	520	101.6	1.2	0.26	1.75
N55	500	101.6	1.3	0.16	0.92
N58	700	101.6	1.4	0.20	1.33

of the type 1 bubbles analysed was in the range from 53 to 85 cm/s. The 'stainless steel' bubbles generated with generator type 2 that are analysed in this study have velocities less than 25 cm/s, see table . These bubbles are observed to condense after detachment which explains the motion of the bubble midpoint towards the wall in Fig. 5. As Fig. 5 shows that both the detachment angle and $\overline{\gamma}$ are strikingly dependent on the liquid velocity in this range. This is explained as follows. The higher the velocity, the more effective the convective heat transfer

from the wall to the liquid prior to bubble initiation, the higher the heat flux required in the stainless steel strip to reach a wall temperature that exceeds saturation temperature by the amount necessary to reach the equilibrium curve for bubble nucleation. In the vicinity of the bubble generator not only the liquid is heated by the small stainless steel strip. The higher the liquid velocity, the higher the wall superheating, Θ_w, before bubbles are initiated.

This is schematised in Fig. 6 that is further discussed below. For type 1 bubbles Θ_w is much larger and so is γ: 0.96 to 1.39 dependent on coil temperature and liquid velocity. See the schematic of Fig. 7. In this case diffusion of heat from the wall towards the growing bubble is not hampering bubble growth since the peripherals of the cavity including the boundary layer are sufficiently superheated. The large superheating is caused by four effects. First of all, the capillary mouth radius is 65 μm as opposed to 1.8 μm of type 2 generators. If V_{bulk} is the same, bubbles have to be larger since the larger surface tension force attracting the bubble towards the wall is proportional to the mouth radius. The bubble surface being at saturation temperature this temperature has to prevail at larger distances from the wall necessitating a larger superheating. Secondly, the heat conduction coefficient is about 1 W /mK for glass and ca. 16 W /mK for stainless steal. To establish the same heat flux the tem-

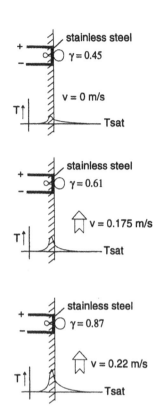

Figure 6: Schematic of the effect of the temperature boundary condition and liquid velocity on bubble generation for generator type 2.

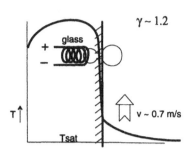

Figure 7: Schematic of the effect of the temperature boundary condition and liquid velocity on bubble generation for generator type 1.

Table 2: Results for the experiments performed with the Pt generator (number 3), T is the Platinum film temperature.

Exp. nr.	V_{bulk} mm/s	T_{sat} (^{o}C)	Θ_{bulk} (K)	T (^{o}C)	$\overline{R_d}$ (mm)	$\overline{\gamma}$
N85	120	101.7	1.6	120.4	0.71	0.43
N86	130	101.7	1.4	124.0	0.65	0.48
N87	130	101.7	1.5	128.1	0.59	0.39

perature gradient in the glass should therefore be about 16 times the one in the stainless steel. In instationary conditions, as during bubble growth, the thermal diffusivity is important, but its values for stainless steel exceeds the one of glass, $6.10^{-7}m^2/s$ by a factor 8 still. The third effect is the larger heating of the peripherals of the coil and the liquid boundary layer in particular due to the positioning of the heating coil away from the wall. The fourth effect is the higher V_{bulk}-values of type 1 experiments. As explained above bubble nucleation only sets in if the superheating of the liquid boundary layer is sufficiently high. If the velocity is higher a higher heating rate is required.

All these effects combine to produce the typical temperature profiles as depicted in Fig's 6 and 7.

The temperature boundary condition is different for the case of the 'Pt bubbles' generated with type 3 generators. The comparison of Fig's 3 and 4 already indicates that diffusion might be more important in this case. This is confirmed by the rather low values of γ : ca. 0.43, see table , which is not much below the value of most nitrogen bubbles having similar shapes as well. This is explained as follows. The wall temperature of the type 3 generators is controlled by the direct current

supply and constant whatever the amount of heat extracted from the platinum. As soon as bubble growth sets in, the heating current in the platinum strip is automatically increased to keep the wall temperature constant. This increased heat flux only reaches the bubble top by diffusion causing the bubble to grow more rapidly at the foot than the top. This might also explain the different shape of Fig. 4. The amount of wall superheating to reach the equilibrium curve for bubble nucleation is relatively small. The temperature drop between the Pt-strip and the bulk liquid is measured to be ca. 17 K, see table .

Further validation of the above interpretation of the effect of the temperature boundary condition has recently been obtained from temperature gradients measured with the aid of Mach-Zehnder interferometry. The thermal conditions thus determined during bubble growth will be reported elsewhere [van Helden et al., 1995b].

FORCE BALANCE ANALYSIS

In an earlier study [van Helden et al., 1995a] the following force balance has been shown to be valid at all times prior to bubble detachment:

$$\frac{dmv_x}{dt} = -C_\sigma R_{cap}\sigma \int_0^{2\pi} \sin\beta(\phi)d\phi + C_{L_1}A\tfrac{1}{2}\rho_l V_m^2 +$$

$$+ C_{L_2} \rho_l V_b V_m \frac{dV_m}{dx} + 2\pi R_{cap} \frac{\partial \sigma}{\partial T} C_{\Delta T} \Delta T +$$

$$+ \pi \rho_l R^2 C_e \{ \frac{3}{2} C_{old} \dot{R}^2 + R\ddot{R} \} \cos(\Phi) +$$

$$+ \left\{ (\rho_g - \rho_l) g H_0 + \frac{2\sigma}{R_{top}} \right\} A_{cap} \qquad (2)$$

The components in x-direction normal to the plane of detachment are selected since the buoyancy force dominates in y-direction. The first term on the RHS is called the surface tension force, the second two are lift forces, the term with $C_{\Delta \sigma}$ a so-called temperature drop force and the other two the expansion and corrected buoyancy force components. For bubbles coming from the Pt-generator (type 3) the temperature drop ΔT between wall and bulk liquid is measured. In other cases the theoretical estimate of the earlier study is used:

$$\Delta T = 0.6 \cdot Q \cdot V_{bulk}^{\xi} \qquad (3)$$

with $\xi = -0.568$. A fitting procedure yields the coefficients in the force balance (2) as thoroughly described in the earlier paper. However, the coefficient C_e was not applied in that paper, only C_{old} was fitted. Repeating the procedure for the nitrogen bubbles studied there yield low values for C_e, 0.0032 typically. In the earlier study C_{old} was selected such by the fitting procedure that the expansion force changed sign during the lifetime of a growing bubble, implying relatively low values for most of the lifetime. This changing sign is physically irrealistic and must be considered as due to lack of fitting capacity due to the absence of C_e. It is not difficult to show that the expansion force must satisfy

$$\frac{1}{1.5 C_e + 1} < \gamma < 1 \qquad (4)$$

in order not to change sign. For the relevant values of γ C_e should exceed 0.82 to satisfy (). However, it did not in the earlier study. In the present study C_{old} has been varied by ± 10 % (values 0.9 and 1.1) without affecting results much. It is therefore concluded that it is inappropriate to fit C_{old}.

It was shown [van Helden et al., 1995a] that the two lift forces are correlated for most of the datasets. It is therefore not attempted here to simultaneously fit C_{L_1} and C_{L_2}. Since the detaching bubbles are nearly spherical most of the streamlines pass by without lowering the pressure on top of the bubble, as was clearly shown by Kenning and Cooper[Kenning, 1966]. Near the bubble foot and near the top pressures will be almost the same. It is therefore not expected that C_{L_1} yields proper estimates of the lift force; this coefficient is put equal to zero.

For 35 nitrogen bubbles with liquid velocity larger than 0 m/s the following coefficient fits are found:

$$C_\sigma = 0.1798 \pm 0.0002 \; ; \; C_e = 0.0032 \pm 0.0001;$$

$$C_{L_2} = 0.01356 \pm 0.00006$$

For 8 nitrogen bubbles with zero liquid velocity the values of C_σ and C_e are about half the above values (and no lift forces can be determined of course). It is noted that C_e should be zero for bubbles that expand without the presence of a wall and also should be zero for bubbles that expand symmetrically around the x-axis. The angles of contact have been measured as a function of size and liquid velocity and the value of $C_\sigma$is is determined from a sufficiently large data set as manifested by scatter plots. For the above value of ca. 0.2 the adhering surface tension force nearly compensates the 'detaching' buoyancy correction term that has opposite sign and is also determined quite accurately. Since the expansion and lift forces are also counteracting, both values could in principle have been higher if the functional relationship of these forces would not forbid.

The value of the C_{L_2} seems rather low and if it would be higher the values of C_σ and C_e would increase. However, at the moment we have no indication whatsoever what C_{L_2} should be. Either dedicated experiments or numerical computations for a turbulent boundary layer should yield an estimate of $C_{L_2}$to improve the results of the fitting procedure.

At this point it is challenging to determine $C_{\Delta \sigma}$ for the bubbles generated with the Pt generator since the driving temperature drop in the force component containing $C_{\Delta \sigma}$ is measured quite accurately in this case. The nine bubbles of N85, N86 and N87 have almost identical liquid velocity (see table) and yield 0.426 for $C_{\Delta \sigma}$ if the above values of C_σ , C_e , and C_{L_2} are implemented and kept constant. However, if C_σ would have been 0.11 approximately, the value of $C_{\Delta \sigma}$ would have dropped to zero. If C_σ is left to be fitted by the programme, there is a tendency to lower the values of both C_σ and $C_{\Delta \sigma}$. This might be the appropriate trend if the observed contact angles would differ from the actual contact angle on microscale since the static contact angle of platinum differs considerably from that of glass used for the nitrogen bubbles, see above. There are indications that in some circumstances the contact angle on microscale might indeed be different from the one usually observed [Chesters, 1994].

Moreover, for the 30 heating coil vapour bubbles presented in the earlier study [van Helden et al., 1995a] different results are found[1]: a negative value of $C_{\Delta \sigma}$ implying attraction to the wall and a tendency to increase both C_σ and $C_{\Delta \sigma}$ if they are fitted simultaneously. The expansion force and the lift force are usually an order of magnitude less than the surface tension, temperature drop and the corrected buoyancy forces for heating coil

---

[1] The fitted values of $C_{\Delta \sigma}$ for the heated coil and the Pt-bubbles are not directly comparable since in the first case the dependance on the water velocity has theoretically been prescribed.

bubbles. There is a clear need of a very accurate value of C_σ for all substrates and because of some lack of knowledge of the lift force the authors doubt if the present value of C_σ is accurate enough to justify the determination of $C_{\Delta\sigma}$.

There is nothing strange about a negative value for $C_{\Delta\sigma}$ since the thermocapillary Marangoni force corresponds to a negative value of $C_{\Delta\sigma}$. In this case a narrow liquid jet is created away from the wall along the bubble interface. The driving surface tension gradient is directed away from the wall, so the force on the bubble, equal but opposite because of Newton's law, is directed towards the wall.

CONCLUSIONS

A bubble generator has been manufactured that enables accurate measurement of the wall superheating during bubble growth on an artificial cavity. Its temperature is constant which is a rather special temperature boundary condition. It has been used to investigate the influence of a possible temperature drop force.

The angle of detachment of bubbles immediately after detachment, Φ, is found to depend on the liquid velocity and on the temperature boundary condition of the surface. This condition also affects the growth rate of a bubble in a manner that could be explained. If the wall temperature is kept constant bubble growth is diffusion controlled with γ -values below 0.5 and a typical wall superheating of 17 K. If the wall superheating is high γ typically exceeds 1.

In the force balance governing bubble growth in the direction normal to the wall the so-called expansion force is less important than described earlier [van Helden et al., 1995a]. Also one of the lift forces, the so-called 'Bernoulli suction', was overestimated in this study. The optimal value of the attaching surface tension force is found to differ for different bubble generators which might be related to different static contact angles. Its value should be known more precisely before even the direction of the, possibly active, temperature drop force can be determined.

If the above findings about bubble growth are to be transfered to practical heater surfaces with uniform mean wall temperature one should bear in mind that during formation on non-artificial cavities often spreading occurs. Apart from this spreading, the results with the stainless steel and heating coil generators can safely be regarded as typical for heating surfaces of common practice.

References

Chesters, A.K., 1994, private communications.

Eckertova, L., 1986, "Physics of thin films", Plenum, New York.

Helden, W.G.J. van, Geld, C.W.M. van der and Boot, P.G.M., 1995a, "Forces on bubbles growing and detaching in flow along a vertical wall", *Int. J. of Heat Mass Transfer*, Vol. 38, No. 11, pp. 2075-2088.

Helden, W.G.J. van, Geld, C.W.M. van der and Boot, P.G.M., 1995b, "The effect of convection on temperature gradients around a cavity-generated boiling bubble", to appear.

Kenning, D.B.R. and Cooper, M.G., 1966, "Flow patterns near nuclei and the initiation of boiling during forced convection heat transfer", *Proc. Instn. Mech. Engrs*, 1965-66, Vol. 180 Pt 3C, paper 11, pp. 112-123.

Meaden, G.T., 1965, "Electrical resistance of metals", Heywood, London.

SUPPRESSION OF NUCLEATION SITES IN FLOW BOILING

J.F. Klausner[1] and R. Mei[2]
[1]Department of Mechanical Engineering
[2]Department of Aerospace Engineering,
Mechanics, and Engineering Science
University of Florida
Gainesville, Florida USA

ABSTRACT

A simple model is presented for estimating the maximum and minimum cavity radius required for ebullition in two-phase flow with heat transfer. It is demonstrated that the nucleation site density in flow boiling is correlated with the ratio of the maximum to minimum cavity radius, r_{max}/r_{min}, and the maximum cavity radius has a controlling influence. As the convective heat transfer associated with bulk turbulence in two-phase flow is enhanced, $r_{max}/r_{min} \to 1$, and the probability of finding surface cavities whose radii lie between r_{max} and r_{min} is reduced. Thus, active nucleation sites become deactivated. The value of r_{max}/r_{min} at which complete suppression of nucleation sites occurs, to a certain extent, depends on the surface/fluid combination and heat flux. For the available data, it is observed that complete suppression of nucleation sites occurs for r_{max}/r_{min} between 50 and 150. Various flow boiling heat transfer data sets have been examined, and it is found that many heat transfer data are not likely in the nucleate boiling regime.

INTRODUCTION

Over the past four decades, the heat transfer rates associated with forced convection boiling have been investigated by hundreds of researchers. Perhaps the first investigators to propose a correlation for predicting flow boiling heat transfer rates were Dengler and Addoms (1956). Since then many other flow boiling heat transfer correlations have been proposed, many of which have been summarized by Gungor and Winterton (1986). The type of correlation which has been used the most was first proposed by Chen (1966) and takes the form

$$h_{2\phi} = Eh_\ell + Sh_b \qquad (1)$$

where $h_{2\phi}$ is the two-phase heat transfer coefficient, h_ℓ is the single-phase heat transfer coefficient based on the liquid fraction

flowing, h_b is the heat transfer coefficient appropriate for pool boiling, E is an enhancement factor to account for increasing bulk turbulence with increasing vapor quality, and S is a suppression factor to account for the effective superheat seen by growing vapor bubbles during flow boiling. Klausner (1989) examined various correlations which take the same form as (1) and found they predict that the macroconvection component of heat transfer, represented by Eh_ℓ, is the dominant heat transfer mechanism, and the microconvection component of heat transfer, represented by Sh_b, typically does not contribute significantly to the total two-phase heat transfer. In contrast, Zeng and Klausner (1993a) measured the microconvection component of heat transfer for horizontal flow boiling of refrigerant R113 and demonstrated that under certain flow and thermal conditions it contributes significantly to the total two-phase heat transfer, as shown in Fig. 1. Others who have reported finding microconvection heat transfer to be significant over a range of flow and thermal conditions include Blatt and Adt (1964), Staub and Zuber (1966), Frost and Kippenhan (1967), Kenning and Cooper (1989), Cooper (1989), Klausner (1989), and Tran et al. (1993). A typical comparison between measured and predicted flow boiling heat transfer coefficients using a correlation which takes the same form as (1) is shown in Fig. 2, in which the Gungor and Winterton (1986) correlation is compared with a sample of the refrigerant R114 flow boiling data of Jallouk (1974). It is seen that a good portion of the data agree reasonably well with the prediction from the empirical correlation while a portion of the data deviate substantially. It is of great interest to understand why the empirical correlations are satisfactory for a certain range of flow and thermal conditions and not others.

In order to understand the discrepancy, it is instructive to consider the saturated vertical upflow boiling experiments by Kenning and Cooper (1989). Over the range of flow and thermal conditions tested, two flow boiling regimes were identified: the apparently convective regime and the apparently nucleate boiling regime. In the apparently convective regime, the measured two-phase heat transfer coefficient appears to be independent of the heat flux and heating surface conditions and can be predicted within an uncertainty of approximately ±10%. In this regime, it appears that the boiling activity has been completely suppressed. In the apparently nucleate boiling regime, the measured heat transfer coefficients are essentially independent of mass flux, vapor quality, and flow regime. They tend not to be repeatable due to variations in surface conditions and gas entrainment, and they are difficult to predict.

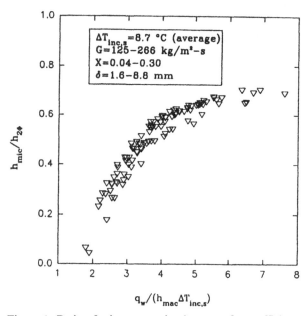

Figure 1 Ratio of microconvection heat transfer coefficient to total two-phase heat transfer coefficient for different flow boiling conditions (from Zeng and Klausner; 1993a).

The importance of the two flow boiling regimes identified by Kenning and Cooper cannot be underestimated. When nucleation sites are completely suppressed and heat transfer is dominated by bulk turbulence and liquid film evaporation, as occurs in the apparently convective regime, the physics governing the heat transfer are completely different than that in the apparently nucleate boiling regime in which heat transfer is mainly controlled by the incipience, growth, and departure of vapor bubbles. In fact, it is the authors' opinion that heat transfer in the apparently convective regime should not be categorized as flow "boiling" heat transfer. As demonstrated in Fig. 1, as soon as incipience occurs, microconvection strongly contributes to the total two-phase heat transfer. Furthermore, when empirical heat transfer correlations are constructed based on data in the apparently convective regime which are interspersed with data in the apparently nucleate boiling regime, they become biased toward one regime or the other if there are significantly more data in one of the regimes. In this work it will be demonstrated that some data sets used for developing flow boiling heat transfer correlations are dominated by data in the apparently convective regime. In order to do so, it is necessary to establish a criterion for predicting when nucleation

sites become completely suppressed, which can serve as a boundary for distinguishing the apparently convective regime from the apparently nucleate boiling regime. In this work a phenomenological model is developed for predicting the complete suppression of nucleation sites in flow boiling; it is calibrated using flow boiling nucleation site density data of Zeng and Klausner (1993b) as well as the heat transfer data of Kenning and Cooper (1989) and Jallouk (1974). It is used to examine existing flow boiling heat transfer data sets to determine what percentage of the data fall in the apparently convective regime.

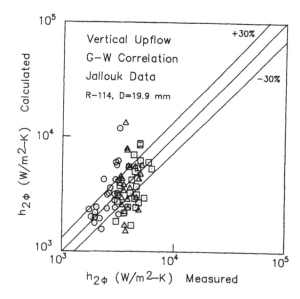

Figure 2 Comparison between measured and predicted flow boiling heat transfer coefficients using correlation of Gungor and Winterton (1986) and typical data from Jallouk (1974).

NUCLEATION SITES IN FLOW BOILING

An extensive experimental investigation of nucleation site density in horizontal flow boiling with refrigerant R113 was reported by Zeng and Klausner (1993b). They demonstrated that

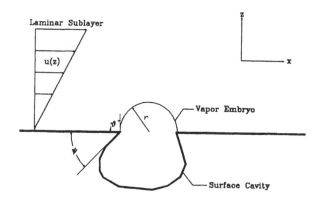

Figure 3 Idealized sketch of a vapor embryo embedded in shear flow protruding a surface cavity.

the behavior of the nucleation site density is significantly different than that found in pool boiling, and although the nucleation site density data show dependence on the critical cavity radius, it is insufficient for correlating the data. They observed a strong dependence on heat flux and mean vapor velocity. As the mean vapor velocity increased, the nucleation site density decreased until complete suppression of nucleation sites occurred. No generalized correlation for the data was suggested. In general the formation of nucleation sites is highly dependent on the size and geometry of the microscopic scratches and pits on the heating surface, the wettability of the fluid, the amount of foreign contaminants on the surface, as well as the material from which the surface was fabricated. Here, the data of Zeng and Klausner (1993b) are reexamined based on theoretical considerations.

It has long been recognized that surface cavities which become active nucleation sites must be capable of trapping vapor (Corty and Foust, 1955). Wang and Dhir (1993) suggested that only those cavities with $\theta > \psi_{min}$ are capable of trapping vapor, in which θ is the liquid-solid contact angle and ψ_{min} is the minimum cavity side angle, or in the case of spherical cavities, it is the mouth angle. Using this vapor trapping criterion and the nucleation site density model proposed by Yang and Kim (1988) for pool boiling, the nucleation site density for either pool or flow boiling may be calculated from,

$$\frac{n}{A} = \frac{N}{A} \int_{r_{min}}^{r_{max}} f_1(r)dr \int_0^\theta f_2(\psi_{min})d\psi_{min} \qquad (2)$$

where n/A is the nucleation site density, N/A is the number of cavities per unit area on the heating surface, f_1 is the probability density function of finding cavities with mouth radius r, f_2 is the probability density function of finding cavities with minimum side angle ψ_{min}, and r_{min} and r_{max} are the respective minimum and maximum mouth cavity radii within which incipience will occur. The use of Eq. (2) assumes a statistically homogeneous distribution of cavities on the heating surface.

Although the applicability of (2) to flow boiling has yet to be experimentally verified, it is quite useful in understanding the nucleation site density data presented by Zeng and Klausner (1993b), in which n/A is observed to decrease rapidly with increasing vapor velocity. Increasing the mean vapor velocity results in enhanced two-phase bulk turbulence and thus reduces the thermal boundary layer thickness. The temperature field which is seen by a vapor embryo protruding a surface cavity, as sketched in Fig. 3, is strongly influenced by the bulk turbulence. A vapor embryo will grow and establish a nucleation site only if there is sufficient energy from the surroundings to sustain growth. If a vapor embryo is too large, which might occur in a cavity with a large diameter, it will protrude into the colder region of the liquid, which reduces the vapor temperature to below the value required by the Clapeyron equation for the equilibrium of the two phases. Therefore, in addition to a minimum cavity radius required for incipience, there also exists a maximum cavity radius, above which the incipience is inhibited. As the bulk turbulence is enhanced, the thermal boundary layer becomes thinner and $r_{max} \to r_{min}$, and thus the nucleation sites become suppressed as suggested by (2). The distinct difference in the formation of nucleation sites between pool and flow boiling lies in the fact that r_{max} in flow boiling is typically much less than that in pool boiling due to a much thinner thermal boundary layer. In fact, Wang and Dhir (1993) were able to adequately predict pool boiling nucleation site density by assuming $r_{max} \to \infty$. As will be demonstrated, such an assumption is not valid for flow boiling.

Now consideration is given to a hemispherical vapor embryo which protrudes a surface cavity as shown in Fig. 3. The static vapor embryo is embedded in a liquid shear flow with a temperature profile governed by macroconvection heat transfer. The embryo radius is typically several orders of magnitude less than the liquid film thickness, and thus the temperature profile in the vicinity of the embryo may be approximated as linear. Following Zeng and Klausner (1993a), the local temperature profile is approximated as,

$$T_\ell(z) = T_w - \frac{Eh_\ell z}{k_\ell}\Delta T_{sat} \qquad (3)$$

where $T_\ell(z)$ is the liquid temperature near the wall, T_w is the mean wall temperature, ΔT_{sat} is the wall superheat, and k_ℓ is the liquid thermal conductivity. In general, during vapor bubble growth, energy is transferred to the vapor through the liquid microlayer which resides beneath the bubble. However, prior to growth no liquid microlayer can be established and thus energy is transferred to the embryo from the surrounding liquid and the solid heating surface beneath the embryo. A physically sound criterion for vapor embryo growth to continue is that its local vapor temperature must always exceed the saturation temperature associated with the vapor pressure. Resolution of the detailed vapor embryo temperature field requires a three-dimensional three-phase heat transfer analysis. Due to the complexity of such an analysis, the incipience criterion proposed by Bergles and Rohsenow (1964) is instead instituted: vapor bubble growth will proceed only when T_ℓ at the top of the hemispherical embryo exceeds the vapor temperature, T_v. T_v is computed from the Clapeyron equation as,

$$T_v - T_{sat} = \frac{T_v T_{sat} R_v}{h_{fg}} \ln(1 + \frac{2\sigma}{r P_{sat}}) \qquad (4)$$

where R_v is the engineering gas constant, σ is the surface tension, P_{sat} is the saturation pressure, and h_{fg} is the latent heat. Thus the maximum cavity radius, r_{max}, which can sustain an active nucleation site may be estimated by setting $T_\ell = T_v$ and $z = r$ in (3) and simultaneously solving (3) and (4) for r. The values of r which satisfy (3) and (4) must be determined iteratively and are denoted by r_{max}. The minimum cavity radius, r_{min}, may be approximated by $r_c = \frac{2\sigma T_{sat}}{\rho_v h_{fg}\Delta T_{sat}}$, where it has been assumed that $\frac{2\sigma}{P_{sat}r_c} \ll 1$. In (3), the macroconvective

heat transfer represented by Eh_ℓ, may be estimated using the correlation proposed by Kenning and Cooper (1989), which is a modified form of that used by Chen (1966),

$$h_\ell = 0.023\frac{k_\ell}{D}Re_\ell^{0.8}Pr_\ell^{0.4} \qquad (5)$$

$$E = 1 + 1.8(\frac{1}{X_{tt}})^{0.87} \qquad (6)$$

where $Re_\ell = \frac{G(1-x)D}{\mu_\ell}$ is the Reynolds number based on the

liquid fraction flowing, and $\frac{1}{X_{tt}} = [\frac{x}{(1-x)}]^{0.9}[\frac{\rho_\ell}{\rho_v}]^{0.5}[\frac{\mu_v}{\mu_\ell}]^{0.1}$ is the

Martinelli parameter.

As suggested by (2), the probability of finding cavities on the heating surface which have mouth radii which fall between r_{max} and r_{min} decreases as $r_{max}/r_{min} \to 1$, and thus r_{max}/r_{min} may serve as a useful correlating parameter for the nucleation site density. In fact, when $r_{max}/r_{min} = 1$ it is not physically possible to maintain heterogeneous boiling, but this is too restrictive a condition to use as a suppression criterion since it is possible for nucleation sites to become completely suppressed for $r_{max}/r_{min} > 1$, depending on the distribution of cavity mouth radii on the heating surface. Furthermore, the Bergles and Rohsenow (1964) incipience criterion used to determine r_{max} is only an approximation. Therefore, the range of r_{max}/r_{min} for which nucleation sites become completely suppressed will be determined empirically.

RESULTS

In Fig. 4 the flow boiling nucleation site density data, measured by Zeng and Klausner (1993b) using R113 on a smooth nichrome heating strip, are displayed as a function of r_{max}/r_{min} for two different heat fluxes, $q_w = 19.3$ and 13.8 kW/m^2. It is seen that for a given heat flux the data are collapsed into a single curve. As r_{max}/r_{min} decreases the nucleation site density appears to decrease proportionally until all nucleation is suppressed. According to (2), the nucleation site density as a function of r_{max}/r_{min} should not be heat flux dependent. The heat flux dependence shown in Fig. 4 is likely due to the fact that (3) and (4) assume a uniform wall temperature. However, in the presence of heterogeneous boiling, the thermal field in the heating element is highly nonuniform with large temperature gradients beneath nucleation sites, which are more pronounced at higher heat fluxes. As the nucleation site density is reduced, the temperature nonuniformities should be less pronounced and the observed convergence of the two curves in Fig. 4 is expected. When nucleation is completely suppressed, the wall

temperature should be uniform and the two curves are expected to intersect at n/A=0. Although they don't exactly intersect at n/A=0, for $q_w = 19.3$ KW/m^2, $(r_{max}/r_{min})_{sup} = 104$ while for $q_w = 13.8$ KW/m^2, $(r_{max}/r_{min})_{sup} = 157$, where the subscript sup denotes the point of suppressed nucleation.

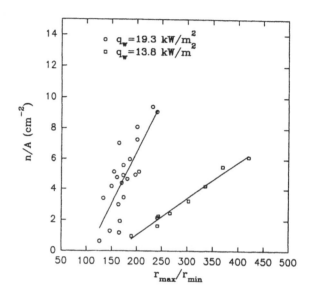

Figure 4 Nucleation site density data of Zeng and Klausner (1993b) shown as a function of r_{max}/r_{min}.

The point of suppressed nucleation can also be determined from heat transfer data as was done by Kenning and Cooper (1989). For a fixed mass flux and vapor quality, the point at which the rate of heat transfer is independent of heat flux is approximately the point of suppressed nucleation. Kenning and Cooper's flow boiling experiments used water on cupronickel tubing. The values of $(r_{max}/r_{min})_{sup}$ based on their measured heat transfer data are summarized in Table 1. Also the R114 heat transfer data measured inside a copper tube by Jallouk (1974) can be used to determine that for $q_w = 19$ kW/m^2, $(r_{max}/r_{min})_{sup} = 112$. These $(r_{max}/r_{min})_{sup}$ data are summarized in Fig. 5 as a function of the heat flux. It is seen that the data demonstrate a slight dependence on heat flux, but are generally scattered between $(r_{max}/r_{min})_{sup} = 50$-$150$. In general, the point of suppressed nucleation will depend on the surface-fluid combination. In this study, both highly wetting fluids (R113 and R114) and an intermediate wetting fluid (water) have been considered, and as indicated in Fig. 5, three regimes may be identified based on r_{max}/r_{min},

Regime I: $r_{max}/r_{min} < 50$, Convective Regime (suppressed nucleation)

Regime II: $50 < r_{max}/r_{min} < 150$, Regime of Uncertainty (nucleation may or may not be sustainable)

Regime III: $r_{max}/r_{min} > 150$, Nucleate Boiling Regime (nucleation is sustained).

Table 1   Point of suppressed nucleation from Kenning and Cooper (1989) heat transfer data.

$(r_{max}/r_{min})_{sup}$	q_w (kW/m^2)	G (kg/m^2-s)
108	50	304
90	100	304
103	100	123
111	150	123
74	200	304
74	200	203
99	250	123
69	300	304
89	300	123
63	350	123

Five different flow boiling heat transfer data sets have been examined to determine what percentage of the data fall in Regimes I-III. The results of this comparison are summarized in Table 2. Klausner (1989) argued that his boiling heat transfer data show a strong dependence on microconvection, and it is seen that 22% of his data fall in the nucleate boiling regime and 66% fall in the regime of uncertainty. For the data of Jallouk (1974), 67% fall in the nucleate boiling regime and 32%

in the regime of uncertainty. For the data of Bennett (1976), Pujol (1968), and Somerville (1962) none fall in the nucleate boiling regime, while 68%, 74%, and 70% respectively fall in the convective regime.

Table 2 Percentage of data within Regimes I,II, and III for typical two-phase heat transfer data sets.

Data Set	Surface-Fluid	# of Data Points	% Data in Regime I	% Data in Regime II	% Data in Regime III
Klausner (1989)	copper/R11	182	12	66	22
Jallouk (1974)	copper/ R114	443	1	32	67
Bennett (1976)	inconel/ water	257	68	32	0
Pujol (1968)	stainless steel/R113	281	74	26	0
Somerville (1962)	stainless steel/ n-butanol	257	70	30	0

DISCUSSION

It is clearly seen that certain two-phase heat transfer data sets which are used to develop flow boiling heat transfer correlations contain very few heat transfer data in the nucleate boiling regime. This is an observation which was also made by Cooper (1989). It is one of the reasons certain correlations which take the form of (1) are biased in favor of macroconvection heat transfer and do not compare favorably with data in the nucleate boiling regime. It is encouraging to note that a large portion of two-phase heat transfer occurs in the convective regime, which is predictable. However, correlations which treat heat transfer in the convective regime should remain distinct from those which treat heat transfer in the nucleate boiling regime since the mechanisms driving the heat transfer are quite different. Thus flow boiling heat transfer correlations should only be compared against data in the nucleate boiling regime.

Kenning and Cooper (1989) have suggested that flow boiling heat transfer in the nucleate boiling regime is very similar to that for pool boiling, and heat transfer rates are essentially independent of the flow conditions. This point of view is not supported by the nucleation site density data shown in Fig. 4 since the microconvection component of heat transfer is highly dependent on the nucleation site density, and it has been demonstrated that the nucleation site density in flow boiling strongly depends on the bulk flow conditions. At present, the prediction of microconvection heat transfer in flow boiling is elusive. Future research efforts should focus on the mechanics of the ebullition process in flow boiling and its role in controlling microconvection heat transfer.

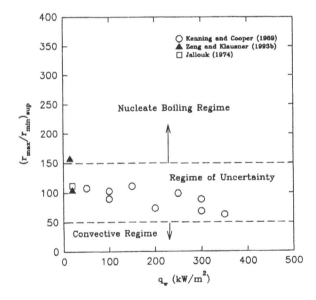

Figure 5 Identification of two-phase heat transfer regimes based on point of suppressed nucleation.

NOMENCLATURE

D	pipe diameter (m)
E	heat transfer enhancement factor
$f_1(r)$	probability density function for finding cavities with mouth radii r
$f_2(\psi_{min})$	probability density function for finding cavities with minimum side angle, ψ_{min}
G	mass flux (kg/m²-s)
h	heat transfer coefficient (W/m²-K)
h_{fg}	latent heat of vaporization (J/kg)
k	thermal conductivity (W/m-K)
n/A	nucleation site density (cm⁻²)
N/A	number of surface cavities per unit area (cm⁻²)
P_{sat}	saturation pressure (Pa)
Pr	Prandtl number
q_w	heat flux (W/m²)
r_{max}	maximum cavity radius required for nucleation (m)
r_{min}	minimum cavity radius required for nucleation (m)
R_v	engineering gas constant (J/kg-K)
Re_ℓ	liquid Reynolds number based on liquid fraction flowing
S	suppression factor
$T_\ell(z)$	liquid temperature near wall (C)
T_w	average wall temperature (C)
T_v	vapor temperature (K)
T_{sat}	saturation temperature (K)
ΔT_{sat}	wall superheat (C)
$\Delta T_{inc,s}$	incipient wall superheat (C)
x	vapor quality
$1/X_{tt}$	Martinelli parameter
z	distance from heating surface (m)

Greek Symbols

δ	liquid film thickness (mm)
θ	liquid/solid contact angle
μ	absolute viscosity (N-s/m²)
ρ	density (kg/m³)
σ	surface tension (N/m)
ψ_{min}	minimum cavity side angle

Subscripts

b	boiling
ℓ	liquid
sup	nucleation suppression point
v	vapor
2ϕ	two-phase

REFERENCES

Bennett, D.L., 1976. "A Study of Internal Forced Convective Boiling Heat Transfer for Binary Mixtures." Ph.D. Thesis, Lehigh University.

Bennett, D.L., Davis, M.W., and Hertzler, B.L., 1980. "The Suppression of Saturated Nucleate Boiling by Forced Convective Flow." *AIChE Symp. Series*, No. 199, Vol. 76, pp. 91-103.

Bergles, A.E., and Rohsenow, W.M., 1964. "The Determination of Forced-Convection Surface-Boiling Heat Transfer." *J. Heat Transfer*, Vol. 86, pp. 365-372.

Blatt, T.A., and Adt, R.R. Jr., 1964. "An Experimental Investigation of Boiling Heat Transfer and Pressure-Drop Characteristics of Freon 11 and Freon 113 Refrigerants." *AIChE Journal*, Vol. 10, No. 3, pp. 369-373.

Chen, J.C., 1966, "Correlation for Boiling Heat Transfer to Saturated Fluids in Convective Flow." *I&EC Proc. Design and Development*, Vol. 5, No. 3, pp. 322-329.

Cooper, M.G., 1989. "Flow Boiling–the 'Apparently Nucleate' Regime." *Int. J. Heat Mass Transfer*, Vol. 32, No. 3, pp. 459-464.

Corty, C., and Foust, A., 1955. "Surface Variables in Nucleate Boiling." *Chem. Eng. Prog. Symp. Series*, No. 51, Vol. 17, pp. 1-12.

Dengler, C.E., and Addoms, J.N., 1956. "Heat Transfer Mechanism for Vaporization of Water in a Vertical Tube." *Chem. Eng. Prog. Symp. Series*, No. 18, Vol. 52, pp. 95-103.

Frost, W., and Kippenhan, C.J., 1967. "Bubble Growth and Heat Transfer Mechanisms in the Forced Convection Boiling of Water Containing a Surface Active Agent." *Int. J. Heat Mass Transfer*, Vol. 10, pp. 931-949.

Gungor, K.E., and Winterton, R.H.S., 1986. "A General Correlation for Flow Boiling in Tubes and Annuli." *Int. J. Heat Mass Transfer*, Vol. 29, No. 3, pp. 351-358.

Jallouk, P.A., 1974, "Two-Phase Flow Pressure Drop and Heat Transfer Characteristics of Refrigerants in Vertical Tubes." Ph.D. Thesis, University of Tennessee at Knoxville.

Kenning, D.B.R., and Cooper, M.G., 1989. "Saturated Flow Boiling of Water in Vertical Tubes." *Int. J. Heat Mass Transfer*, Vol. 32, No. 3, pp. 445-458.

Klausner, J.F., 1989, "The Influence of Gravity on Pressure Drop and Heat Transfer in Flow Boiling." Ph.D. Thesis, University of Illinois at Urbana-Champaign.

Pujol, L., 1968. "Boiling Heat Transfer in Vertical Upflow and Downflow in Tubes." Ph.D. Thesis, Lehigh University.

Somerville, G.F., 1962. "Downflow of n-Butanol in a Uniformly Heated Tube." M.S. Thesis, University of California at Berkeley.

Staub, F.W., and Zuber, N., 1966. "Void Fraction Profiles, Flow Mechanisms, and Heat Transfer Coefficients for Refrigerant 22 Evaporating in a Vertical Tube." *ASHRAE Trans.*, Vol. 72, Part 1, pp. 130-146.

Tran, T.N., Wambsganss, M.W., France, D.M., and Jendrzejczyk, J.A., 1993. "Boiling Heat Transfer in a Small, Horizontal, Rectangular Channel." *AIChE Symp. Series*, Vol. 89, pp. 253-261.

Wang, C.H., and Dhir, V.K., 1993. "On the Gas Entrapment and Nucleation Site Density During Pool Boiling of Saturated Water." *J. Heat Transfer*, Vol. 115, pp. 670-679.

Yang, S.R., and Kim, R.H., 1988. "A Mathematical Model of the Pool Boiling Nucleation Site Density in Terms of Surface Characteristics." *Int. J. Heat Mass Transfer*, Vol. 31, No. 6, pp. 1127-1135.

Zeng, L.Z., and Klausner, J.F., 1993a, "Heat Transfer, Incipience, and Hysteresis in Saturated Flow Boiling. " *AIChE Sypmp. Series*, Vol. 115, pp. 215-221.

Zeng, L.Z., and Klausner, J.F., 1993b, "Nucleation Site Density in Forced Convection Boilng." *J. Heat Transfer*, Vol. 115, pp. 215-221.

INVESTIGATION OF BUBBLE DEPARTURE MECHANISM IN
SUBCOOLED FLOW BOILING OF WATER USING HIGH-SPEED PHOTOGRAPHY

Satish G. Kandlikar, Viktor R. Mizo, and Michael D. Cartwright
Department of Mechanical Engineering
Rochester Institute of Technology
Rochester, New York

ABSTRACT

The bubble growth rates under flow boiling conditions are obtained experimentally for subcooled flow of water in a rectangular channel 3 mm x 50 mm wide and over a 10 mm diameter heater surface in the bulk temperature range of 60 to 80 °C, and a liquid Re range of 1290 to 5160, and near atmospheric pressure. A high speed camera capable of obtaining images at a frame rate of up to 6000 fps and a microscope capable of providing up to 1350X magnification are employed. The effects of flow rate, subcooling, heater surface temperature, and cavity size are investigated and specific trends observed in the growth rate behavior are reported. The growth rates in flow boiling are compared with the ones for pool boiling case, and a need for an improved model for predicting bubble growth rates in flow boiling is established.

INTRODUCTION

Early work on nucleation in flow boiling presented by Bergles and Rohsenow (1964) extended the nucleation model for individual cavities in pool boiling (Hsu, 1962) to the case of flow boiling. Another area of research under focus since 1960's is the prediction of departure bubble diameter of individual bubbles as a function of flow parameters. Kandlikar and Stumm (1993) provide a summary of the work done by previous investigators and propose a new model for small diameter bubbles (less than 500 micrometers).

As seen from the above discussion, bubble nucleation and bubble departure in flow boiling have been investigated to some extent in literature. However, the important region of bubble growth lying between bubble nucleation and its departure has not been looked into. The bubble growth of individual bubbles occurs in a flowing liquid under the presence of strong velocity and temperature fields in the vicinity of the wall. The temperature field is altered due to flowing liquid and the transient heating model of pool boiling is no longer directly applicable.

The present work is part of an ongoing effort to understand the mechanism of nucleation in flow boiling. The focus here is to obtain experimental data for the bubble growth rates in flow boiling.

OBJECTIVES OF THE PRESENT WORK

The main objective of the present work is to obtain experimental data on bubble growth rate from individual cavities of known sizes in subcooled flow boiling of water near atmospheric pressure. The growth rate curves will be obtained for different cavity sizes as a function of flow velocity (liquid Reynolds number), subcooled water temperature, and heater surface temperature. The curves will be compared with the bubble growth rate curve predicted for the pool boiling case using Mikic and Rohsenow's (1969) model, and the need for a separate model to account for the changes in the temperature field adjacent to the bubble caused by liquid flow will be established. The work on the theoretical development of bubble growth rates is currently under progress in the Thermal Analysis Laboratory at RIT.

EXPERIMENTAL SETUP

The experimental setup is described by Kandlikar, Cartwright and Mizo (1995) which appears under another session at the same conference. Further details of the experimental setup may be found in Kandlikar (1992, 1994) and Kandlikar and Stumm (1993).

EXPERIMENTAL PROCEDURE

The experimental procedure for the preparation of the setup prior to obtaining data is same as that described by Kandlikar, Cartwright and Mizo (1995) appearing elsewhere in the same conference proceedings. The bubble growth rates of individual bubbles were obtained by maintaining water temperature and flow rates at the desired values and gradually increasing the heat input to the heater in small steps starting from a very low value. Signs of bubble activity were carefully monitored through microscope on the video screen. When any bubble activity was suspected at any sites, video pictures were recorded and replayed at a slower rate to detect bubbles. Certain sites were nucleated as the heat flux was increased, and further observations at increasing heat flux and flow rates were focused on these sites. As the bubble activity became more rapid, it was impossible to see any activity directly without the high-speed camera recording and replay. The maximum magnification used for regular camera (30 fps) was 1350X, and that with the high-speed camera (up to 6000 fps) was 430X. The maximum frame rate employed in the current investigation was 4000 fps.

EXPERIMENTAL RESULTS AND DISCUSSION

The experimental results are presented for the following ranges of parameters: bulk temperature: 60 to 80 °C; Reynolds number: 1290 to 5160; saturation temperature of water: 100 °C; and heater surface temperature: 100 to 120 °C.

Before reviewing the experimental data, it would be useful to see the theoretically predicted growth curve for pool boiling case using Mikic and Rohsenow's (1969) model. Figure 1 shows one such plot drawn for T_b=80 °C, T_w=108 °C, and a cavity radius r_c=5.53 μm. These conditions are comparable to those employed in the flow boiling experiments.

The growth rate curve shown in Fig. 1 indicates that for the stated conditions under pool boiling, a bubble growing on this cavity grows to a size of 90 μm in about 75 μsec. The bubble growth is extremely rapid and in order to visually observe say 10 frames during the bubble growth, a frame rate of 140,000

fps is needed. It may be noted that in pool boiling, nucleation is obtained over much larger cavity sizes which corresponds to a somewhat slower bubble growth. Small cavities of the size used in Fig. 1 will generally not be activated as they may be covered by larger bubbles. These cavities therefore may not play an important role in pool boiling. In the following paragraphs the flow boiling data obtained in the present investigation is presented.

Figure 2 shows the bubble growth curves for r_c=3.2 μm, T_w=80 °C, and Re=1664 for three different wall temperatures of 106.4, 107.2 and 107.9 °C. A change of 0.5 °C in wall temperature causes significant changes in the bubble growth curves. The growth time reduces from 130 ms to 22 ms, and further down to 5 ms with increasing wall temperatures. Note however that the departure bubble radius remains approximately the same around 40 to 45 μm. This data was obtained using high speed camera at 1000 fps.

With further increase in wall temperature, the growth rate becomes even faster. Figure 3 corresponds to the same conditions as for Fig. 2, except the wall temperature being higher at 109.6 and 110.4 °C. The bubble growth time is reduced drastically to about 1.6 ms, while the departure bubble radius became only slightly smaller to about 34 μm. Data in Fig. 3 was obtained using the high-speed camera at 3000 and 4000 fps for the two sets. The increasing growth rate with increasing wall temperature is still evident at these wall superheats.

A very important point may be made at this time regarding the visual observations. For all bubbles presented in Figs 2 and 3, direct observation of the heating surface revealed no nucleation activity and one may conclude the complete absence or "suppression" of nucleation. However, the high speed camera reveals significant bubble activity as shown in these figures. Further increase in surface temperature caused the bubbles to grow even faster, and their images soon started to become a blur on the screen. At the next step increase in surface temperature, no bubble activity could be observed even at 6000 fps (the highest speed available). The trends observed here make us believe that the bubble activity continued with increasing wall temperatures, perhaps in excess of 10,000 bubbles per second. The presence of nucleate boiling was further confirmed through the heat transfer coefficient values, which were higher than the single phase convection values. This point is further discussed by Kandlikar, Cartwright and Mizo (1995) in an accompanying paper presented at another session in this conference.

Figure 4 shows a similar trend as in Figs. 2 and 3 confirming a faster growth rate with an increase in wall superheat. Here, the cavity radius is somewhat larger, 4.2 μm. This causes the departure bubble radius to be correspondingly larger to about 66 μm. Note however that for both cases of wall superheat values in Fig. 4, the departure bubble radius is almost same, only the growth rate is affected.

The effect of cavity size on the departure bubble radius is shown in Fig. 5. For the same flow rates, and wall and bulk temperatures, cavities of radii 4.6 μm and 5.1 μm yield bubbles departing at 73 and 90 μm radii respectively. The growth rate curves are essentially parallel.

The effect of cavity radius on bubble growth rate is seen in Fig. 6. As the Reynolds number is increased to 5068, smaller cavities of 2.8 and 3.7 μm became active. The growth rates for the bubbles growing over these cavities were much faster. The larger cavity yielded a larger bubble, although the departure bubble sizes were also considerably reduced for both cavities to 18 and 22 μm respectively.

Fig. 7 presents two sets of conditions at the same flow rate yielding identical bubble growth rate curves. In the first case, a lower wall superheat combined with larger cavity size yielded the same curve as that for a higher wall superheat with lower cavity size. Again note that the departing bubbles are extremely small, reaching only 12 μm radius.

Effect of flow velocity is seen in Fig. 8. Here same cavity yields larger bubbles at smaller flow velocities under identical set of conditions; at Re=2280, departure radius is 27 μm, while at Re=1267, the departure radius increases to 38 μm. The bubble growth curves in the two cases are almost identical.

Figure 9 shows a similar plot as Fig. 8, however here the flow velocities are quite different. At Re=1568, the bubble growth rate curve is much slower and yields a departure bubble radius of 89 μm. For the case of Re=5643, the growth rate is much faster with the departure radius being only 15 μm. The cavity radii in the two cases are somewhat different, 5.5 and 5.1 μm respectively.

Based on the trends observed in the data collected and the high-speed visual observations, following specific comments can be made.

Effect of Flow Velocity

As the flow velocity increases, larger cavities stop nucleating, and bubbles start nucleating from smaller cavities. The departing bubble radii become smaller with increasing flow velocity. Higher flow rate makes the temperature gradient steeper near the wall, and larger cavities can no longer be activated. This effect is described in detail by Kandlikar (1990). Detailed experimental data on nucleation are presented by Kandlikar, Cartwright and Mizo (1995) in an accompanying paper, and a paper is under progress on the theoretical aspect of nucleation under flow conditions.

Effect of Wall Superheat

The effect of wall superheat is to increase the bubble growth rate from a nucleating cavity. The growth rate increases very rapidly near a certain value of wall superheat, dependent on the operating conditions. With as little as a 0.5 °C change of wall superheat, the growth rate increases ten-fold. The bubble frequency also increases as is to be expected. Bubble activity under these conditions could be traced only up to a certain point, beyond which even 6000 fps camera speed could not capture the rapid growth and departure. It is believed that the bubble activity continued at a very high frequency at least for some further increases in wall superheat values. It is not clear at this time what exactly happens in this region of bubble growth.

Effect of Subcooling

The effect of subcooling is quite interesting. At lower bulk temperatures (high subcooling), bubble growth was slow and was essentially in the thermally controlled region due to rapid condensation occurring at the top of the bubble exposed to the bulk flow. With higher bulk temperatures, the bubble growth was rapid, departure bubble radii became smaller, and bubble frequency increased. The bubble growth was so strongly dependent on the bulk temperature that at and above 85 °C, no bubble activity could be traced in spite of careful monitoring of the surface temperatures in small steps up to 116 °C. It is possible that bubbles smaller than 5 μm are ejected at high speeds (in excess of 6000 bubbles/second) and go undetected with the present imaging capabilities.

Comparison with Pool Boiling Bubble Growth Rates

The data presented in Figs. 2-9 can be compared with the pool boiling growth rates presented in Fig. 1 and experimental data reported by Hsu (1962). From this comparison, following observations can be made. (i) The bubble growth rate curves are significantly slowed down with increasing flow velocity under flow boiling. (ii) The sizes of cavities nucleating in flow boiling are much smaller as compared to pool boiling. (iii) Departure bubble radii are much smaller in flow boiling than in pool boiling.

CONCLUSIONS

Based on the present experimental study, the following conclusions are made:

1. The bubble growth rates in flow boiling are significantly different than those under the pool boiling conditions. Cavity sizes and departure bubble radii are also smaller in flow boiling.

2. An increase in flow velocity causes a shift in the nucleation activity to smaller sized cavities. The departure bubble radii also become smaller with increasing flow velocity.

3. An increase in wall superheat causes a rapid increase in the bubble growth rates, makes the departure bubble radii smaller, and causes smaller sized cavities to nucleate.

4. A decrease in subcooling (an increase in bulk temperature)

causes the bubble growth rates to increase, increases the bubble frequency and reduces the bubble radii. The bubble frequency becomes extremely large and no bubble activity could be observed (due to higher frequency than the camera speed), although the presence of bubbles is strongly suspected under some of the conditions. No nucleation activity could be observed for bulk temperatures above 90 °C.

ACKNOWLEDGEMENTS

The authors gratefully acknowledge the support provided by the Mechanical Engineering Department at RIT. The help provided by David Hathaway and Tom Locke in the Machine Shop is sincerely appreciated.

NOMENCLATURE

T_w - wall temperature, °C
T_b - bulk temperature, °C
r_c - cavity radius, m
Re - Reynolds number

REFERENCES

Bergles, A.E, and Rohsenow, W.M., 1964, "The Determination of Forced-Convection Surface Boiling Heat Transfer," Journal of Heat Transfer, Vol. 86, pp. 365-372.

Hsu, Y.Y., 1962, "On the Size Range of Active Nucleation Cavities on a Heating Surface," Journal of Heat Transfer, Vol.84, pp.207-216.

Kandlikar, S.G., 1990a, "A Mechanistic Model for Flow Boiling Heat Transfer," Paper presented at the 1990 ASME Winter Annual Meeting, Dallas, Nov., Single and Multiphase Convective Heat Transfer, eds. M.A. Ebadian, K. Vafai, and A. Levine, eds., ASME HTD-Vol. 145, pp. 61-69.

Kandlikar, S.G., 1992,"Bubble Behavior and Departure Bubble Diameter of Bubbles Generated Over Nucleating Cavities in Flow Boiling," Pool and External Flow Boiling, Proceedings of The Engineering Foundation Conference on Pool and External Flow Boiling, March 22-27, ASME.

Kandlikar, S.G., 1994,"Measurement of Departure Bubble Diameter and Advancing and Receding Contact Angles in Subcooled Flow Boiling of Water," Paper presented at the joint ASME and ISHMT Conference, Jan. 5-8, BARC, Bombay, India.

Kandlikar, S.G., Cartwright, M.D., and Mizo V.R., 1995, "A Photographic Study of Nucleation Characteristics of Cavities in Flow Boiling," Proceedings of International Conference on Convective Flow Boiling, ASME Foundation Conference, Banff, Canada, April 30-May 5.

Kandlikar, S.G., and Stumm B.J., 1993, "A Control Volume Approach for Investigating Forces on a Departing Bubble under Subcooled Flow Boiling," ASME National Heat Transfer Conference, Colorado Springs. Also accepted for publication in Journal of Heat Transfer, to appear in Nov. 95 issue.

Mikic, B.B., and Rohsenow, W.M., 1969, "Bubble Growth Rates in Non-uniform Temperature Field," Progress in Heat and Mass Transfer, Vol. II, pp. 283-293.

Nukiyama, S., 1934, "The Maximum and Minimum Values of Heat q Transmitted from Metal Surface to Boiling Water under Atmospheric Pressure," Journal of Society of Mechanical Engineers (Japan), Vol. 37, pp. 367-374, 553-554.

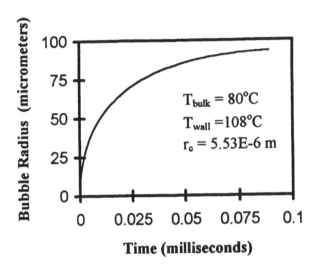

Figure 1. Pool Boiling Bubble Growth Curve from Mikic and Rohsenow's (1969) Model

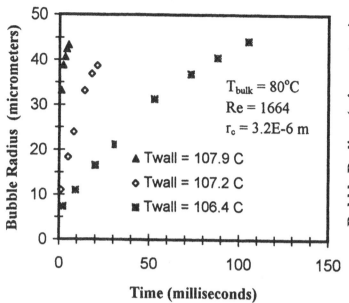

Figure 2. Effect of Wall Temperature on the Bubble Growth Rate for Tb = 80 °C, Re = 1664, and Cavity Radius = 3.2 μm (1000 fps).

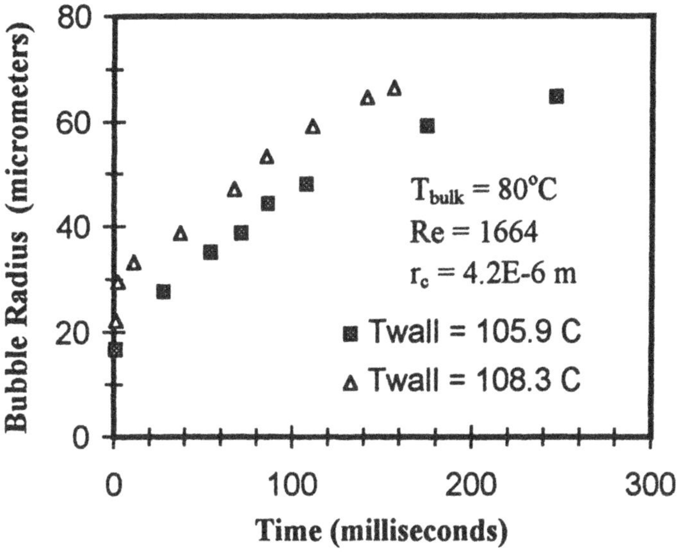

Figure 4. Effect of Wall Temperature on the Bubble Growth Rate for Tb = 80 °C, Re = 1664, and Cavity Radius = 4.2 μm (1000 fps).

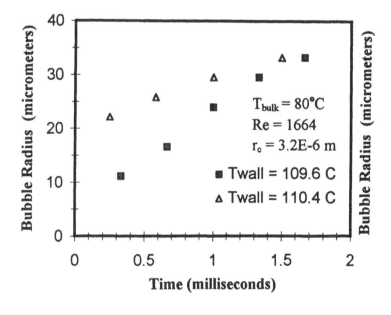

Figure 3. Effect of Wall Temperature on the Bubble Growth Rate for Tb = 80 °C, Re = 1664, and Cavity Radius = 3.2 μm (3000 and 4000 fps).

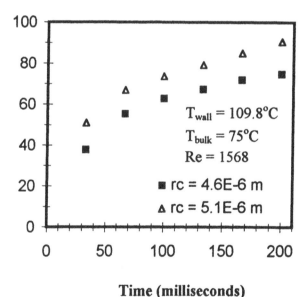

Figure 5. Effect of Cavity Size on the Bubble Growth Rate for Tb = 75 °C, Re = 1568, (1000 fps).

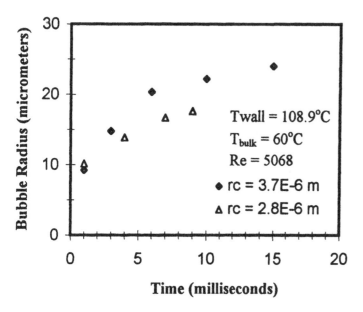

Figure 6. Effect of Cavity Size on the Bubble Growth Rate for Tb = 60 °C, Re = 5068, (1000 fps).

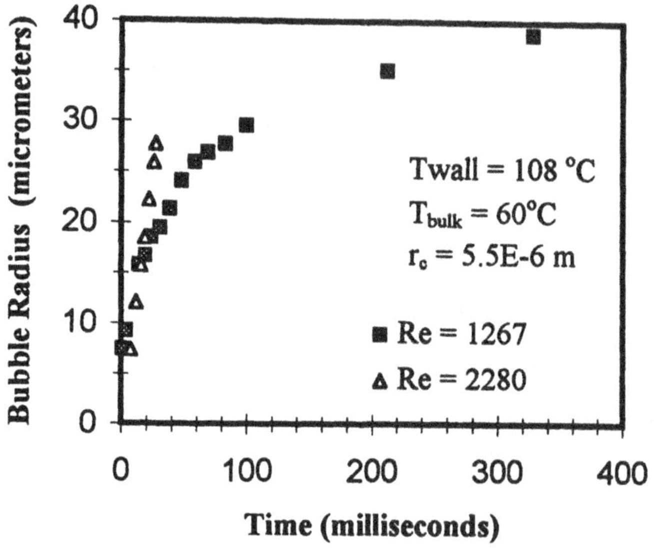

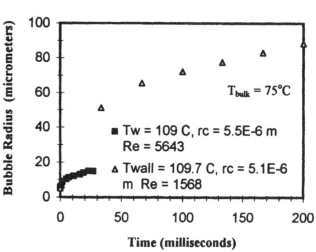

Figure 8. Effect of Reynolds Number on the Bubble Growth Rate for Tb = 60 °C, Cavity Radius = 5.5 μm, (1000 fps).

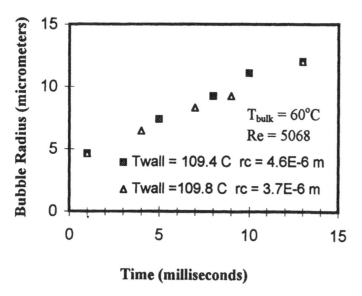

Figure 7. Effect of Cavity Size and Wall Temperature on the Bubble Growth Rate for Tb = 60 °C, Re = 5068, (1000 fps).

Figure 9. Effect of Reynolds Number on the Bubble Growth Rate for Two Identical Cavities at Tb = 75 °C, (1000 fps).

THE PHYSICAL DIMENSION IN CONVECTIVE BOILING

Keith Cornwell and I A Grant
Department of Mechanical and Chemical Engineering
Heriot-Watt University,
Edinburgh, EH14 4AS, Scotland

ABSTRACT

The aim of this paper is to **critically review** the use of the physical dimension and the perceived influence of geometry in convective boiling. It is confined to saturated, single component pool and flow boiling although many of the conclusions are more general. The case is made for representation of convective boiling in a similar way to single phase forced convection with dependence on a physical dimension. Much of the experimental evidence is based on work at Heriot-Watt University using water and Refrigerants in rigs providing a wide range of geometries. Initial results from new work on boiling in small diameter tubes is included.

INTRODUCTION

Much of the early work on the design of tube-side and shell-side boilers and evaporators showed some appreciation of the need to allow escape of the vapour and thus the influence of geometry. Generally however the heat transfer resistance of the boiling side was so low compared to the bon-boiling side that the resistance was effectively ignored and no serious study of boiling was made until the 1930's when Nukiyama (1934) published the so-called boiling curve and Jacob (1949) and others showed the importance of the bubble site density and the bubble departure size and frequency.

In retrospect it is unfortunate for engineering heat transfer development that there was so much preoccupation with certain aspects of nucleate boiling. This diverted attention away from the need to develop engineering correlations of the type used in say forced convection. These diversions may be outlined as follows:-

Diversion 1, Nucleation Sites

The intrinsic attraction of studying the incipience of boiling, the seeding of the sites, the bubble growth, the local heat transfer mechanism at the site, the bubble departure size, the site density, interaction between sites and so on all at low heat flux is undeniable. The physics is interesting, the experiments are fairly simple, the cost is low and the number of variables is enormous. However most industrial equipment operates at heat fluxes which are about a quarter to a third of the critical heat flux where there is a high intensity of bubbles on the surface and strong convective currents in the liquid. Representations of advancing liquid fronts at a known contact angle flowing into neat conical cavities simply do not apply under these conditions. The surface itself will always ensure that the micro-mechanism is intangible.

Diversion 2, Pool Boiling

Pool boiling is deemed to occur when there is no **imposed** liquid velocity in the potential field around the boiling surface. However the **actual** liquid velocity at the surface is a strong function of the vapour flow, the geometry and the liquid properties even under these so-called pool boiling conditions. An obvious example is pool boiling on a horizontal tube as shown in Figure 1 where most of the heat transfer occurs in the bubbly flow layer formed at the sides and is thus dependent on the tube diameter.

Diversion 3, Flat Plates and Wires

The relative ease of conducting laboratory experiments with these two geometries has led to most of the data being obtained for geometries which are far removed from those

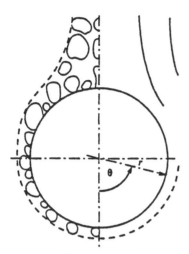

Fig 1. The Bubble Layer

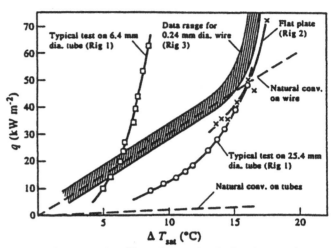

Fig 2. Pool Boiling of R113 in Similar Geometries, Cornwell et al (1982)

used in industrial processes. This may not matter at high heat fluxes where geometry is less important but is important at the intermediate values used in most processes. Furthermore the horizontal flat plate is an anomalous geometry in that no buoyancy bubble layer develops and the wire is thinner than the departing bubble, thus disallowing formation of a consistent liquid micro-layer under the bubble. These factors lead to large differences in the lower regions of the boiling curves as shown in Figure 2 for the boiling of R113 on various geometries in our laboratory, Cornwell et al (1982).

The following Sections review the mechanisms and correlations for nucleate boiling at low to intermediate heat flux and the influence of geometry in external and internal flow boiling.

CONVECTIVE MECHANISM AND CORRELATION

The well-known models of boiling on a flat plate by Mikic and Rohsenow (1969) and Judd and Hwang (1976) essentially divide the heat transfer into a natural convection (nc), enhanced boiling convection (bc) around the bubble and microlayer evaporation (me) under the bubble:

$$q = q_{nc} + q_{bc} + q_{me}$$

The latter authors present the contributions from each for water at 1 atmosphere. Under intermediate nucleate boiling conditions the heat flux is divided between the enhanced convection at the bubble (bc) and microevaporation (me).

Essentially therefore saturated nucleate boiling is dependent on conduction in a boundary layer which is either bounded by liquid in potential flow at T(sat) or a two-phase boundary at T(sat) under a bubble. This is not dissimilar to turbulent forced convection with a driving force of T(w) − T(sat) and boiling under these conditions may well be correlated by forced convection relationships. A difficulty arises with this approach however with the identification of an effective velocity and we argue here that nucleate boiling

is a function of two Reynold numbers; the **Boiling Reynolds Number** Re_b dependent on the vapour flow and the **Flow Reynolds Number** Re_f dependent on the imposed liquid flow as shown diagramatically in Figure 3. Convective boiling may then be correlated in dimensionless form as:-

$$Nu = f\left(Re_b, Re_f, Pr_f\right) \qquad (1)$$

The definition of Re_f is clearly the conventional expression used in turbulent forced convection. The definition of Re_b may be found by considering pool boiling on a horizontal surface when the Re_f term has no influence as there is no imposed flow. Under such conditions the driving force per unit area for the liquid flow is taken as the product of the vapour mass fluid G_g and the velocity u. The viscous force resisting this flow is μ_g u/d where the physical dimension d is conventionally taken as the bubble departure diameter d_b. If the Boiling Reynolds Number is the ratio of these forces it becomes

$$Re_b = \frac{G_g u d_b}{\mu_f u} = \frac{q d_b}{h_{fg}\mu_f} \qquad (2)$$

where saturated conditions are assumed. Equation (1) can then be written in the following form:-

$$\frac{\alpha d_b}{k_f} = C\left(\frac{q d_b}{h_{fg}\mu_f}\right)^m Pr_f^{\,n} \qquad (3)$$

Fig 3. The Imposed Liquid Flow and Vapour Flow Reynolds Numbers

This is the classical Rohsenow correlation in which m = 0.667 and n = 0.7. Kuloor and Radhakrishnan (1966) list four other correlations of this form in which m varies from 0.625 to 0.72 and n is about 0.3.

However, as was mentioned earlier, this case is anomalous as the geometry dictates that no transverse bubbly flow develops. For other cases such as outside tubes, inside tubes and non-horizontal surfaces the vapour removal is strongly dependent on the geometry and the characteristic dimension d is a feature of this geometry as shown in the following Sections.

THE PHYSICAL DIMENSION AND EXTERNAL FLOW BOILING

Given that forced-convective-type correlations may be used for nucleate boiling it is reasonable to attempt correlation of the type shown in equation (3).

$$Nu_b = C \, Re_b{}^m \, Pr^n \qquad (4)$$

for geometries other than a flat plate (where there is no imposed liquid velocity). Cornwell et al (1982) first developed such a correlation for a limited range of fluids and conditions and this has recently been developed by Cornwell and Houston (1994) into a general correlation for nucleate boiling on tubes.

The plotting of Nu against Re_b with the tube diameter D as the physical dimension was particularly revealing. Figures 4,5 and 6 (from Cornwell and Houston, 1994) show respectively the data from 6 experimental studies for water, 16 for R113 and 6 for R11 mainly at 1 atmosphere except for a few points in Figure 6 which are at higher pressure. The route to generalising the equation was less obvious, but use of the Mostinski (1963) pressure factor and the critical pressure to take account of pressure and fluid property effects, yielded a final correlation as:-

$$Nu_b = AF(p) \, Re_b{}^{0.67} \, Pr^{0.4} \qquad (5)$$

where $A = 9.7p_c{}^{0.5}$ (with p_c in bar)

$$F_{(p)} = 1.8p_r{}^{0.17} + 4p_r{}^{1.2} + 10p_r{}^{1.0}$$

$$p_r = p / p_c$$

(The lines in Figures 4, 5 & 6 give this correlation for 1 atmosphere). The correlation applies to tubes in the diameter range 8-50mm with as-machined surface, heat fluxes in the range 10-80% of critical heat flux, pressures of 0.001-0.8 of critical pressure and ordinary fluids such as water, refrigerants, organics, but not liquid metals and cryogenics. Data fit to the general correlation is shown in Figure 7 and while there are rogue results and considerable scatter it is no greater than that for boiling correlations which are more complex than equation 5.

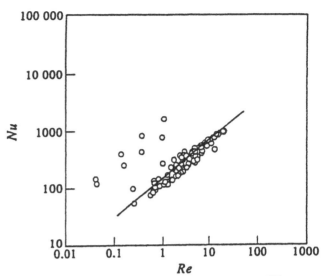

Fig 4. The Nu-Re_b Plot for Data on Water

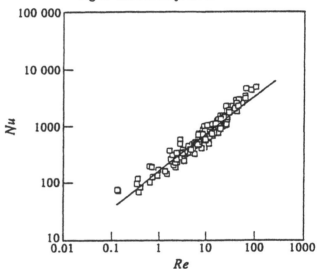

Fig 5. The Nu-Re_b Plot for Data on R113

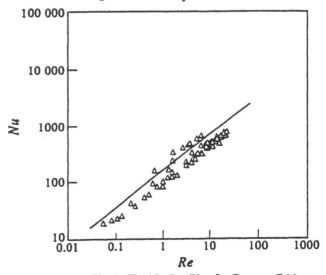

Fig 6. The Nu-Re_b Plot for Data on R11

The correlation implies that:

$$\alpha = f\left(D^{-0.33}\right) \qquad (6)$$

and this trend has indeed been shown by Cornwell et al (1982) and Hahne and Shi (1993). Data on refrigerants reviewed by the latter authors shown (in simplified form) in Figure 8 where the ordinate is the ratio of are acual α to its value on a tube of 8mm diameter. (The line shown actually indicates a power of -0.5 in equation 6 but the data is too scattered to be precise).

Interestingly, both these papers indicate a fall in x at diameters less than 6-8mm to values approaching half the maximum for wires as $D \rightarrow O$. This is to be expected as the additional heat transfer due to the evaporation under the sliding bubbles in the bubble layer (see Fig 1) is absent for wires and also for flat horizontal plates which have a similar α - value to wires.

The influence of imposed flow in the correlation is to multiply equation 5 by a flow Reynolds number such that:-

$$Nu_{fb} = Nu_b \, Re_f^n \qquad (7)$$

There is limited precise data in the literature but analysis of that in Cornwell and Emersson (1990) indicates values of n = 0.20 to 0.25. Here

$$Re_f = \frac{\rho_f U D}{\mu_f}$$

and is taken as unity when velocity U becomes less than that of the mean bubble stream velocity with no imposed flow (pool boiling). An interesting case worthy of investigation would be that of a downward flow on a horizontal tube.

Horizontal plates provide a unique boiling geometry as mentioned earlier. However non-horizontal plates exhibit many of the features of a tube with a bubble layer commencing at the leading edge this being the lowest. Cornwell and Einarsson (1990) show that if a tube is considered as a series of flat plates a similar peripheral heat transfer distribution is obtained. Small horizontal surfaces tend to yield higher heat transfer and Nishikawa and Fujita (1990) make allowance for this by inclusion of a 'representative length of heating surface' ℓ. Their correlation implies.

$$\alpha = f\left(\ell^{-0.45}\right) \text{ for laminar flow}$$
$$\alpha = f\left(\ell^{-0.86}\right) \text{ for turbulent flow}$$

These variations have not been independently assessed but arise from the general correlation and should therefore be treated with caution.

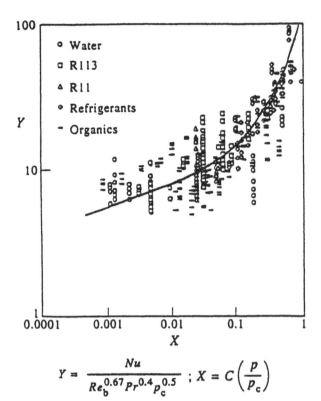

$$Y = \frac{Nu}{Re_b^{0.67} Pr^{0.4} p_c^{0.5}} \; ; \; X = C\left(\frac{p}{p_c}\right)$$

Fig 7. General Correlation for All the Fluids

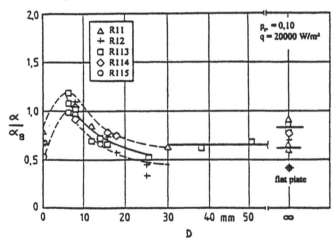

Fig 8. Effect of Tube Diameter on Heat Transfer Relative to an 8mm Tube, Hahne & Shi (1993)

THE PHYSICAL DIMENSION AND INTERNAL FLOW

Large Diameter Tubes

The Chen (1966) correlation and more recent forms such as Gungor and Winterton (1986) and Liu and Winterton (1991) based on a similar approach have been satisfactory for prediction purposes. However the Shah (1976) correlation, which also gives satisfactory prediction, is conceptually more appropriate to our argument here. For the case where gravitation effects are small it may be expressed as:-

either $Nu = C\,Bo^m Nu_{\ell_o}$ (8)

or $Nu = F\,Nu_{\ell_o}$ (9)

whichever is the larger (Note that the Boiling Number, is the ratio of liquid to vapour mass flow rate and F, the two-phase enhancement factor, is a function of quality and densities). At high quality the second equation applies and in this region both the Chen and Shah methods are similar.

The first equation which applies to nucleate boiling at low quality is of particular interest here. Shah gave m = 0.5 and the equation may therefore be written as:-

$$Nu = C\left[\frac{q}{Gh_{fg}}\right]^{0.5}\left[\frac{Gd(1-x)}{\mu_f}\right]^{0.8}Pr_f^{0.4} \qquad (10)$$

Rearrangement yields

$$Nu_b = C\left[\frac{qd}{h_{fg}\,u_f}\right]^{0.5}\left[\frac{Gd}{\mu_f}\right]^{0.3}[1-x]^{0.8}\,Pr_f^{0.4}$$

and at low quality

$$Nu = C\,Re_b^{0.5}\,Re_f^{0.3}\;Pr_f^{0.4} \qquad (11)$$

Thus essentially, at low quality (x<0.1), heat transfer is an identified function of the *vapour flow Reynolds number* and the *liquid flow Reynolds number* as indicated in Figure 3. Note that Bo is simply the ratio of these two numbers.

$$Bo = \frac{Re_b}{Re_f} = \frac{q}{h_{fg}G} \qquad (12)$$

and that the power of Re_f is not dissimilar to that in Equation 7. Once again G has a minimum value dictated by its effective local value with no imposed flow.

It would appear that the Shah correlation indicates that both in the nucleate boiling and convective evaporation ranges

$$\alpha = f\!\left(d^{-0.2}\right)$$

The Chen correlation yields no variation with d in nucleate boiling and more recent correlations exhibit a complex variation owing to inclusion of Re_f in the suppression factor.

Small Diameter Tubes and Channels

In this context small-diameter tubes are deemed to be those that yield flow and heat transfer characteristics which differ from those predicted by correlations based on large tubes. Recent experimental work conducted by the authors (Cornwell, Kew and Grant, 1995) has examined the heat transfer of R141b   in tube sizes around the transition

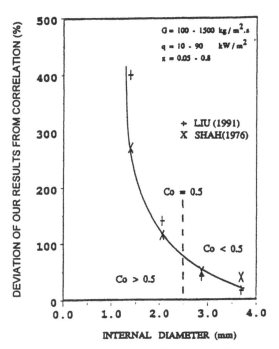

Fig 9. Influence of Diameter on Fit to
Correlations for Flow Boiling

range from large to small diameter and some typical results are shown in Figures 9-12.

In our work on compact channels (Kew and Cornwell, 1994) we have shown that a Confinement Number Co may be used to represent the influence of confinement on the heat transfer coefficient and also to indicate the transition which appeared to occur at Co = 0.5 where

$$Co = \frac{1}{d}\left[\frac{\sigma}{g(\rho_f - \rho_g)}\right]^{0.5} \qquad (13)$$

For R141b at 1 atmosphere this yields a critical diameter of about 2.5mm below which confinement has an effect. Figure 9 appears to confirm that departure from large diameter correlations occurs around this point.

Figures 10, 11 and 12 indicate that for a 'small' tube the heat transfer may be greater or less than that predicted by the 'large' tube correlations depending on the mass flow rate. Reasons for this variation lie in different flow regimes which occur under the same conditions in small tubes. At low flow rate bubbly and slug flow yields a thin rapidly evaporating liquid layer on the surface which may enhance the heat transfer as shown in Figure 10.   At high flow rate observation showed that small slugs of liquid were spat intermittently through the tube at high velocity rather than the smooth annular flow presumed in the predictions. This yielded a heat transfer more akin to nucleate boiling than convective evaporation and thus very much lower as shown

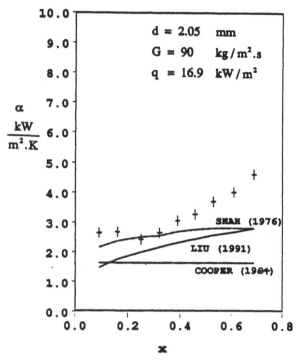

Fig 10. Variation of α with Quality at low Flow-Rate

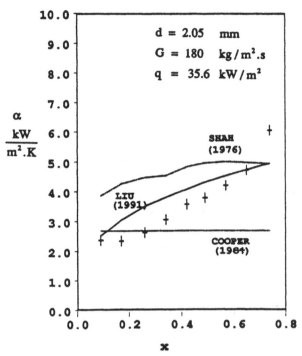

Fig 11. Variation of α with Quality at Medium Flow-Rate

in Figure 12. Some corroboration of this hypothesis is provided by Damianeides and Westwater (1988) who show an extension of the intermittent flow range at smaller diameters. Wambsganss et al (1993) examine boiling of R113 in a 2.92mm tube for flow rates of 50-300 kg/m²s and find that for this range the large tube correlations work well. Obviously there is a great need for further study of this matter.

At the lower flow rates of interest in compact heat exchangers with 'small' channels, Kew and Cornwell (1994) have shown that the enhancement in the nucleate boiling regime may be correlated using Co and from the limited data in the literature have tentatively suggested a correlation based on equation 11 in the form:-

$$Nu = C\,Re_b^{0.5}\,Re_f^{0.3}\,Co^{0.5}\,Pr_f^{0.4} \qquad (14)$$

This yields the dependence at constant G

$$\alpha = f\!\left(d^{-0.7}\right)$$

which, as expected, is a much stronger dependence than for large tubes. Whether correlations developed from flow in between plates and in small channels (where liquid may linger in the corners) may be applied to round tubes is questionable.

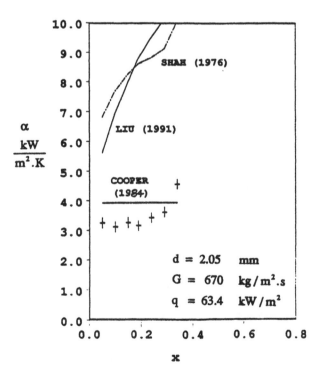

Fig 12. Variation of α with Quality at High Flow Rate

CONCLUSIONS

1. Unlike forced convection the influence of the physical dimension in nucleate boiling has not been widely studied although it is implicit in some studies and the resulting correlations.

2. The case of pool boiling outside horizontal tubes is the only geometry where detailed studies have been conducted.

3. The indications are that for pool and flow nucleate boiling inside tubes, outside tubes and over finite flat plates the physical dimension has a fairly strong influence. In the relationship:

$$\alpha = f(d')$$

the power r is generally in the range -0.2 to -0.9.

4. Work is continuing on boiling relationships of the forced convection type, particularly on imposed flow effects on the outside of tubes and over flat plates at various angles.

REFERENCES

Chen J G (1966) 'Correlation for Boiling Heat Transfer to Saturated Fluids in Convective Flow", I & EC Proc. Des & Dev. 5, 322-333.

Cooper, M G (1984), 'Heat flow rates in saturated nucleate pool boiling - a wide-ranging examination using reduced properties', In Advances in Heat Transfer, Vol. 16, pp 157-239. Academic Press, New York (1984).

Cornwell, K and Einarrson, J G (1990) 'The influence of fluid flow on nucleate boiling from a tube', Exp. Heat Transfer (Hermisphere), 3, 101-116.

Cornwell K, Schuller R B and Einarsson J G (1982) 'The Influence of diameter on nucleate boiling outside tubes, Proc. 7th. Int. Heat Transfer Conf. 4, 47-53, Munich.

Cornwell, K and Houston S D (1994), 'Nucleate Boiling on Horizontal Tubes : A Convection-Based Correlation', Int J Heat & Mass Transfer, 37, (Sup 2) 303-309.

Cornwell, K, Kew P A & Grant I A G (1995) - To be published.

Diamiandes C A & Westwater J W (1988), 'Two-Phase Flow Patterns in a Compact Heat Exchanger and in Small Tubes', Proc 2nd UK Nat H T Conf, Glasgow, Scotland, 2, 1257-1268.

Gungor R E, & Winterton, R H S (1986) ' A general correlation for flow boiling in tubes and annuli', Int J Heat Mass Transfer 29, 351-353.

Hahne E & Shi, K (1993) 'How to obtain pool boiling heat transfer coefficients from tubes from wire experiments' Int. Commun. Heat Mass Transfer 20, 361-371.

Jakob, M (1949) Heat Transfer, Vol 1 & 2, Wiley, New York.

Judd, R L and Hwang, K S (1976), 'A comprehensive model for nucleate pool boing heat transfer including microlayer evaporation', ASME, J Heat Transfer, 9B, 623-629.

Kew, P A and Cornwell K (1944) 'Confined Bubble Flow end Boiling in Narrow Spaces' Heat Transfer 1994 (10th Int Heat Transfer Conf), 7, 473-478.

Kuloor, N R and Radhakrishnan, V N (1966) 'Effect of surface roughness on nucleate boiling', Chem Proc Eng 47(6), 276-286.

Liu Z and Winterton R H S (1991) 'A General Correlation for Saturated and Subcooled Flow Boiling in Tubes and Annuli, Int J Heat Mass Transfer, 34, 2759-2766.

Mikic, B B and Rohsenow, W M (1969), 'A new correlation of pool boing data including the effect of heating surface characteristics' ASMEJ J Heat Transfer, 91, 245-250.

Mostinski. I L (1963), 'Calculation of Heat Transfer and Critical Heat Flux in Boiling Liquid based on the Law of Corresponding States, Teploenergetika' 10(4), 66-71.

Nishikawa K and Fujita Y (1990) 'Nucleate Boiling and its Augmentation', Adv. Heat Transfer, 20, 1-82.

Nukiyoma, S (1934) 'The maximum and minimum values of the heat transmitted from metal to boiling water under atmospheric pressure', Tr. by C J Lee form J. Jap. Soc. Mech. Engr, 37, 367-374, (1934) in Int J Heat Mass Transfer, 9, 1419-1433 (1966).

Shah, M M (1976) 'A new correlation for heat transfer during boiling through pipes' Trans. ASHRAE 82, 66-86.

Wambsganss M W et al (1993) 'Boiling Heat Transfer in A Horizontal Small-Diameter Tube', J Heat Transfer 115, 963-972.

THE PRINCIPAL MECHANISMS FOR BOILING CONTRIBUTION IN FLOW BOILING HEAT TRANSFER

Victor V.Yagov
Moscow Power Engineering Institute
(Technical University)
Moscow, Russia

ABSTRACT

The nucleate boiling heat transfer mechanisms play the basic role in the flow boiling at the wide range of controllig parameters. The paper presents the mechanistic model of the fully developed nucleate boiling which accounts for the two main ways of the heat removal from the wall. The predictive correlation obtained on the base of the model is found to be in good agreement with numerous experimental data for both pool and flow boiling. At the relatively high flow velocities the simple interpolation correlations are recommended for the flow boiling heat transfer computations.

INTRODUCTION

The present paper is devoted to that mode of flow boiling which may be identified as the nucleate boiling regime. This regime extends from the incipience boiling channel cross-section till that axial position where the annular flow with rather thin vaporizing liquid film on the tube wall is established. According to Labuntsov (1960) the upper border of nucleate boiling regime under forced flow conditions may be defined by the void fraction value $\varepsilon \approx 0.7$. Under well developed nucleate boiling the heat transfer intensity doesn't depend on such hydrodynamical and thermodynamical effects as mass flow rate and liquid subcooling, the gravitational acceleration and electrical field superposition. The main problem of the convective flow boiling heat transfer is consequently to reveal the internal mechanisms of the nucleate boiling itself. Many of the correlations for flow boiling heat transfer (Kutateladze, 1979; Labuntsov, 1960; Chen, 1966; Bennet and Chen, 1980) are built as two-terms formulas reflecting the simultaneous effects of single-phase convection and pool boiling on the total heat transfer coefficient. However all the correlations for nucleate boiling heat transfer used in these works are empirical.

There are some reasons which don't allow to create a strict theory of nucleate boiling heat transfer. Firstly, the closed mathematical description of the nucleate boiling phenomenon is practically impossible because it ought to include (besides the ordinary set of the conservation equations for both phases) the conjugation conditions at the interface. But the shape and the space and temporal coordinates of the interface are unknown before and could be found only as a decision result. In addition it's necessary to bear in mind that the motion both of liquid and vapour phases is turbulent and so the ordinary problem of the mathematical description also exists here.

Secondly, the above mentioned independence of nucleate boiling heat transfer on such "strong" factors as mass flow rate, liquid subcooling and gravitational acceleration combines with very essential influence on the nucleate boiling of some "weak" factors, such as small surfactant admixture, dissolved gases, wettability, roughness of the heated wall and so on. It's doubtful that applicable to practice theory could take into account such a diversity of the different effects. Nevertheless there exists a possibility to find out some (not many) main features of nucleate boiling process which determine the principle mechanisms of heat transfer.

On this base an approximate theory of nucleate boiling heat transfer may be worked out and it's intended to describe HTC dependence on the controlling parameters under some "average conditions". Namely this way was chosen by Labuntsov (1959, 1963, 1972) and then was developed by the present author (Yagov, 1988, 1995). D.Labuntsov (1959) expressed a profound idea that the processes essential for the boiling heat transfer occurred in that region of the boundary layer near the heated surface where in the single-phase flows any hydrodynamical perturbations weren't felt. This approach had its natural continuation in the "approximate theory of the developed nucleate boiling heat transfer" (Labuntsov, 1963) which was the first successful attempt to get the correlation for nucleate boiling HTC directly on the ground of a physical model. According to this model at the isolated bubbles boiling regime heat from the solid surface is transferred to the cold liquid core by the conduction through the liquid sublayer which is to some degree analogous to the viscous sublayer in the single phase convection. The effective thickness of such a conductive sublayer in boiling depends on the nucleate sites density and the bubbles growth velocity. For the nucleation sites density (n_f) Labuntsov used the very simple law:

$$n_f = C_1 R_*^{-2} \tag{1}$$

with the unknown coefficient C_1 of an order of 10^{-8} - 10^{-7} The critical vapour bubble radius is determined as

$$R_* = 2\sigma / \Delta p, \tag{2}$$

where for small temperature differences $\Delta T = T_W - T_S$

$$\Delta p = \frac{\Delta T h_{lg} \rho_g}{T_S}. \tag{3}$$

The formula (1) is based on the dimensions analysis and on the qualitative consideration of boiling conditions on the commercial surfaces (Labuntsov,1963). The vapour bubble growth rate according to Labuntsov is determined by the heat flux from the wall through the thin liquid layer in the base of the bubble (microlayer). This model gives for the current bubble radius:

$$R = \sqrt{\frac{2\beta\lambda\Delta Tt}{h_{lg}\rho_g}} \tag{4}$$

This formula was experimentally affirmed at Jakob numbers $Ja<10$ (relatively high pressures) at the empirically determined coefficient $\beta=6$. The conducted sublayer average thickness is governed by the action of a number of vapour bubbles simultaneously growing on the wall at the different stages of their evolution. To estimate their influence Labuntsov has calculated the kinetic energy of liquid agitated by the vapour bubbles. This kinetic energy value determines the mean fluctuating liquid velocity w_*:

$$w_* = \frac{C_2 R \dot{R}}{R_*}, \tag{5}$$

where $\dot{R} = dR/dt$, C_2 is a constant of an order of 10^{-2}. So using the simplest definition of "convective" heat flux at boiling which is transferred by the heat conduction through the liquid sublayer:
$q_1 \approx \lambda\Delta T / \delta_1,$

where $\delta_1 \approx 10v / w_*$, one can get from the expressions from (1) to (5)

$$q_1 = \frac{C_3\lambda^2\Delta T^3}{v\sigma T_S}, \tag{6}$$

where C_3 is the numerical factor of an order of 10^{-3}.

In the discussed work (Labuntsov, 1963) the additional heat flux q_2 connected with the liquid "direct evaporation" near the wall was also introduced. Unfortunately the expression for this heat flux q_2 was not grounded so thoroughly as for q_1 was. Later using the new experimental results upon the nucleate boiling mechanism (in particular, Gaertner, 1965; Iida and Kobayasi, 1969) Labunntsov (1972) showed that heat flux density through the liquid macrofilm could be expressed with the same formula (6) which was obtained for "convective" heat flux at the isolated bubbles regime. The resulting correlation of the paper of Labuntsov, (1972) includes the empirical corrective function of vapour/liquid densities relation and has a form

$$\alpha = 0.075(\frac{\lambda^2 q^2}{v\sigma T_S})^{1/3}[1+10(\frac{\rho_g}{\rho_l-\rho_g})^{2/3}] \tag{7}$$

This simple formula is practically convenient and demonstrates rather good agreement with experimental values of nucleate boiling HTC. However, the empirical expression in the square brackets causes some objections not only from the position of the scientific purity but from the side of practical applications because the accuracy of the correlation (7) can be insufficient in some cases (for example at nucleate boiling of some refrigerants). The problem is to reflect adequately the role of direct evaporation of the liquid on the heated wall.

THE NEW MODEL OF NUCLEATE BOILING HEAT TRANSFER

As it's based in the present author's work (Yagov, 1988) there exist the two simultaneously acting mechanisms which determine the total heat flux from the heated surface. So

$$q = q_1 + q_2, \tag{8}$$

where q_1 is the heat flux by conduction through the liquid macrofilm of the mean thickness δ_0, the second component q_2 is connected with intensive evaporation in the dry patches vicinity. The macrofilm mean thickness was calculated by the analogy with the boundary layer thickness in the single- phase flow. The characteristic scales of the velocity and the length were correspondingly the average evaporation rate $u_e = q /(h_{lg} r_g)$ and the average distance l_* between nucleation sites i.e. dry spots (see Fig.1). It's assumed that at high heat flux regime any vapour stem in the macrofilm may be considered as the nucleation site. Nucleation sites density is supposed to be determined by the Labuntsov's correlation (1). At the uniform dry spots distribution on the surface

$$l_* = n_f^{-1/2} = C_4 R_* \tag{9}$$

with the constant C_4 of an order $10^3 - 10^4$. Thus the mean macrofilm thickness is determined as follows

$$\delta_0 = C_5 (\nu R_* h_{lg} \rho_g / q)^{1/2}, \tag{10}$$

with the constant C_5 being of an order of 10^2. This follows from the previous estimations of the constants C_1 and C_4 and satisfactorily agrees with the experimental measurements of δ_0 in the works Gaertner, (1965); Bhat et al, (1986); Rajvanshi et al., (1992). (In the calculations of δ_0 the expressions (2) and (3) were used). The first ("convective") component of the heat flux q_1 is determined consequently as

$$q_1 = C_5^{-1} \lambda \Delta T (\frac{q}{\nu R_* h_{lg} \rho_g})^{1/2}. \tag{11}$$

It's easy to show that at the conditions which were used in the paper of Labuntsov (1972) $(q = q_1, R_* = 2\sigma T_s / (h_{lg} \rho_g \Delta T))$ the expression (6) is obtained again. But now q_1 depends on the total heat flux q.

From the general consideration it seems to be obvious that near the boundary of any dry spot (nucleation site or vapour stem in the macrofilm) there must be an area of the very intensive evaporation. Wayner and his co-workers (Wayner, 1973, 1978, 1982; Wayner et al, 1976, Potash and Wayner, 1972; Renk and Wayner, 1979) showed that such an area was placed in the part of the intrinsic meniscus region which bordered with the evaporating thin liquid film. The initial thickness δ_0 of the evaporating thin liquid film at the interline is of an order of a molecular diameter ($\sim 10^{-9}$ m). In this part of liquid film the evaporation rate is limited by both the London - van der Waals dispersion forces and the heat conduction resistance δ / λ. The dispersion forces influence diminishes but the heat conduction resistance increases with the liquid film thickness increasing. Wayner found out that the maximum heat flux through the liquid film (and the maximum evaporation rate) took place at that region of the extended meniscus where the London - van der Waals forces were negligible. Some results of Wayner and his coworkers were used in the recent paper of Lay and Dhir (1994) where a rather successfull attempt has been made to get the theoretical decision of the fully developed nucleate boiling heat transfer problem. However, according to the model used in this paper the nucleation sites density strongly depends on the contact angle. But this characteristic even for a

given surface-liquid combination is rather indefinite in practical applications, in any case the contact angle is a value of poor reproducibility. So the approach discussed is rather the parametrical study than the predicting one. Besides, the very large calculated vapour stem diameters (0.9 - 4.7 mm at one atmosphere pressure water boiling) cause the great doubts and contradict to the Gaertner's (1965) measurements and especially to the results of Dhuga and Winterton (1985) who have established that dry spots fraction on the heated surface has been very small during nucleate boiling regime (less than 1-2% of the whole surface area). In the work of Pasamehmetoglu et al (1993) where numerical investigations of nonsteady heat transfer in the vicinity of a vapor stem in the liquid macrofilm are performed the empirical correlations are used for the nucleation sites density and for void fraction near the wall. Besides "the triple-point mass evaporation coefficient" is introduced as an ajustable parameter in calculations.

In the present author's opinion in this situation it's reasonable to confine oneself to consider only the approximate model of the heat transfer in the vicinity of a dry spot. Here the principal ideas of such a model, firstly presented in the previous paper (Yagov, 1988) are developed. Fig.1 depicts the extended meniscus near a nucleation side (dry patch) and shows the linear scales which are used in the analysis. It's accepted that the liquid inflow into the intensive evaporation area is caused by the meniscus curvature gradient. The maximum curvature at the contact line is assumed to be equal to $K_* = 1 / R_*$. This estimation agrees with the Wayner's results for the liquid meniscus with the non-zero apparent contact angle (Wayner, 1973, 1982; Potash and Wayner, 1972). For linear decrease of the meniscus curvature and linear increase of the film thickness with radius the meniscus curvature gradient may be estimated as

$$dK / dR = \frac{C_6}{R_* \delta_m} \tag{12}$$

where δ_m is the liquid layer thickness at the region of the maximum evaporation rate, C_6 is the constant of an order of 10^{-2}. For radial flow in the thin liquid film of the

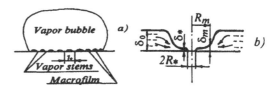

Fig.1. The models of a macrofilm (a) and a dry spot (b)

uniform thickness δ_m the mass flow rate per the film width unit is calculated as

$$\Gamma_R = 1/3 \frac{dP}{dR} \frac{\delta_m^3}{\nu}.$$

Assuming that this type correlation is true for the extended meniscus and calculating the pressure gradient as

$$dP/dR = -\sigma dK/dR$$

one gets (bearing in mind the eq. (12))

$$\Gamma_R = C_6' \frac{\sigma \delta_m^2}{\nu R_*} \qquad (13)$$

where C_6' has the same order of magnitude as C_6. For steady conditions the balance exists between arrived and evaporated liquid. So the heat flux per the film width unit may be written as follows

$$q_R = C_6' \frac{\sigma h_{lg} \delta_m^2}{\nu R_*}. \qquad (14)$$

On the other hand the heat flux in the interline region is determined by the heat conductance across the liquid layer. For the linear approximation of the meniscus shape it's easy to show that the total heat flux from the circle of the radius $R = R_m$ (i.e. the heat flux per one nucleation site) is expressed as

$$Q_{ns} = C_7 \lambda \Delta T \delta_m \qquad (15)$$

where C_7 is a constant of an order of 10^{-3} - 10^{-4}. The formula (15) is based additionally on the assumption $\delta_m \gg \delta_*$. So for the linear heat flux at $R = R_m$ one gets

$$q_R = C_8 \lambda \Delta T \qquad (16)$$

with the constant C_8 of an order of 10^1 - 10^2. From the correlations (14) and (16) it follows that the liquid film thickness in the maximum evaporation rate area is determined as

$$\delta_m = C_9 \left(\frac{\lambda \Delta T R_* \nu}{\sigma h_{lg}} \right)^{1/2}, \qquad (17)$$

where C_9 is the constant of an order of 10^2.

Substituting this result into the correlation (15) and using the formula (1) for the nucleation sites density one obtains the expression for the 2-nd constituent of the heat flux:

$$q_2 = C_{10} \left(\frac{\lambda \Delta T}{R_*} \right)^{3/2} \left(\frac{\nu}{\sigma h_{lg}} \right)^{1/2}. \qquad (18)$$

An order of the constant C_{10} is 10^{-2}. Substituting the expressions (11) and (18) into the equation (8) one gets the quadratic equation for q. Its physically meaningful root includes two unknown factors which are a combination of the constants C_5 and C_{10}, so these factors are determined with the accuracy of an order of magnitude. The below cited numerical values of these two factors were determined on the base of the comparison with the experimental data. Based on the Clapeyron - Clausius equation the linear dependence (3) between pressure and temperature drops along the saturation line is valid only at small ΔT. By means of keeping the second term in the $p_S - T_S$ dependence expansion the more exact correlation for rather high temperature drops may be obtained. After some simplifications this correlation takes a form

$$\Delta p = \frac{h_{lg} \rho_g \Delta T}{T_S} (1 + \frac{h_{lg} \Delta T}{2 R_i T_S^2}), \qquad (19)$$

where R_i is the gas constant for a given fluid. Bearing in mind these remarks the resulting equation for the nucleate boiling heat transfer may be presented in the following form

$$q = 3.43 \cdot 10^{-4} \frac{\lambda^2 \Delta T^3}{\nu \sigma T_S} (1 + \frac{h_{lg} \Delta T}{2 R_i T_S^2})(1 + \sqrt{1 + 800 B} + 400 B), \qquad (20)$$

where

$$B = \frac{h_{lg} (\nu \rho_g)^{3/2}}{\sigma (\lambda T_S)^{1/2}}. \qquad (21)$$

For the heat transfer coefficient this equation may be rearranged as follows

$$\alpha = 0.07 [\frac{\lambda^2 q^2}{\nu \sigma T_S} (1 + \frac{h_{lg} \Delta T}{2 R_i T_S^2})(1 + \sqrt{1 + 800 B} + 400 B)]^{1/3}. \qquad (22)$$

DISCUSSION

In the paper (Yagov, 1988) the calculations according to the equation (22) were compared with about 3000 experimental points for nucleate boiling (mainly pool boiling) of water, ethanol, methanol, benzene, methane, ethane, ethylene, nitrogen, oxygen and different refrigerants at the reduced pressure range $p_s / p_{cr} = 1.8 \times 10^{-4}$ - 0.94. More than 91% of experimental points deviate from the predicted values less than $\pm 35\%$.

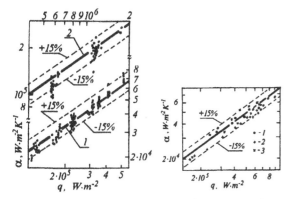

Fig.2. The predictive and experimental HTC values for water flow boiling: 1 - p_s=4.0 MPa (Morozov, 1964), 2 - p_s=9.8 MPa (Tarasova, Orlov, 1964)

Fig.3. The comparison of the predictive correlation (22) with the experimental data of Aihara et al (1993) for LN_2 inpinging jet flow boiling at flow rates 1 - 0.29 g/s, 2 - 0.39 g/s, 3 - 0.5 g/s

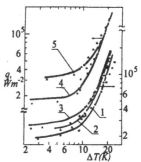

Fig.4. The comparison of the predictive correlation (23) with the experimental data by Hasan et al (1990) for subcooled flow boiling of R113. 1, 2, 3 - p_s=0.184, 0.219, 0.253 MPa at G=801 kg/(m²s); 4, 5 - G=579 and 1102 kg/(m²s) at p_s=0.219 MPa.

As it's discussed before under developed nucleate boiling the basic events occur in the closed vicinity of the heated wall, this region not being experienced any essential influence of the main liquid flow. A very simple rule to define the developed nucleate boiling boundary follows from the Labuntsov's recommendations for HTC calculation at flow boiling. According to Labuntsov (1960)

$\alpha = \alpha_q$ if $\alpha_q \geq 2\alpha_u$, $\alpha = \alpha_u$ if $\alpha_q \geq 2\alpha_u$,

$\alpha = \alpha_u(4\alpha_u + \alpha_q) / (5\alpha_u - \alpha_q)$ if $0.5\alpha_u \leq \alpha_q \leq 2\alpha_u$

where α_q, α_u and α are respectively heat transfer coefficients at pool nucleate boiling, at single-phase convection at a given number and total HTC at flow boiling. When $\alpha_q \geq 2\alpha_u$ the heat transfer is controlled by the nucleate boiling internal mechanisms only and so this unequality may be considered as the boundary mentioned above. As for the correlation (22) it was specially investigated (after the paper (Yagov, 1988) publication) by means of the wide comparison with the experimental data on heat transfer under flow boiling of different liquids. The good agreement is established if $\alpha_q \geq 2\alpha_u$. Fig.2 demonstrates such an agreement of the experimental and predicted HTC for water flow boiling in the tube of the inner diameter 32 mm and in the annular channels with the gap size 1.06-3 mm at the relatively high pressures - 4.0 and 9.8 MPa (Morozov, 1964; Tarasova and Orlov, 1964).The proposed correlation (22) compares also well to the recent experimental results of Aihara et al (1993) on liquid nitrogen (LN_2) boiling on the inside surface of the copper cryoprobe tip onto which LN_2 jet is impinged. These experimental data for flat and hemispherical tips

are shown in Fig.3. The most of the experimental points for the mass flow rates G = 0.29, 0.39 and 0.5 g/s lie within ± 15% of the prediction if the distance between the jet mouth and the probe tip doesn't exceed 0.5 mm (there is no effect of the turning vapour-liquid flow). These results are noticeable because of the very high heat fluxes for LN_2 at near atmospheric pressure (0.109 MPa). The ordinary critical heat flux for LN_2 pool boiling (which is about 200 KW m⁻²) is exceeded here more than four times due to high liquid velocities and good vapour evacuation from the heated wall. Obviously the main features of the nucleate boiling mechanism which are reflected by the present model remain at these unusual conditions.

As for the transition from the single-phase convection to the nucleate boiling heat transfer which occurs at rather small void fractions the simple interpolation formulas of the type of the mentioned above Labuntsov's one or the analogous Kutateladze (1979) corelation are probably suitable for practical calculations. The latter may be easy rearranged for subcooled flow boiling conditions. Bearing in mind that for the single-phase convection the natural reference temperature is liquid bulk temperature $\bar{T}_b$ (the mean temperature in the given channel cross-section) and doing interpolation between corresponding heat flux densities one gets

$$\alpha = [\alpha_q^2 + \alpha_u^2(1 + \Delta T_{sub} / \Delta T)^2]^{1/2}. \qquad (23)$$

This correlation was applied for the comparison with the experimental data of Hasan et al (1990) on subcooled flow boiling of refrigerant R113. Fig.4 reflects rather good agreement between the experimental and predicting values. The nucleate boiling HTC α_q was calculated on the formula (22), the single phase convection HTC α_u was computed on the Petukhov's correlation (Petukhov et al, 1986).

NOMENCLATURE

h_{lg}	latent heat of evaporation
p	pressure
q	heat flux
T	temperature
t	time
α	heat transfer coefficient
λ	thermal conductivity
μ, ν	dynamic and kinematic viscosity
ρ	density
σ	surface tension

Subscripts

g	vapour
l	liquid
s	saturation
w	wall

REFERENCES

Aihara T., Kim J.-K., Suzuki K., Kasahara K., 1993. "Boiling heat transfer of a micro-impinging jet of liquid nitrogen in a very slender cryoprobe", *Int J. Heat Mass Transfer*, Vol. 36, pp.169-175.

Bennet D.L., Chen J.C., 1980, "Forced convective boiling in vertical tubes for saturated pure components and binary mixtures", *AIChE Journal*, Vol. 26., pp.454-461.

Bhat A.M., Saini J.S., Prakash R., 1986, "Role of macrolayer evaporation in pool boiling at high heat flux", *Int J. Heat Mass Transfer*, Vol. 29, pp. 1953-1961.

Chen J.C., 1966, "Correlation for boiling heat transfer to saturated fluids in convective boiling", *Ind Engng. Chem. Proc. Des. Develop.*, Vol.5, pp.322-329.

Dhuga D.S., Winterton R.H.S., 1985, "Measurement of surface contact of transition boiling", *Int J. Heat Mass Transfer*, Vol. 28, pp. 1869-1880.

Gaertner R.F., 1965, "Photographic study of nucleate pool boiling on a horizontal surface", *Trans of ASME. J. Heat Transfer*, Vol.87, pp. 17 - 29.

Hasan A., Roy R.P., Karla S.P., 1990, "Experiments on subcooled flow boiling heat transfer in a vertical annular channel", *Int J. Heat Mass Transfer*, Vol. 33, pp. 2285-2293.

Iida Y., Kobayasi K., 1969, "Distribution of void fraction above a horizontal heating surface in pool boiling", *Bull. of JSME*, Vol. 12, pp. 283 - 290.

Kutateladze S.S., 1979, Heat transfer theory fundamentals. Moscow (in Russian).

Labuntsov D.A., 1959, "Heat transfer at nucleate boiling in liquids", *Teploenergetika*, N12, pp. 19 - 26 (in Russian).

Labuntsov D.A., 1960, "Generalized correlations for liquid nucleate boiling heat transfer," *Teploenergetika*, N5, pp. 76- 81 (in Russian).

Labuntsov D.A., 1963, "The approximate theory of the developed nucleate boiling heat transfer," *Izvestiya*

Academii nauk of the USSR, Energetika i Transport, N1, pp. 58- 71 (in Russian).

Labuntsov D.A., 1972, "Nucleate boiling heat transfer issues", *Teploenergettika*, N9, pp. 14 - 19 (in Russian).

Lay J.H. and Dhir V.K., 1994, "A nearly theoretical model for fully developed nucleate boiling of saturated liquids", *Proc. of 10-th Int. Heat Transfer Conf.,.* Brighton, UK, 1994., Vol. 5, pp. 105 -110.

Morozov V.G., 1964, "Investigations of water boiling heat transfer in tubes", *Convektivnaya teploperedacha v dvukhphasnom i odnophasnom potokakh*, Moscow-Leningrad, pp. 130 - 139 (in Russian).

Pasamehmetoglu K.O., Chappidi P.K., Unal C., Nelson R.A., 1993, "Saturated pool nucleate boiling mechanisms of high heat fluxes", *Int J. Heat Mass Transfer*, Vol.36, N15, pp.3859-3868.

Petukhov B.S., Genin L.G., Kovalev S.A., 1986, "Heat transfer in nuclear power plants", Moscow, Energoatomizdat, (in Russian).

Potash M., Jr., Wayner P.C., Jr., 1972, "Vaporation from a two dimensional extended meniscus", *Int J. Heat Mass Transfer*, Vol. 15, pp. 1851-1863.

Rajvanshi A.K., Saini J.S., Prakash R.,1992, "Investigation of macrolayer thickness in nucleate pool boiling at high heat flux", *Int J. Heat Mass Transfer*, Vol. 35, pp. 343-350.

Renk F.J., Wayner P.C., Jr., 1979, "An evaporating ethanol meniscus, Part I. Experimental studies. Part II. Analytic studies", *Trans of ASME. J. Heat Transfer*, Vol. 101, pp. 55 - 58, 59 - 62.

Tarasova N.V., Orlov V.M., 1964, "Heat transfer and pressure drop at subcooled flow water boiling in annular channels", *Convektivnaya teploperedacha v dvukhphaznom i odnophaznom potokakh*, Moscow-Leningrad, pp. 162 - 187 (in Russian).

Wayner P.C., Jr., 1973, "Fluid flow in the interline region of the evaporating non-zero contact angle meniscus", *Int J. Heat Mass Transfer*, Vol. 16, pp. 1777-1783.

Wayner P.C., Jr., 1978, "A constant heat flux model of the evaporating interline region", *Int J. Heat Mass Transfer*, Vol. 21, pp. 362-364.

Wayner P.C., Jr., 1982, "Adsorption and capillary condensation at the contact line in change of phase heat transfer", *Int J. Heat Mass Transfer*, Vol. 25, N5, pp. 707-713.

Wayner P.C., Jr., Kao J.K., LaCroix L.V., 1976, "The interline heat transfer coefficient on an evaporating wetting film", *Int J. Heat Mass Transfer*, Vol. 19, pp. 487-492.

Yagov V.V., 1988, "Heat transfer with developed nucleate boiling of liquids", *Therm. Eng.*, Vol.35, pp.65-70.

Yagov V.V., 1995, "The scientific legacy of Labuntsov and modern ideas on nucleate boiling process", *Therm. Eng.*, Vol. 42, pp. 181-189.

CONTRIBUTIONS OF CONVECTION AND BOILING
TO CONVECTIVE FLOW BOILING

John C. Chen and Kemal Tuzla
Institute of Thermo-Fluid Engineering and Science
Lehigh University
Bethlehem, Pennsylvania   USA

ABSTRACT

This work is an attempt to experimentally delineate between the contributions of convection and boiling to saturated convective flow boiling. This was accomplished by measuring heat transfer coefficients for two-phase convection with gas-liquid flow (no boiling) and two-phase saturated boiling with superheated wall. Results indicate that two-phase flow enhances convective heat transfer (beyond that for liquid phase convection) and suppresses the contribution of nucleate boiling (below that for comparable pool boiling).

INTRODUCTION

It is widely recognized that in convective flow boiling both convective and boiling mechanisms contribute to the total heat transfer process. Beginning in the 1950's, various models have been proposed suggesting that the effective heat transfer coefficient for saturated convective boiling is some functional combination of the individual heat transfer coefficients for convection and for nucleate boiling,

$$h_{cb} = f \left(h_c, h_b \right) \tag{1}$$

where all heat transfer coefficients are based on the temperature difference $(T_w - T_s)$. There has been no consensus

$$h_{cb} = F_{DA} \cdot h_c \tag{2}$$

regarding the correct functional form for eqn. 1. One group of models suggest a geometric combination of the individual coefficients whereby h_{cb} is some product function of h_c and h_b. An example is the early model of Dengler and Addoms (1952), where F_{DA} is a correction factor representing the nucleate boiling contribution. Other models of this

multiplicative nature include that of Guerrieri and Talty (1956), Bennett et al. (1959) and Shrock and Grossman (1962).

Another group of models suggest a basic additive superposition of the contributions. The first model proposed by Rohsenow (1952) used a simple addition of the individual coefficients

$$h_{cb} = h_c + h_b \tag{3}$$

The Chen Correlation of 1966 follows the same form but suggests that the individual components are modified by interactions between the two mechanisms,

$$h_{cb} = h_{mac} + h_{mic} \tag{4}$$

here h_{mac} = macro-convection with two-phase flow
$\qquad\quad = F \times h_l$
$\quad h_{mic}$ = micro-convection with nucleate boiling
$\qquad\quad = S \times h_{pb}$

Chen's F function is a two-phase multiplier to account for enhanced convection due to co-current flow of vapor and liquid; the S function is a "suppression" factor to account for the possible reduction of nucleate boiling heat flux due to bulk convection. A variation on this superposition formulation is the "power type" addition models of Kutateladze (1961) and Steiner and Taborek (1992). These use an expression formalized by Churchill (1974) to represent transitional addition between diffeent regimes,

$$h_{cb} = \left[h_c^n + h_b^n \right]^{\frac{1}{n}} \tag{5}$$

where n = an empirical constant.

181

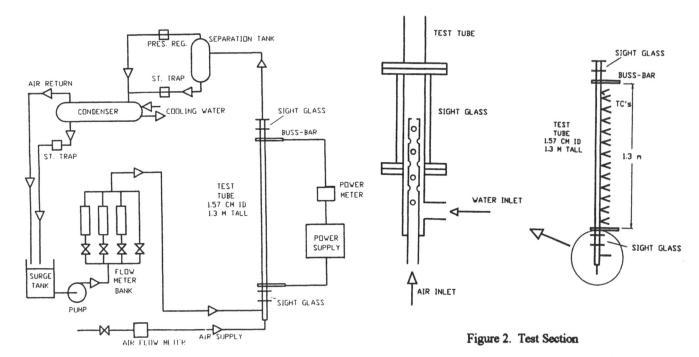

Figure 1. Multi-Phase Flow Loop

Figure 2. Test Section

This diversity of approach reflects a basic uncertainty regarding the fundamental mechanisms. To quote Steiner and Taborek (1992): "In flow boiling, the nucleate and convective components are superimposed by a very complex mechanism which so far is not well understood."

In addition to uncertainty regarding the right form for combining convective and boiling contributions in eqn. 1, there is also uncertainty about the models for calculation of the individual coefficients h_c and h_b . The development of such models have always been hampered by the fact that experimental data for convective boiling provide values only for the final effective heat transfer coefficient, thereby necessitating some guess work to assign relative contributions by each of the individual mechanisms. This approach raises a number of phenomenological issues:

- Is the effective convective coefficient greater than that for single phase convection?

- If so, can the convective multiplication factor be correlated in a manner similar to that for the multiplicative factor of two-phase pressure drop?

- Is the contribution of nucleate boiling, as measured by the effective boiling coefficient h_b, equal to, greater than, or less than the coefficient for pool nucleate boiling at same wall superheat?

- If the effective h_b is different than h_{pb}, how does the correction factor correlate with operating variables?

The present research is an attempt to experimentally delineate between the contributions of convection and boiling. This is accomplished by measuring heat transfer coefficients for two matched series of tests:

a.   Two-phase convection with gas-liquid wherein heat is transferred to the two-phase fluid with all temperatures below saturation (no boiling);

b.   Two-phase saturated boiling with superheated wall at various vapor-liquid flow qualities.

The concept of this approach is that coefficients measured in test (a) represent only the convective contribution, h_c. Coefficients measured in test (b) represent the total convective boiling coefficient, h_{cb}. The difference then is taken to represent the effective nucleate boiling contribution h_b. The individual contributing coefficients (h_c, h_b) are then compared to single-phase convective coefficeints for the liquid phase flow (h_l) and the pool boiling coefficient (h_{pb}), respectively.

EXPERIMENT

Experiments were carried out in a multi-phase loop, with the capability for operation with air-water two-phase flow, steam-water two-phase flow and air-steam-water three-phase flow. Figure 1 shows a schematic diagram of the loop. A pair of variable-speed positive displacement pumps delivered liquid (water) to the bottom of a vertical test section. At the test section inlet, provision was made for injection of inert gas (air) at a metered and controlled rate. The multi-phase fluid flowed vertically upward through the test section where heat can be added at variable heat fluxes. After exiting from the top of the vertical test section, the multi-phase fluid flowed to a condenser where gas was separated and discharged, vapor was condensed and the liquid was subcooled for return to to the pumps.

As shown in Fig. 2, the test section was a vertical stainless steel tube of 1.57 cm ID, heated by passage of AC current directly through the tube wall. The high-current,

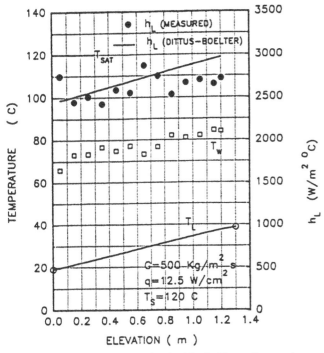

Figure 3. Sample Results for Single-Phase Convection

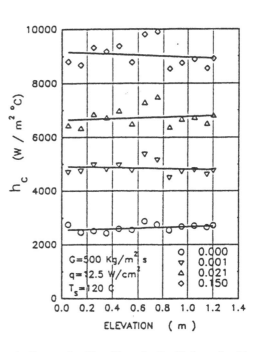

Figure 4. Convective Heat Transfer Coefficients for Air-Water Flow ($Re_l = 10,000$)

low-voltage power supply was capable of generating heat fluxes up to 350,000 W/m² at the tube inside surface. In runs with air injection, the air was fed to the bottom inlet of the test section through a concentric nozzle as indicated in Fig. 2. A length of 20 cm between the point of air injection and the start of the heated test section aided in two-phase mixing and flow development. Thirteen thermocouples were attached to the outside surface of the heated test section for measurement of wall temperatures along the axial length. All thermocouple junctions were electrically isolated from the wall metal to prevent stray EMF from contact with the electrically powered tube wall. Fluid temperatures were measured by thermocouples immersed in the flowing fluid at test section inlet and outlet, as well as at the discharge from the condenser. A pressure transducer at the test section outlet measured absolute fluid pressure, used to determine saturation temperature in the experiments. Separate flow meters were used to measure the individual flow rates of water and air. Flow rate of vapor generated in the test section was obtained by energy balance, using the measured heat flux of the test section. It was estimated that for a typical run with $(T_w\text{-}T_s)$ of 20°C or greater, the effective heat transfer coefficient at the tube ID could be determined with an accuracy of better than ± 4%.

In a typical experimental run, the inlet fluid flows (water and air) would be established at selected values by adjustment of pumps and throttle valves. With steady-state flow, power would then be applied to the test section and cooling water started through the condenser. Adjustment of the balance between heat input at the test section and heat removal at the condenser allowed control of the test section

pressure and saturation temperature. Data would be recorded after obtaining steady-state in flow, pressure, heat flux and wall temperatures along the test section.

A total of 58 experimental runs were carried out. Ten runs were with single-phase subcooled liquid, taken for the purpose of verifying the experimental methodology and instrumentation. Twenty-three runs were carried out with air-water with subcooled temperatures; these runs represented two-phase heat transfer without nucleate boiling. Finally, 25 runs were carried out with convective flow boiling of water, progressing from subcooled boiling to saturated boiling along the test section.

RESULTS AND DISCUSSION

Figure 3 shows sample results for a run with liquid mass flux of 500 kg/m²s, corresponding to a liquid Reynold number of 10,000. At a steady-state heat flux of 12.5 W/cm², the axial profile of measured wall temperatures are shown along with the measured inlet and exit fluid temperatures. It was deemed necessary to first verify the experimental methodology by tests with single-phase heat transfer. Twenty-three rus were carried out with subcooled water at different flow velocities, giving measurements of the local heat transfer coefficients along the test section length. The resulting local heat transfer coefficient corresponding to each measured wall temperature is plotted and compared to the Dittus-Boelter correlation (indicated by solid line). It is seen that the measured local coefficients had a scattered band of approximately ten percent but were in excellent agreement with the Dittus-Boelter correlation. Similar agreement was

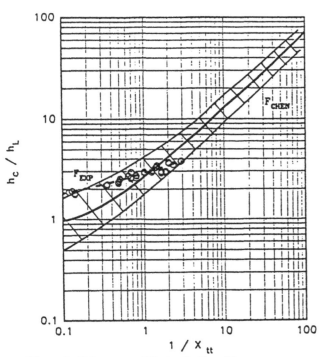

Figure 5. Enhancement Factor for Two-Phase
Convection

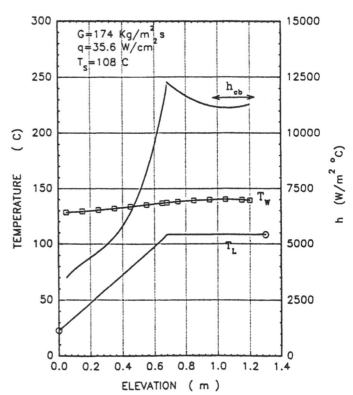

Figure 6. Sample Results for Convective Boiling

obtained for all other single-phase runs with turbulent liquid flow, thus confirming the validity of the experimental methodology and measurement instrumentation. Using the same experimental method, convective heat transfer coefficients were measured for two-phase flow of air-water. Results for four cases are presented in Fig. 4. These data represent tests at a constant liquid flow rate (500 kg/m²s) and a constant wall heat flux (12.5 W/cm²). Each parametric set of data represents the local heat transfer coefficient measured along the test section length for a given air flow rate, representing a different flow quality. It is seen that for these two-phase convection cases, there is little axial variation of local coefficients, indicating quick development of flow and thermal conditions at the beginning of the test section. These results show the very significant increase of heat transfer coefficient with increasing gas quality. Compared to the value for single-phase convection, (zero percent quality), it is seen that the magnitude of the two-phase coefficients increased by factors of 1.9, 2.7 and 3.6 as the flow quality of gas increased to 0.1 percent, 2.1 percent and 15.0 percent, respectively. Clearly there is an enhancement of the turbulent convective heat transfer mechanism with addition of a gaseous phase. The heat transfer coefficients measured for air-water flow represent the two-phase convective coefficient (h_c). There is no ambiguity about contribution of

boiling (by h_b) since all tests were carried out with subcooled fluid and wall temperatures. These data may then be compared with the single-phase convective coefficient for corresponding liquid flow alone (h_l). A related question is whether or not this two-phase enhancement can be correlated with some general two-phase-flow parameter, as has been achieved with pressure-drop multipliers for two-phase flow. Figure 5 shows a plot of the ratio (h_c/h_l) against the inverse Martinelli parameter ($1/X_{tt}$). Since the ratio $_ch/_lh$ is by definition equivalent to Chen's F function, this latter function as proposed by Chen (1966) is also shown in this figure for comparison. Several points are worth noting. First, the experimental data indeed show a tight correlation with the inverse Martinelli parameter. This confirms that the enhancement of convective heat transfer by two-phase flow is governed by the same parameters that effect enhancement of momentum transfer (pressure drop). Secondly, it is seen that these new data fall within the uncertainty band indicated by Chen in 1966, but show a discernable difference in trend. For the conditions of this experiment we can represent the convective enhancement factor by a more accurate F_{exp} as indicated by the dashed line. This will be utilized subsequently to extract the convective contribution in saturated convective boiling.

The final series of runs involve convective boiling without gas injection. Figure 6 shows typical conditions for

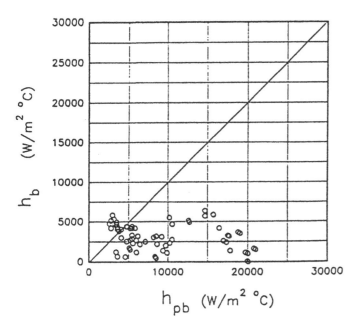

Figure 7. Comparison of Boiling Contribution to
Forster-Zuber's Correlation for Pool
Boiling

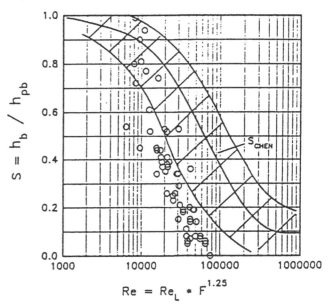

Figure 8. Suppression Factor for Boiling Contribution

one such run. With a given liquid flow rate, test section heat flux and pressure, the axial variation of fluid temperature (T_l), wall temperature (T_w) and local heat transfer coefficient (h_{cb}) are plotted as functions of elevation from start of the heated test section. For this run, the liquid entered with approximately 75°C subcooling and attained saturation temperature midway along the test section. The local heat transfer coefficients started in the range corresponding to single-phase convection and increased steeply as subcooled boiling occurred, before the advent of saturated convective boiling in the upper half of the test section. There appears to have been a short entry region immediately downstream of the saturation point wherein the local heat transfer coefficient decayed slightly from subcooled boiling to developed saturated boiling. For our subsequent analysis of saturated boiling results, only data from the developed saturated-boiling regime (near the upper end of the heated test section, as illustrated in Fig. 6) were utilized.

The convective boiling coefficients thus measured were used to extract an "effective boiling contribution" by the following relationships,

$$h_b = h_{cb} - h_c \qquad (7)$$

$$h_c = h_l \cdot F_{exp} \qquad (8)$$

were h_l is calculated by the Dittus-Boelter relationship for liquid convection alone and F_{exp} is obtained from the empirically determined function representing two-phase convection as measured in the air-water runs. By this logic, the heat transfer associated with the boiling contribution (h_b) is that amount above and beyond the heat transfer contributed by equivalent two-phase convection at the same gas quality.

The results of such analysis can be compared to the pool boiling heat transfer (without flow or net vapor quality) at the same wall superheat. An unsuccessful attempt was made to directly measure the pool boiling heat transfer coefficients in the test tube; due to the small channel cross-section and the vertical tube placement, steady-state pool boiling could not be sustained. The next best assessment was to compare the boiling contribution (h_b) to that calculated by standard pool boiling correlations (h_{pb}). Obviously, there are numerous pool boiling correlations which are suitable candidates for such a comparison. Figure 7 shows a parity plot comparing h_b to h_{pb} calculated by the well known Forster-Zuber pool-boiling correlation (1955). It is seen that for almost all test conditions, the effective contribution of nucleate boiling to convective flow boiling is significantly less than the magnitude anticipated in normal pool boiling at the same wall superheat.

The ratio h_b/h_{pb} is the suppression function (S) defined by Chen in 1966. The values of the S function, obtained from measurements in this investigation and using the Forster-Zuber correlation as pool boiling basis, are plotted in Fig. 8 against the two-phase convective Reynolds number (Re). Note that for this analysis, the two-phase Reynold number was determined using the experimental values of the convective multiplication function (F_{exp}) shown in Fig. 5. Thus the data in Fig. 8 represent the experimentally-determined boiling contribution compared to the heat transfer that would normally be anticipated for pool boiling at the same wall superheat. It is seen that the S values were all found to be less than unity, indicating a net "suppression" of the nucleate boiling contribution. It is seen that these new results follow the same trend, but are of lower magnitude (more suppression) than that suggested by Chen in 1966.

SUMMARY AND CONCLUSIONS

By utilizing a combination of experiments with two-phase convection (no boiling) and saturated convective boiling, it was possible to delineate the relative contributions of the convective heat transfer and boiling heat transfer mechanisms. The results indicate that in comparison to single-phase convection and pool boiling, two-phase convection significantly enhances convective heat transfer but reduces the contribution of nucleate boiling heat transfer.

NOMENCLATURE

h Heat transfer coefficient
F Enhancement factor for two-phase convection
S Suppression factor for boiling contribution
T Temperature
X_{tt} Martinelli-Lockhard parameter

Subscripts
b Boiling contribution
c Convective contribution
cb Convective boiling
ℓ Liquid alone
pb Pool boiling
s Saturation
w wall

REFERENCES

Bennet, J.A.R., Collier, J.G., Pratt, H.T.C., Thornton, J.D., 1959, "Heat Transfer to Two-Phase Gas Liquid Systems,: At. Energy Res. Establishment, Rept. AERE-R3159.

Chen, J.C., 1963, "A Correlation for Boiling Heat Transfer to Saturated Fluids in Convective Flow," *Ind. Eng. chem. Process Design Develop,* V. 5, No. 3, pp. 322-329.

Churchill, S.W., 1974, *The Interpretation and Use of Rate Data,* Hemisphere, New York.

Dengler, C.E., Addoms, J.N., 1956, *Chem. Eng. Progr. Symp. Ser. 52,* No. 18, pp. 95-103.

Forster, H.K., Zuber, N., 1955, *AIChE J.,* 1, No. 4, pp. 531.

Guerrieri, S.A., Talty, R.D. 1956,, *Chem. Eng. Progr. Symp. Ser. 52,* No. 18, pp. 69-77.

Kutateladze, S.S., 1961, "Boiling Heat Transfer," *Int. J. Heat Mass Transfer,* V. 4, pp. 3-45.

Schrock, V.E., Grossman, L.M., 1962, *Nuclear Sci. Eng.* 12, pp. 474-481.

Steiner, D. And Taborek, J., 1992, "Flow Boiling Heat Transfer in Vertical Tubes Correlated by an Asymptotic Model", *Heat Transfer Engineering,* Vol. 13, No. 2. pp 43

AN ALTERNATIVE MODEL FOR FLOW BOILING HEAT TRANSFER

Vishwas V. Wadekar
Heat Transfer and Fluid Flow Service (HTFS),
AEA Technology, Harwell,
Oxfordshire OX11 0RA, U.K.

ABSTRACT

A previously presented model for the addition of the convective and nucleate boiling components is explored further. The model is based on the concept of suppression of convective heat transfer due to the presence of nucleate boiling. It is demonstrated that this alternative model of flow boiling can provide reasonably good predictions of measured data. The comparisons of published experimental data for R12 with the predictions are presented here. Some areas of further work are described.

INTRODUCTION

In flow boiling two mechanisms of heat transfer, namely, nucleate boiling and convective heat transfer are generally treated as additive with some appropriate interaction between them. All flow boiling correlations take into account the contributions of these two mechanisms to the overall heat transfer. However, the correlations may differ from each other in the way they account for the interaction between nucleate boiling and convective heat transfer. The most widely used model for this interaction is that of suppression of nucleate boiling due to flow (Chen, 1966).

Recently a flow boiling correlation based on the concept of suppression of two-phase convective heat transfer was presented (Wadekar, 1992, 1994). The concept of suppression of convective heat transfer is a radical departure from the conventional approach of the suppression of nucleate boiling. The previous work (Wadekar, 1992, 1994) has shown that such an alternative approach to the conventional norm does provide good predictions.

In the initial development (Wadekar, 1992), the main focus of the work was to reproduce the trends of the asymptotic addition of the nucleate boiling and convective heat transfer components as given by the following equation.

$$\dot{q}_T = \left(\dot{q}_c^{\,n} + \dot{q}_{nb}^{\,n} \right)^{1/n} \qquad (1)$$

In a subsequent paper (Wadekar, 1994), further development of the model was carried out beyond merely reproducing the trends of Equation (1). In the present paper the model of suppression of convective heat transfer is further examined and tested against some more published experimental data.

BACKGROUND

In this section, the individual component correlations for nucleate boiling and convective heat transfer are described. These component correlations are needed to translate the alternative model of flow boiling into a full correlation

Nucleate Boiling Component Correlation

Cooper (1984) investigated various nucleate pool boiling correlations and groups of fluid physical properties used in them. He concluded that most of the physical property variations can be well represented by the reduced pressure and molecular weight of a fluid. The following equation, based on the Cooper correlation, is used to obtain the nucleate boiling component of flow boiling heat transfer:

$$\dot{q}_{nb}^{0.33} = C \, \Delta T \, p_r^{0.12} \left(-log_{10} p_r \right)^{-0.55} M^{-0.5} \qquad (2)$$

This correlation has been extensively tested for water. The value of the constant C is typically in the range of 35 to 55.

Two Phase Convective Component Correlation

A correlation based on the work of Kenning and Cooper (1989) is used here. From the measured flow boiling data, they carefully separated those data which fall into the apparently convective heat transfer region. Using those data, they developed the following correlation for the F factor which is a ratio of two-phase to single phase heat transfer coefficient.

$$F = 1.8 \left(1/\chi_{tt} \right)^{0.87} + 1 \qquad (3)$$

where χ_{tt} is the Martinelli parameter. This correlation is used in the following manner to obtain the convective component of flow boiling heat transfer.

Single phase liquid fraction coefficient, obtained from the Dittus-Boelter equation, is firstly used to obtain the two-phase convective heat transfer coefficient, a_c.

$$\alpha_c = F \, \alpha_1 \qquad (4)$$

where, $\alpha_1 = 0.023 \dfrac{\lambda_l}{D} Re_l^{0.8} Pr_l^{0.4} \qquad (5)$

The convective heat flux, $\dot{q}_c$ is then calculated by the following equation:

$$\dot{q}_c = \alpha_c \left(T_w - T_{sat} \right) \qquad (6)$$

Note that in the previous work (Wadekar, 1992, 1994), a reduced pressure based correlation was used for obtaining the convective heat flux, $\dot{q}_c$.

MODEL OF SUPPRESSION

In this section the basis of the model is discussed. This is followed by a discussion on how the model provides an alternative view of flow boiling.

Basis of the Model

The suppression of the convective component due to the presence of nucleate boiling is a simple concept. A variety of nucleate boiling heat transfer mechanisms that arise from bubble related activities, such as bubble growth and bubble departure, occur within or very close to the laminar sublayer. Thus nucleate boiling can be viewed as a heat transfer mechanism internal to the laminar sublayer. In contrast to this, convective heat transfer arising from flow can be viewed as a heat transfer mechanism which is external to the laminar sublayer. For example, increasing flow velocity will have an effect of thinning of the laminar sublayer or increase in the number of turbulent bursts acting on the sublayer. It can, therefore, be argued that wherever the internal heat transfer mechanism of nucleate boiling is present, the externally imposed mechanism of convective heat transfer is likely to have little or no effect. The concept of suppression of the convective component, thus described, is expressed as:

$$\dot{q}_T = \left(1 - A_{nb} \right) \dot{q}_c + \dot{q}_{nb} \qquad (7)$$

Here A_{nb} represents the time averaged fraction of the total heat transfer area over which nucleate boiling is effective. Note that the nucleate boiling component is not multiplied by A_{nb} because pool boiling correlations, used in obtaining the nucleate boiling heat flux, are based on the overall area rather than the effective area. In other words A_{nb} is already built-in in the calculation of the nucleate boiling heat flux, $\dot{q}_{nb}$. Decrease in the nucleate boiling heat flux with increasing mass flux or vapour quality,

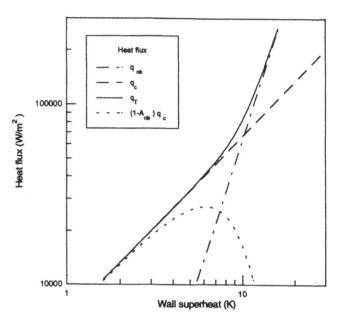

Figure 1 A plot of heat flux against wall superheat showing various heat flux curves (Wadekar, 1994).

although not apparent from Equation (7), occurs because of cooling effect of flow which reduces the wall superheat. This reduction in the nucleate boiling heat flux, as a result of reduction in the wall superheat is proposed as the equivalent of the conventional suppression of nucleate boiling. As the relevant wall superheat is based on the heating surface temperature in contact with the liquid, i. e. the inside wall temperature, the reduction in nucleate boiling can occur for electrically heated as well as fluid heated test-sections.

Wadekar (1992) found that the following relationship for A_{nb} is required to successfully predict the trends for typical flow boiling data.

$$A_{nb} = 1 - e^{-\theta} \qquad (8)$$

where, $\theta = \dfrac{k \, \Delta T^{2.5} \, p_r}{Re_{2p}}$

It was argued that the wall superheat, ΔT, represents a combined effect of bubble parameters such as their population density, growth rates, departure diameters, frequencies, and their interactions with each other. The two-phase Reynolds number is included because it models the decrease in A_{nb} resulting from decreasing bubble departure diameters with increasing flow velocity. The following expression, proposed by Chen (1966), is used for the two-phase Reynolds number

$$Re_{2p} = Re_1 \, F^{1.25} \qquad (9)$$

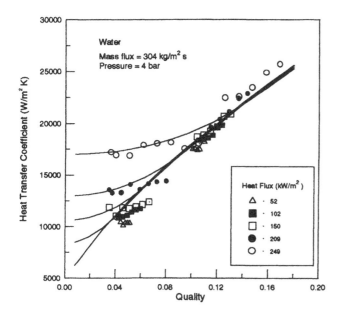

Figure 2 Comparison of predictions with published data for water at 304 kg/m²s mass flux and 4 bar pressure using $C= 35$ in Equation (2) (Wadekar, 1994)

In the initial development a value of 6200 was recommended for the empirical constant, k. Subsequently the constant k was related to the constant C in Equation (2) by the following expression (Wadekar, 1994).

$$k = 177 \, C \qquad (10)$$

It should be noted that similar to the constant C in Equation (2), the constant k is a dimensional constant and therefore is not a universal constant.

Alternative View of Flow Boiling

Figure 1 illustrates the basic heat flux curves in flow boiling heat transfer, namely, the convective curve, the nucleate boiling curve and the flow boiling curve. The curves are prepared for an example of water flowing in a tube of a given diameter, at a fixed pressure, mass flux and quality. The convective curve is obtained from Equations (3) to (6). The nucleate boiling curve is obtained from Equation (2) with $C = 35$. Experimental measurements show that the flow boiling data are generally bound by these two asymptotes and can typically be represented by the solid line shown in Figure 1. This line, referred to as the flow boiling curve, represents the total heat flux, $\dot{q}_T$. It shows that at low wall superheats, $\dot{q}_T$ approaches the convective heat flux, $\dot{q}_c$, and at high wall superheats, $\dot{q}_T$ approaches the nucleate boiling heat flux, $\dot{q}_{nb}$.

Thus, in order to obtain the flow boiling curve from the nucleate boiling and convective components, there should be a suppression of nucleate boiling heat transfer with decreasing wall superheat and a suppression of convective heat transfer with increasing wall superheat. This behaviour of suppression of each

mechanism is well represented by Equation (1) which can reproduce a flow boiling curve with an appropriate value of n greater than unity. In Figure 1 a curve for $\dot{q}_c \, (1 - A_{nb})$ is also plotted. This represents a reduced convective heat flux because of the area based suppression of the convective heat transfer. It can be seen that this reduced convective heat flux is close to $\dot{q}_c$ at a low wall superheat, as A_{nb} is nearly zero at a low wall superheat. With increasing wall superheat, $\dot{q}_c (1 - A_{nb})$ starts to deviate from $\dot{q}_c$ and, in this particular case, falls rapidly after a wall superheat of about 6 K since A_{nb} rapidly approaches unity with further increase of the wall superheat. As the flow boiling curve can be reproduced by adding $\dot{q}_c \, (1-A_{nb})$ and $\dot{q}_{nb}$, this represents an alternative interpretation of the observed flow boiling curve. As noted before the suppression or reduction in the nucleate boiling heat flux, $\dot{q}_{nb}$, takes place because of the cooling influence of the flow on the temperature of heating surface in contact with the liquid.

The asymptotic model shown in Equation (1) correlates the observed flow boiling data, but does not provide the magnitudes of the individual components. The suppression of convective heat transfer model provides the magnitudes of the individual components and, thus, can potentially provide a way for verification with a detailed parametric study. This point is further elaborated in a later discussion.

COMPARISON WITH SOME PUBLISHED DATA

Figures 2 to 4 show comparisons of the predictions of the present method, based on the suppression of convective heat transfer, with some published data. Comparisons shown in Figures 2 and 3 are from previous work (Wadekar, 1994). These comparisons are made on heat transfer coefficient against quality plots for a given mass flux and pressure. In each plot different

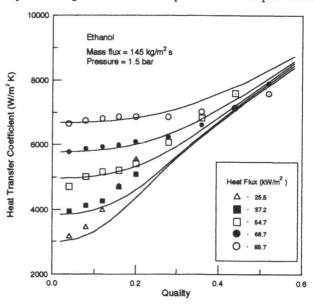

Figure 3 Comparison of predictions with published data for ethanol at 145 kg/m²s mass flux and 1.5 bar pressure using $C = 48$ in Equation (2) (Wadekar, 1994)

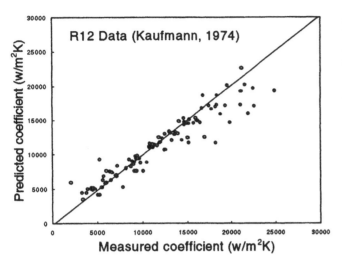

Figure 4 Comparison of measured R12 data (Kaufmann, 1974) with the current method using $C = 57$ in Equation (2).

symbols show the measured data at various heat fluxes and the solid lines represent the predictions for the corresponding heat fluxes. It can be seen that the trends exhibited by the measured data are well predicted for typical water data (Kenning and Cooper, 1989) shown in Figure 2 and for typical ethanol data (Robertson and Wadekar, 1988) shown in Figure 3.

Figure 4 compares the data reported by Kaufmann (1974) for Refrigerant 12 with the predictions of the present method. Kaufmann obtained these data from a 20.5 mm diameter tube at pressures ranging from 11 to 33 bar, mass fluxes ranging from of 658 to 4850 kg/m²s and heat fluxes ranging from 9.5 to 111 kW/m². With these conditions, he covered a range of vapour qualities from 0.02 to 0.24. It can be seen from Figure 4 that this data-set covering an extensive range of pressures, mass fluxes, heat fluxes and qualities are well predicted by the present method. Table 1, showing the comparisons of the same data with the predictions of various other methods, indicates that the current method appears to predict the data relatively well.

FURTHER DISCUSSION OF DATA

Some special features and trends exhibited by the ethanol and R12 data are discussed here.

Ethanol Data

Figure 3 shows that there is some amount of cross over of the measured data at different heat fluxes, for example at the heat fluxes of 54.7 and 85.7 kW/m². This can be explained, at least partly, by the pressure changes occurring in a 3 m tall test-section used by Robertson and Wadekar. Although each experiment at a given heat flux is conducted at a nominal pressure of 1.5 bar, at the downstream end of the test-section, where high qualities occur, the pressure can be significantly lower than 1.5 bar. This is a result of pressure loss associated with the two-phase flow.

The effect of the lowering of local pressure on the local measured heat transfer coefficient at the exit end of the test-

section will depend on the relative magnitudes of the nucleate boiling and convective heat transfer. If convective heat transfer is a dominant mechanism, which is likely at a lower heat flux, the result will be an additional increase in heat transfer coefficient because the convective coefficient increases with decreasing pressure. Such additional increase is likely to be absent at a higher heat flux, because nucleate boiling and convective heat transfer can be present in comparable magnitudes. It should be pointed out that the pressure changes alone can not fully explain the observed cross over of the heat transfer coefficients. This issue is discussed again in the next section.

R12 Data

In spite of extensive range of parameters covered by Kaufmann, his R12 data can not be meaningfully plotted as heat transfer coefficient against quality as in Figure 2 or 3; the reason being that at each heat flux the data covers only two vapour qualities. Therefore, the data are presented, in Figures 5 and 6, as a ratio of the measured to predicted heat transfer coefficient against two chosen parameters.

Figure 5 shows a graph of this ratio against A_{nb}, the time average fractional area over which nucleate boiling is active. It shows that the data cover a range of values of A_{nb} from nearly zero to 0.6. Figure 5 also shows that the predictions are better at higher values of A_{nb}. The fact that the maximum value of A_{nb} is 0.6 suggests that the data are never completely dominated by nucleate boiling. This could probably result from the relatively high mass fluxes used in the experiments. A larger scatter in the ratio of coefficients at low values of A_{nb}, when the convective mechanism dominates, is some what surprising. Generally, flow boiling data are much more reproducible and hence show less scatter when they are dominated by convective heat transfer rather than nucleate boiling.

The reason for the observed scatter in the ratio of heat transfer coefficient at low values of A_{nb} becomes clear when the same data are plotted against measured wall superheat as shown in Figure 6. It can be seen that, barring two data points with low ratio of heat transfer coefficients at about 4.5 to 5.5 K wall superheat, the greater scatter in the data corresponds to lower wall superheat where the accuracy of measuring temperature differences is at the lowest. It can be seen that beyond a wall superheat of about 4 K there is generally less scatter in the data.

Table 1 Prediction of R12 data (Kaufmann, 1974) by various methods as reported by Steiner and Taborek (1992) and the present method

Method	% Avg. Dev	% Mean Dev
Shah (1976)	- 40.5	41.9
Kandlikar (1987)	- 17.9	22.8
Gungor and Winterton (1987)	- 29.3	34.9
Steiner and Taborek (1992)	9.2	19.1
Present method	0.7	11.6

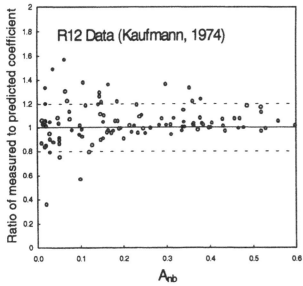

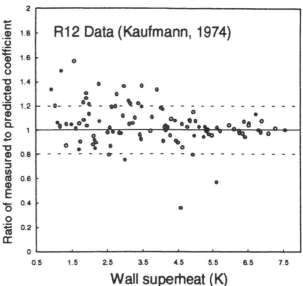

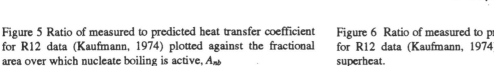

Figure 5 Ratio of measured to predicted heat transfer coefficient for R12 data (Kaufmann, 1974) plotted against the fractional area over which nucleate boiling is active, A_{nb}

Figure 6 Ratio of measured to predicted heat transfer coefficient for R12 data (Kaufmann, 1974) plotted against the local wall superheat.

GENERAL REMARKS AND SCOPE FOR FURTHER WORK

Although the asymptotic model, expressed mathematically by Equation (1) can predict the overall trends of flow boiling reasonably well, there are some peculiar trends in flow boiling which cannot be correlated by an asymptotic model. As discussed earlier, there appears to be a small cross over of the measured heat transfer coefficients at different heat fluxes in the high quality region (see Figure 3). Only a part of this behaviour can be attributed to the effect of decrease in pressure along the length of a test-section as discussed before. With a model based on the suppression of convective heat transfer, it may be possible to explain and correlate this trend. This, however, may require some change in the reasoning of suppression of convective heat transfer. At a lower wall superheat the heat transfer area based suppression may arise from the fact that nucleate boiling is a better mechanism of heat transfer than convective heat transfer. With increasing wall superheat, the number of vapour bubbles growing from a given surface area increases as smaller nucleation sites are activated. This can lead to coalescence of some growing bubbles with a part of the heating area being covered by a vapour patch for a length of time. On this area, nucleate boiling heat transfer will not be very effective and yet the area will be physically unavailable for the convective heat transfer mechanism, at least part of the time. Thus at a high wall superheat part of the suppression of convective heat transfer may occur not because nucleate boiling is a better mechanism but because it is physically obstructing the two-phase convective heat transfer. This may explain why the overall coefficient is sometimes lower at a higher heat flux in the high quality region as shown by some data in Figure 3. Further work is required to develop these ideas.

As discussed earlier, the current model provides an alternative view of flow boiling. For example, referring to Figure 1, at a wall superheat of about 7 K, where the curves for $\dot{q}_c (1-A_{nb})$ and $\dot{q}_{nb}$ cross, the model predicts that the overall heat flux consists of equal amounts of convective and nucleate boiling heat flux. Normally such a data point falling very close to the line for $\dot{q}_c$ will be classified as one falling into the purely convective regime. It should be possible to change the relative magnitudes of nucleate boiling and convective heat transfer by varying the pressure whilst holding all other experimental conditions constant. This will provide a systematic approach to verify whether this data point is in the purely convective region or in a mixed region, with equal magnitudes of convective and nucleate boiling heat transfer. Detailed parametric studies can thus enable us to assess the validity of the model.

CONCLUSIONS

It appears that the correlation, based on the proposed alternative model for flow boiling, predicts the published Refrigerant 12 data well. A method, identified here for the possible experimental verification of the model, is worth following in the future work. Further development of the model may provide explanations for some of the peculiar trends exhibited by flow boiling data.

ACKNOWLEDGEMENTS
The author gratefully acknowledges the permission given by Heat Transfer and Fluid Flow Service (HTFS) to publish this work.

NOMENCLATURE

A_{nb}	fractional time average area for nucleate boiling
C	constant in Equation (2)
D	diameter (m)
F	ratio of two phase to single phase transfer coefficient
G	mass flux(kg/m^2s)
k	constant in Equation (11) ($K^{2.5}$)
M	molecular weight
n	exponent in Equation (1) (typically 2 to 4)
p_r	reduced pressure ($p/p_{critical}$)
Pr_l	liquid Prandtl number
Re_l	liquid fraction Reynolds number, $DG(1-x)/\mu$
Re_{2p}	two-phase Reynolds number
$\dot{q}$	heat flux (W/m^2)
T	temperature (K)
x	vapour quality
α	heat transfer coefficient (W/m^2K)
χ_{tt}	Martinelli parameter

$$= \left(\frac{1-x}{x}\right)^{0.9}\left(\frac{\rho_g}{\rho_l}\right)^{0.5}\left(\frac{\mu_l}{\mu_g}\right)^{0.1}$$

ΔT	wall superheat, T_w-T_{sat} (K)
μ	viscosity (Pa s)
λ	thermal conductivity (W/m K)
θ	constant in Equation (11)

subscripts

c	convective
l	liquid phase
nb	nucleate boiling
sat	saturation
T	total or overall
w	wall

REFERENCES

Chen, J. C., *Correlation for boiling heat transfer to saturated liquids in convective flow*, Ind. Eng. Chem. Process Design and Develop., Vol. 5, pp 322-329, 1966.

Cooper, M. G., *Heat flow rates in saturated nucleate pool boiling - a wide ranging examination using reduced properties*, Advances in Heat Transfer, Vol. 16, pp. 157-239. 1984.

Kaufmann W. D. *Heat transfer and pressure drop in two phase flow of R12 in a vertical tube*, PhD Thesis Diss Nr 5196, ETH Zurich, 1974.

Kenning D. B. R. and Cooper, M. G., *Saturated flow boiling of water in vertical tubes*, Int. J. Heat Mass Transfer, Vol. 32, pp. 445-458, 1989.

Robertson, J. M. and Wadekar, V. V., *Vertical upflow boiling of ethanol in a 10 mm diameter tube*, Proc. 2nd UK National Heat Transfer Conference, Glasgow, Vol. 1, pp. 67-77, 1988.

Steiner, D. and Taborek, J., *Flow boiling heat transfer in vertical tubes correlated by an asymptotic model*, Heat Transfer Engineering, Vol. 13, pp. 43-69,1992.

Wadekar, V. V., *A model for addition of convective and nucleate boiling heat transfer in flow boiling*, Proc. of 3rd UK National Heat Transfer Conference, Birmingham, Vol. 1, pp. 181-187, 1992.

Wadekar, V. V., *A flow boiling model based on suppression of convective heat transfer*, Paper No TPF-23, First ISHMT-ASME Heat and Mass Transfer Conference, Bombay 1994.

A GENERAL HEAT TRANSFER MODEL FOR TWO-PHASE ANNULAR FLOW

G. Sun, W. H. G. T. Chan and G. F. Hewitt

Department of Chemical Engineering and Chemical Technology
Imperial College of Science, Technology and Medicine
London SW7 2BY, UK.

ABSTRACT

A new model is presented for the analysis of hydrodynamics and heat transfer in two-phase annular flow which can be used to predict heat transfer coefficients over the full region from the onset of annular flow to the dryout point.

The model follows the traditional interpretation of evaporative heat transfer, namely that of forced convective film heat transfer which may be enhanced by bubble nucleation processes and growth.

For the forced convective component, a theoretical model for turbulent film heat transfer is used together with the analytical method for annular flow hydrodynamics developed by Owen and Hewitt (1987). The effect of damping turbulence near the interface is also considered, using the analysis of Levich (1962).

For the component of nucleate boiling in the liquid film, the whole hydrodynamic and heat transfer process is considered, from the bubbles growing from active sites to their ultimately bursting through the surface. The model is shown to be consistent with data.

1. INTRODUCTION

Two-phase annular flow patterns occur in many process and power generating systems.

The liquid flows on the channel wall as liquid film, and the gas or vapour flows in the channel core together with the rest of the liquid which entrained as liquid droplets. At low liquid velocities the gas-liquid interface is relatively smooth having only small amplitude ripples. At increasing liquid velocity the interface becomes covered with large amplitude waves. This dominant form of wave on the interface is the so-called disturbance wave which occurs above a critical liquid flow rate in the film and which increases the interfacial friction between the core and the film and gives rise to the droplet entrainment from the tips of the waves.

In the annular flow regime, two complementary mechanisms coexist for heat transfer, namely heat transfer resulting from nucleate boiling in the liquid film and heat transfer resulting from convection from the tube surface to the interface, where evaporation occurs. As the film thins along the tube, forced convection becomes more dominant since, with the increasing heat transfer coefficient, the wall temperature falls and the intensity of nucleate boiling decreases.

However, it is shown that, though a rational explanation of the observed behaviour is possible in some case, there is, at present, significant larger uncertainty in the correlations for two-phase annular flow heat transfer, compared to those for single-phase flow. This may be due to inadequacies in the correlations and/or in the experimental data. The assumptions behind Chen-type correlations have been strongly questioned in recent years. Mesler (1973), one of the questioners, pointed out that experiments performed only with axially uniform heat flux (as most boiling experiments are) cannot separate the effects of upstream history and local quality. Mesler also argued that conventional deep pool nucleate boiling is irrelevant for the very thin liquid films of annular flow regime.

The challenge faced in the work described in the paper was to try to develop a complete and rational basis for predicting boiling in annular flow, taking account of the nucleation and of the forced convection. The processes involved are extremely complex and a detailed prediction of all the local phenomena is certainly beyond the scope of current analytical capability. However, a *phenomenological model* of the combined processes can be constructed and this is what is described here. Though, as with many phenomenological models for multiphase systems, empirical relationships have to be invoked for some of the local processes, such a model can form a rational basis for interpretation and (more importantly) prediction of data.

In what follows below, modelling of forced convection in annular flow will be discussed in Section 2. This is followed by a description of a new nucleate boiling model in Section 3. Comparisons between experimental data and predictions based on the superposition of forced convection and nucleate boiling are shown in Section 4.

2.   MODELLING OF CONVECTIVE HEAT TRANSFER IN ANNULAR FLOW

For annular flow with evaporation taking place at the vapour-liquid interface, the temperature profiles within the liquid film can be calculated from the following expression:

$$\dot{q} = -\left(\lambda_L + \varepsilon_H c_{pL}\rho_L\right)\frac{dT}{dy} \qquad (2\text{-}1)$$

where ε_H is the eddy diffusivity for heat, λ_L, c_{pL} and ρ_L are the thermal conductivity, specific heat and density of the liquid. It is convenient to express eq. (2-1) in a dimensionless form as follows:

$$1 = \left(\frac{1}{Pr_L} + \frac{\varepsilon_H}{\nu_L}\right)\frac{dT^+}{dy^+} \qquad (2\text{-}2)$$

where Pr_L is the liquid Prandtl number, ν_L the liquid kinematic viscosity and

$$T^+ = \frac{c_{pL}\rho_L u^*}{\dot{q}}(T_W - T) \qquad (2\text{-}3)$$

where u^* is the friction velocity, T_W the wall temperature. If the film thickness δ and average shear stress are known, then eq. (2-3) can be integrated to give:

$$T^+ = \int_0^{\delta^+} \frac{1}{1/Pr_L + \varepsilon_m/\nu_L}dy^+ \qquad (2\text{-}4a)$$

where the non-dimensional variables defined as follows:

$$y^+ = yu^*/\nu_L \ , \ u^* = \sqrt{\tau_{avg}/\rho_L} \ , \ \tau_{avg} = \frac{1}{\delta}\int_0^{\delta}\tau dy \ ,$$

$$r_o^+ = \frac{r_o u^*}{\nu_L} \ , \ \delta^+ = \frac{\delta u^*}{\nu_L}$$

The heat transfer coefficient can be computed from the relationship

$$\alpha_c = \frac{c_{pL}\rho_L u^*}{T^+} \qquad (2\text{-}5)$$

In general, the heat transfer coefficient predicted from eq. (2-5) is somewhat higher than the observed one; this might be attributable to the existence of a reduction in eddy diffusivity in the region adjacent to the interface. A further explanation may be found in the transient nature of the flow. The existence of the large disturbance waves opens up the possibility that the average shear stress calculated from the mean pressure gradient is not the one which should be used in calculating u^*. In the intervals between the disturbances the shear stress may be much lower. Here Levich (1962) suggestion is adopted by introducing a damping factor as a multiplier to the eddy diffusivity. Its general form is expressed as $\left(1 - y/\delta\right)^n$, where the exponent n is not a universal constant; its value has been varied in different analyses to fit experimental data

(Mudawwar and El-Masri, 1986).   A new correlation of damping factor derived from regression method in terms of convective experimental data is given by the present authors as follows:

$$n = c_1\left(Re_{LF} - Re_{LFC}\right)^{c_2}\left(\frac{\delta\tau_i}{\sigma}\right)^{c_3} \qquad (2\text{-}6)$$

where Re_{LF} is liquid film Reynolds number, Re_{LFC} the critical liquid film Reynolds number. $c_1 = 11.94$, $c_2 = -0.04$ and $c_3 = 0.71$. As a result, eq. (2-4) is modified to:

$$T^+ = \int_0^{\delta^+} \frac{1}{1/Pr_L + \left(\varepsilon_m/\nu_L\right)\left(1 - y/\delta\right)^n}dy^+ \qquad (2\text{-}4b)$$

In order to solve eq. (2-5), we need to know δ and τ_{avg}. To calculate δ and τ_{avg} we must invoke the equivalent hydrodynamic model of annular flow and this will be described in the following section.

2.1.   Calculation of Film Thickness and Shear Stress Distribution

To calculate film thickness and shear stress distribution we employ the Owen and Hewitt model (1987). In this model, new relationships were proposed for both the liquid film and the gas-droplet core.

The triangular relationship in liquid film

This relationship, so called because it relates the three principle dependant variables in annular flow (i.e. film thickness, film flowrate and the pressure gradient), is described by Hewitt and Hall-Taylor (1970). If the accelerational pressure gradient is small and can be neglected, a momentum balance results in the following expression:

$$\tau = \tau_i\left(\frac{r_i}{r}\right) - \frac{1}{2}\left(\frac{dp}{dz} + \rho_L g\right)\left(\frac{r^2 - r_i^2}{r}\right) \qquad (2\text{-}7)$$

where τ_i is the interfacial shear stress and r_i the interfacial radius.

Taking account of the eddy damping in the liquid film, the shear stress distribution in the liquid film can be expressed as

$$\tau = \rho_L\left[\nu_L + \varepsilon_m\left(1 - \frac{y}{\delta}\right)^n\right]\frac{du}{dy} \qquad (2\text{-}8)$$

It is convenient to write eq. (2-8) in non-dimensional form

$$\frac{du^+}{dy^+} = \frac{\tau}{\tau_{avg}}\left[\frac{1}{1 + \left(\varepsilon_m/\nu_L\right)\left(1 - y/\delta\right)^n}\right] \qquad (2\text{-}9)$$

In eq. (2-9), ε_m is the single-phase eddy diffusivity, which can be derived, for instance, from von Karman's universal velocity profile:

$$\varepsilon_m/\nu_L = 0 \qquad\qquad y^+ < 5$$
$$\varepsilon_m/\nu_L = y^+/5 - 1 \qquad 5 < y^+ < 30 \qquad (2\text{-}10)$$
$$\varepsilon_m/\nu_L = y^+/2.5 - 1 \qquad 30 < y^+$$

The film flowrate can be calculated from the velocity profile by the following expression:

$$\dot{M}_{LF} = \int_{r_i}^{r_o} 2\pi r \rho_L u \, dr \qquad (2-11)$$

The core model

In the core region, ignoring the change of momentum and assuming a homogeneous core, a force balance gives the following expression:

$$\tau_i = -\frac{r_i}{2}(\rho_C g + \frac{dp}{dz}) \qquad (2-12)$$

where ρ_C is the homogeneous density in the core region and defined as:

$$\rho_C = \frac{\dot{m}_E + \dot{m}_G}{\dot{m}_E/\rho_L + \dot{m}_G/\rho_G} \qquad (2-13)$$

where $\dot{m}_E$ and $\dot{m}_G$ are the liquid entrainment mass flux which can be calculated from Owen model (1986) and gas mass flux, respectively. The interfacial friction factor can be defined in the conventional form as:

$$f_i = \frac{\tau_i}{\frac{1}{2}\rho_C(U_C - u_i)^2} \qquad (2-14)$$

where u_i is the liquid-vapour core interfacial velocity, U_C is a homogeneous velocity in the core region

$$U_C = \frac{\dot{m}_G}{\rho_G} + \frac{\dot{m}_E}{\rho_L} \qquad (2-15)$$

Because of the presence of interfacial waves, which act like a rough pipe wall on the gas core flow, f_i is greater than for a smooth tube.

Owen (1986) modified Nikuradse's single-phase frictional factor equation in deriving the following expression for calculating the interfacial friction factor, which is consistent with eq. (2-14):

$$\sqrt{\frac{2}{f_i}} = \frac{1}{K_{TP}} log\left(\frac{r_i}{e_E}\right) + A_r^* - \frac{1.5}{K_{TP}}$$
$$+\frac{1}{K_{TP}}\left(\frac{2\delta}{r_o} log\left(\frac{r_o}{\delta}\right) - \frac{\delta}{r_o}\right) \qquad (2-16)$$

where K_{TP} is the two-phase von Karman constant and e_E is the effective roughness which was correlated empirically.

From eq. (2-7), it can be seen that the film model requires values of pressure gradient and interfacial shear stress, which can be calculated from the core model. Equally, eq. (2-14) and eq. (2-15) show that the core model requires values of film thickness and interfacial velocity, which are calculated from the film model. Thus, the two models have to be solved simultaneously and iteratively for a given liquid film flowrate.

3   MODELLING OF NUCLEATE BOILING HEAT TRANSFER IN ANNULAR FLOW

In this section a physical processes of bubble growth will be described. Then equations and correlations will be presented for

calculating various parameters which are related the heat transfer process.

Considering a single active wall nucleation site from the moment when a bubble starts to grow, the rate of growth of a vapour bubble is determined by the surface tension, the forces produced by the ambient flowing liquid and the difference between the pressure within the bubble and the ambient, or external pressure. As the bubble grows larger, the forces, imposed by the liquid flow on the bubble, increase. The bubble will detach when the forces reach the imbalance point. The detached bubble continues to grow when it travels up the liquid film, convected by the liquid flow and the moving bubble eventually bursts at the interface. By transient conduction, a temperature profile over the active nucleation site starts to develop in the liquid that replaces the detached bubble. When this temperature profile satisfies the required superheat for incipience of nucleation, a new bubble starts to grow from the site. In addition, nucleation centres may be created by the entrainment of bubbles at the gas liquid interface.

Heat transfer within one nucleation cycle (the time from the growth of a potential bubble in a given site to the bursting of the bubble) over a single active nucleation site is modelled. The time average values of the variables involved are considered. For calculating the entire heat transfer coefficient, the parameters of bubble cycle frequency, bubble departure and burst sizes and the cumulative size distribution of potential nucleation sites must be known.

From an energy balance the heat transfer coefficient associated with latent heat removal by bubble growth in the film is given by:

$$\alpha_n = \frac{\left(\frac{4}{3}\pi R_b^3\right) h_{fg}\rho_G f_b}{(T_w - T_{SAT})} \qquad (3-1)$$

where f_b is the frequency per unit peripheral area at which bubbles burst through the interface. R_b is the bubble burst diameter. h_{fg} is the latent heat of evaporation, ρ_G the vapour density, T_w the wall temperature, T_{SAT} the saturation temperature. The estimation of the parameters in equation (3-1) will be discussed next.

The incipient criterion of nucleate boiling on the wall

An equilibrium bubble will grow when the superheat of the surrounding liquid exceeds the critical (metastable equilibrium) value. The critical superheat temperature for bubble growth can be calculated from the following equation

$$T_b - T_{SAT} = \frac{RT_{SAT}^2}{Mh_{fg}} ln\left(1 + \frac{2\sigma}{R_E p_L}\right) \qquad (3-2)$$

If a hemispherical shape is assumed for the original equilibrium bubble, the bubble height is equal to the bubble radius and the mouth radius of the cavity, i.e. $y_b = R_E = r_c$. Following the usual assumption, it is supposed that the incipient bubble will grow if the temperature around its surface is everywhere greater than the metastable equilibrium temperature

T_b. Since heat is being transferred into the fluid, there is a temperature gradient from the wall. Thus, bubble growth will occur if the temperature at y_b (at the top of the bubble) exceeds T_b.

Bubble frequency

The bubble frequency f_b can be written as follows:

$$f_b = f_s n_{aw} \qquad (3\text{-}3)$$

where n_{aw} is the number of active sites on the wall and f_s is the frequency of bubble departure from each site. Suppose that the maximum site is r_{cmax} and r_c is a specified site radius with $r_c < r_{cmax}$. The number of potentially active sites n_s within the average r_c to r_{cmax} may be calculated from the form proposed by Brown (1967) and adopted later by Mikic and Rohsenow (1970) is used for nucleate boiling on the wall.

$$n_s = \left(r_{cmax} / r_c \right)^s \qquad r_c < r_{cmax} \qquad (3\text{-}4)$$

For a given surface, r_{cmax} and s are the characterising factors. In some boiling incipience analyses, the values of r_{cmax} were considered to be in the order of a few microns (1×10^{-6} m). Using the distribution given by eq. (3-4), and using the incipience eq. (3-2), n_{aw} may be calculated if r_{cmax} and s are known; of course, r_{cmax} and s are not known and have to be obtained by fitting the data. f_s, the frequency of bubble departure from a given sites given by

$$f_s = \frac{1}{t_w + t_s} = \frac{1}{t_{wg}} \qquad (3\text{-}5)$$

where t_w is the "waiting time" between the departure of a bubble and the initiation of the next one, t_s is the bubble growth time before departure and $t_{wg} = t_w + t_s$. The bubble departs when it grows to a radius R_D at which the forces on it are in imbalance.

Bubble departure diameter

There is substantial body of literature which presents methods to estimate R_D. In this study, the analysis of Klausner et al. (1993) was adopted. This analysis was for horizontal stratified flow. The situation being considered here is somewhat different since our analysis is for a vertical upwards annular flow regime. However, the analysis of Klausner et al. (1993) is still applicable since we are concerned mainly with the various forces acting on the bubble.

Bubble burst diameter

For the bubble burst diameter at the vapour-liquid interface there are no published correlations (or even data) in annular flow nucleate boiling. However, Chan (1990) reports the analysis of some early Harwell cine films which showed that

R_b was surprisingly large (several times greater than the liquid film thickness). The prediction of bubble burst diameter is based on a phenomenological analyses of the characteristics of bubble burst at vapour-liquid interface in flow boiling. There are two dimensionless parameters which are identified to be important to bubble burst. One is Jakob number, Ja, which represents heat transfer driving force affecting the rate of bubble growth. Another is Weber number, We, which represents the ratio of disruptive force to cohesive force acting on a bubble. Through a regression analysis the present author give the following correlation

$$R_B = c_1 \left(Ja \right)^{c_2} \left(We \right)^{c_3} \qquad (3\text{-}6)$$

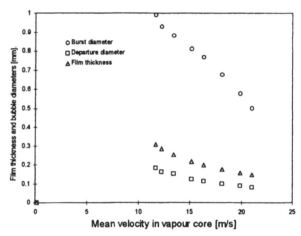

Figure 3-1 Typical parameters of film thickness and bubble diameters against mean velocity in the vapour core.

where R_B is the bubble burst radius, We ($= \rho_L \Delta V^2 R_{eq} / \sigma$) is Weber number. Here $\Delta V = V_G - V_i$. V_G and V_i are gas core and interfacial velocities, respectively. c_1 to c_3 are the constant regression factors which are given below

Media	c_1	c_2	c_3
H₂O	1.18×10^{-2}	-1.29	-0.04
R12	2.71×10^{-3}	-1.2	-0.21

4. THE NEW SUPERPOSITION MODEL FOR TWO-PHASE ANNULAR FLOW AND DATA ANALYSIS

Having obtained the forced convection and nucleate boiling heat transfer coefficients, the superposition model can be written as

$$\alpha_{tp} = \alpha_c + \alpha_n \qquad (4\text{-}1)$$

where α_{tp} is the heat transfer coefficient for two-phase annular flow.

The experimental data analysed are listed in table 1. Fig. 4-1 shows one example of the contribution to the overall heat transfer coefficient from forced convection and nucleate boiling. From the data analysis, we will see that nucleate boiling is gradually suppressed whereas the forced convection becomes dominate with quality increasing. This may indicate that a forced convective mechanism of the conventional type in annular flow rather than the alternative thin film boiling mechanism as suggested by Mesler (1977). Fig. 4-2 shows the comparisons between some typical water data in table 1 and the model predictions. Most of the data are correlated within ±25% .

Table 1. Range Of Data Used In Testing The Model

Ref.	Fluid	Quality	Pressure bar	Mass flux $kg/m^2 s$	Heat flux KW/m^2
Kenning (1989)	water	0.035-0.5	1.6-3.9	123-304	53-300
Sun (1995)	water	0.1-0.5	0.9-1.2	125-240	80-985
Jung (1989)	R12	0.08-0.85	3.9	254-532	10-17

5. CONCLUSIONS

The work described in this paper shows that a rational model for heat transfer can be constructed for annular flow taking account of:

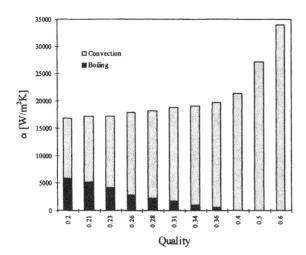

Figure 4-1 The contribution to the overall heat transfer coefficient from nucleate boiling and forced convection.

(1) The hydrodynamics of the liquid film, including interfacial shear stress and entrainment, principally using the analysis of Owen and Hewitt (1987) with later modifications (Hewitt and Govan, 1990).

(2) The convection of heat through the turbulent liquid film, including the effect of interface damping.

(3) The growth and departure of bubbles at and from wall nucleation sites, and the subsequent slide growth and bursting of the bubbles as they slid along the heat transfer surface.

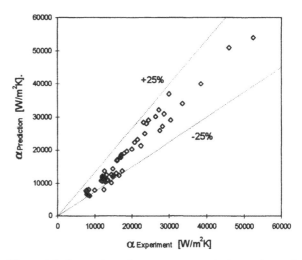

Figure 4-2 Comparisons between some typical experimental data and the present model.

The model shows that a reasonable phenomenological representation of the evaporation process can be achieved taking account of the mechanisms. The model does, of course, require empirical fits (for instance for the distribution of number and size of potential wall nuclei and for the bubble burst diameter). Thus, the model should in no way be regarded as a general predictive tool. However, the fact that it can be constructed and that it is consistent with the phenomena occurring may encourage further developments in this direction.

6. REFERENCES

Chan W. H. G. T., 1990, "Evaporation and condensation in annular vertical upward flow of water-steam", PhD. Thesis. Imperial College.

Chen J. C., 1966, "Correlation for boiling heat transfer to saturated fluids in convective flow", Ind. Eng. Chem. Proc., Des. Dev. Vol. 5, pp. 322-329

Hewitt G. F. and Govan A. H., 1990, "Phenomena and prediction in annular two-phase flow", Invited Lecture, Symposium on Advances in Gas-liquid Flows, ASME Winter Annual Meeting, Dallas, November 1990 (FED-Vol. 99, HTD-Vol. 155)

Hewitt G. F. and Hall-Taylor N. S., 1970, "Annular two-phase flow", Pergamon Press.

Jung D. S., 1989, "Horizontal-flow boiling heat transfer using refrigerant mixtures", EPRI Rep. ER-6364.

Kenning D. B. R. and Cooper M. G., 1989, "Saturate flow boiling of water in vertical tubes", Int. J. Heat Mass Transfer, Vol. 32, No. 3, pp. 445-458.

Klausner J. F., Mei R., Bernhard D. M. and Zeng L. Z., 1993, "Vapour bubble departure in forced convection boiling", Vol. 36, No. 9, pp. 2271-2279

Levich V. G., 1962, "Physicochemical hydrodynamics", Chap. XII, Prentice-Hall Inc., Englewood Cliffs, New Jersey

Mesler R. B., 1973, "An alternative to the Dengler and Adams convective concept of forced convective boiling heat transfer", AIChE. Jl Vol. 27, pp. 448-453

Mudawwar A. and El-Masri M. A., 1986, "Momentum and heat transfer across freely-falling turbulent liquid film", Int. J. Multiphase Flow, Vol. 12, No. 5, pp. 771-790

Owen D. G., 1986, "An experimental and theoretical analysis of equilibrium annular flows", PhD. thesis, University of Birmingham

Owen D. G. and Hewitt G. F., 1987, "An improved annular two-phase flow model". Proc. 3rd Int. Multiphase Flow Conference, The Hague, Netherlands, Paper No. C1.

Sun G. and Hewitt G. F., 1995, "Experimental studies on heat transfer in annular flow". 2nd European Thermal-Sciences and 14th UIT National Heat Transfer Conference, Rome, 29-31 May 1996 (waiting for publish)

CRITICAL HEAT FLUX AND TRANSITION-FILM-BOILING

UPSTREAM CHF OF FORCED CONVECTION BOILING IN AN UNIFORMLY HEATED VERTICAL TUBE

Sadao Yokoya, Makoto Watanabe and Masahiro Shoji

Department of Mechanical Engineering
The University of Tokyo
Tokyo, Japan

ABSTRACT

Critical heat flux condition of forced convection boiling inside an uniformly heated tube usually takes place at the tube exit end but it is sometimes detected at an upstream location under high mass velocity at high system pressure. To make clear the characteristics of this anomalous CHF phenomenon, systematic experiments were conducted by employing Freon-115 as a test fluid for wide experimental ranges of system pressure, mass velocity, heated tube length and inlet subcooling, using three kinds of test tube which have the same inner diameter but different inner surface roughness. Spatial and temporal wall temperature excursion at and post CHF was investigated in detail to detect correctly the upstream CHF. From the experiments, the followings are found: (1)upstream CHF is an autonomous phenomenon; (2)upstream CHF condition is likely to occur in a rough tube; (3)upstream CHF is likely to be detected at low inlet subcooling at high pressure under high mass velocity; (4)upstream CHF is higher than regular downstream CHF for the same inlet subcooling. CHF data are correlated as a function of local quality χ_c and it is shown that the data of upstream CHF branch from an unique CHF-χ_c relation.

INTRODUCTION

In forced convective boiling inside an uniformly heated tube, the temperature of liquid increases in a flow direction in subcooled region and the quality of two-phase flow increases in saturated or superheated region. So it is quite reasonable to expect that critical condition takes place at the exit end of the heated tube. Actually, under usual experimental conditions, CHF is detected at the tube exit end. It has been known, however, that when the system pressure is high and the mass velocity is also extremely high, CHF is sometimes detected at an upstream location. This anomalous CHF is often designated as "upstream CHF" against regular downstream CHF.

The upstream CHF phenomena has been investigated by many researchers [Waters et al. (1964),Matzner et al. (1965), Peskov et al. (1969), Groeneveld (1972, 1974), Merilo (1977), Merilo and Ahmad (1979), Katto and Yokoya (1982) and Katto and Ashida (1982)]. In spite of their brave efforts, however, the aspect and the mechanism have not been made fully clear.

In the author's earlier study[Yokoya et al.(1994)], systematic experiments were conducted by employing Freon-115 as a test liquid for wide parameter ranges of system pressure, mass velocity, test tube length and inlet subcooling, and several important features of upstream CHF were made clear. One of the findings in it was that upstream CHF is accompanied by downstream CHF under some parameter conditions. It was also recognized concerning the occurrence of CHF that the rise of tube wall temperature declines noticeably with increasing system pressure and mass velocity and decreasing inlet subcooling. These facts may imply that upstream CHF is of subtle phenomenon and care must be taken to detect it. In the present study, with an aim of making clear the situation of occurrence of upstream CHF, spatial and temporal wall temperature excursion at and post CHF is investigated in detail. Other characteristics and parameter effects on upstream CHF are also discussed.

EXPERIMENTS

Experimental apparatus is illustrated schematically in Fig.1. Freon-115 is employed as a test liquid. A part of the test liquid flowing out of a circulating pump, passes through flow meters and an electric preheater to enter the test section with a prescribed subcooling. The vapor and liquid leaving the test section flow into a pressurizer. Another part of the subcooled liquid leaving the circulating pump passes through a cooler (counter flow heat exchanger cooled by water), and then the greater part of it returns to the circulating pump through a filter while the rest is sprayed into the pressurizer to condense the vapor coming from the test section. A water-cooled condenser and an electric heater equipped in the pressurizer is used to control the temperature of the saturated liquid to realize a prescribed system pressure. The liquid flowing out of the

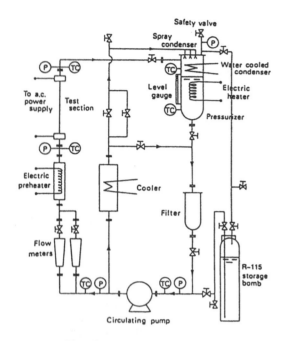

Fig.1 Experimental apparatus
(P: pressure gauge, TC: Thermocouple)

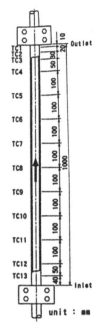

Fig.2 Composition of test section
(unit of length: mm, T: thermocouple)

Fig.3 Thermocouples on a test
tube (a case of L=1000mm)

Table 1 Positions and numbering of thermocouples on the outer surface of the test tube

L/d	Distance from outlet x (mm)													
	Thermocouple No.													
	1	2	3	4	5	6	7	.....	11	12	13	14	.....	31	32
50	10	30	60	110	160	210								
200	10	30	60	110	210	310	410	.....	810	910	960			
600	10	30	60	110	210	310	410	.....	810	910	1010	1110	.....	2810	2910

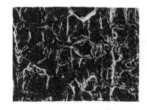

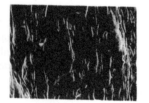

(a) Type-B test tube   (b) Type-C test tube
Fig.4 Photomicrographs of the inner surface of test tubes

pressurizer joins the subcooled liquid coming from the cooler and returns to the circulating pump.

The test tube is a stainless steel tube of inner diameter 5 mm with wall thickness of 1 mm. The length of the tube is chosen as 250 mm, 1000 mm and 3000 mm, which correspond to the length-to-inner diameter ratio of 50, 200 and 600 respectively. As the space of the test section is taken 3000 mm in height, an additional tube is set at an upstream part of the short test tube such as shown in Fig.2 and used as a preheater for adjusting the inlet subcooling. The test tube is heated directly by A.C. current. As is shown in Fig.3, which is a case of the tube length of 1000 mm, 0.1 mm diameter Chromel-Alumel thermocouples are spot-welded to the outer surface of the test tube in order to measure

the wall temperature and to detect CHF. Table 1 summarizes the location and numbering of the thermocouples. According to the results of the author's previous experiment[Yokoya et al.(1994)], upstream CHF differs noticeably depending on the tube inner surface finish. So three test tubes, a smooth tube (named as Type-A) and two kinds of rough tube (named as Type-B and Type-C) are prepared:

- Type-A tube: an annealed tube with polished inner surface, having maximum roughness of 5 μm.
- Type-B tube: a commercial tube with inner surface washed by acids, having maximum roughness of 38 μm.
- Type-C tube: a commercial tube produced by pulling out from the ingot with inner surface washed by acids, having maximum roughness of 30 μm.

The inner surface roughness of Type-B and Type-C tubes is almost the same but the surface appearance (minute structure) is different between the two as seen in photomicrographs of Fig.4.

In every boiling test, the tube was heated gradually step by step and the heat flux when the wall temperature raised sharply was taken for CHF. The wall temperature at every thermocouple location was monitored every ten seconds. Not to burnout the test tube, a CHF-detector was set to operate automatically for shutting off the electric input power when the thermocouple detect the wall temperature rise higher than the

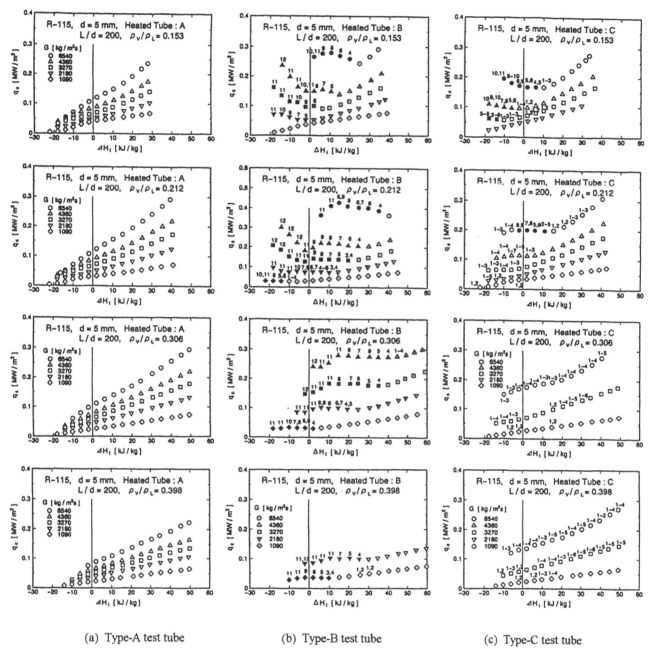

(a) Type-A test tube (b) Type-B test tube (c) Type-C test tube

Fig.5 Raw CHF data obtained using three types of test tube. Tube length is L=1000mm and inner diameter is d=5mm. Open symbols represent regular CHF and solid symbols upstream CHF. Numeral at the side of datum point denotes the thermocouple location with which critical condition was detected first.

temperature plus about 50 K. As shown in Fig.2, a thermocouple and a pressure gauge were equipped near the entrance of the test tube to find inlet condition of test liquid. Since the pressure drop through the test tube was at most 2.2% of the absolute pressure in the present experimental range, the reading of the pressure gauge was taken for the value of system pressure. The experiment was carried out in the following parameter ranges:

- system pressure p: 1.4 -3.0 MPa (corresponding vapor-to-liquid density ratio: $\rho_v/\rho_L = 0.1 - 0.4$)
- mass velocity G: 545 - 6540 kg/m²s

- tube length L: 250 - 3000 mm(corresponding length-to-inner diameter ratio: L/d = 50 - 600)
- inlet enthalpy (inlet subcooling) ΔH_i = -30 - 60 kJ/kg.

EXPERIMENTAL RESULTS
Upstream CHF

Figure 5 are the raw data of CHF obtained using separately the Type-A, Type-B and Type-C tubes under the same parameter conditions: The tube length was fixed as 1000 mm (L/d=200) and system pressures was varied as 1.88, 2.27, 2.68 and 3.12 MPa

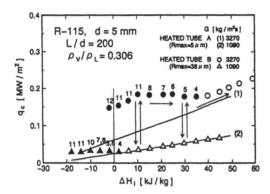

Fig.6 Two sets of data showing the difference of regular
and upstream CHF.   Arrow symbols indicate the
changing of experimental condition.

(ρ_V/ρ_L=0.153, 0.212, 0.306 and 0.398), mass velocity from 545
to 6540 kg/m^{2}s and inlet enthalpy ΔH_i from -30 to 60 kJ/kg.
Open symbols represent the data of regular downstream CHF
detected at the tube exit end and the solid symbols the data of
upstream CHF. The numeral at the side of each symbol indicates
the location of the thermocouple (see Table 1) with which the
critical condition is detected. We can see in Fig.5 that only
regular downstream CHF is obtained by Type-A tubes whereas
upstream CHF is detected in a certain experimental range of
parameters in cases of Type-B and Type-C tubes. It should be
also noted that in a case of Type-C tube, upstream CHF is mostly
accompanied by regular downstream and pure upstream CHF is
detected only in a case of ρ_V/ρ_L=0.212. Data of CHF for other
tube lengths were reported elsewhere[Yokoya et al.(1994)].
     From the results of Fig.5, the followings are found:
(1) Upstream CHF phenomenon is likely to take place in a rough
     tube.
(2) The difference of CHF among test tubes becomes marked as
     the mass velocity increases.
(3) Upstream CHF phenomenon is more likely to take place for
     small ΔH_i under a fixed mass velocity.
(4) The region of ΔH_i where upstream CHF is detected becomes
     wide as the mass velocity increases.

Regular Downstream CHF

     As seen in Fig.5, in a region of subcooled inlet condition,
regular downstream CHF varies linearly with inlet enthalpy ΔH_i.
By extrapolating the linear relation, the value of CHF at saturated
inlet condition, q_{co}, can be obtained and compared with Katto and
Ohno's correlation. The result of comparison is quite the same to
the previous ones [Yokoya et al.(1994)] and it is found that
(1) q_{co} agrees well in general with Katto and Ohno's correlation;
(2) The data at high pressure (high ρ_V/ρ_L) and high mass
     velocity deviate from Katto and Ohno's correlation;
(3) The deviation becomes large as the pressure increases and
     also mass velocity increases.
Regular CHF data for positive inlet enthalpy(ΔH_i) can be
compared in a general form with Katto and Ohno's correlation if
we employ "boiling length, ℓ_b" in place of actual tube length.
The result is also the same to the previous one that are
(1) CHF data for ℓ_b/L less than 200 deviates from Katto and
     Ohno's correlation;

(2) The deviation becomes marked as ℓ_b/L is reduced.

Autonomy of Upstream CHF Phenomenon

     Figure 6 is a typical set of CHF data obtained at a density
ratio of ρ_V/ρ_L=0.306 for two mass velocities of 1090 and 3270
kg/m^{2}s using Type-B tube. The CHF data obtained for the same
experimental conditions using smooth Type-A tube is shown by
solid lines for comparison.   The arrow symbols in the figure
indicate the changes of experimental condition made to see the
reproducibility of the data and to see whether or not the upstream
CHF phenomenon is autonomous.   The test was conducted by
varying ΔH_i from a higher value to a lower value under a fixed
mass velocity of 3270 kg/m^{2}s. During the test, at the condition of
ΔH_i=10 kJ/kg, the mass velocity G was changed to 1090 kg/m^{2}s
and then changed back to the value of G=3270 kg/m^{2}s. On the
other hand, the inlet enthalpy ΔH_i was changed in the increasing
direction of ΔH_i under a fixed mass velocity of G=3270 kg/m^{2}s,
and, at ΔH_i=30 kJ/kg, mass velocity was changed in the same
manner as was mentioned above. From these tests, it is found
that
(1) upstream CHF phenomenon is autonomous, that is, the
     value of upstream CHF is uniquely determined depending
     on the parameter conditions;
(2) upstream CHF is higher than the estimated regular CHF.

Wall Temperature Excursion

     Typical sets of CHF data obtained using the Type-C tube
with three different tube lengths are shown in Fig.7, in which the
regular as well as upstream CHF data are included and shown by
open and solid symbols respectively.   The datum "a" of this
figure is a typical upstream CHF with downstream CHF and data
of "b" and "c" are typical upstream CHF.   Numerals beside data
points of a, b and c denote the thermocouple location (see Table
1) with which critical condition was detected first.   The temporal
and spatial wall temperature excursions at and post CHF are
shown in Fig.8 and Fig.9, which are the cases of data a, b and c
of Fig.7 respectively. Fig.8 shows the wall temperature rise with
time after changing heat flux slightly beyond CHF.   It is clear
that wall temperature no longer rises after a few minutes pass
and becomes quasi-steady state.   Fig.9 shows the longitudinal
distribution of wall temperature at the quasi-steady state when
two minutes pass after changing the heat flux (at the time shown

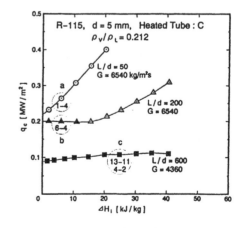

Fig.7 Typical sets of CHF data relating to Figs 8 and 9.

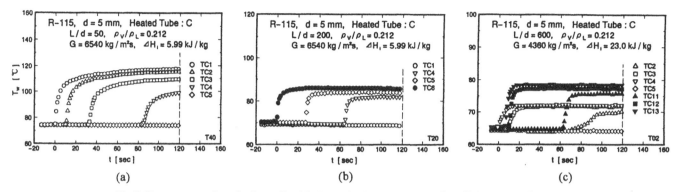

(a)　　　　　　　　　　　　　(b)　　　　　　　　　　　　　(c)

Fig.8 Temperature rise of tube wall with time after increasing heat flux slightly beyond CHF.
Figures of (a),(b) and (c) correspond to CHF data of a, b and c in Fig.7 respectively.

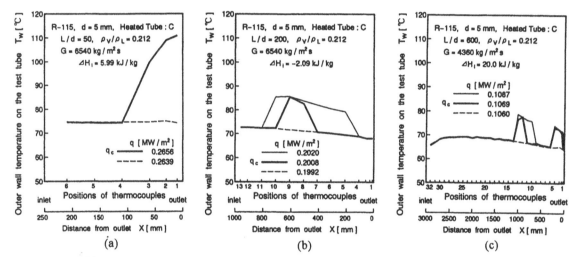

(a)　　　　　　　　　　　　　(b)　　　　　　　　　　　　　(c)

Fig.9 Quasi-steady state longitudinal wall temperature distribution at and post CHF.
Figures of (a),(b) and (c) correspond to CHF data of a, b and c in Fig.7 respectively.

by a dotted line in each figure of Fig.8). It should especially be noted here that upstream CHF condition are detected at two upstream locations as shown in (c) of Fig.8 and Fig.9.

Generally, in cases of usual regular downstream CHF, the wall temperature rises sharply only at the tube exit even when a few minutes pass after changing heat flux slightly beyond CHF. Contrary, in cases of upstream CHF accompanied by regular downstream CHF, wall temperature rises first at the tube exit end and then the temperature rise propagates toward the upstream direction in a short time. In this sense, the terminology such as "upstream CHF accompanied by downstream CHF" or " upstream CHF with downstream CHF " is impertinent. This kind of CHF may be classified into regular downstream CHF.

Effect of System Pressure and Mass Velocity

As is seen in Fig.5, the ΔH_i region where upstream CHF is detected widen with increasing system pressure and mass velocity. Namely, it is said that upstream CHF is more likely to take place when the system pressure is high and the mass velocity is also high. However, at extremely high system pressures, the wall temperature excursion at critical condition is small and it becomes rather difficult to detect CHF. This type of CHF, sometimes designated as slow CHF, can be recognized also

at extremely high mass velocity. These results agree with the report of Groeneveld (1972,1974) and of Peskov et al. (1969).

Effects of Length-to-Inner Diameter Ratio

Data for different tube length is available elsewhere[Yokoya et al.(1994)] in which we found that

(1) upstream CHF is more likely to be detected when the tube is long;

(2) CHF of short tube is generally regular CHF but mostly of the type of " upstream CHF with downstream CHF "

(3) upstream CHF for long tube occasionally take place at two upstream locations at high pressures.

Typical example of the wall temperature excursion concerning with the item (3) above is given in figure (c) of Fig.8 and Fig.9.

CHF-LOCAL QUALITY RELATION

All data of CHF including regular as well as upstream CHF obtained for positive inlet enthalpy (ΔH_i) in the present experiments are correlated as a function of quality χ_C at an onset location of CHF. For cases of upstream CHF, the quality at an upstream location where critical condition was detected first is taken for the value of χ_C. One typical set of the result is shown in Fig.10. As was discussed in detail also in the previous paper

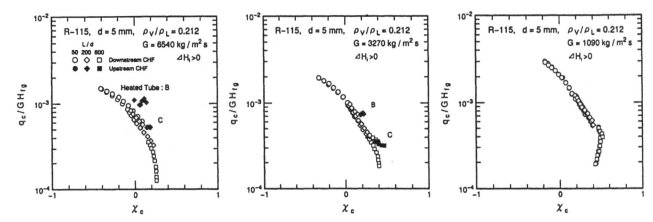

Fig.10 A typical result of correlation of CHF for $\Delta H_i > 0$ as a function of local quality.
Open and solid symbols represent regular CHF and upstream CHF respectively.

[Yokoya et al.(1994)], it seems that an unique relation exist between regular CHF(q_c) and local quality(χ_c) and that data of upstream CHF deviate from the unique q_c-χ_c relation. The dependence of upstream CHF on tube inner surface finish appears as the difference of the position of deviation such as shown by points of B and C in Fig.10. Though CHF is well correlated as a function of local quality as seen in Fig.10, the physical basis is not yet revealed and remains as a future problem.

CONCLUSIONS

Upstream CHF phenomenon was studied by carrying out systematic experiments for wide ranges of experimental parameters such as system pressure, mass velocity, tube length-to-inner diameter ratio, inlet subcooling for three types of test tube with different inner surface finish. Special attention was paid to the situation at critical condition, and tube wall temperature excursion was investigated in detail. From the experiments, the following results are obtained:
(1) upstream CHF phenomenon is an autonomous phenomenon.
(2) upstream CHF is affected noticeably by inner surface finish of test tube.
(3) upstream CHF phenomenon is likely to takes place at low inlet subcooling at high pressure and high mass velocity.
(4) upstream CHF is higher than the estimated regular CHF .
(5) A type of upstream CHF which is accompanied by downstream CHF should be classified into a kind of regular downstream CHF.
(6) Regular downstream CHF is well correlated as a function of local quality, and the upstream CHF deviates and branches from the unique CHF-local quality relation.

NOMENCLATURE:

d   inner diameter of test tube [mm]
G   mass velocity [kg/m2s]
H_{fg} latent heat of evaporation [kJ/kg]
ΔH_i inlet enthalpy [kJ/kg]
L   tube length [mm]
ℓ_b boiling length [mm]
p   system pressure [Pa]
q   heat flux [W/m2]
q_c critical heat flux [W/m2]
q_{co} critical heat flux when ΔH_i=0 [W/m2]

T_w wall temperature of test tube [K]
χ_c local quality at an onset point of CHF [-]
ρ_L density of liquid [kg/m3]
ρ_V density of vapor [kg/m3]

REFERENCES

Groeneveld,D.C., 1972, "The thermal behavior of a heated surface at and beyond dryout", Atomic Energy of Canada Ltd., AECL-4309.

Groeneveld,D.C., 1974, "The occurrence of upstream dryout in uniformly heated channels", Heat Transfer IV, pp.265- 269.

Katto,Y. & Yokoya,S., 1982, "CHF of forced convection boiling in uniformly heated vertical tubes: Experimental study of HP-regime by the use of refrigerant 12", Int.J.Multi-phase Flow, Vol.8, pp.165-181.

Katto,Y. & Ashida,S., 1982, "CHF in high-pressure regime for forced convection boiling in uniformly heated vertical tubes of low length-to-diameter ratio", Proc. 7th Int. Heat Transfer Conf., Vol.4, pp.291-296.

Katto,Y. & Ohno,H., 1984, "An improved version of the generalized correlation of critical heat flux for the forced connective boiling in uniformly heated vertical tubes", Int. J. Heat Mass Transfer, Vol.27, No.9, pp.1641-1648.

Matzner,B. et al., 1965, "Critical heat flux in long tubes at 1000psi with and without swirl promoters", ASME-paper No.65-WA/HT-30.

Merilo,M., 1977, "Critical heat flux experiments in a vertical and horizontal tube with both Freon-12 and water as coolant", Nucl. Engng Design, Vol. 44, pp.1-16.

Merilo,M. & Ahmad,S.Y., 1979, "Experimental study of CHF in vertical and horizontal tubes cooled by Freon-12", Int. J. Multiphase Flow, Vol.5, pp.463-478.

Peskov,O.L. et al., 1969, "The critical heat flux for the flow of steam-water mixtures through pipes", In Progress of Heat Transfer and Hydraulics of Two-phase Media, Edited by Kutateladze, Pergamon Press Oxford, pp.48-62.

Waters,E.D. et al., 1964, "Experimental observations of upstream boiling burnout", Chem.Engng Prog.Symp.Ser. Vol.61,No.57, pp.230-237.

Yokoya S. et al., 1994, "Upstream Critical Heat Flux in Flow Boiling inside a Tube", Proc. of German-Japanese Symposium on Multi-Phase Flow, Karlsruhe, Germany, August, pp.75-88.

CRITICAL HEAT FLUX OF SUBCOOLED FLOW BOILING IN TUBE WITH AND WITHOUT INTERNAL TWISTED TAPE UNDER UNIFORM AND NON-UNIFORM HEAT FLUX CONDITIONS

Hideki Nariai

Institute of Engineering Mechanics

University of Tsukuba

Tsukuba, Ibaraki, Japan

Fujio Inasaka

Nuclear Technology Division

Ship Research Institute

Mitaka, Tokyo, Japan

ABSTRACT

Critical heat flux (CHF) of subcooled flow boiling with water in tube with internal twisted tape under circumferentially non-uniform heat flux condition was experimentally investigated using direct current heating of stainless steel tube. The critical heat flux in tube with internal twisted tape under non-uniform heat flux condition was higher than those under uniform heat flux condition. It was explained with the disruption of bubble boundary layer along the twisted tape flow under non-uniform heat flux condition. The boiling curve of high velocity subcooled flow was derived and the net vapor generation heat flux was confirmed to be almost coincided with the Levy model. By assuming the disruption of bubble boundary layer at the lower heat flux part where the heat flux was lower than the net vapor generation heat flux, the increase of the CHF with twisted tape flow under non-uniform heat flux condition was well explained.

INTRODUCTION

High heat flux components of nuclear fusion reactors are subjected to very high heat flux and the cooling channel is under non-uniform heat flux condition since heat by irradiation comes in, in general, from one side of the tube wall of the cooling channel. Subcooled flow boiling with water has the high possibility to remove high heat load, and the critical heat flux (CHF) is the most important factor to determine the upper limit of the coolability. In order to increase the CHF, internal twisted tape is usually inserted in the cooling channel. However, the researches on the CHF of subcooled flow boiling with water with internal twisted tape under non-uniform heat flux condition are very scarce.

Ornatskii et al (1965) and Leontiev et al (1981) have conducted experiments using direct current heating of the test tube. By using electron beam apparatus, several experimental researches with internal twisted tape have recently been conducted by Koski et al (1988), Schlosser et al (1993), Araki et al (1992) and some other researchers. The heat flux distribution around the cooling channel was under non-uniform heat flux condition, and the CHF was estimated to be higher than the predictions by the existing CHF correlations under uniform heat flux condition. Nariai et al (1991, 1992, 1993, 1994, 1995) have recently continued the experiments

on the CHF of subcooled flow boiling with internal twisted tape under uniform and non-uniform heat flux conditions.

Present paper shows the CHF experimental results in tube with and without internal twisted tape under uniform and non-uniform heat flux conditions. Then, the boiling characteristics of high velocity subcooled flow are presented. Finally, the higher CHF in tube with internal twisted tape under non-uniform heat flux condition is explained assuming the disruption of bubble boundary layer at the lower heat flux part.

EXPERIMENTAL APPARATUS AND METHOD

Experiments were conducted with vertical test tube made of stainless steel, which was directly heated by a DC power source. Figure 1 shows the schematic drawing of water loop used to circulate water from water tank to the test tube. Experiments consisit of two parts. One is the CHF experiments with and without internal twisted tape under uniform and non-uniform heat flux conditions. The other is to obtain the boiling curve with and without internal twisted tape under uniform heat flux condition.

CHF Experiment

Figure 2 shows a schematic drawing of the test tube. In order to simulate circumferentially non-uniform heat flux condition, a part of the test tube was made thin by electrochemical machining. The cross section in the figure is the case of a thinned part angle $\beta = 180°$. Figure 3 shows the cross section of the test tube under uniform heat flux and six cases of non-uniform heat flux conditions. Dimensions of the test tubes were 10 cm in heated length, 6 mm in inside diameter and 0.35 mm in wall thickness of thick part and 0.25 mm or 0.15 mm in thin part. For the experiments with internal twisted tape, a twisted tape made of zirconia was inserted in the test tube. The gap between the tape and the tube wall was 0.2 mm in average. Experimental conditions are as follows;

Inlet temperature of water : about 40°C
Mass velocity of water : 5,000, 8,700 kg/m²s
Pressure at the tube outlet: 0.1, 0.6, 1.1, 1.5 MPa
Cut thickness (α) : 0.1 , 0.2 mm
Cut angle (β) : 90° , 180° , 270°
Twisted tape ratio(y) : 2.64, 3.47, ∞

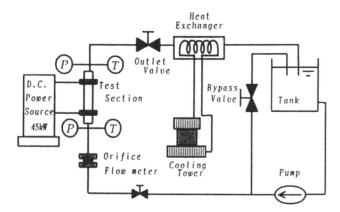

Fig. 1 A schematic drawing of test loop

(Twisted tape ratio y is defined as the axial length of 180° tape twist devided by the tube inside diameter, and y=∞ means the straight tube without internal twisted tape.)

Boiling Curve Experiment

In order to confirm the boiling characteristics of high velocity subcooled flow, boiling curves were derived both with and without internal twisted tape under uniform heat flux condition. The experimental conditions are as follows;

Inlet temperature of water: about 40℃

Mass velocity of water: 2,000 to 9,000 kg/m²s

Pressure at the tube outlet: 0.6 MPa

Twisted tape ratio(y) : 3.47, ∞

CHF CHARACTERISTICS
CHF In Straight Tube

Figure 4 shows the CHF against the qualities of tube outlet in the case of α = 0.1 mm and 0.2 mm with mass velocity of 5,000 kg/m²s, and outlet pressure of 1.1 MPa. For non−uniform heat flux conditions with β=90° ,180° and 270° , the symbols $q_{c,H}$, $q_{c,L}$ and $q_{c,av}$ mean the heat flux of thicker part or higher heat flux at burnout, thinner part or lower heat flux at burnout, and the average heat flux, respectively. The symbol $q_{c,unif}$ is the CHF under uniform heat flux condition. The dotted line shows the prediciton by the Modified Tong correlation which Inasaka−Nariai (1987) proposed for straight tube under uniform heat flux condition. The solid line shows the relation between the heat flux and the average qualities resulting from the heat balance at the inlet temperature 40 ℃. The figure shows the decrease of the average qualities with the increase of non−uniformity or the increase of α and/or β. The CHF in straight tube under non-uniform condition agrees with the Modified Tong prediction, though $q_{c,L}$ is different between α = 0.1 mm and 0.2 mm. This result means that even if the ratio of higher to lower heat flux is different, the higher heat flux under non−uniform heat flux condition can be predicted by the correlation for uniform heat flux condition using actual average outlet qualities.

CHF In Tube With Internal Twisted Tape

Figure 5 shows the CHF against the average qualities at tube outlet for outlet pressure of 0.6 MPa with mass velocity 5,000 kg/m²s and twisted tape ratio 2.64. The dotted line shows the prediction by the Nariai−Inasaka correlation which Nariai et al

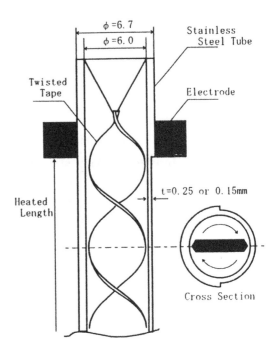

Fig.2 A schematic drawing of the test tube

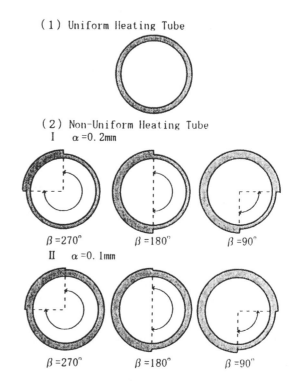

Fig. 3 Cross sections of the test tube

(1993) proposed for tubes with internal twisted tape under uniform heat flux condition. As shown in the figure, though average qualities of tube outlet are almost equal or larger than those under uniform heat flux, the increasing rate of $q_{c,H}$ from Nariai−Inasaka correlation or the CHF under uniform heat flux condition becomes larger with the increase of the non−uniformity. In the figure, $q_{c,L}$

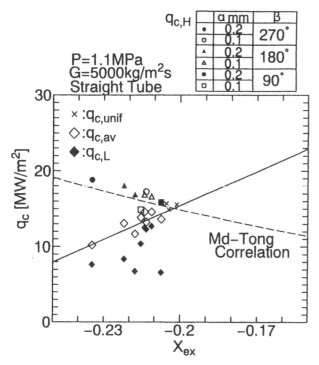

Fig. 4 CHF in straight tube under non−uniform heat flux condition

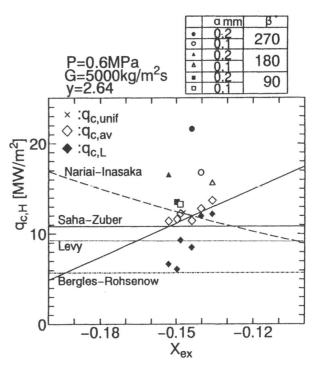

Fig. 5 CHF with internal twisted tape uder non−uniform heat flux condition

or lower heat flux for each datum is shown. The boiling inception heat flux by Bergles−Rohsenow and the net vapor generagtion heat flux by Saha−Zuber and Levy are also presented for each experimental condition.

BOILING CURVE

In order to derive the boiling curve, outer wall temperatures of the test tube near the outlet of the test section were measured using CA thermocouple with 0.25 mm diameter. The heat flux was evaluated from the input power, and the inner wall temperature was calculated with heat conduction equation.

Straight Tube

Figure 6 shows the boiling curve in straight tube without internal twisted tape for five mass velocities. The symbol ΔT_s means the wall superheat temperature. In the figure, the boiling inception correlation by Bergles−Rohsenow, nucleate boiling correlations by Thom and Jens−Lottes are indicated. The symbol T_{kk} means the homogeneous nucleation temperature at the inner tube wall.

In the figure, the gradient of the nucleate boiling curve coincides with Thom at lower mass velocities. However, with the increase of the mass velocities, the gradient becomes more gentle than those of Jens−Lottes and Thom. The forced convection effect governs particularly single phase and bubble attached regions. Further for very high mass velocities, the effect seems to affect on the boiling characteristics at high heat flux and the inner wall temperature approaches to the homogeneous nucleation temperature.

Figures 7 and 8 show the heat transfer coefficients at mass velocities of 3,000 and 9,000 kg/m²s, respectively. Sieder−Tate correlation predicted well the single phase heat transfer coefficients. Bergles−Rohsenow correlation also predicts well the boiling inception heat flux. For net vapor generation heat flux, Levy and Saha−Zuber predict similar heat flux at low mass

velocities. However, at high mass velocities, Levy model predicted well in our experimental condition. The decrease of the heat transfer coefficient at high heat flux may be closely related to the mechanism of high velocity subcooled flow boiling.

Tube With Internal Twisted Tape

Figures 9 to 10 show the boiling curve, and heat transfer coefficients in tube with internal twisted tape, respectively. The general trend is also very similar to the case for straight tube. Single phase heat transfer coefficients are predicted well by Gambill correlation and the Levy model predicted in general well the net vapor generation heat flux.

When we compare Fig.8 and Fig.10, the boiling heat transfer coeficients are, as already well known, almost the same for both cases with very small increase in the case of internal twisted tape. On the other hand, the CHF increases with the twisted tape insertion.

CHF ENHANCEMENT UNDER NON−UNIFORM HEAT FLUX CONDITION

The increase of CHF with internal twisted tape under non−uniform heat flux condition occurs due to the disruption of bubble boundary layer. For that, the bubble boundary layer has to be disrupted at the lower heat flux part along the twisted tube flow. We estimated that when $q_{c,L}$ was lower than the net vapor generation heat flux, then the disruption of bubble boundary layer might occur. The thinned part angle is also very important to disrupt the bubble boundary layer. According to the results of boiling curve, we adopted the Levy model as net vapor generation heat flux. In our experimental condition, $q_{c,L}$ with $\alpha=0.2$ mm are almost always below Levy, and $q_{c,L}$ with $\alpha=0.1$ mm above Levy.

Figure 11 shows the comparison of $q_{c,H}$ and $q_{c,unif}$ for $\alpha=0.2$ mm.

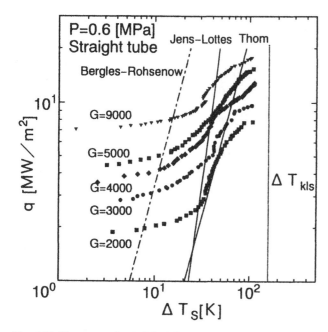

Fig. 6 Boiling curve in straight tube

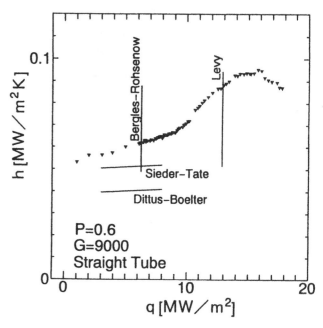

Fig. 8 Heat transfer coefficient in straight tube

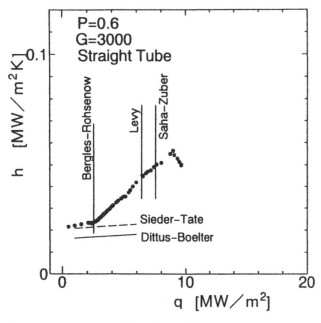

Fig. 7 Heat transfer coefficient in straight tube

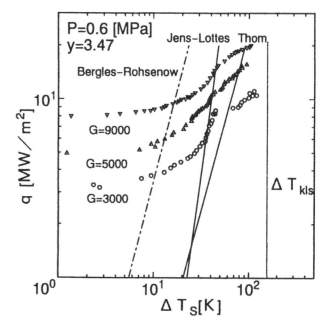

Fig. 9 Boiling curve with internal twisted tape

The CHF at higher part $q_{c,H}$ becomes almost 1.7 times of $q_{c,unif}$ at $\beta = 270°$. With the decrease of β, the ratio of $q_{c,H}$ to $q_{c,unif}$ approached to unity.

Figure 12 shows the case for $\alpha = 0.1$ mm. In this case the ratio is not so much high. Figure 13 shows the comparison of $\alpha = 0.2$ and 0.1 mm for $\beta = 270°$. We can clearly show that in this case the disruption of bubble boundary layer is effective under low $q_{c,L}$ condition. However, when $\beta = 90°$, the disruption of bubble boundary layer is not effective even when low $q_{c,L}$ condition.

CONCLUDING REMARKS

Critical heat flux of subcooled flow boiling with and without internal twisted tape under non-uniform and uniform heat flux conditions were experimetally investigated. Main results are as folows;

(1) The CHF with internal twisted tape under non-uniform heat flux condition is higher than that under uniform heat flux condition. The enghancement increases with the increase of the non-uniformity.

(2) The boiling curve for high mass velocity was derived, which

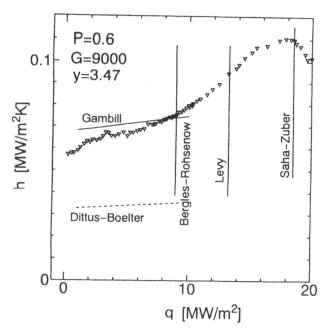

Fig.10 Heat transfer coefficient with internal twisted tape

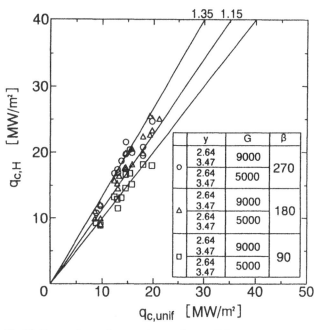

Fig.12 Comparison of $q_{c,H}$ and $q_{c,unif}$ for $\alpha = 0.1$ mm

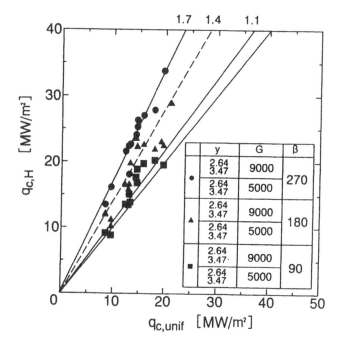

Fig.11 Comparison of $q_{c,H}$ and $q_{c,unif}$ for $\alpha = 0.2$ mm

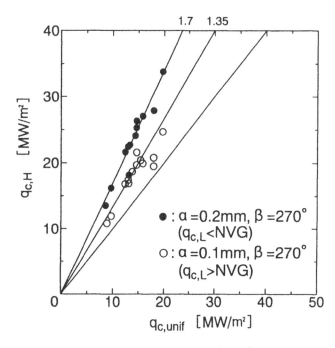

Fig.13 Comparison of $\alpha = 0.2$ and 0.1 for $\beta = 270°$

showed that the net vapor generation heat flux was almost coincided with the Levy prediction.

(3) The increase of the CHF with twisted tape flow under non-uniform heat flux condition was well explained by assuming the disruption of bubble boundary layer at the lower heat flux part where the heat flux was lower than the net vapor generation heat flux.

NOMENCLATURE

G	Water mass velocity
h	Heat transfer coefficient
P	Pressure
q	Heat flux
ΔTs	Wall superheat
X	Quality
y	Twisted tape ratio
α	Cut thickness
β	Cut angle

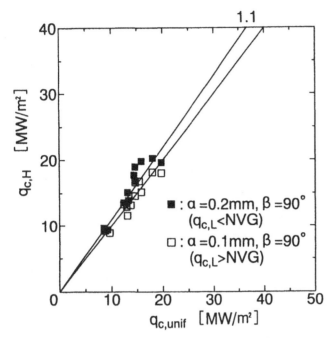

Fig.14 Comparison of α=0.2 and 0.1 mm for β=90°

Subscripts
av   Average heat flux
c   Critical heat flux
ex   Tube outlet
kls   Homogeneous nucleation temperature
L   Lower part heat flux
H   Higher part heat flux
unif   Uniform heat flux

REFERENCES

Araki, M., Ogawa, M., and Akiba, M., 1992,"Experimental Evaluation of Critical Heat Flux under One Sided Heating Condition," Proc. of the 2nd Specialist's Workshop on High Heat Flux Components Thermal–Hydraulics, Rome, Italy.

Inasaka, F., and Nariai, H., 1987 "Critical Heat Flux and Flow Characteristics of Subcooled Flow Boiling in Narrow Tubes," JSME International,Vol.30, p.1595–1600.

Leontiev, A. I., Mostinsky, I. L., Polonsky, V. S., Styrikovich, M. A., and Chernika, I. M. , 1981,"Experimental Investigation of the Critical Heat Flux in Horizontal Channels with Circumferentially Variable Heating," Int. J. Heat Mass Transfer, Vol. 24, No. 5, pp. 821–828.

Nariai, H., Inasaka, F., Fujisaki, W., and Ishiguro, H., 1991,"Critical Heat Flux of Subcooled Flow Boiling in Tubes with Internal Twisted Tape," 7th Proc. of Nuclear Thermal Hydraulics, ANS, pp.38–46.

Nariai, H., Inasaka, F., Ishikawa, A., and Fujisaki, W., 1992, "Critical Heat Flux of Subcooled Flow Boiling in Tube with Internal Twisted Tape under Non–Uniform Heating Condition," Proc. of the 2nd JSME–KSME Thermal Engineeering Conf., Vol. 3, pp.285–288.

Nariai, H., and Inasaka, F., 1993, "Critical Heat Flux of Subcooled Flow Boiling in Tubes with and without Internal Twisted Tape under Non–Uniform Heating Condition," Proc. of the 6th Int. Topical Meeting on Nuclear Thermal Hydraulics, Grenoble, Vol. 1, pp.715–722.

Nariai, H., Inasaka, F., Ishikawa, A., and Kinoshita, H., 1994, "Effefct of Internal Twisted Tape on Critical Heat Flux of Subcooled Flow Boiling under Non–Uniform Heating Condition," Heat Transfer 1994, Vol. 7, pp.521–526.

Nariai, H., Kinoshita, H., and Inasaka, F., 1995, "The Critical Heat Flux of Subcooled Flow Boiling under Non–Uniform Heating Condition in Tube with Internal Twisted Tape," Proc. of the ASME–JSME Thermal Engineering Joint Conference, Maui, Hawaii

Ornatskii, A. P., and vinyarskii, L.S., 1965, "Het Transfer Crisis in a Forced Flow Underheated Water in Small–Bore Tubes," High Temperature, Vol. 3, pp.400–406.

Schlosser, J., and Boscary, J. 1993, "Thermal Hydraulics Tests at NET/ITER Relevant Conditions on Divertor Targets Using Swirl Tubes," Proc. of the 3rd Specialist's Workshop on High Heat Flux Component Cooling, Cadarache, France, p.5a–1.

CRITICAL HEAT FLUX AND MINIMUM HEAT FLUX OF FILM BOILING OF BINARY MIXTURES FLOWING UPWARDS IN A VERTICAL TUBE

Hein Auracher [1], Andreas Marroquin [2]
[1] Institut für Energietechnik
Technische Universität Berlin
Berlin, Germany
[2] Institut für Technische Thermodynamik
und Thermische Verfahrenstechnik
Universität Stuttgart
Stuttgart, Germany

ABSTRACT

The effects of heated length, local quality or subcooling and mole fraction at constant pressure (1.1 MPa) and mass flux ($\dot{m}$ = 1000 kg/(m^{2}s)) on the critical heat flux (CHF) and the minimum heat flux of film boiling (MHF) have been investigated for the refrigerant mixture $CBrF_3$ (R13B1)/$C_2Cl_2F_4$ (R114). The study is restricted to the DNB type of the boiling crisis. It was found that CHF has a maximum in a CHF versus mole fraction plot. The higher the outlet subcooling the more pronounced is the maximum. Only in long heated test sections and at an outlet quality of $\dot{x}$ = 0.2 no maximum was observed. There is also a heated length effect. For the pure components : $(l/d)_{min} \approx$ 10, above which the CHF does not change for fixed outlet subcooling or quality. The corresponding limit for mixtures was found to be significantly larger: $(l/d)_{min} \approx$ 30. All above mentioned effects are of minor importance for MHF.

INTRODUCTION

Only few studies on CHF of mixtures in forced flow are known. To the authors best knowledge data on MHF are not available. Previous results for CHF have been summarized recently by Collier and Thome (1994). It was found that CHF has a maximum in a CHF versus mole fraction plot where the mole fraction difference between vapor and liquid of the more volatile component: (y-x), reaches a maximum. These experiments are, however, limited to low pressures, test sections with a small ratio l/d and low quality or subcooled outlet conditions. Celata et al. (1994) presented CHF data for the refrigerant mixture CCl_2F_2 (R12) and $C_2Cl_2F_4$ (R114) in forced convective upflow in a vertical stainless steel uniformly heated tube, 2.1m long and 7.57mm i.d.. The pressures were 1.2 and 2.8 MPa and the inlet subcooling was 23 K, but the local quality at CHF was always positive. In most of the test runs CHF due to dryout occurred. It was found that no significant maximum exists for CHF versus mole fraction. A similar study to Celata et al. has been performed by Mori et al.

(1990) (vertical stainless steel tube, 9mm i.d., 2m long) with $CHClF_3$ (R22) and $C_2Cl_2F_4$ (R114) mixtures at relative high pressures (2.3 + 4.8 MPa). Again no significant influence of mole fraction on CHF was found.

There is obviously a strong contradiction about the behaviour of CHF versus mole fraction which could be due to the different experimental systems: short test sections with low quality or subcooled outlet conditions on the one hand and long test sections with higher outlet qualities on the other. The present study is focused on this problem. The effects of heated length, local quality or subcooling and mole fraction on CHF and MHF are investigated for the refrigerant mixture $CBrF_3$ (R13B1)/$C_2Cl_2F_4$ (R114). Main emphasis is laid on precise experimental data rather than on a physical interpretation and corresponding correlations. This will be the topic of a later report.

A technique has been developed to carry out steady-state experiments of the entire boiling curve with an electrically heated system. On the vertical test tube a cylindrical copper block is soldered which is divided into several short sections. Each of these sections can be heated separately. It is also possible to control the temperature or the heat flux of each section by feedback systems. Hence, CHF and MHF can be measured under steady-state conditions without the disturbance or dangerous effect of an instability. It is furthermore possible to vary the heated length of the tube.

EXPERIMENTAL APPARATUS

Fig. 1 shows the test section. It consists of a vertical tube with an inner diameter of 17mm. The first part is a three-stage pre-evaporator (1,2,3) followed downstream by the main evaporator (4). The pre-evaporators are used to establish a certain quality or subcooling at the main evaporator inlet. Temperature and pressure at the test section inlet, where the fluid is always subcooled, are measured by a resistance thermometer (W_1) and a pressure transducer (p_1), resp..

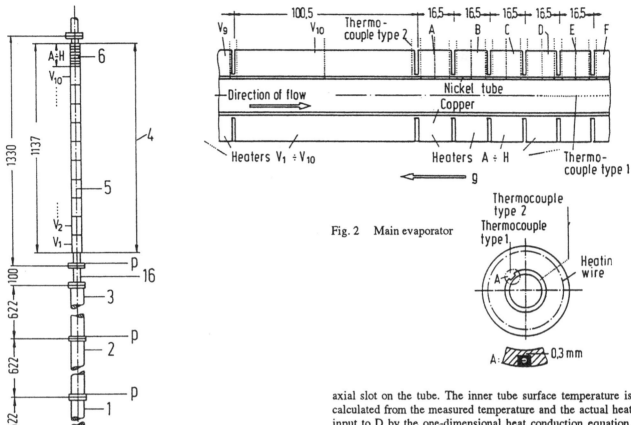

Fig. 1   Test section

1-3 Pre-evaporator
4   Main evaporator
5,6 Sections of the main
       evaporator V_1-V_{10}, A-H

Fig. 2   Main evaporator

Furthermore, a small amount of liquid is removed at the inlet region (x_1) and is supplied to a gas-chromatograph to determine the mole fraction of the mixture. The pressure drop along the test section (p) is measured by three transducers for pressure differences.

Fig. 2 shows cross-sections of the main evaporator. The test tube is made of nickel. Cylindrical cooper blocks are soldered onto the test tube. 10 copper sections (V_1-V_{10}) with 100.5mm length and 47mm o.d. each, and 8 copper sections (A-H) with 16.5mm length and 34mm o.d. each, are used. The copper sections are separated by small slots of 0.45mm width so that only a thin copper bridge (1mm) remains between the sections. Each of these copper sections can be heated separately by a sheathed resistance wire, which is rolled and brazed into a coiled channel on the copper surface. Quasi local heat transfer measurements are carried out in the short sections (A-H), which are DC-heated.

Section D serves as the main heater. All data and parameters of measurements presented later refer to this section. The others (A-C and E-H) are used as guard heaters. Maximum surface heat fluxes of $1MW/m^2$ can be established along the main evaporator. The wall temperature in D is measured by a thermocouple (type 1, Fig. 2), soldered into an axial slot on the tube. The inner tube surface temperature is calculated from the measured temperature and the actual heat input to D by the one-dimensional heat conduction equation. In each section three thermocouples (type 2) are installed radially to measure the temperature distribution along the entire test section. In particular, they are used to smooth out the inner wall temperature of the main heater D and the adjacent guard heaters. The thermocouples in the middle of each section serve as sensors to give the deviation between set-point and actual temperature separately for each section.

All radially installed thermocouples are soldered into holes of 0.7mm diameter with their tip just beneath (0.3mm) the inner surface. Due to the non-isothermal installation of such thermocouples, the measured temperature deviates slightly from the real one. This is, however, not relevant, since only the axial temperature distribution is determined and the error is approximately the same for each sensor of type 2.

Each of the copper sections is heated and controlled separately. It is therefore possible to vary the temperature or the heat flux distribution along the test section. It is in particular possible to establish a constant temperature along the tube wall regardless of the given boiling regime. For control of the short sections (A-H), the dither-technique is applied which has been presented by Auracher and Marquardt (1986). The longer copper sections (V_1-V_{10}) are AC-heated and controlled by a packet method. The maximum power of each long section is 7kW. Hence, the total power in these sections is 70kW. In addition 8kW are supplied in the short sections (1kW each). The maximum possible system pressure is 35bar and the maximum fluid temperature in the test section 130°C. The lowest possible temperature at the test section inlet is -18°C.

EXPERIMENTAL PROCEDURE

Main emphasis of the present study was to find out the influence of mole fraction, quality (or subcooling) and heated length on CHF and MHF. The investigation is limited to CHF due to departure from nucleate boiling (DNB). The experimental technique easily permits to detect the transition from DNB to dryout. In the latter case the dryout front moves forth and back in the region of heater D, which results in temperature oscillations with low frequency and hence large amplitudes. In contrast, DNB is characterized by temperature oscillations of higher frequency and small amplitudes due to the inertia of wall and thermocouple. It was found that up to a local quality of $\dot{x} = 0.2$ always DNB occurs in heater D. The test program, summarized in table 1, reflects the above mentioned aims.

TABLE 1: TEST PROGRAM	
Pressure p, MPa	1.1 ($p / p_{cr,C_2Cl_2F_4} = 0.34$; $p / p_{cr,CBrF_3} = 0.28$)
Mass flux $\dot{m}$, kg / (m²s)	1000
Local quality $\dot{x}$,-	-0.4, -0.2, 0, 0.2
Mole fraction $CBrF_3, x_1$, %	0, 27, 54, 78, 100
Heated length l	variable

Due to experimental limits, $\dot{x} = -0.4$ could only be realized at mole fractions x_1 of 0 and 0.27.

CHF and MHF could be measured by either increasing or decreasing wall temperature. Mostly the CHF was determined by starting at fully developed nucleate boiling and stepwise moving up to a wall temperature higher than the one at CHF ($T_{W,CHF}$). For MHF the wall temperature was stepwise reduced around $T_{W,MHF}$. The control system prevents an instability once CHF or MHF are reached. For some system conditions the entire boiling curves were measured, by stepwise increasing and/or decreasing the wall temperature along all boiling regions.

Pressure p and quality $\dot{x}$ change downstream the test section. Therefore, if these quantities are assigned to a test run or a boiling curve, it holds for the center of section D. Furthermore, in the case of positive $\dot{x}$ the liquid mole fraction x decreases downstream depending on the given system parameters (see next chapter). Because x is directly measured at the test section inlet (x_1,Fig.1) we assign always this x_1 to the results presented later. During a test run the heat input to the heated part of the test section varies significantly. Consequently the heat input to the pre-evaporators had to be adjusted to ensure a constant local $\dot{x}$ in section D for each point of the boiling curve.

EVALUATION OF LOCAL QUALITY

The fluid enters the test section (z = 0) always in subcooled condition with an enthalpy h_1. Then - assuming thermodynamic equilibrium - the local quality at location z (center of section D in the main evaporator, Fig. 2) follows from an energy balance

$$\dot{x} = \frac{1}{\Delta h_v} \left(\frac{\dot{Q}\ 4}{\dot{m}\,\pi d^2} + h_1 - h_{ls} \right) \quad . \qquad (1)$$

Given the pressure p_z at location z the saturation enthalpies in Eq. (1), h_{ls} and $\Delta h_v = h_{lv} - h_{ls}$, are fixed in the case of pure fluids. Q denotes the total heat flow rate supplied to the test section between z = 0 and z. $\dot{Q}, \dot{m}, h_1$, and p_z are known from measurements. Especially at small and large values of $\dot{x}$ the assumption of thermodynamic equilibrium is not fulfilled. Nevertheless, as usual in comparable studies, $\dot{x}$ was always determined by Eq. (1), since no reliable information is available to account for non-equilibrium.

The first two terms in the brackets of Eq. (1) represent the spec. enthalpy h of the fluid at location z. In subcooling: $h < h_{ls}$, hence a negative quality is obtained. Introducing the temperature difference of subcooling at location z.

$$\Delta T_{sub} = T_{ls} - T \qquad (2)$$

and taking into account that $h - h_{ls} = -c_p\,\Delta T_{sub}$, we obtain from Eq. (1)

$$\Delta T_{sub} = -\frac{\dot{x}\,\Delta h_v}{c_p} \qquad (3)$$

where the spec. heat capacity c_p can be determined for pressure p_z and a mean temperature between T_{ls} and T.

Eq. (1) holds also for mixtures. However, the properties are now in addition dependent on the mole fraction: Δh_v, (p_z,x_z), $h_{ls}(p_z,x_z)$, $h_1(p_1,x_1,T_1)$, where x_z and x_1 denote the liquid mole fraction of the more volatile component ($CBrF_3$) at z and z = 0, resp.. The local subcooling can again be calculated by Eq. (3) with the spec. heat capacity c_p of the liquid mixture at location z.

EVALUATION OF LOCAL MOLE FRACTION

Under subcooling conditions, the mole fraction x_z in the center of section D is taken to be equal to x_1. At positive qualities: $x_z < x_1$, because the fraction of the less volatile component ($C_2Cl_2F_4$) in the liquid increases downstream. x_z cannot be measured directly without disturbing the flow. A reasonable estimate is, however, possible if a closed system evaporation and thermodynamic equilibrium is assumed, as shown in VDI-Wärmeatlas (1994). This model yields x_z as follows

$$x_z = \frac{x_1 \left[1 + \dot{x}(M_l / M_v - 1) \right]}{1 + \dot{x} \left[(y_z / x_z)(M_l / M_v) - 1 \right]} \qquad (4)$$

where y_z denotes the equilibrium vapor mole fraction of the more volatile component at location z. M_l and M_v are the molar masses of the liquid and vapor fractions, resp., given by

$$M_l = x_z M_1 + (1 - x_z)M_2 \quad , \qquad (5)$$
$$M_v = y_z M_1 + (1 - y_z)M_2 \quad , \qquad (6)$$

with the molar masses M_1 and M_2 of the pure components.

Eqs. (1) and (4) to (6) have to be solved iteratively. A useful start is $x_z = x_1$. Under the present system conditions (p = 1.1 MPa, $\dot{m} = 1000$ kg / (m²s)) the correspondences summarized in table 2 are obtained for $\dot{x} = 0.2$.

TABLE 2. LOCAL MOLE FRACTIONS AND SARURATION TEMPERATURES AT $\dot{x} = 0.2$
$x_1 = 0.27$ ($T_{sat,x_1} = 55.01°C$) → $x_z = 0.21$ ($T_{sat,x_z} = 61.23°C$)
$x_1 = 0.54$ ($T_{sat,x_1} = 33.56°C$) → $x_z = 0.48$ ($T_{sat,x_z} = 37.57°C$)
$x_1 = 0.78$ ($T_{sat,x_1} = 20.33°C$) → $x_z = 0.74$ ($T_{sat,x_z} = 22.29°C$)

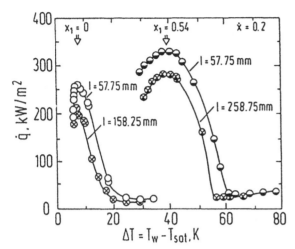

Fig. 3   Boiling curves for pure fluids and mixtures of mole
fraction $x_1 = 0.54$ at different heated lengths. Quality:
$\dot{x} = 0.2$

Obviously the downstream decrease in the mole fraction
cannot be neglected especially if we look at the corresponding
increase of the saturation temperature of several degrees.

The thermophysical properties, needed for data evaluation,
were taken from Kältemaschinenregeln (1981) for the pure
components and for the mixtures equations developed by
Lee/Kesler (1975) and Plöcker (1977) were used.

RESULTS

Figs. 3 to 8 give a survey of the most important results in
terms of mole fraction, quality/subcooling and heated length
effect on CHF and MHF.

Typical Boiling Curves

In Fig. 3 boiling curves for $C_2Cl_2F_4$ ($x_1 = 0$) and of a
mixture with $x_1 = 0.54$ are compared for different heated
lengths. Except for other effects, which are discussed later, it
is quite obvious that boiling curves of mixtures are located at
significantly higher temperature differences ΔT than those of
pure substances . The reason is well known: The liquid mole
fraction of the less volatile component ($C_2Cl_2F_4$) near the wall
is higher with respect to the bulk. Hence, the saturation
temperature at the vapor-liquid interface of a bubble is also
higher. Since a minimum supersaturation is required for vapor
generation the wall temperature increases to reach the
necessary ΔT (note that for the mixture curves the real T_{sat,x_z}
according to table 2 were introduced). The mixture CHFs are
considerably larger than the ones of the pure substances, but
the heat transfer coefficients α are much smaller, e.g. at a
heated length of $l = 57.75mm$:

$$CHF_{x1=0} = 259.5kW/m^2 \rightarrow \alpha_{CHF} = 34kW/m^2K \, ;$$

$$CHF_{x1=0.54} = 329.8kW/m^2 \rightarrow \alpha_{CHF} = 7.8kW/m^2K \, .$$

CHF and MHF Versus Heated Length

The design of the test section (Fig. 1) permits a stepwise
increase of the heated length l. The smallest possible l is
57.75mm (A to center of D, see Fig. 2). Smaller heated

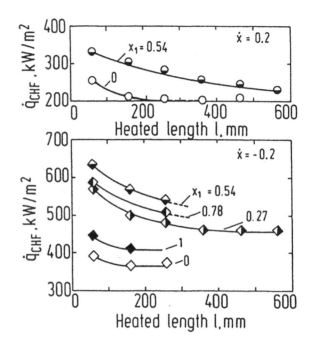

Fig. 4.   CHF vs. heated length

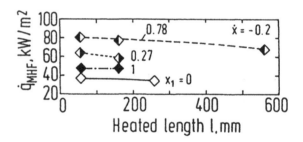

Fig. 5   MHF vs. heated length

lengths are not possible, since the heaters A etc. are needed as
guard-heaters. l can be increased in steps of $\Delta l = 100.5mm$
(heaters V10 etc.). Fig.4 shows the results for CHF at two
different local qualities $\dot{x} = 0.2$ and $\dot{x} = -0.2$ (note that
"local" is identical   to "outlet", since these quantities
correspond to section D at the outlet of the test section). For
the pure components no effect of l was observed at $l_{min} \geq$
170mm regardless of the local quality (in the range of the
measurement accuracy). Very short test sections ($l < l_{min}$),
however, show an increase in CHF at the same local
conditions. This is due to the complex effects and interactions
of the inertia of boiling phenomena and the local thermodynamic non-equilibrium, which we cannot quantify. The
$(l/d)_{min} \approx 10$ above which no heated length effect on CHF was
observed for the pure components in this study may therefore
change, especially if other fluid velocities exist. An important
result of the present investigation is that the $(l/d)_{min}$ of
mixtures is always significantly larger than the one of pure
substances. A first order of magnitude estimate shows (Fig. 4)

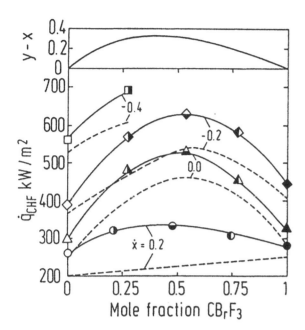

Fig. 6   CHF vs. mole fraction
        — Heated length l = 57.75mm
        --- (l/d) ≥ (l/d)$_{min}$, i.e. asymptotic CHF values

that (l/d)$_{min}$, above which no heated length effect is observed, is independent of local quality and mole fraction x_1 and amounts to (l/d)$_{min}$ ≈ 30. It is reasonable to assume that the larger heated length limit of mixtures is due to a mass transfer effect, but a direct proof is not possible by the present experiments. In the previous studies, summarized in Collier and Thome's book (1994), mostly the (l/d) was smaller than the limit of 30 found here for the mixtures. The strong maxima for CHF observed in these studies may therefore be partly due to the heated length effect.

According to Fig. 5 MHF is not considerably affected by the heated length. It slightly decreases with increasing l for all mole fractions x_1 and all local subcoolings or qualities. One has to take into consideration, however, that the absolute changes are small compared with changes of CHF in the region of small heated lengths.

CHF and MHF Versus Mole Fraction

Fig. 6 summarizes the mole fraction effect on CHF. The full lines represent the result for the smallest possible heated length, according to a l/d = 3.4. The dotted lines present the results for l/d ≥ (l/d)$_{min}$ taken or extrapolated from the lines drawn in Fig. 4. In the following we call these data asymptotic CHF values. Obviously we observe both effects claimed in the literature. All asymptotic values for subcooled outlet conditions - also at $\dot{x}$ = 0 the liquid is subcooled since thermodynamic equilibrium does not really exist - and in particular the short heated length data show a pronounced maximum between the pure fluid data in accordance with previous findings. Only if l/d ≥ (l/d)$_{min}$ and at positive outlet qualities the CHF of mixtures changes linearly between the pure substance data. This confirms qualitatively the more

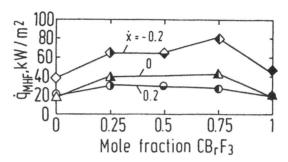

Fig. 7   MHF vs. mole fraction
        Heated length l = 57.75mm

recent results by Mori et al. (1990) and Celata et al. (1994), where similar system conditions have been applied. The maximum does not always occur precisely at the maximum of the mole fraction difference (y-x), which is normally claimed in the literature. This discrepancy may be a result of CHF-measurement inaccuracies.

Fig. 7 shows the mole fraction effect on MHF for a short heated test section. The MHF of mixtures is obviously a little higher than the one obtained by linear interpolation between the pure component data. The effect is, however, not very pronounced, if we look at the scale and take into account that the experimental reproducibility is for sure not better than ± 10kW/m^2.

CHF Versus Local Quality

Fig. 8 presents CHF data versus the local quality at the test section outlet. The asymptotic CHF values ((l/d) ≥ (l/d)$_{min}$) are plotted there. In Table 3 the subcoolings corresponding to the plotted qualities are summerized for the different mole fractions. The qualitative characteristics of CHF versus quality observed in the present experiments is well confirmed by numerous measurements (see e.g. Collier and Thome (1994), chap. 8) The mixtures show a stronger increase of CHF with increasing subcooling than the pure fluids, which is clearly due to the mole fraction effect on CHF (Fig. 6). Also the characteristics of the change is different. We do not claim, however, that there is a distinct physical reason for this effect. We would need more data for different mixtures and system conditions to come to a more general conclusion.

TABLE 3. LOCAL QUALITY AND CORRESPONDING SUBCOOLING							
x_1,-	0	0	0.27	0.27	0.54	0.78	1
$\dot{x}$,-	-0.2	-0.4	-0.2	-0.4	-0.2	-0.2	-0.2
ΔT_{sub}, K	19.9	41.2	23.8	52.6	28.6	29.4	25.4

CONCLUSIONS

1. CHF and MHF of mixtures occur at higher supersaturations compared with the corresponding quantities of the pure fluids. This is due to the enrichment of the less volatile component near the liquid-vapor interface.
2. A heated length effect was observed. For pure substances a limit ratio of (l/d)$_{min}$ ≈ 10 was found. At smaller ratios CHF increases with decreasing (l/d) for fixed outlet

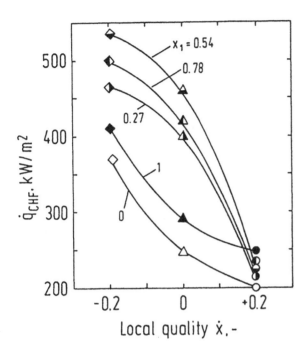

Fig. 8   CHF vs. local quality
         Asymptotic values

subcooling or quality, at larger ratios CHF remains constant. The corresponding limit ratio for mixtures was found to be significantly larger: $(l/d)_{min} \approx 30$.

3. In the range of the present experiments, which are restricted to the DNB type of the boiling crisis, CHF shows a maximum in a CHF versus mole fraction plot. This maximum becomes more pronounced as the outlet subcooling is increased. Only in long heated test sections and at an outlet quality of $\dot{x} = 0.2$ no maximum was observed.

4. CHF increases with increasing outlet subcooling for both the pure fluids and the mixtures.

5. All effects studied in the present project are of minor importance for the MHF.

ACKNOWLEDGEMENT

This work has been sponsored by the "Arbeitsgemeinschaft Industrieller Forschungsvereinigungen (AIF)" whom the authors wish to thank for financial support. Our thanks are also due to the "Hoechst AG" for providing us with the test fluids. Last but not least we acknowledge the contribution of A. Miyara, Saga University, Japan.

NOMENCLATURE

c_p spec. heat capacity [kJ/(kgK)]
d inner diameter of test tube [m]
h spec. enthalpy [kJ/kg]
Δh_v heat of vaporization, [kJ/kg]
l length [m]
M_1 molar mass of more volatile component [kg/kmol]
M_2 molar mass of less volatile component [kg/kmol]
$\dot{m}$ mass flux [kg/(m²s)]

p pressure [MPa]
$\dot{Q}$ heat flow rate [kW]
$\dot{q}$ heat flux [kW/m²]
T temperature [K]
ΔT_{sub} temperature difference of subcooling [K]
x mole fraction of more volatile component, liquid,-
x_1 mole fraction of more volatile component at the test section inlet,-
$\dot{x}$ quality,-
y mole fraction of more volatile component, vapor,-

Subscripts
1   inlet of test section
cr   critical
l   liquid
s, sat   saturation
v   vapor
z   at location z

REFERENCES

Auracher, H., and Marquardt, W., 1986, "The Dither-Technique For Steady-State Transition Boiling Measurements", *Proceedings 8th Interantional Heat Transfer Conference*, Hemisphere Publishing Corp., Washington, D.C., Vol. 2, pp. 501-506.

Celata, G.P.,Cumo, M., and Setaro, T., 1994, "Critical Heat Flux in Upflow Convective Boiling of Refrigerant Binary Mixtures", *Int. J. Heat Mass Transfer*, Vol. 37, pp. 1143-1153.

Collier, J.G., and Thome, J.R., 1994, "Convective Boiling and Condensation", 3rd ed., Clarendon Press, Oxford, UK.

Kältemaschinenregeln, 1981, 7th ed., Verlag C.F. Müller, Karlsruhe, Germany.

Lee, B.I., and Kesler, M.G., 1975, "A Generalized Thermodynamic Correlation Based on Three-Parameter Corresponding States", AICHE *Journal*, Vol. 21, pp. 510-526.

Mori, H., Yoshida, S., Ohno, M., Kusumoto, K., and Itoh, T, 1990, "Critical Heat Flux for Non-Azeotropic Binary Mixtures at High Pressures", *Proceedings of* JSME, No. 908-2, Saga, pp. 210-214.

Plöcker, U., 1977, "Berechnung von Hochdruck-Phasengleichgewichten mit einer Korrespondenzmethode unter besonderer Berücksichtigung asymmetrischer Gemische", Diss. TU Berlin.

VDI-Wärmeatlas, 1994, "Strömungssieden gesättigter Flüssigkeiten", Hbb21-Hbb25, 7th ed., VDI-Verlag, Düsseldorf, Germany.

THEORETICAL MODELING OF CHF FOR NEAR-SATURATED POOL BOILING AND FLOW BOILING FROM SHORT HEATERS USING THE INTERFACIAL LIFT-OFF CRITERION

Issam Mudawar[†], Jesse E. Galloway[††],
Christopher O. Gersey[††], and Stanley J. Reed[††]
Boiling and Two-Phase Flow Laboratory
School of Mechanical Engineering
Purdue University
West Lafayette, Indiana, U.S.A.

ABSTRACT

Pool boiling and flow boiling were examined for near-saturated bulk conditions in order to determine the critical heat flux (CHF) trigger mechanism for each. Photographic studies of the wall region revealed features common to both situations. At fluxes below CHF, the vapor coalesces into a wavy layer which permits wetting only in wetting fronts, the portions of the liquid-vapor interface which contact the wall as a result of the interfacial waviness. Close examination of the interfacial features revealed the waves are generated from the lower edge of the heater in pool boiling and the heater's upstream region in flow boiling. Wavelengths follow predictions based upon the Kelvin-Helmholtz instability criterion. Critical heat flux in both cases occurs when the pressure force exerted upon the interface due to interfacial curvature, which tends to preserve interfacial contact with the wall prior to CHF, is overcome by the momentum of vapor at the site of the first wetting front, causing the interface to lift away from the wall. It is shown this interfacial lift-off criterion facilitates accurate theoretical modeling of CHF in pool boiling and in flow boiling in both straight and curved channels.

1. INTRODUCTION

Predicting CHF has been the focus of a considerable body of research spanning over three decades. For pool boiling, the well known CHF model of Zuber *et al.* (1961) has maintained its popularity because of its theoretical appeal and use of well-established hydrodynamic instability hypotheses in determining limits on liquid access to a horizontal surface. Many attempts have been made to alter this model in order to account for effects the original model did not address. They include finite heater size, heater geometry, and surface orientat -

ion. The latter of these effects, particularly the case of a vertical surface, is of special interest to the present study.

While the model by Zuber *et al.* predicts zero CHF for vertical surfaces, experimental evidence proves CHF for this orientation and all orientations between horizontal and vertical are only slightly smaller than for a horizontal surface. Not only does this negate the suitability of this model for vertical surfaces, but it also raises questions concerning the CHF trigger mechanism on which the model is founded, even for horizontal surfaces. It is one of the key objectives of the present study to develop a new model for pool boiling CHF from vertical surfaces.

With regard to flow boiling, several types of CHF models have been proposed. The present study concerns the *interfacial lift-off* model developed by Galloway and Mudawar (1993a,b) and Gersey and Mudawar (1995a,b).

The present study will first explore the trigger mechanism for CHF from a short vertical surface in near-saturated flow boiling using an apparatus which lends itself to high resolution photographic study of interfacial features. A model is proposed the validity of which will be tested for both straight and curved flow boiling. The curved flow configuration facilitates an assessment of the accuracy of the model in predicting the enhancement effects due to curvature (Gambill and Greene, 1958; Galloway and Mudawar, 1992). This study will then treat pool boiling on a vertical surface simply as a limiting condition of flow boiling corresponding to zero liquid velocity. It will then be shown the proposed model reduces to a simple expression for pool boiling CHF.

2. EXPERIMENTAL METHODS
Straight and Curved Flow Boiling Visualization Facility

A CHF flow visualization apparatus was designed to maximize photographic access to interfacial features in close proximity to the heater surface in both straight and curved flows. The apparatus consisted of a curved flow channel having a 4.19-cm outer radius of curvature, which was located downstream from a

[†] Professor and Director of the Purdue University Boiling and Two-Phase Flow Laboratory; author to whom correspondence should be addressed
[††] Graduate Student

219

straight channel. A 1.27-cm long heater was inserted in each of the straight and curved regions of the channel. As shown in Fig. 1, the flow channel was formed by milling a 0.16 cm × 0.64 cm slot into a transparent polycarbonate plastic plate. A second plate was clamped onto the first plate trapping an o-ring seal. The centerline of the curved heater was positioned at a 135-degree angle relative to the inlet flow. Both the straight and curved heaters were constructed from copper and heated by a thick-film electrical resistor silver soldered to the outer protruding surface. The heaters were inserted into insulating flanges made from fiberglass plastic. An o-ring was pressed between the base of each heater assembly and the flow channel plate providing a leak proof seal. FC-87, a 3M dielectric fluid, was tested at a pressure of 1.37 bars (T_{sat} = 39 °C) with 8 °C inlet subcooling. Tilting the entire flow channel module allowed all tests to be conducted in an upflow configuration with respect to the tested heater. Only one heater was operated at a time.

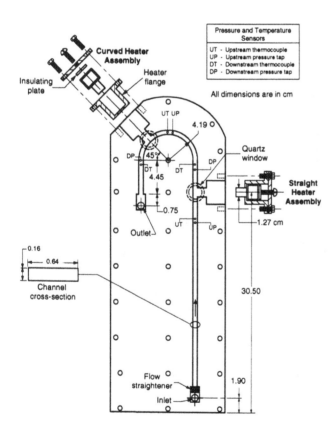

Fig. 1   Flow boiling apparatus.

Pool Boiling Facility

Figure 2 shows a schematic of the pool boiling test chamber which was constructed of fiberglass plastic and fitted on the front and back with polycarbonate plastic windows. The chamber was equipped with two condensers and three cartridge heaters. The condenser coiled inside the chamber effectively recovered all of the vaporized liquid. An external reflux condenser connected to the chamber's vent acted as a final barrier to any escaping vapor during both deaeration and testing. To prevent the boiling on the cartridge heaters from influencing CHF on the primary test heater, the cartridge heaters

were placed at the back of the test chamber, isolated from the test heater by a baffle plate. The test heater consisted of a 12.0 × 62.0 mm^2 copper block which was heated by a resistive heater pressed against its back. The heater was mounted on an angular rotation platform to facilitate testing at different surface orientations.

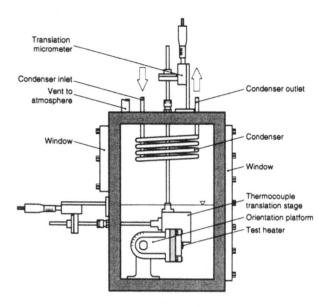

Fig. 2   Pool boiling test chamber.

3. PHOTOGRAPHIC STUDY OF VAPOR LAYER INTERFACIAL FEATURES

The flow and pool boiling facilities showed similar vapor production behavior at conditions leading to CHF. At a heat flux of about 85% of CHF, large coalescent bubbles were observed sliding over the heater surface. The length of these coalescent bubbles increased with increasing heat flux until, eventually, a fairly continuous wavy vapor layer was formed over the heater surface at heat fluxes below CHF. Boiling was sustained by liquid entrainment near the lower edge of the heater in pool boiling and the heater's upstream edge in flow boiling, as well as in *wetting fronts*, where the liquid-vapor interface made contact with the heater surface. Experimental evidence supporting this wetting front description is also available from studies by Hino and Ueda (1985a,b), and Galloway and Mudawar (1992); all of whom measured fluctuations in the heater surface temperature synchronous with the passage of vapor slugs.

Figure 3 shows the vapor layer formation on the curved heater in flow boiling corresponding to heat fluxes equal to 99% of CHF. A violent surge of small bubbles in the wetting fronts was observed to be feeding the vapor layer. The wavelength for both the straight and curved heaters decreased with increasing velocity. However, the curved heater exhibited significant differences in the shape of the wavy vapor layer as compared to the straight heater. For equal inlet velocities, the interfacial wavelength was greater for the straight heater than for the curved heater and, occasionally, at inlet velocities greater than 1.25 m/s, the curved heater projected vapor away from its surface in the form of vapor slugs which protruded from the wave peaks. No such behavior was observed with the straight heater.

Vapor layer mean thickness and wavelength were measured from high-speed video images captured by a 6000 partial frames per second Kodak Ektapro 1000 motion analyzer, which were later analyzed on a 55-cm wide screen; only still photography was used in the pool boiling tests. Thirty measurements were made for each inlet velocity to quantify the randomness of the interfacial features.

4. CHF MODEL
Model Assumptions

The proposed CHF model is built upon physical observations from extensive high speed video imaging studies as discussed in the previous section and illustrated in Fig. 4: (1) at heat fluxes approaching CHF, vapor coalesces to form a fairly continuous wavy vapor layer; (2) liquid is entrained at wetting fronts where the liquid-vapor interface contacts the heater surface; (3) vigorous boiling persists near the leading edge of the heater and in the wetting fronts while regions between neighboring wetting fronts dry out; (4) CHF commences when the liquid-vapor interface separates from the heater surface at the location of the most upstream wetting front; and (5) remaining wetting fronts are separated, in succession, after separation of the upstream wetting front.

As shown in Fig. 4, the first wetting front is established at a distance $z*$ from the leading edge and then propagates along the heater surface at a speed c_r. The vapor layer interfacial wavelength, $2\lambda_c$, was determined from hydrodynamic instability theory (discussed below) and observations made using the high-speed video imaging. When the liquid-vapor interface is unstable, a disturbance having a wavelength equal to λ_c is assumed to touch the heater surface at $z = z*$ ($z*$ is slightly greater than λ_c, the difference being a negligible distance z_0 over which the vapor velocity just exceeds the liquid velocity as discussed by Galloway and Mudawar, 1993b), enabling liquid to contact the heater over a localized region. A short time later, at $t = \lambda_c/c_r$, another disturbance approaching the heater surface will be forced away by the momentum of vapor emanating from residual liquid at $z = z*$ left after the passage of the previous wetting front. Not until a later time $t = 2\lambda_c/c_r$, after the residual liquid has been consumed at the location of the first wetting front, will a new wetting front be established on the heater surface. Wetting is, therefore, skipped every other cycle and wetting fronts are separated by $2\lambda_c$ wavelengths.

Surface Energy Balance at CHF

The photographic study and data base revealed the span (length) of each wetting front was one-fourth the separation distance between wetting fronts; i.e., CHF is about one-fourth the heat flux concentrated in the wetting fronts. A surface energy balance detailed by Galloway and Mudawar (1993b) reduces equation (1) to the following expression for CHF:

$$q_m = \left[1 - \frac{\lambda_c}{16(L - z^*)}\right]\frac{q_l}{4}, \quad (1)$$

where the coefficient in the brackets is close to unity for most operating conditions and accounts for continuous wetting in the region $0 < z < z*$ and any partial wetting fronts in the downstream region, and q_l is the heat flux required to cause lifting of the most upstream wetting front.

CHF Trigger Mechanism: Lift-off Criterion

The lift-off heat flux will develop when the normal momentum of vapor generated in the wetting front just exceeds the pressure

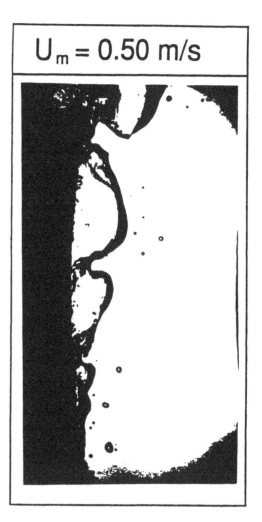

$U_m = 0.50$ m/s

Fig. 3   Wavy vapor layer development on curved heater in flow boiling just prior to CHF.

force exerted upon the interface as a result of interfacial curvature.

$$\rho_g \left[\frac{q_l}{\rho_g h_{fg}\left(1 + \frac{c_{p,f} \Delta T_{sub}}{h_{fg}}\right)}\right]^2 = \overline{P_f - P_g}, \quad (2)$$

where the average pressure difference across the interface, $\overline{P_f - P_g}$, is calculated by integrating the pressure difference over the span of the most upstream wetting front.

Combining equations (1) and (2) yields an expression for CHF which is applicable to both straight and curved heaters.

$$q_m = \frac{1}{4}\rho_g h_{fg}\left(1 - \frac{\lambda_c}{16(L - z^*)}\right)\left(1 + \frac{c_{p,f}\Delta T_{sub}}{h_{fg}}\right)\left[\frac{\overline{P_f - P_g}}{\rho_g}\right]^{1/2} \quad (3)$$

The lift-off criterion alone can explain why, in flow boiling, CHF is greater for the curved, concave heater than the straight heater. For an assumed sinusoidal wave with wavelength λ and amplitude η_0, curvature at the wave peak is proportional to $(\eta_0 / \lambda)^2/\eta_0$. Although the straight and curved heaters produced waves with fairly equal amplitudes, these waves acquired greater curvature over the curved heater than they did over the straight heater. This increased curvature resulted in a greater pressure force exerted upon the interface and, consequently, increases both the lift-off heat flux in the wetting fronts and CHF relative

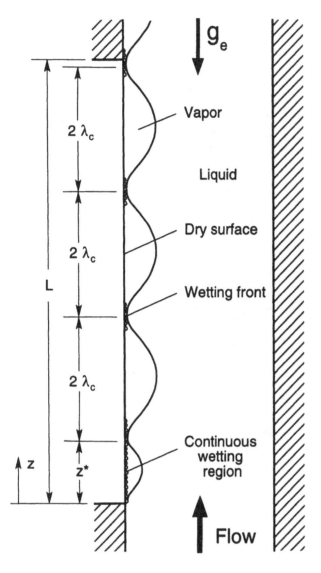

Fig. 4 Wetting front propagation along a vertical surface.

to the straight heater. As will be shown later, the curved heater produced an average enhancement of 23% compared to the straight heater.

Interfacial Instability of Vapor Layer

The interfacial waviness illustrated in Fig. 4 can be idealized as a hydrodynamic instability of an interface between a vapor layer of velocity $\bar{u}_{g,m}$ and height H_g and a liquid layer of velocity $\bar{u}_{f,m}$ and height H_f. Using classical instability theories, the interfacial pressure difference resulting from a sinusoidal disturbance of the form $\eta(z,t) = \eta_o\,e^{ik\,(z-c\,t)}$ perpendicular to the unperturbed interface can be expressed as (Galloway and Mudawar, 1993b)

$$P_f - P_g =$$
$$-\eta\,k\left[\rho_f^{''}\left(c-\bar{u}_{f,m}\right)^2 + \rho_g^{''}\left(\bar{u}_{g,m}-c\right)^2\right] - \left(\rho_f-\rho_g\right)g_n\,\eta = \quad (4)$$
$$-\sigma\,k^2\,\eta\,,$$

where g_n is the body force per unit mass perpendicular to the unperturbed interface ($g_n = 0$ for vertical upflow over a straight

heater), and the modified density terms for a straight channel are expressed as (Galloway and Mudawar, 1993b)

$$\rho_f^{''} = \rho_f \coth\left(k\,H_f\right), \qquad (5a)$$
$$\rho_g^{''} = \rho_g \coth\left(k\,H_g\right). \qquad (5b)$$

Solving equation (4) yields the following equation for c:

$$c = \frac{\rho_g^{''}\,\bar{u}_{g,m} + \rho_f^{''}\,\bar{u}_{f,m}}{\left(\rho_g^{''} + \rho_f^{''}\right)}$$
$$\pm \sqrt{\frac{\sigma k}{\left(\rho_g^{''}+\rho_f^{''}\right)} - \frac{g_n\left(\rho_f-\rho_g\right)}{k\left(\rho_g^{''}+\rho_f^{''}\right)} - \frac{\rho_g^{''}\,\rho_f^{''}\left(\bar{u}_{g,m}-\bar{u}_{f,m}\right)^2}{\left(\rho_g^{''}+\rho_f^{''}\right)^2}}\,. \qquad (6)$$

The critical wavelength is defined as the wavelength that produces a neutrally stable wave. This wavelength can be calculated by setting the argument of c in equation (6) equal to zero.

$$\frac{2\pi}{\lambda_c} = k_c = \frac{\rho_f^{''}\,\rho_g^{''}\left(\bar{u}_{g,m}-\bar{u}_{f,m}\right)^2}{2\,\sigma\left(\rho_f^{''}+\rho_g^{''}\right)}$$
$$+ \sqrt{\left[\frac{\rho_f^{''}\,\rho_g^{''}\left(\bar{u}_{g,m}-\bar{u}_{f,m}\right)^2}{2\,\sigma\left(\rho_f^{''}+\rho_g^{''}\right)}\right]^2 + \frac{g_n\left(\rho_f-\rho_g\right)}{\sigma}}\,. \qquad (7)$$

This classical hydrodynamic instability model is not applicable to curved flow. Recently, Galloway and Mudawar (1995) developed a new model for hydrodynamic instability along a curved interface. Interestingly, the above instability model was proven equally valid for curved flow, provided the liquid and vapor velocities in equations (4), (6) and (7) are calculated along the interface, and the modified density terms are replaced by the following:

$$\rho_f^{'''} = \rho_f\left[\frac{\left(\frac{R_0}{R_1}\right)^{2k\,R_0}+1}{\left(\frac{R_0}{R_1}\right)^{2k\,R_0}-1}\right], \qquad (8a)$$

$$\rho_g^{'''} = \rho_g\left[\frac{\left(\frac{R_0}{R_2}\right)^{2k\,R_0}+1}{1-\left(\frac{R_0}{R_2}\right)^{2k\,R_0}}\right]. \qquad (8b)$$

CHF Model Predictions for Straight and Curved Flow Boiling

A separated flow model given by Galloway and Mudawar (1993b; 1995) was employed to predict local mean values of vapor layer thickness and velocities of the liquid and vapor layers in terms of inlet velocity, subcooling, and heat flux. These local values are required in order to predict the interfacial wavelength and $\overline{P_f - P_g}$.

Figure 5 shows the CHF model predicts the experimental data for the straight and the curved heaters with mean absolute errors of 7% and 14%, respectively. The accuracy of the model predictions is proof of the validity of the assumptions used in constructing the surface energy balance and of the lift-off criterion. The CHF enhancement obtained with the curved heater over the straight heater is a direct consequence of the increased curvature of the individual interfacial waves causing an increase in the net pressure force exerted upon the interface in the wetting fronts.

5. THEORETICAL MODEL FOR POOL BOILING CHF FROM A VERTICAL SURFACE
Growth of Vapor Layer

Figure 6 shows pool boiling data for water for orientations

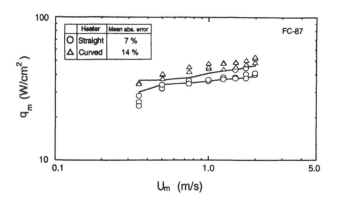

Fig. 5   Comparison of model predictions and flow boiling CHF data for straight and curved heaters (Galloway and Mudawar, 1995).

ranging from horizontal ($\theta = 0$ degrees) to vertical (90 degrees), compared with predictions of the model by Zuber *et al.* (1961), modified by replacing g_e by $g_e \cos\theta$. CHF decreases with increasing angle of orientation following, to some degree, the predicted trend. However, while the model predicts zero CHF for the vertical orientation, the data show an increase in CHF between 75 and 90 degrees. Obviously, the mechanism proposed by Zuber *et al.* is not suitable for near-vertical orientations.

In pool boiling on a vertical surface, the mean velocity of liquid outside the vapor layer is zero. Equations for the mean vapor velocity, $\bar{u}_{g,m}$, and vapor layer thickness, δ, can be derived by applying mass, momentum, and energy conservation for a control volume of the vapor layer of length Δz. Combining both mass and energy conservation for this control volume and integrating with respect to z gives

$$\rho_g \bar{u}_{g,m} \delta = \frac{q_m z}{h_{fg} \left[1 + \dfrac{c_{p,f} \Delta T_{sub}}{h_{fg}} \right]}. \quad (9)$$

A momentum balance on the same control volume yields

$$\frac{d}{dz} \left[\rho_g \bar{u}_{g,m}^2 \delta \right] = (\rho_f - \rho_g) g_e \delta. \quad (10)$$

Combining equations (9) and (10) gives a differential equation relating $\bar{u}_{g,m}$ to z. The solution for this equation is

$$\bar{u}_{g,m} = \left[\frac{2}{3} \left(\frac{\rho_f - \rho_g}{\rho_g} \right) g_e z \right]^{1/2}. \quad (11)$$

Substituting for $\bar{u}_{g,m}$ using equation (11) in equation (9) gives the variation of δ with z.

$$\delta = \frac{q_m}{\rho_g h_{fg} \left[1 + \dfrac{c_{p,f} \Delta T_{sub}}{h_{fg}} \right]} \left[\frac{2}{3} \left(\frac{\rho_f - \rho_g}{\rho_g} \right) g_e \right]^{-1/2} \sqrt{z}. \quad (12)$$

The critical wavelength corresponding to the onset of instability can be derived from equation (7) by setting g_n and $\bar{u}_{f,m}$ equal to zero; further simplification is also possible because, for the conditions of the pool boiling study, $\rho_g'' \cong \rho_g$.

$$\lambda_c = \frac{2\pi}{k_c} = \frac{2\pi \sigma (\rho_f + \rho_g)}{\rho_f \rho_g'' \bar{u}_{g,m}^2}. \quad (13)$$

The mean interfacial pressure force over the wetting front for a sinusoidal wave with $k = 2\pi / (2\lambda_c)$ becomes

$$\overline{P_f - P_g} = 2\sqrt{2} \, \pi \frac{\sigma \delta}{\lambda_c^2}. \quad (14)$$

Substituting the above expression in equation (3), neglecting the coefficient due to the upstream continuous wetting zone and the downstream partial wetting front, gives

$$q_m = \frac{1}{4} \rho_g h_{fg} \left[1 + \frac{c_{p,f} \Delta T_{sub}}{h_{fg}} \right] \left[(2\sqrt{2} \, \pi) \frac{\sigma \delta}{\rho_g \lambda_c^2} \right]^{1/2}. \quad (15)$$

Since the first wetting front is centered at $z = \lambda_c$, the critical wavelength, equation (13), can be expressed in terms of $\bar{u}_{g,m}$, equation (11), corresponding to $z = \lambda_c$. This gives

$$\lambda_c = \left[3\pi \left(\frac{\rho_f + \rho_g}{\rho_f - \rho_g} \right) \frac{\sigma}{\rho_f g_e} \right]^{1/2}. \quad (16)$$

Substituting the expressions for δ (based on $z = \lambda_c$), equation (12), and λ_c, equation (16), in equation (15) yields the following CHF relation

$$q_m =$$
$$\frac{1}{8} \left(\frac{\pi}{3} \right)^{1/4} \left[1 + \frac{c_{p,f} \Delta T_{sub}}{h_{fg}} \right] \left[\frac{\rho_f}{\rho_f + \rho_g} \right]^{3/4} \rho_g h_{fg} \left[\frac{\sigma (\rho_f - \rho_g) g_e}{\rho_g^2} \right]^{1/4} \quad (17)$$

The above equation can be further simplified for saturated conditions, and pressures much smaller than critical to

$$q_m = 0.126 \, \rho_g h_{fg} \left[\frac{\sigma (\rho_f - \rho_g) g_e}{\rho_g^2} \right]^{1/4}. \quad (18)$$

Interestingly, equation (18) is identical in form to the model by Zuber *et al.*, although the mechanisms proposed in the individual models are distinctly different. Also, while the model by Zuber *et al.* predicts zero CHF for vertical surfaces, the present model, as shown in Fig. 6, predicts a CHF value for the same orientation only 11% smaller than the data.

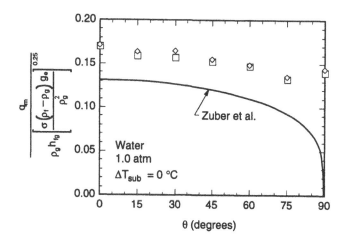

Fig. 6   Comparison of pool boiling CHF data for different orientations with predictions of the modified model by Zuber *et al..*

6. CONCLUSIONS

Experiments involving pool and flow boiling from vertical surfaces were performed to ascertain the CHF trigger mechanism for each. The flow boiling experiments included both straight and curved surfaces. Key conclusions from the study are as follows:

(1) A fairly continuous wavy vapor layer engulfs the heater surface at heat fluxes smaller than CHF in both pool and flow boiling. Boiling remains active in wetting fronts, where the

interface of the vapor layer contacts the heater surface. CHF is triggered when the normal momentum of the vapor produced in the wetting front exceeds the pressure force exerted upon the interface due to interfacial curvature.

(2) A CHF model constructed from these observations predicts flow boiling CHF data for the straight and curved heaters with mean absolute errors of 7% and 14%, respectively.

(3) In flow boiling, CHF is higher for the curved heater than for the straight heater because of a greater pressure resistance to interfacial separation in the case of the curved heater.

(4) For pool boiling, the present model reduces to an expression identical in form to the model by Zuber *et al.*, although the mechanisms proposed in the individual models are distinctly different. However, while the model by Zuber *et al.* predicts zero CHF for vertical surfaces and is therefore unsuitable for vertical surfaces, the present model predicts a CHF value for the same orientation only 11% smaller than the data.

ACKNOWLEDGMENT

The authors are grateful for the support of the Office of Basic Energy Sciences of the U.S. Department of Energy (Grant No. DE-FE02-93ER14394).

NOMENCLATURE

c	wave speed
c_p	specific heat at constant pressure
c_r	real component of wave speed
g_e	Earth gravity
g_n	component of body force per unit mass normal to liquid-vapor interface
H	channel height, $H_f + H_g$, $R_2 - R_1$
H_f	liquid layer thickness
H_g	vapor layer thickness
h_{fg}	latent heat of vaporization
k	wave number, $2\pi/\lambda$
k_c	critical wave number, $2\pi/\lambda_c$
L	heater length
P	pressure
$\overline{P_f - P_g}$	mean interfacial pressure difference in wetting front
q_l	heat flux (at CHF) corresponding to wetting front separation (lift-off)
q_m	critical heat flux (CHF)
R_0	radius of unperturbed interface in curved flow
R_1	inner radius of curved channel
R_2	outer radius of curved channel and curved heater surface
t	time
T	temperature
ΔT_{sub}	liquid subcooling
$\bar{u}_m$	mean velocity across liquid or vapor layer
U_m	mean liquid velocity upstream of heater in flow boiling
z	spatial coordinate in the stream-wise direction
z^*	distance from leading edge of heater to center of first wetting front, $z_0 + \lambda_c(z^*)$
z_0	position from leading edge where the liquid and vapor velocities become equal

Greek Symbols

δ	mean vapor layer thickness ($= H_g$)
η_0	interfacial displacement
η	amplitude of interfacial displacement
θ	surface angle of orientation
λ	wavelength of interfacial perturbation
λ_c	critical wavelength corresponding to onset of instability
ρ	density
ρ''	modified density defined in equations (5a) and (5b)
ρ'''	modified density defined in equations (8a) and (8b)
σ	surface tension

Subscripts

f	saturated liquid
g	saturated vapor
m	mean
sub	subcooling.

REFERENCES

Galloway, J. E. and Mudawar, I., 1992, "Critical Heat Flux Enhancement by means of Liquid Subcooling and Centrifugal Force Induced by Flow Curvature," *Int. J. Heat Mass Transfer*, Vol. 35, pp. 1247-1260.

Galloway, J. E. and Mudawar, I., 1993a, "CHF Mechanism in Flow Boiling from a Short Heated Wall - Part 1. Examination of Near-Wall Conditions with the aid of Photomicrography and High-Speed Video Imaging," *Int. J. Heat Mass Transfer*, Vol. 36, pp. 2511-2526.

Galloway, J. E. and Mudawar, I., 1993b, "CHF Mechanism in Flow Boiling from a Short Heated Wall - Part 2. Theoretical CHF Model," *Int. J. Heat and Mass Transfer*, Vol. 36, pp. 2529-2540.

Galloway, J. E. and I. Mudawar, I., 1995, "A Theoretical Model for Flow Boiling CHF from Short Concave Heaters," *J. Heat Transfer*, in press.

Gambill, W. R. and Greene, N. D., 1958, "Boiling Burnout with Water in Vortex Flow," *Chem. Eng. Prog.*, Vol. 54, pp. 68-76.

Gersey, C. O. and Mudawar, I., 1995a, "Effects of Heater Length and Orientation on the Trigger Mechanism for Near-Saturated Flow Boiling CHF - Part 1. Photographic and Statistical Characterization of the Near-Wall Interfacial Features," *Int. J. Heat Mass Transfer*, Vol. 38, pp. 629-642.

Gersey, C. O. and Mudawar, I., 1995b, "Effects of Heater Length and Orientation on the Trigger Mechanism for Near-Saturated Flow Boiling CHF - Part 2. CHF Model," *Int. J. Heat Mass Transfer*, Vol. 38, pp. 643-654.

Hino, R. and Ueda, T., 1985a, "Studies on Heat Transfer and Flow Characteristics in Subcooled Flow Boiling-Part 1. Boiling Characteristics," *Int. J. Multiphase Flow*, Vol. 11, pp. 269-281.

Hino, R. and Ueda, T., 1985b, "Studies on Heat Transfer and Flow Characteristics in Subcooled Flow Boiling-Part 2. Flow Characteristics," *Int. J. Multiphase Flow*, Vol. 11, pp. 283-297.

Zuber, N., Tribus, M. and Westwater, J. M., 1961, "The Hydrodynamic Crisis in Pool Boiling of Saturated and Subcooled Liquid," *Int. Dev. in Heat Transfer*, Boulder, CO, pp. 230-236.

CRITICAL HEAT FLUX IN FORCED CONVECTIVE SUBCOOLED BOILING WITH ONE AND MORE IMPINGING JETS

M. Monde and Y. Mitsutake

Department of Mechanical Engineering,

Faculty of Science and Engineering, Saga University

1, Honjo, Saga, 840 Japan

ABSTRACT

Critical heat flux during forced convective subcooled boiling on a rectangular heated surface, being supplied with subcooled liquid through a single round jet or multiple round jets which impinge on the surface has been studied experimentally employing three different liquids of water, R113, and R22 at a pressure from 1 to 25 bar at a high subcooling condition up to 115 K.

According to observation on a variation of flow aspect on the heated surface with an increase in heat flux, characteristics for both the single jet and multiple jets are similar up to the CHF, when focusing on the region controlled by each individual jet. A criterion is proposed that the CHF in the saturated region can be determined from the existing equation for the CHF for the saturated liquid, while for the same heat flux in the subcooled region, an equation can be derived from a simple heat balance on an assumption that the subcooled liquid is heated up at the saturation temperature at the leading edge of the saturation region. By eliminating the length of the subcooled region from the two equations and after revising it slightly, a generalized correlation can be derived, yielding the CHF data not only for the single jet but also for the multiple jets with the same accuracy.

INTRODUCTION

Boiling provides an efficient heat transfer process for applications such as the cooling of electric components(Bar-Cohen, 1991, and Mudawar and Wadsworth, 1991) and fusion components(Boyd, 1983 and Dhir and Scott, 1987). The critical heat flux(CHF), which defines the upper limit of the efficient range of boiling heat transfer, limits the maximum heat flux which may be dissipated from the components. An enhancement of the CHF becomes important to utilize the boiling heat transfer properly. An impinging liquid jet(Monde, 1991), for example, as shown in Fig.1, can be considered very simple and attractive as a means of the CHF enhancement. Grambill and Lienhard (1989) pointed out that the impinging jet is one of the potential coolings for the components with high heat fluxes.

In the past two decades, the critical heat flux in saturated convective boiling with an impinging jet and multiple jets, has been studied rather extensively (Monde and Katto, 1978, Sharan and Lienhard, 1985, Monde and Inoue, 1991, and Monde, 1991),

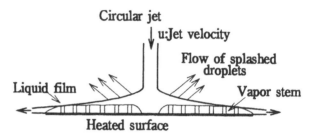

Fig.1 Boiling near CHF with an impinging jet

so that the characteristics of the CHF can be made clear. According to Monde(1991), its characteristics may be divided into four regimes. In addition, generalized correlations, which are applicable for the CHF within a wide range of the density ratio ρ_l/ρ_v = 5.3 to 1603 and the reciprocal of the Weber number, $2\sigma/\rho_l u^2(D - d) = 2 \times 10^{-7}$ to 10^{-3}, have been derived by Monde (1991). The generalized correlation for one(i.e., V-regime) of the four regimes, for example, is given by the following equation:

$$\frac{q_{co}}{\rho_v H_{fg} u} = 0.221 \left(\frac{\rho_l}{\rho_v} \right)^{0.645} \left(\frac{2\sigma}{\rho_l u^2(D - d)} \right)^{0.343} (1 + D/d)^{-0.364} \qquad (1)$$

The CHF predicted by Eq.(1) has a feature such that the CHF increases with an increase in the jet velocity, some liquid flows out from the heated surface just before the CHF takes place, and the CHF is related closely to one determined from the concept of Critical Film Thickness proposed by Haramura and Katto(1983).

Monde and Inoue(1991) show that equation(1) is also applicable for the CHF for multiple impinging jets when a maximum flow length in the domain controlled by one jet is employed as a characteristic length. In addition, they suggest that equation(1) can be applied to the CHF for the jet impinging on any configuration other than a disc, if a maximum flow length is chosen appropriately in place of diameter D.

On the other hand, there are few experiments employing a subcooled liquid as the impinging jet except for the Monde and Katto experiment(1978) within our knowledge about this boiling system, although the CHF would be expected to be improved by using a subcooled liquid. They proposed the following equation,

Table 1   Experimental range

	Water		R113	R22
N	1	2, 4	1	1
L [mm]	40, 60	45.1, 46.5	40, 60	40, 60
d [mm]	2		2	2
u [m/s]	5 - 16	5 - 25	4.2 - 18.4	6.6 -33.9
ΔT_{sub} [K]	0 - 115	0 - 80	0 - 80	0 - 30
P [bar]	1, 3		1 - 5	16 - 25
ρ_l/ρ_v	658 - 1605		40.1 - 202	8.8 - 16.1
Ja	0 - 210	0 - 170	0 - 30	0 - 3.1

yielding an effect of the subcooling on the CHF within a limited range of a low subcooling level.

$$q_c/q_{co} = 1 + 2.7(\rho_l/\rho_v)^{-1.5}Ja^2 \qquad (2)$$

In order to make it clear whether Eq.(2) can be extended to a high subcooling level and to show the effect of subcooled liquid on enhancement of the CHF, one would need an advanced study in this field.

In the present experiment, the CHF for one or multiple jets will be measured for water, R113, and R22 over a wide experimental range given in Table 1. A generalized correlation will be derived for the CHF up to a high subcooling of 115 K, predicting the CHF with a good agreement.

EXPERIMENTAL APPARATUS AND PROCEDURE

Experimental Apparatus

Figure 2 shows only details of the main part of experimental apparatus, because the whole system is similar to one in Monde(1987).

At the top of the pressure vessel, both ends of a stainless steel foil of thickness 0.3 mm are attached to two copper blocks to which D.C. power is automatically supplied from a large D.C. transformer controlled by a personal computer. This foil heated thereby, serves as the heated surface. The foil except for downward facing surface is totally packed with an epoxy adhesive to prevent heat loss from the back side of the foil, thereby Joule heat generated in the foil being properly transferred to the test liquid with a heat loss less than 1 % of the electrical input, because nucleate boiling heat transfer is very high compared with the heat conduction to the backside. The error of q calculated by assuming Joule heat to be transferred into the liquid, has been estimated to be less than 1 %.

The test liquid, which is pressurized by a regenerative pump and then adjusted with the heater up to a designated temperature, flows out through one, two or four holes of inner diameter d = 2 mm, and impinges on the heated surface vertically. Figure 3 shows the arrangement of the impinging jets on the heated surface, where in the hatched area, the length of L/2 corresponds to the maximum flow length which is controlled by one jet. According to a calibration of mass flow rate done before the experiment, mean velocity at the nozzle exit for the different holes, u can be determined within an error of 3 - 5 % from the pressure difference between the inlet and outlet of the nozzle measured by means of a differential pressure transducer. In addition to this, the calibration shows that there is little difference between the mass flow rates discharged through each hole.

The lower half of the inside of the vessel is filled with the test liquid (about 0.4 to 0.5 m deep) and the vapor above the liquid is condensed by a cooling coil. The temperature inside the vessel is

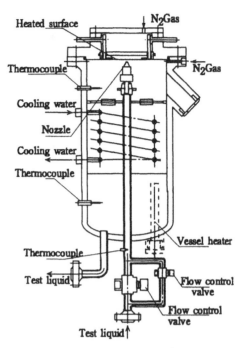

Fig.2 Essential part of experimental apparatus

maintained by controlling the vessel heater as well as the cooling coil.

The pressure inside the vessel is controlled at a designated value by using N_2 gas in order to get a high subcooling level. An effect of the N_2 gas on the CHF will be discussed later.

Procedure of Experiment and Measuring CHF

The jet velocity, the liquid temperature, and the system pressure are first fixed at each designated value. Then, the electric power to the heated surface is automatically increased at a time interval of about 3 minutes in increments that are less than 5 % of each preceding heat flux, and finally a point is reached where the heated surface temperature runs away to burn out. At this point, the CHF is determined with an uncertainty of 0 to 5 %. Incidentally, the temperature change of the heated surface can be detected from the change in the resistance of the foil heater. But, the temperature of the heated surface for water, increases too fast to protect it from its damage while the rise in its temperature for R113 and R22, can be stopped before a failure of the heated surface because of a relatively slow increase in the temperature compared with that for water.

Effect of N_2 Gas on CHF

In order to get a high subcooling up to $\Delta T_{sub} = 115$ K, we employed the N_2 gas in the experiment to raise the system pressure. Before starting the measurement of the CHF with the impinging jet, we measured the CHF on a horizontal wire of d = 1 and 2 mm in a subcooled pool boiling during which the system pressure is raised or controlled by means of N_2 gas.

The experimental result shows that the existence of the N_2 gas would not have severe effect on the CHF in the subcooled pool boiling. It should be noted that most of the experiment of subcooled pool boiling were done about the CHF under the condition that the test liquid was surrounded with atmosphere (for example, Elkassabgi and Lienhard, 1988). The fact that the inert N_2 gas has little effect on the CHF, may be natural from such a situation that

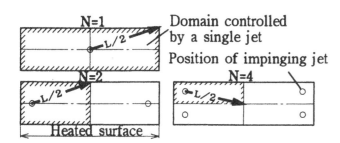

Fig.3 Geometrical arrangement of impinging jets on the
       heated surface

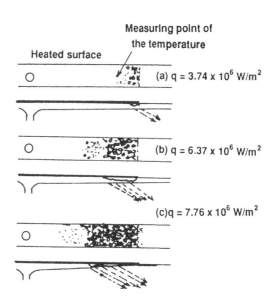

Fig.4 Variation of flow state with an increase in heat flux

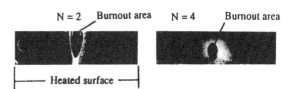

Fig.5 Burnout point of the heated surface for N = 2 and 4

most of dissolved gases have been taken out from the liquid near
the heated surface together with a great deal of vapor produced by
a very violent nucleate boiling near CHF.

EXPERIMENTAL RESULTS AND DISCUSSION

Variation of Fluid Behavior with Increase in Heat flux

Figure 4 is a sketch for variation of the flow configuration with
an increase in heat flux at a velocity of u = 10 m/s and for N = 1
based upon a visual observation.

Figure 4 depicts that the flow situation can be divided into subcooled and saturated regions. In the subcooled region, after impingement of the jet, the liquid film flows radially over the heated
surface. Weak boiling appears along the heated surface, but the
surface is soon cooled down by the subcooled liquid. The vapor
generated on the heated surface, however, is hardly condensed
near a downstream part in the subcooled region, from which the
saturated region starts. In the saturated region, most of the liquid
flowing over the heated surface is splashed out as droplets, but a
residual liquid still remains as a very thin liquid film totally covering on the heated surface and insuring heat transfer from the
heated surface due to its evaporation. With increasing the heat
flux, the saturation region gradually extends from the outer of the
heated surface toward the center and inversely the subcooled
region is shrunk back. For the case of multiple jets, it is observed
that when one focuses on the domain controlled by each jet, the
variation of the flow due to an increase in heat flux is similar to
that observed in the single jet except for the interference between
radial flows formed on the heated surface after the impingement
of jet at low heat fluxes, but finally, this interference fades due to
the increase of the splashed droplets. .

Finally, it may be of importance to note that when the saturated
region is focused, the flow situation for the subcooled liquid is
very similar to that for the saturated liquid and to emphasize that
near the CHF point, the flow features of both multiple jets and
single jet become totally similar in the domain controlled by the
single jet.

Burnout of Stainless Steel Foil due to Occurrence of CHF

Figure 5 shows representative photographs of the heated surface for both N = 2 and 4 when the electric power is shut down
immediately after the occurrence of the CHF. The case of N = 1 is
omitted here because the CHF always takes place at the edge of
the heated surface at any cases.

Comparing the CHF point in the heated surface and the do-
main(see the hatched area in Fig.3), we may notice that the CHF
takes place at the point which is farthest from the position of the
impinging jet within the domain. The same evidence is observed
for the saturation condition. The variation of the flow aspect
when increasing heat flux up to the CHF and the point of the CHF
occurrence may allow us to conclude that the CHF phenomena are
not different in the case of the multiple jets versus the single jet,
provided that the heated surface can be divided into a subdivision
where the liquid flow formed on the heated surface after impingement of the jet is influenced only by that jet.

Flow Model for Behavior of Liquid and Vapor

Figure 4 may allow us to propose a flow model as shown in
Fig.6. A domain, A, in Fig.6 corresponds to the saturated region
not only where a thin liquid film being like a critical film thickness proposed by Haramura and Katto(1983), is formed on the
heated surface but also where a violent boiling appears. In addition, the residual liquid flowing over the heated surface, is still
enough to cool it by its evaporation, although most of the liquid is
splashed out. On the other hand, the liquid in B and C parts is still
kept subcooled. In the part of B, a weak boiling would take place
on the heated surface but be soon cooled down so that the
splashed droplets is not produced.

It is necessary to say, finally that the flow model as shown in
Fig.6 is substantially identical to the one for a disk surface, although it shows the rectangular surface used in the experiment
which is cut off as a part of the disk.

Effect of Jet Velocity on CHF

Figure 7 represents the CHF for water plotted against the jet
velocity at a different subcooling up to ΔT_{sub} = 80 K. A straight
line in Fig.7, for reference, is the CHF, which is calculated from
Eq.(1) for the saturated liquid at a corresponding pressure. The
similar result for R113 and R22 as well as for the multiple jets is
also obtained.

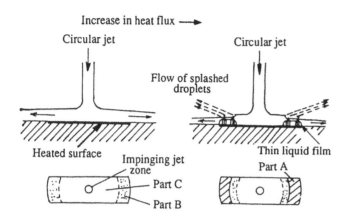

Fig.6 Flow model of liquid and vapor

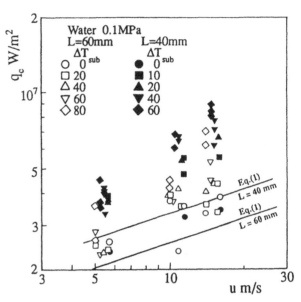

Fig.7 Relationship between CHF and velocity for water

Figure 7 shows that an effect of the subcooling on the CHF depends on the jet velocity, that is the enhancement of the CHF at the same subcooling is enlarged with an increase in the velocity. In addition to this effect, enhancement of the CHF due to the subcooling becomes larger for the shorter length of the heater surface when comparing them for L = 40 and 60 mm. In other words, the enhancement of the CHF owing to the subcooling tends to disappear either at a low velocity or for a long surface because the liquid radially flows being heated and approaches the saturation condition near the impinging zone. In the case of L = 60 mm, for example, the CHF at about u = 5 m/s, may be slightly influenced by the subcooling while the CHF at u > 8 m/s is significantly increased.

These results can be understood qualitatively using the flow model in Fig.6: for a low velocity and long heated surface, the length, L', of the subcooled area occupied by the parts of B and C near the CHF, becomes relatively small comparing with the total length, L.

Effect of Subcooling on CHF

The effect of the subcooling on the CHF is made clearer by normalizing the measured CHF data with Eq.(1). Figure 8 shows the normalized value, $\Phi = q_c/q_{c\infty}$, only for water plotted against the Jakob number Ja, because for the other liquid, a similar trend is obtained. A solid line in Fig. 8 is Eq.(2), for reference.

The effect of the subcooling on the CHF depends on the velocity and the length of the heated surface. This dependence significantly appears at low velocity of about u = 5 m/s on comparison between L = 40 and 60 mm in Fig.8. The CHF for the same velocity and the same length is seen from Fig.8 to increase with increasing the Jakob number. In addition to this effect, it should be necessary to say that the CHF enhancement due to the subcooling for the same liquid becomes larger as the system pressure rises.

CORRELATION OF CHF

Theoretical Analysis and Correlation of CHF

The flow aspect on the heated surface near the CHF can be divided in two different regions namely the saturated region, A and the subcooled region, (B+C) as shown in Fig. 6. The violent boiling takes place in the region A as given in Fig.6 so that the CHF may be reasonable to be governed by a thin liquid film formed on the heated surface, which plays an essential role in determining the CHF in saturated boiling. Before applying Eq.(1) to the CHF of this boiling, one has to know the velocity, u_1 at

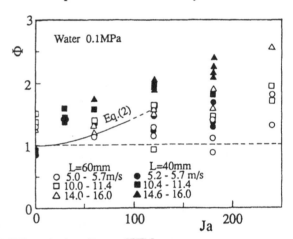

Fig.8 Effect of subcooling on CHF for water

which the liquid radially spreading on the heated surface after the jet impingement, flows into the thin liquid film. The velocity at the leading edge of the liquid film is smaller than the jet velocity, due to the shear force acting on the liquid radially flowing through the length of the subcooled region. However, there is a great difficulty in determining the effect of the shear force on the velocity which is needed to calculate it. Therefore, we assume here for the sake of simplicity that the velocity is proportional to the jet velocity, namely $u_1 = k_o u$ where the factor k_o mainly depends on the length of the subcooled region and the nature of what was brought out on the heated surface, for example the nature of a weak boiling. After substituting $u_1 = k_o u$ into Eq.(1), one can derive easily as:

$$\frac{q_c}{\rho_v H_{fg} k_o u} = 0.221 \left(\frac{\rho_l}{\rho_v}\right)^{0.645} \left(\frac{2\sigma}{\rho_l (k_o u)^2 (L - L')}\right)^{0.343}$$
$$(1 + L/L')^{-0.364} \qquad (3)$$

According to Monde and Inoue(1991), the length of the boiling area across which the liquid runs, is significantly important in determining the CHF and the difference in configuration between rectangular and circular surfaces has little effect on the CHF. Therefore, the shape of the heated surface will be considered as a disk to make the later discussion simpler.

The heat balance in the region (B+C) can be easily derived as:

$$k_1 \rho_l \frac{\pi}{4} d^2 u c_p \Delta T_{sub} = q_c \frac{\pi}{4} L'^2 \qquad (4)$$

where the constant k_1 is the fraction of the liquid supplied by the jet that is heated up to the saturation temperature, and its value is generally a function of the Reynolds number and the Prandtl number for a liquid flow. Substituting the length, L' in Eq.(4) into Eq.(3), one can get an equation, predicting the CHF. However, the equation derived thereby becomes implicit and rather complicated. Therefore, before substituting Eq.(4) into Eq.(3), one can reform Eq.(3) by dividing it by Eq.(1) as follows:

$$\frac{q_c}{q_{co}} = k_o^{0.314} \left(\frac{1+d/L}{1+L'/L} \right)^{0.021} \left(\frac{1-(d/L)^2}{1-(L'/L)^2} \right)^{0.343} (L'/d)^{0.364} \qquad (5)$$

and then Eq.(5) can be rearranged by using relation such that L'/L <1, d/L << 1, and d/L' < 1, make the second power of them ignore and then by setting {1+0.0021(d/L)(1-L'/d)} near equal to unity, yielding as:

$$\frac{q_c}{q_{co}} = k_o^{0.314} \left\{ 1 + 0.343(d/L)^2 \right\} \left\{ (L'/d)^2 - 1 \right\} (L'/d)^{0.364} \qquad (6)$$

One may notice that $k^{0.314} \times (L'/d)^{0.364}$ in Eq.(6) approaches unity because k, L'/d, and $q_c/q_{co} \to 1$ for the saturation condition, that is Ja $\to 0$ and then the k value always becomes smaller than unity, while the value L'/d inversely becomes larger than unity.

Incidentally, it may be of interest to discuss about behavior in the radial spread of liquid after the jet impingement. According to Watson(1964) who theoretically studied the radial spread, we can get an important information on the k_o value: in laminar flow, the k_o value linearly decreases with any increase in the radial distance for $d/2 < r < r_o$ and then is about proportional to the reciprocal of r for $r_o < r < 2r_o$, where r_o is the point at which the boundary layer reaches the free surface and is given as $r_o/(d/2) = 0.367 (ud/v)^{1/3}$. Consequently, what we assume $k_o^{0.314} \times (L'/d)^{0.364} = 1$ here, seems to be reasonable, although it is very difficult to prove it.

Eq.(6) finally becomes as

$$q_c/q_{co} = 1 + 0.343(d/L)^2 (L'/d)^2 \qquad (7)$$

Substituting Eq.(4) into Eq.(7) to eliminate the length, L', one can get

$$\frac{q_c}{q_{co}} = 1 + 0.343(\frac{d}{L})^2 \frac{\rho_v k_1 u H_{fg}}{q_c} Ja \qquad (8)$$

Consequently, Eq.(8) being rearranged explicitly for the CHF in the subcooled liquid, it becomes as:

$$\frac{q_c}{q_{co}} = \frac{1 + \sqrt{1 + 4CJa}}{2} \qquad (9)$$

where

$$C = \frac{0.343 k_1 (d/L)^2 (1 + L/d)^{0.364}}{0.221 (\rho_l/\rho_v)^{0.645} (2\sigma/\rho_l u^2 (L-d))^{0.343}} \qquad (10)$$

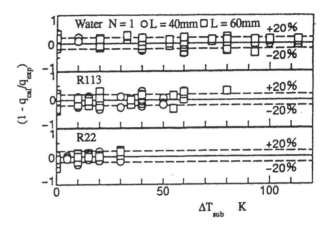

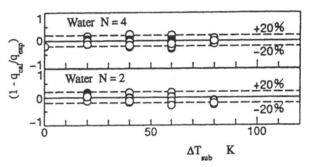

Fig.9 Comparison of CHF data with Eq.(9) with Eq.(11)

The unknown value, k_1, however is too complicated to estimate its value which may be greatly influenced by the weak boiling in the region, B, and amount of the splashed droplets. An alternative way is to keep the functional form given in Eq.(10) and correct the constants included in C such that the CHF data agree with those calculated from Eq.(9). We can get the final form of C as:

$$C = \frac{0.95(d/L)^2 (1 + L/d)^{0.364}}{(\rho_l/\rho_v)^{0.43} (2\sigma/\rho_l u^2 (L-d))^{0.343}} \qquad (11)$$

Comparing Eqs.(10) and (11), one easily notices that the exponent of the density ratio slightly changes from 0.645 to 0.43 and the constant does with its change. This fact means that the effects of the velocity as well as the heated length are appropriately predicted on the CHF by Eq.(9) with Eq.(11). Equation(9) shows the non-linear relation between the CHF and the subcooling, although most of the researches to evaluate the effect of the subcooling on the CHF in both pool and forced convective subcooled boiling gave the subcooling effect on the CHF as linear.

Figure 9 shows the CHF measured with comparison of the CHF value predicted by Eq.(9) with Eq.(11) against the subcooling.

It is found from Fig.9 that the CHF data are in good agreement with the predicted values. For the single jet, 82 % of the CHF data for the subcooled liquid can be predicted with an accuracy of ± 20 % and then 95 % of the total data falls within a range of ± 30 %. For the multiple jets, 85 % of the CHF data can be predicted with an accuracy of ± 20 %, while 95 % of the total data falls within a range of ± 30 %. Finally, statistical results for all the CHF data points are listed in Table 2.

Table 2   Accuracy of the CHF values predicted

	No. of data	E_1	E_2	E_3
Single jet	271	-0.06	0.124	0.152
Multiple jets	121	-0.019	0.115	0.160

Average error	$:E_1 = \Sigma(\,1 - q_{cal}/q_{exp}\,)/N$
Mean variation	$:E_2 = \Sigma\,|\,1 - q_{cal}/q_{exp}\,|/N$
Standard variation	$:E_3 = \sqrt{\Sigma(\,1 - q_{cal}/q_{exp}\,)^2/N}$

Length of the Subcooling Area

The length of the subcooling area can be estimated provided that the CHF and the fraction k_1 in Eq.(4) are known. The CHF can be determined from Eqs.(1), (9), and (11), while it is very difficult to get the fraction k_1 because it is related to the temperature profile over the thickness of the liquid film and amount of the splashed droplets. On the basis of the visual observation, the length L' has a tendency such that the value of L'/L becomes larger with an increase in subcooling and velocity of the impinging jet and with a decrease in the density ratio.

CONCLUSIONS

Critical heat fluxes for not only single but also multiple circular impinging jets have been measured by employing water, R113 and R22 at subcoolings up to $\Delta T_{sub} = 115$ K and velocities up to 25 m/s.

1. A flow model for the subcooled boiling with the jet impingement is proposed and the fully-developed boiling within the saturation zone is identical to that in the saturated boiling.

2. Characteristics of CHF for both the single jet and the multiple jets are similar, when focusing on the region controlled by each individual jet.

3. The generalized correlation (9) with Eq.(11) derived on the basis of the flow model, can predict the CHF data for the single jet as well as the multiple jets with the same accuracy as listed in Table 2.

NOMENCLATURE

C   :constant defined by equation(3)
c_p :specific heat of liquid,
d :diameter of an impinging jet,
H_{fg} :latent heat of evaporation,
Ja^s :Jakob number $((\rho_l/\rho_v)(c_p\Delta T_{sub}/H_{fg}))$,
L   :two times of maximum distance (Fig.3),
N   :number of impinging jets,
q_{co} :critical heat flux for saturated boiling,
q_c :critical heat flux for subcooled boiling,
ΔT_{sub} :subcooling temperature of jet($= T_{sat} - T_{liq}$)
u   :velocity of a liquid jet at the nozzle exit,
ρ_l,ρ_v :density of saturated liquid and vapor,
σ :surface tension.

REFERENCES

Bar-Cohen, A., 1991, Thermal Management of electronic Components with Dielectric Liquids, *Proc of 3rd ASME-JSME Thermal Engineering Proceedings* (Edited by Lloyd, and Kurosaki), Vol.2, pp.xv-xxxix.

Boyd, R. D., 1983, Review of Subcooled Boiling Critical Heat Flux (CHF) and Its Application to Fusion Systems, Part I and II, *Fundamentals of Heat Transfer in Fusion Energy Systems* (edited by Kazimi, M. S. and Jones, O. C.), 21st National Heat Transfer Conf.(1983, Seattle, Washington), HTD-Vol.24, pp.19-42.

Dhir, V. K. and Scott, J. H., 1987, On the Superposition of Injection Induced Swirl during Enhancement of Subcooled Critical Heat Flux, *Int. J. Heat Mass Transfer*, Vol.30, No.10, pp.2013-2022.

Elkassabgi, Y. and Lienhard, J. H., 1988, The Peak Pool Boiling Heat Flux from Horizontal Cylinders in Subcooled Liquids, ASME *Journal of Heat Transfer*, Vol.110, pp.479-486.

Gambill, W. R. and Lienhard, J. H., 1989, An Upper Bound for the Boiling Heat Flux, ASME *Journal of Heat Transfer*, Vol.113-3, pp815-818.

Haramura, Y. and Katto, Y., 1983, A New Hydrodynamic Model of Critical Heat Flux Applicable Widely to Both Pool and Forced Convection Boiling Submerged Bodies in Saturated Liquids, *Int. J. Heat Mass Transfer*, Vol.26, pp.389-399.

Monde, M. and Katto, Y., 1978, Burnout in a High Heat-Flux Boiling System with an Impinging Jet, *Int. J. Heat Mass Transfer*, Vol.21-3, pp.295-305.

Monde, M., 1987, Critical Heat Flux in Saturated Forced Convective Boiling on a Heated Disk with an Impinging Jet, ASME *Journal of Heat Transfer*, Vol.109-4, pp.991-996.

Monde, M. and Inoue, T., 1991, Critical Heat Flux in Saturated Forced Convective Boiling on a Heated Disk with Multiple Impinging Jets, ASME *Journal of Heat Transfer*, Vol.113-3, pp.722-727.

Monde, M., 1991, Critical Heat Flux in Saturated Forced Convective Boiling on a Heated Disk with One or Multiple Impinging Jets, *Trends in Heat, Mass & Momentum Transfer*, 1, pp.33-44.

Mudawar, I. and Wadsworth, D. C., 1991 "Critical Heat Flux From a Simulated Chip to a Confined Rectangular Impinging Jet of Dielectric Liquid", *Int. J. Heat Mass Transfer*, Vol.34-6, pp.1465-1479.

Sharan, A. and Lienhard, J. H., 1985, On Predicting Burnout in the Jet-Disk Configuration, ASME *Journal of Heat Transfer*, Vol.107-2, pp.398-401.

Watson, E. J., 1964, The Radial Spread of a Liquid Jet over a Horizontal Plane, *Journal of Fluid Mech.*, Vol.20, Part 3, pp.481-499.

EFFECT OF WALL THERMAL CONDITIONS ON CROSS FLOW FILM BOILING HEAT TRANSFER

Osama A. Montasser
Department of Power Engineering
Ain Shams University
Cairo, Egypt

Masahiro Shoji
Department of Mechanical Engineering
The University of Tokyo
Tokyo, Japan

ABSTRACT

Theoretical and experimental studies have been carried out to investigate the effect of the wall heat flux conditions on upward cross flow film boiling heat transfer from a horizontal cylinder. The Extreme condition of uniform wall heat flux was first considered and theoretically investigated. The predicted overall average Nusselt number was concluded to be higher than that reported in the literature for the uniform wall temperature condition. Besides, the results showed that a minimum limit of the wall heat flux is required to sustain the vapor film and that the separation angle is always encountered at an angle $\theta \approx \pi/2$ regardless of liquid subcooling or wall superheat.

Experimental investigation followed to check the angular distribution of local wall heat flux and local wall temperature around a horizontal cylinder. Both of the uniform input heat flux and the uniform wall temperature conditions were applied. The measured distributions were utilized to estimate the heat transfer contribution of the back side surface down stream the separation points which is neglected by the theory.

Theoretical study was finally performed to develop a mathematical model for the uniform input heat flux condition, the nearest condition to the actual applications in which electrically heated surfaces are used. The model aimed at studying the effect of the heat transfer by conduction through the heated wall on the heat transfer characteristics. Results of the model showed a good agreement with the present experimental data.

INTRODUCTION

In the past, not a few studies have been done in the field of forced convection film boiling heat transfer from a cylinder or a sphere. But generally it can be said that all of the previous theoretical studies considered only the uniform wall temperature condition despite of the fact that in most of the previous experimental works rather than some actual fields of application, electrically heated surfaces are used, or in other words a uniform heat flux technique is applied, and thus the surface acquire some angular temperature distribution. The temperature difference between the topmost and stagnation points was reported to reach

such a high values as 100-200 K, see Bromley (1953) and Sankaran (1990).

It is also noted that the heat transfer contribution of the back side surface downstream the separation points was neglected by the previous mathematical models. In spite of that, this neglected contribution was not experimentally estimated in the literature. A big gap was also observed between the predicted results for the average heat transfer coefficient and the experimental data.

It is therefore one of the objectives of the present work is to theoretically check how the results of the uniform wall heat flux condition differ from those reported in the literature for the uniform wall temperature one. Another target is to experimentally estimate the back side heat transfer contribution neglected by the theory to clear up whether it is really of negligible value or it plays some role in increasing the overall heat flux. The experimental investigation aimed also at evaluating the theoretical prediction for the front side heat transfer as well as to check how the flow film boiling heat transfer alter when each of the uniform wall temperature or the uniform input heat flux conditions is applied.

Note: Regarding the difference between the uniform wall heat flux and the uniform input heat flux expressions which are frequently used in this study, it is noted that while the first refers to a uniform heat flux dissipation from the outer surface adjacent to the boiling medium, the later means that either the heated surface is supplied by a uniform input heat flux at its inner surface or a uniform heat is generated in its wall.

A theoretical investigation of the uniform input heat flux condition is also aimed by this study to check the effect of the heat transfer by conduction through the heated wall on switching the case between the uniform wall temperature and the uniform wall heat flux extreme conditions.

THE UNIFORM WALL HEAT FLUX MODEL

Cess (1962) has reported that considering a flat plate of uniform wall heat flux resulted in predicting a higher average Nusselt number by a factor of 1.41 compared to that of the constant wall temperature solution. It is therefore the previous uniform wall temperature model developed by Witte and Orozco (1984) for cylindrical and spherical surfaces was modified here to the uniform wall heat flux condition in an attempt to get higher predictions for the average heat transfer coefficient so that the big gap between theoretical results and experimental data might be recovered. Their governing equations were modified to the new condition by applying a heat balance on the wall surface and considering a local wall heat flux of constant value around the surface. The resulted governing equations are as follows:

Cylinder, q_w-constant model

$$\frac{d\Delta}{d\theta} = \frac{\frac{1}{2} A^* - \sqrt{\frac{2}{\pi}} \, C \cos\frac{\theta}{2} - \Delta \cos\theta - \frac{2}{3} B \, \Delta^3 \cos 2\theta}{\left[1 + 2 B \, \Delta^2 \cos\theta\right] \sin\theta} \quad (1)$$

Sphere, q_w-constant model

$$\frac{d\Delta}{d\theta} = \frac{\frac{2}{3} A^* - \frac{4}{\sqrt{\pi}} C \, \dfrac{\cos^2\frac{\theta}{2}}{\sqrt{2+\cos\theta}} - 2\Delta \cos\theta - \frac{1}{2} B \, \Delta^3 \left(3 \cos^2\theta - 1\right)}{\left[1 + \frac{3}{2} B \, \Delta^2 \cos\theta\right] \sin\theta} \quad (2)$$

where:

$$A^* = \frac{Re_{v,q}}{Re_v}, \quad Re_{v,q} = \frac{\left(q_w/\rho_v h_{fg}\right) D}{v_v}, \quad B = Re_v^x, \quad C = \frac{Ja_l^x}{\sqrt{Pe_l}} \quad (3)$$

Results of The Uniform Wall Heat Flux Model

The governing equations of both the uniform wall heat flux and the uniform wall temperature models were numerically solved for water and freon-113 under the atmospheric pressure and over a wide ranges of wall superheat and liquid subcooling

The model showed the following results:

(1) Minimum limit was found for the applied wall heat flux under which film boiling can not be established. The wall heat flux should be greater than the maximum value, at stagnation point, of the heat flux transferred to a subcooled liquid due the temperature difference between saturation temperature at the liquid-vapor interface and the liquid bulk temperature. The minimum wall flux is thus a function of the liquid subcooling and is equal to zero at the saturation condition. The same result is concluded for the uniform input heat flux model, presented later, as well.

(2) Angular distribution of Nusselt number, presented in Fig. 1, shows a very high local Nusselt number at the stagnation point, $\theta = 0$, and decreases sharply with the angle θ to intersect with the corresponding distribution of the uniform wall temperature model at an angle less than 90 degrees and continue to decrease until the separation angle is encountered at $\theta = 90$.

(3) Separation angle is always fixed at about 90 degrees regardless of liquid subcooling or wall superheat, see Fig. 9.

(4) Overall average Nusselt number is higher than that of the uniform wall temperature model within a ratio ranged between about 30% at saturation condition to about 200% at the highest applied liquid subcooling, 25 [K] for freon-113 and 70 [K] for water, at which the predicted values was found to approach the level of the corresponding experimental data, see Fig. 10.

(5) In both of the uniform wall heat flux and the uniform input heat flux models, the back side heat transfer contribution could be indirectly compensated. This is explained as follows: The heat transfer coefficient of the back side surface is still neglected by the models, however in calculating the average wall superheat, the

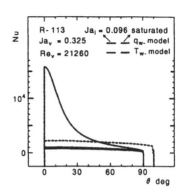

Fig. 1 Nusselt Number Angular Distribution, Comparison Between The Tw-constant And qw-constant Models.

whole surface is considered of an active wall heat flux instead of considering a dead back side surface. This results in calculating a higher value for the average wall superheat and thus leads to grading up the average heat transfer results (the value obtained for the average heat transfer coefficient is plotted against an average wall superheat higher than that calculated for zero heat transfer from the back surface).

(6) The numerical solution data was utilized to suggest the following dimensionless groups. The groups were used to present and correlate the heat transfer results in the present study.

$$Ja_v^{**} = \frac{Ja_v^x}{Pr_v} \qquad Ja_l^{**} = \frac{Ja_l^x}{\sqrt{Pr_l \dfrac{\mu_v}{\mu_l}}} \quad (4)$$

$$Nu^{**} = \frac{Nu}{\sqrt{Re_v^x}} \left(Ja_v^{**}\right)^{n_{sub}}, \qquad n_{sub} = 0.356 \, \frac{Ja_l^{0.24x}}{\left(\sqrt{Pr_l \dfrac{\mu_v}{\mu_l}}\right)^{0.24}} \quad (5)$$

THE EXPERIMENTAL INVESTIGATION.

No estimation for the back side heat transfer contribution was detected in the previous research work. Chou's work (1992) is the only exception. She estimated numerically the heat transfer from the back side of a cylindrical heater by introducing an angularly symmetric wake boundary resembled a cosine curve. A conclusion was reported that the heat transfer in the wake region was as high as 20% of the overall heat transfer and should not be neglected especially at high subcooling.

In the present experimental study, the angular variation of local Nusselt number and local wall temperature around a 3.3 cm horizontal heated cylinder was investigated in upward cross flow film boiling of Freon-113 near the atmospheric pressure. The investigation aimed at estimating how is great the back side heat transfer as well as to evaluate the prediction of the mathematical models for the heat transfer from the front side surface. Both of the uniform wall temperature and the uniform input heat flux conditions were applied to check how the flow film boiling heat transfer alters with each condition.

Results of the uniform wall heat flux model, presented above for the average Nusselt number and the separation angle, were hoped to be checked by the present experimental investigation. However this hope could not be achieved since the heat transfer by conduction through the test heating block was of so high level that supplying it with a uniform input heat flux did not result in a uniform heat flux distribution transferred from its outer surface.

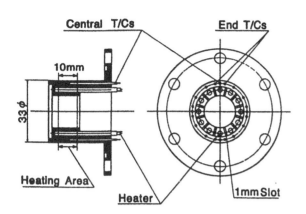

FIG. 2 Cutaway View In The Test Heating Block.

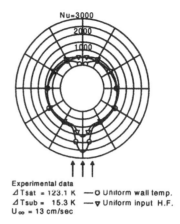

Experimental data
ΔT_{sat} = 123.1 K   —O Uniform wall temp.
ΔT_{sub} = 15.3 K   —▽ Uniform input H.F.
U_∞ = 13 cm/sec

FIG. 3 Typical Results Of Angular Distribution Of Local Nusselt Number.

Experimental Apparatus And Instrumentation

A test flow loop was used to produce steady low temperature flows over an electric heater. A test heating block is installed cross to the liquid flow direction. The heating block, Fig. 2, was designed in such a way that measurements of the distribution of local heat flux and local temperature around the surface can be achieved as well as to enable to switching the film boiling from the uniform wall temperature to the uniform input heat flux conditions or vice versa.

It is a brass cylindrical tube of 33 mm outer diameter and 35.5 mm long, of which 10 mm lies inside the test duct and serves as the active boiling surface. Electric power was supplied to twelve small heating elements inserted in the block and distributed uniformly around the cross section. The elements are separated by 12 slots 1 mm in width to reduce the heat conduction within the heating block. The angular distribution of the surface temperature at the center of the boiling surface is measured by 12 Chromel-Alumel thermocouples. Another 8 thermocouples are inserted to measure the surface temperature at four points (right, up, left and down) around each end section of the boiling surface and used to estimate the heat loss through the ends. A universal scanner-multimeter instrumentation system connected to a personal computer was used to read and save the data and to demonstrate the angular distribution of both the surface temperature and the local heat flux on the computer screen.

Experimental Conditions

The experiments covered subcooling and velocity ranges of 5-25 °C and 5-20 cm/sec respectively at wall superheat of about 120 °C. At the lowest velocity level, the wall superheat effect was investigated over around 90-120 °C range and the effect of a high subcooling at about 35 °C was also checked.

Results And Discussion

After the film boiling was established, each of the uniform wall temperature or the uniform input heat flux condition could be achieved by controlling the input heat power to each heating element using accurate variable voltage transformers while observing the angular distribution of surface temperature or local input heat flux on the computer screen. Heat conduction between each heating element and the nearby others was obtained using the angular temperature distribution. The net wall heat flux was calculated for each heating element from the values obtained for input power, heat conduction and the estimated end losses.

Angular Distribution of Local Wall Temperature And Local Heat Flux.

Wall temperature, in the present uniform input heat flux runs, was found to be of almost uniform distribution around the heating surface. The maximum temperature difference attained between the topmost point behind the surface and the stagnation point was as low as 7% of the average wall temperature. The reason is due to the high thermal conductivity of the test heating block. No matching was therefore observed between the present experimental data and the results of the present uniform wall heat flux model and thus the model's results could not be checked by the present experimental investigation.

Typical results for the angular distribution of the local Nusselt number are shown in Fig. 3. A comparison between the uniform wall temperature and the uniform input heat flux conditions is presented in the figure. The figure shows that although each condition is of a distinct distribution, both are of almost the same average Nusselt number. The reason is explained by what is mentioned above that the uniform input heat flux runs were rather near to the uniform wall temperature condition than to the uniform wall heat flux one.

The predicted results of the uniform wall temperature model for the local Nusselt number distribution were found to take the same trend as the measured data but the later is of higher values and of distinguished high local Nusselt number at the stagnation point. On the contrary, the uniform wall heat flux model shows a fully different results for both the value and trend, see Fig. 8.

Stagnation Effect.

Figure 3 also shows an interesting and newly found result, that is the unexpected great value of local Nusselt number observed at the stagnation point compared to the nearby points. The reason may be due to a solid-liquid contact caused by removing the thin vapor film at the stagnation point by the moving liquid stream. This phenomenon was observed in all the present experimental runs and is referred to, in this study, as the stagnation effect. It is a new parameter added to the other parameters neglected by the theoretical models and result in the observed gap between the experimental data and the predicted results. The stagnation effect was found to be proportional to the liquid velocity and to the ratio between liquid subcooling and wall superheat. Stagnation effect was estimated by the present data to reach a maximum ratio of 10% of the overall average heat transfer.

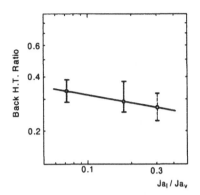

FIG. 4 An Estimation For The Back Side To Overall
Average Heat Transfer Ratio.

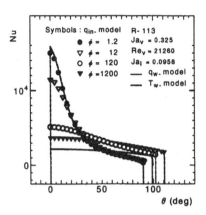

Fig. 5 Angular Distribution Of Nusselt Number, Results Of
The qin-constant Model.

Wake Formation Angle. The measured values for the wake formation angle was found to take almost the same trend as that of the separation angle results predicted by the uniform wall temperature model but higher in value, see Fig. 9. The higher measured values are logically expected since the predicted values are for the angle at which the vapor flow reversal starts which is an early stage of the formation of an observable vapor wake.

Back Side Contribution. The back side heat transfer contribution could be estimated, for the first time, by the present experimental investigation. The measured distribution for local wall heat flux around the surface was utilized to determine the heat transfer from the back surface area downstream the separation angle predicted by the theory. Separation angle predictions by the uniform wall temperature model were used to estimate the actual ratio of the back side heat transfer neglected by the model. Fig. 4. Shows that the back side contribution to the overall average heat transfer ratio may reach more than 35% at low liquid subcooling and decreases as the ratio between liquid subcooling and wall superheat, Ja_l/Ja_v, is increased. The back side heat transfer ratio is therefore of a considerable great value and should be taken into consideration in the theoretical studies.

Average Heat Transfer Results. The overall average Nusselt number results in both conditions of the present experimental runs were found to be of almost the same level, see Fig. 10. Results for the front average Nusselt number were obtained from the present experimental data by excluding the estimated back side heat transfer. A comparison between these results and those predicted by models showed that although the back side effect has already been excluded, the experimental data is always underestimated by the uniform wall temperature model. The reason may be due to the laminar vapor flow assumption in the theory and also to the stagnation effect foregoing mentioned. Different results were observed for the uniform wall heat flux condition. The experimental data was found to be overestimated by the model results as the liquid subcooling and liquid velocity are increased. A comparison between the present and the previous experimental data showed a rather good agreement.

THE UNIFORM INPUT HEAT FLUX MODEL

In most of the previous experimental investigations rather than some actual applications, a uniform heat flux is supplied to the heated surface at its inner surface. The boiling case is therefore changed between the uniform wall temperature and the uniform

wall heat flux extreme conditions as the level of heat transfer by conduction through the heating block is increased from zero to a considerably high value. The wall conduction heat transfer was thus theoretically considered through a cylindrical heating block supplied by a uniform input heat flux to develop the uniform input heat flux mathematical model. The energy conservation principle within the heater's wall is applied to get relations for the temperature distribution around the surface. These relations were numerically solved simultaneously with the vapor film thickness distribution of the uniform wall temperature model using a finite-difference scheme. The size and thermal conductivity values of the heating block in the present experimental investigation were used to predict corresponding theoretical results for comparison.

The level of heat conduction within the heater was classified using the parameter ϕ which was extracted from the governing equations and defined as:

$$\phi = \frac{1}{\left(1 - \frac{t_h}{D}\right)} \frac{\lambda_h \, t_h}{\lambda_v \, D} \qquad (6)$$

Results of The Uniform Input Heat Flux Model

The uniform input heat flux model pointed out the main following results:

(1) Results at low wall heat conduction, $\phi \le 120$, were found to coincide completely with that corresponding of the uniform wall heat flux model. As the level of wall heat conduction is increased, Fig. 5, it was observed that:

(A) Wall temperature becomes gradually of uniform distribution around the surface.

(B) Angular distribution of local Nusselt number changes gradually form the typical one of the uniform wall heat flux model to become virtually of the same trend as that of the uniform wall temperature model but higher in values.

(C) Separation angle becomes a function of the ratio between liquid subcooling and wall superheat but of higher values and of different trend than that of the uniform wall temperature model, see Fig. 6.

(D) Average Nusselt number decreases to approach a minimum limit of a value higher than that of the uniform wall temperature model by a ratio inversely proportional to the liquid subcooling, see Fig. 7.

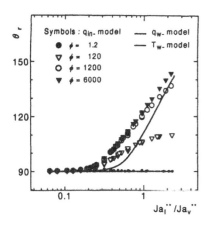

Fig. 6 Separation Angle, Comparison Between The Results
Of qin-constant Model And Results Of
The Other Models.

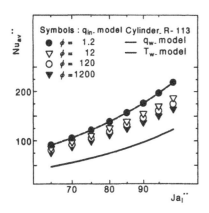

Fig. 7 Average Nusselt Number Results, Comparison
Between The qin-constant Model And The Tw-constant And
qw-constant Extreme Models.

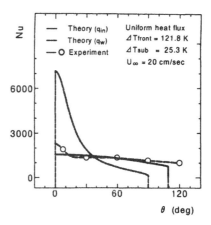

Fig. 8 Angular Nusselt Number Distribution, Comparison
With The Present Experimental Data At High Subcooling-
High Velocity Conditions, $\phi \approx 1500$.

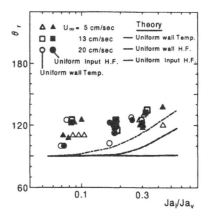

Fig. 9 Separation Angle, Comparison Between The
Mathematical Models Results And The Present
Experimental Data.

(2) At high levels of wall heat conduction, $\phi \geq 1200$, at which the uniform wall temperature distribution is resulted, as liquid subcooling is increased from the saturation condition to an extremely high liquid subcooling, it was found that:

(A) Separation angle approaches the values and the trend of that predicted by the uniform wall temperature model so that they become almost the same, see Fig. 6.

(B) Average Nusselt number changes from only 5% lower than that of the uniform wall heat flux model to become virtually the same as that of the uniform wall temperature model, see Fig. 7.

(3) A comparison with the present experimental data and the results of the other theoretical models shows that:

(A) The angular distribution of local Nusselt number at high subcooling-high velocity conditions becomes much nearer to the experimental data see Fig. 8.

(B) The predicted results for the separation angle are higher than those of the uniform wall temperature model and nearer to the experimental data see Fig. 9.

(C) The average Nusselt number results are of better agreement with the present experimental data and of almost the same level as that of Chou's model (1992), see Fig. 10.

CONCLUSIONS

In this work, the effect of wall thermal conditions on upward cross flow film boiling heat transfer from a horizontal cylinder has been theoretically and experimentally investigated. Both of the uniform wall heat flux and the uniform input heat flux conditions were considered and compared the previous condition of the uniform wall temperature. The main conclusions of this study are:

(1) With the uniform wall heat flux condition, a minimum limit for the wall heat flux is required to sustain the vapor film. The separation angle was fixed at a value of about $\pi/2$ regardless of liquid subcooling or wall superheat. Results of the model for the average heat flux were found to overestimate the experimental data as liquid subcooling and liquid-velocity are increased.

(2) Angular distributions for local wall heat flux and local wall surface temperature around the heated surface were experimentally investigated for the first time. Measured value of the local heat flux at the stagnation point showed unexpected great value compared to the nearby points. Back side heat transfer was experimentally estimated for the first time and found to have a considerable great value at low subcooling.

(3) Effect of heat transfer by conduction through the heated wall was considered by the present uniform input heat flux model. The

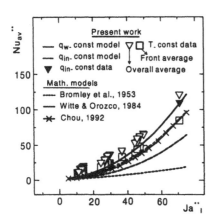

Fig. 10 Average Nusselt Number Results, Comparison Between The Present Study And The Previous Mathematical Models.

model's results show a good agreement with the present experimental data.

(4) Conducting experiments using a heating surface of thin wall and of low thermal conductivity is recommended to allow experimentally attaining the uniform wall heat flux condition so that the results of the present model can be checked.

NOMENCLATURE

Cp	Specific heat [J/kg.K]
D	Outside diameter [m]
deg	degree
h	Heat transfer coefficient [W/m^2.K]
h_{fg}	Latent heat of vaporization [J/kg]
h'_{fg}	Modified latent heat of vaporization, [J/kg] Eq. 7
Ja_l	Liquid Jakob number [Cp ΔT_{sub} / h'_{fg}]
Ja_v	Vapor Jakob number [Cp ΔT_{sat} / h'_{fg}]
Nu	Nusselt number [hD/λ]
Pe	Peclet number [DU$_\infty$ /α]
Pr	Prandtel number [ν /α]
q_w	Wall heat flux [W/m^2]
Re	Reynolds number [DU$_\infty$ /ν]
$Re_{v,q}$	Modified Reynolds number, defined be Eq. 3
t_h	Heater wall thickness [m]
T	Temperature [K]
ΔT_{sat}	Wall super heat, T_w - T_{sat} [K]
ΔT_{min}	Minimum film boiling wall superheat [K]
ΔT_{sub}	Liquid subcooling, T_l - T_{sat} , [K]
U_∞	Liquid approaching velocity [m/s]

Greek Symbols

α	Thermal diffusivity [m^2/s]
δ_v	vapor film thickness [m]
Δ	Dimensionless vapor film thickness, δ_v /D = 1/Nu
ϕ	Parameter used to classify the wall conduction Eq. 6
λ	Thermal conductivity [W/m.K]
ν	Kinetic viscosity [m^2/s]
π	Constant, 3.1415
θ	Angle [deg]

Δ	Angle step [deg]
θ	
θ_r	Separation angle [deg]
ρ	Density [kg/m^3]

Subscripts

h	Heater
l	Liquid
v	Vapor
sat	Saturation
sub	Subcooled
w	Wall

Superscripts

x	Denotes multiplying by ρ_l / ρ_v
**	Denotes the overall correlation parameters Eqs. 4, 5

$$h'_{fg} = h_{fg} \left[1 + 0.4 \frac{Cp_v \, \Delta T_{sat}}{h_{fg}} \right]^2 \qquad (7)$$

REFERENCES

Bromley, L. A., LeRoy, N. R. and Robbers, J. A., 1953, "Heat Transfer in Forced Convection Film Boiling", Ind. Eng. Chem., Vol. 45, p. 2639.

Cess, R. D., 1962, "Forced-Convection Film Boiling on a Flat Plate with Uniform Surface Heat Flux", J. Heat Transfer, Vol. 84, p. 395.

Chang, K. H. and Witte, L. C., 1988, "Liquid-Solid Contact During Flow Boiling of Freon-113", ASME HTD, Vol. 96, p. 659.

Chou, X. S., 1992, "Subcooled Flow Film Boiling Across a Horizontal Cylinder", Ph. D. Dissertation, University of Houston, Houston, Texas.

Epstein M. and Hauser G. M., 1980, "Boiling in the Forward Stagnation Region of a Sphere or Cylinder", Int. J. Heat Mass Transfer, Vol. 23, p.179.

Jacobson, R. N. and Shair, F. H., 1970, "Film Boiling from a Sphere During Forced Convection of Subcooled Water", Ind. Eng. Chem. Fundam., Vol. 9(1), p. 183.

Montasser, Osama, A., 1994, "Effect of Wall Heat Flux Conditions On Cross Flow Film Boiling Heat Transfer", PH. D. Dissertation, The University of Tokyo, Tokyo, Japan.

Motte, E. I. and Bromley, L. A., 1957, "Film Boiling of Flowing Subcooled Liquids", Ind. Eng. Chem., Vol. 49, p. 1921.

Sankaran, S. and Witte L. C., 1990, "Highly Subcooled Flow Boiling of Freon-113 over Cylinders", ASME HTD, Vol. 136, p. 29.

Sideman, S., 1966, "The Equivalence of the Penetration and Potential Flow Theories", Ind. Eng. Chem., Vol. 58(2), p. 54.

Witte, L. C. and Orozco, J., 1984, "The Effect of Vapor Velocity Profile Shape on Flow Film Boiling From Submerged Bodies", J. Heat Transfer, Vol. 106, p. 191.

Witte, L. C., 1968, "Film Boiling from a Sphere", Ind. Eng. Chem.Fundam., Vol. 7, p. 517.

Witte, L. C., Baker, L. and Haworth, R. R., 1968, "Heat Transfer from Spheres into Subcooled Liquid Sodium During Forced Convection", J. Heat Transfer, Vol. 90, p. 397.

Yilmaz, S. and Westwater, J. W., 1980, "Effect of Velocity on Heat Transfer to Boiling Freon-113", J. Heat Transfer, Vol. 102, p. 26.

STUDY OF TRANSITION BOILING HEAT TRANSFER IN QUENCHING OF A FLAT SURFACE USING SURFACE TEMPERATURE AND HEAT FLUX MICROSENSORS

Jason J. Xu, Kamal Adham-Khodaparast and M. Kawaji
Department of Chemical Engineering and Applied Chemistry
University of Toronto
Toronto, Ontario Canada

ABSTRACT

Instantaneous, surface temperature and heat flux have been measured during quenching of a flat copper plate with a flow of Freon-113 using 2-micron thick sensors. The frequency of direct liquid-solid contact was obtained from the power spectrum of the surface heat flux and temperature sensor data. There exists a peak in the frequency with decreasing wall superheat in transition boiling regime and the peak frequency increased with increasing flow rate.

INTRODUCTION

Transition boiling starts at the onset of rewetting and ends as the surface is completely rewetted and nucleate boiling is established. The physical mechanisms responsible for quenching and transition boiling heat transfer are, however, still unclear due to the complex heat transfer processes involved and the difficulties encountered in experimental studies. In recent years the interest in this boiling regime has increased mainly in connection with the safety analysis of nuclear reactors. In hypothetical loss of coolant accidents the transition region is traversed in a transient process. Other quenching processes, e.g. in material processing, also go through the same boiling curve. Reliable prediction methods for transition boiling heat transfer are also required to design high-performance evaporators heated by a liquid or a condensing fluid. Such heat exchangers can be operated in the transition region without the danger of instabilities because the heat transfer process is temperature controlled.

The past experimental studies on the transition boiling mechanisms and heat transfer have shown that the heating surface is alternately in contact with the liquid and vapor. Since the rate of heat transfer to the liquid is higher than that to the vapor, the processes at the points of wall-liquid contact are dominant in transition boiling.

Kalinin *et al.* (1987) and Auracher (1990) have presented full reviews of liquid-solid contact in transition pool boiling. Quantitative studies on liquid-solid contact in transition boiling regime have mostly used either conductivity probes or analysis of surface temperature fluctuations. Using an electrical conductance probe, Yao and Henry (1978) have confirmed liquid-solid contacts in transition pool boiling. Lee *et al.* (1985) studied the wall temperature fluctuations by using a microthermocouple flush-mounted at the boiling surface. Their results showed that the time-averaged local liquid-solid contact fraction increased with decreasing surface superheat. The frequency of liquid-solid contact reached a maximum of ~50 Hz at a surface superheat of ~100 °C and decreased gradually to 30 Hz near the critical heat flux.

Dhuga and Winterton (1985) measured the impedance between a thin, electrically-insulating layer coated on the heating surface and the boiling liquid in transition boiling. They found that the fraction of the wetted area increased with decreasing wall superheat. Using this technique, Alam Rajabi and Winterton (1988) found that the heat flux during the liquid contact periods is not constant but falls with increasing temperature.

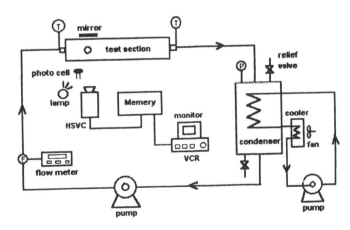

Figure 1. The schematic of flow loop.

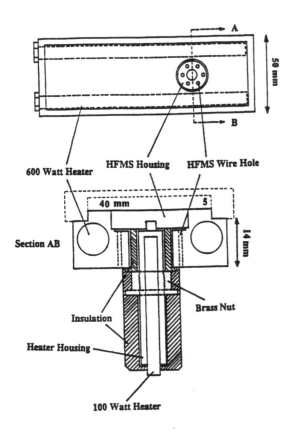

Figure 2. The schematic of the test section

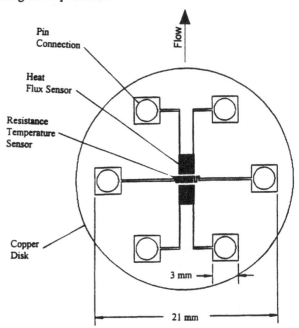

Figure 3. The schematic of heat flux sensor.

Ragheb and Cheng (1978, 1979) used a platinum-tipped zirconium probe to measure the liquid-solid contacts in quenching of a hot tube and observed larger fluctuations in the probe output in the transition boiling regime.

None of those studies involved direct measurement of the surface heat flux in boiling heat transfer. In the present study, the instantaneous surface temperature and heat flux have been measured during quenching of a flat copper plate with a 2-micron thick heat flux sensor. The surface heat flux and temperature data were analyzed to find the effects of wall superheat and flow conditions on liquid-solid contact characteristics in transition boiling.

EXPERIMENTAL SET UP

Figure 1 shows the schematic of the experimental setup. Freon-113 was circulated in a flow loop consisting of the test section, pump, a two-phase condenser equipped with a cooling water loop, and the connecting Teflon tubing. Real time data acquisition was implemented using an IBM-compatible 486DX-33. The flow rate of liquid Freon entering the test section was measured by an infrared flow sensor. Two type-T thermocouples were used to measure the inlet and outlet temperatures of the test section. The rate of cooling water circulation in the condenser and a pressure relief valve setting were used to stabilize the pressure in the condenser at about 120 kPa

Figure 2 shows the schematic of the test section which was designed for studying flow boiling of R-113 on a flat surface. It was built in the shape of a 200 mm long flow channel with a 5 mm by 40 mm rectangular cross section comprised of a stainless steel heated wall and an aluminum casing. Two 600 Watt cartridge heaters were used to heat up the bottom plate of the test section, while a 100 Watt, 3 mm diameter heater could be used to locally heat the heat flux microsensor region. The test section design allowed different flow boiling and quenching experiments to be performed on the same apparatus, by changing the heating procedure and flow conditions.

The sensor was a Vatell Corp. heat flux microsensor equipped with two thermopiles and one resistance temperature sensor (RTS), see Figure 3. It was mounted on a 25.4 mm diameter, 6.4 mm thick copper disk with a 100 μm thick plasma sprayed aluminum oxide base insulating layer and a 1 μm thick overcast of sputtered aluminum oxide preventing any oxidation of the sensor surface and providing a moderate abrasion resistance and electrical insulation. The sensor was fabricated by direct metal deposition technique and was only 2-micron thick. The uncertainty in the surface temperature measurement by RTS was estimated to be ±1.0 °C. The micro heat flux sensors were also calibrated by the manufacturer and the uncertainty in the heat flux measurements was estimated to be ±5% of the reading.

In order to verify that the heat flux sensor gives reliable measurement, transient one-dimensional heat conduction problem was modelled by Adham-Khodaparast

(1995) to determine if the surface temperature response could be simulated using the measured heat flux at the sensor surface as a boundary condition. The result showed good agreement between the one-dimensional model predictions and the RTS surface temperature measurements, indicating reliable response of the heat flux sensors.

For the data acquisition system, the signal conversion time was 8.5 μs for DAS1402 A/D board, 15 μs for EXP-16 signal conditioning board and less than 250 μs for Ectron 687 amplifier. The response time of the MHFS and RTS was reported as 10 μs and 1 ms, respectively by the manufacturer. So, the total response time for the heat flux measurement was estimated to be 273.5 μs and 1.024 ms for the surface temperature measurement.

A high speed video camera was used to study the flow and boiling conditions on the heat flux sensor during the experiments. A mirror at 45° angle was used to direct the image from the top circular window of the test section to the camera. To synchronize the data acquisition with video recording, a 0.002 second flash light signal was detected by a photo cell. The flash light was triggered just after the video recording started. Work is still underway to obtain detailed synchronized correspondence between the heat flux data and the observed events on the heat flux sensor during transition boiling.

For transition boiling experiments, the temperature of the heated wall of the initially voided test section was increased to about 230 °C, using two 600-Watt heaters. After reaching the desired temperature, heating was stopped and the flow of Freon-113 with subcooling of about 25 °C was initiated. The instantaneous temperature and heat flux on the surface of the HFMS were recorded at a frequency of 900 Hz.

More detailed description of the experimental apparatus, instrumentation and procedures can be found elsewhere (Adham-Khodaparast *et al.*, 1995).

RESULTS AND DISCUSSION

The first mode of heat transfer detected by the heat flux sensor was film boiling which continued until the surface temperature fell to a sufficiently low value, usually referred to as the rewetting temperature. During this film boiling period, the sensor surface was entirely covered by a vapor film with a wavy liquid-vapor interface as shown in Figure 4. After rewetting, the boiling mode changed to transition boiling, and the video pictures showed large bubbles and vigorous vaporization in the form of sudden vapor bursts on the sensor surface. It was clear that the surface is alternately wetted by the liquid and covered by the vapor as evident in a picture shown in Figure 5. As the surface temperature further decreased, the surface was completely rewetted and nucleate boiling was observed on the surface. Figure 6 shows the view of the heat flux sensor area in nucleate boiling which was accompanied by regular releases of small size bubbles from the nucleation sites on the sensor surface.

Figure 7 shows the history of transient heat flux and wall superheat $(T_w - T_{sat})$ taken at a rate of 100 Hz for one of the rewetting experiments. Before pumping the liquid into the test section, the heat flux and surface temperature signals were constant. After the liquid was injected into the test section, a short period of dispersed flow film boiling of liquid over the sensor surface was observed followed by inverted annular flow, in which a vapor film separates the liquid and the surface. In this period, the instantaneous heat flux data showed small amplitude fluctuations when the quench front was far away from the sensor, and the surface temperature decreased gradually until the quench front came sufficiently close to the sensor. Occasionally, large heat flux peaks were observed indicating that some liquid may have contacted the solid surface, but the surface rapidly dried out due to the high thermal capacity of the hot plate. The surface temperature data showed only small fluctuations (about 1 - 2 °C). There was an apparent decrease in heat flux just before the quench front reached the heat flux sensor and large fluctuations in heat flux began to appear.

Figure 4. The view of film boiling regime.

Figure 5. The view of transition boiling regime.

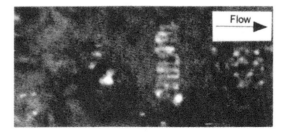

Figure 6. The view of nucleate boiling regime.

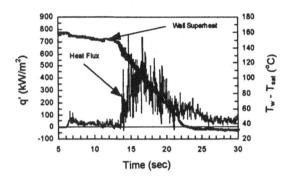

Figure 7. A typical quenching curve.

At the onset of rewetting, the surface temperature profile exhibited a knee in the slope (for some runs it was sharp, but for other runs more gradual), and the temperature decreased much more rapidly through the transition boiling and nucleate boiling regimes. Large amplitude fluctuations appeared in the heat flux data from the onset of rewetting through the maximum heat flux, and surprisingly to the stage involving low wall superheat, conventionally regarded as nucleate boiling. As shown in Figure 8, the surface temperature and heat flux fluctuated almost synchronously during transition boiling despite a slight difference in the positions of the RTS and MHFS, which reveals the size of the rewetted area to be of the order of the size of temperature and heat flux sensors and the distance between them.

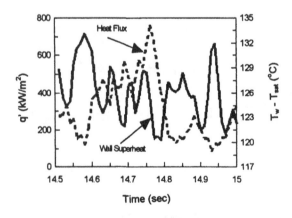

Figure 8. Synchronized response shown by heat flux and temperature sensors.

Direct and simultaneous measurements of the local surface temperature and surface heat flux during rewetting have, to our knowledge, been measured for the first time in several decades of boiling heat transfer research. Previous measurement methods using thermocouples embedded in the heated plate to obtain heat flux data by inverse conduction calculations, could not reliably yield such rapid and large

fluctuations in the surface heat flux as the present heat flux microsensor. The data obtained in this work point to the usefulness of the heat flux microsensor in boiling heat transfer research to elucidate the microscopic phenomena as discussed below.

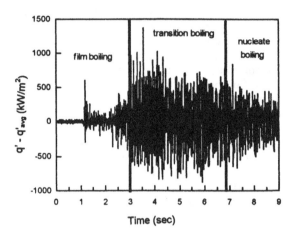

Figure 9. The magnitude of heat flux fluctuations in three boiling regimes.

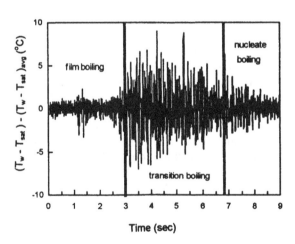

Figure 10. The magnitude of temperature fluctuations in three boiling regimes.

The magnitudes of heat flux and wall superheat fluctuations were obtained by subtracting their respective mean values, obtained by 100-point moving average, from their instantaneous values, thus leaving only the fluctuating components. The amplitude of heat flux fluctuations first increased but remained nearly constant throughout the transition boiling regime as shown in Figure 9. The surface temperature fluctuations also increased in amplitude initially, but reached a maximum amplitude and then gradually decreased, as shown in Figure 10. The amplitude of heat flux fluctuations was of the order of several hundred kW/m^2 while the amplitude of surface temperature fluctuations was less than

5°C. The peak values of the heat flux oscillation in transition boiling are greater than the steady-state CHF values for pool boiling, and do not appear to be directly related to the extension of nucleate boiling curves obtained in the present work and reported by Xu et al. (1995).

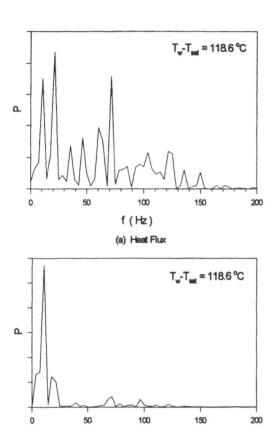

(a) Heat Flux

(b) Surface Temperature

Figure 11. Typical power spectra of heat flux
and surface temperature fluctuations.

It is noted that the amplitudes of fluctuations in heat flux and temperature are still quite large when the wall superheat has decreased below the maximum heat flux point. This is probably the first measurement to yield such heat flux and surface temperature data in quenching experiments. In steady boiling experiments, the local maximum in heat flux is the Critical Heat Flux and the lower wall superheat region corresponds to nucleate boiling regime in which the surface is presumably wetted and fluctuations in heat flux should be quite small. Gaertner (1965) photographically studied the high heat flux region in nucleate boiling below the Critical Heat Flux in pool boiling and observed that dry patches appear in this region of the boiling curve. He called it as second transition region of nucleate boiling regime. Whether the large heat flux fluctuations appearing in our heat flux data correspond to this

second transition remains to be determined more clearly in the future.

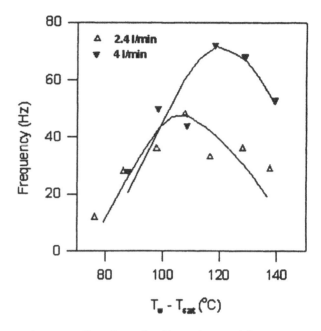

Figure 12. The effects of wall superheat and flow rate on
liquid-solid contact frequency.

Next, the frequency of direct liquid-solid contact during transition boiling was obtained from the power spectra of the surface temperature and heat flux data using the FFT analysis. For each spectrum, at least 512 data points were processed over a given interval of time during which the surface temperature dropped by a maximum of about 10 °C. Typical power spectra for the surface temperature and heat flux data during transition boiling are shown in Figure 11. For the temperature data, a dominant frequency at about 10 Hz was found for many runs under different conditions. However, more than one dominant peak frequency were observed for the heat flux data, one at a low frequency of about 10 Hz and others at higher frequencies that changed with wall superheat and flow rate. The lack of high frequency peaks in the temperature sensor data may be attributed to the larger area of the temperature sensor compared to the heat flux sensor and small amplitude of temperature fluctuations. If the high frequency peaks are associated with small-scale rewet and dryout events, the temperature sensor would not be able to capture these events, so it would show only the low frequency peak.

In observing the boiling processes on the heat flux sensor by a high speed video camera, it was found that a large area encompassing the entire heat flux sensor could be clearly seen when the surface was rewetted as previously shown in Figure 5. More often, however, smaller parts of the heat flux sensor could be seen clearly which indicates small-scale or partial wetting of the surface. A clear view of the full sensor was obtained about 8 to 10 times per second which corresponds to the low peak frequency in the power spectrum data. Partially clear views of the sensor were obtained more often and may correspond to high peak frequencies in transition boiling heat flux fluctuations. It seems that rewetting and dryout of the surface involve different length and time scales, with the large-scale rewet/dryout occurring less frequently than the small-scale rewet-dryout phenomena.

Figure 12 shows the effects of wall superheat and flow rate on the highest liquid-solid contact frequency detected by the present heat flux sensor. With decreasing wall superheat, the peak frequency first increases and then decreases, so there exists a local maximum in the contact frequency. Also, the curve drifts to higher wall superheats with increasing flow rate. The dominant frequencies of about 50 to 70 Hz obtained in this work at wall superheats of about 110 to 120 °C are somewhat higher than those of about 40 Hz at wall superheats of about 40 °C for pool boiling of R-113 with similar liquid subcooling (Kalinin et al., 1987).

CONCLUSIONS

The boiling heat transfer and liquid-solid contact characteristics during quenching of a hot surface have been studied using a micro heat flux sensor deposited on a flat copper surface. Large fluctuations in surface heat flux and temperature have been observed not only in transition boiling regime, but also in nucleate boiling regime near the maximum heat flux region. There also exists a peak in the liquid-solid contact frequency as the wall superheat changes.

REFERENCES

Adham-Khodaparast, K., J.J Xu and M. Kawaji, "Flow Film Boiling Collapse and Surface Rewetting in Normal and Reduced Gravity Conditions", *Int. J. Heat Mass Transfer*, **38**, No.15, 2749-2760 (1995)

Adham-Khodaparast, K., "Theoretical Modeling and Experimental Investigation of Film Boiling and Rewetting at Different Gravity and Flow Conditions", Ph.D. thesis, Department of Chemical Engineering and Applied Chemistry, University of Toronto, (1995)

Auracher, H., "Transition Boiling", *Proc. 9th Int. Heat Transfer Conf.*, Jerusalem, Israel, 69-90 (1990)

Dhuga, D.S. and R.H.S. Winterton, "Measurement of Liquid-Solid Contact in Boiling", *J. Phys. E. Sci. Inst.*, **19**, 69-75 (1986)

Gaertner, N.F., "Photographic Study of Nucleate Pool Boiling on a Horizontal Surface", *J. Heat Transfer*, **15**, 17-29 (1965)

Kalinin, E.K., I.I. Berlin and V.V. Kostiouk, "Transition Boiling Heat Transfer", *Advances in Heat Transfer*, **18**, 241-323 (1987)

Lee, L.Y.W., J.C. Chen and R.A. Nelson, "Liquid-Solid Contact Measurements Using a Surface Thermocouple Temperature Probe in Atmospheric Pool Boiling Water", *Int. J. Heat Mass Transfer*, **28**, 1415-1423 (1985)

Ragheb, H.S., S.C. Cheng and D.C. Groeneveld, "Measurement of Transition Boiling Boundaries in Forced Convective Flow", *Int. J. Heat Mass Transfer*, **21**, 1621-1624 (1978)

Ragheb, H.S. and S.C. Cheng, "Surface Wetted Area during Transition Boiling in Forced Convective Flow", *Trans. ASME J. Heat Transfer*, **101**, 381-383 (1979)

Rajabi, A.A.A. and R.H.S. Winterton, "Liquid-Solid Contact in Steady-State Transition Pool Boiling", *Int. J. Heat and Fluid Flow*, **9**, 215-219 (1988)

Xu, J.J, K. Adham-Khodaparast and M. Kawaji, "Boiling Heat Transfer in Reduced Gravity during Quenching of a Hot Surface with R-113", *Proceedings of the 30th National Heat Transfer Conference*, **3**, 39-47, (1995)

Yao, S. and R.E. Henry, , "An Investigation of the Minimum Film Boiling Temperature on Horizontal Surfaces", *Trans. ASME J. Heat Transfer*, **100**, 260-267, (1978)

FORCED CONVECTIVE HEAT TRANSFER TO SUPERHEATED STEAM IN ROUND TUBE

Haiyan Chen and Yuzhou Chen
China Institute of Atomic Energy
P.O.Box 275-59, Beijing 102413, China

ABSTRACT

Heat transfer coefficient in steam is essential for modelling the film boiling of water. The experiment of forced convective heat transfer to superheated steam in vertical round tube of 12mm ID has been performed, covering the range of pressure 1.5 to 60.9 bar, Reynolds number 1800 to 58000, steam temperature 450 to 750K, and surface temperature 590-910K. From the results the change of flow from turbulent region to transition region is evident, which exhibits a dramatic change in the heat transfer trend, and occurs at substantially different Reynolds number for different pressure. The results are compared with six correlations. Reasonable agreement is found with the Heineman and Hadaller correlations for turbulent region, but a significant deviation from the data exists when these correlations are extended to the transition region.

INTRODUCTION

Convective heat tranfer to pure steam is of great interest to the investigators over all of the world due to the significance in variety of engineering practices. In film boiling of water it is the major heat transfer mode, and the calculation of the heat transfer coefficient in steam is essential for modelling the film boiling regime (Groeneveld 1984).

A number of experiments of heat transfer in steam have been carried out by many investigators (McAdams, Kennel et al. 1950, Heineman 1960, Bishop, Krambeck et al 1964, Hadaller and Banerjee 1969). These works generally dealt with the range of higher Reynolds number (Re>20000). While for the range of Re<20000, which is typical in dispersed flow film boiling, the data are limeted. Only the McAdams experimet in an annulus covered a lower range of Reynolds number (Re=6000-40000). In dispersed flow film boiling the correlations obtained from these experiments are usually extended to Re>2300 for the calculation of steam convective heat transfer component due to lack of appropriate correlation. In recent studies on the film boiling the effect of the entrained droplets on steam convective heat transfer is particularly concerned. (Hewitt 1984, Chen 1994). Precise calculation of the heat transfer to pure steam flow is needed as a

basis for evaluation of this effect. Therefore, it is imperative to evaluate the uncertainties of the correlations caused by extending to the outside range.

In addition to the correlations from the experiments of steam some correlations from experiments of gas or liquid, typically those suggested by Kutateladze (1966), Gnielinski (1976) and the well-known Dittus-Boelter correlation (1930), are also widely used for the heat transfer to steam over Re>2300 (Hartnett 1994).

In the present experiment the data of heat transfer to steam convective flow were obtained in a round tube to study the heat transfer behaviour in the range of Re>1800,and were compared with various correlations to examine their applicability in this region.

EXPERIMENTAL APPARATUS AND METHOD

A schematic diagram of the experimental apparatus is shown in Fig.1. The subcooled water coming from the pump was heated into two-phase in the preheater I. In the separator a part of steam was separated from steam-water mixture. It was superheated to some degree by the preheater II before entering the test section. The two-phase mixture from the bypass was cooled into liquid before returning to the pump. On the downstream of the test section the superheated steam was condensed into liquid, then flowed through the coil flowmeters. The flow rate was controlled by the outlet valve and bypass valve.

The test section shown in Figure 2 was a vertical Inconel tube with 12mm ID 15mm OD and heating length of 1.5m. On the upstream of the test section there was a starting section of 1m in length with the same diameter, which ensured a fully developed velocity profile to be established at the inlet of heating section. To measure the wall temperatures of the test section, 28 calibrated thermocouples were discharge-welded to the outer surface. At each position of length-to-diameter ratio L/D=10, 27, 46 and 70 four thermocouples were placed $90°$ apart on the surface. An alternative current was connected to the test section. Approximately uniform heating can be provided to the test section due to weak temperature-dependence of electric

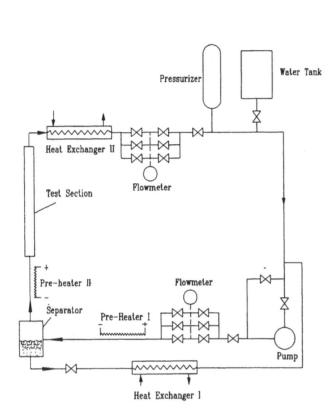

Fig. 1 The diagram of test loop

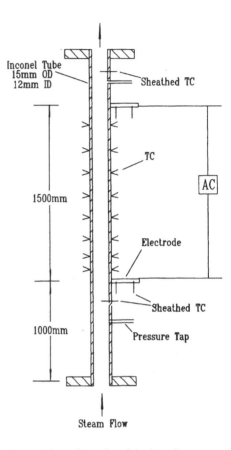

Fig. 2 The schematic of test section

resistance of the material .

The test section was thermally-insulated with glass fibre, and the heat loss was deliberately calibrated in an auxiliary test. The heat loss coefficient was expressed as a function of surface temperature of the test section. To evaluate the conduction heat loss through the electrodes from the test section, a sheathed thermocouple was mounted at each end of the electrodes (copper bars).

The measuremets included: the inlet pressure, the voltage and current to the test section, the surface temperatures, the inlet and outlet steam temperatures, the flow rate, and the temperatures in the electrodes. All the readings were scanned and recorded by a PC.

By keeping carefully the flow rate and power at a fairly stable condition for sufficient time to reach stabilization (40-120 min for a run), good heat balances were attained: all runs were within 4.5% (most within 3%). Totally 47 runs were performed, covering the range of pressure 1.5 to 60.9 bar, Reynolds number 1800 to 58000, steam temperature 450 to 750K, and wall temperature 590 to 910K.

EXPERIMENTAL RESULTS AND DISCUSSIONS

Data Reduction

Local heat transfer coefficient was determined based on the equation,

$$h = \frac{q_0 - q_l - q_c}{T_w - T_g - \Delta T} \qquad (1)$$

where q_0 is the local gross wall heat flux, which is obtained from the average wall heat flux, taking into account the effect of temperature on the electric resistance. q_l is the heat loss to the surroundings, evaluated from the local surface temperature and the calibrated loss coefficient. q_c is the axial heat conduction evaluated from the measured temperature distribution on the basis of heat conduction equation. T_w is local outer surface temperature. T_g is local steam temperature evaluated from the heat balance equation. ΔT is the temperature difference across the wall, evaluated from the correlation of heat conduction in a hollow cylinder with heat source.

For comparison the radiation component, a minor fraction of the total heat transfer coefficient, is not excluded from the h as in other correlations obtained from steam experiments.

The experimental results were expressed by dimensionless groups: Nusselt number Nu, Reynolds number Re and Prandtl number Pr. The properties of superheated steam in these groups were calculated by the equations given in the code TRAC-BD1/MOD1 (Shumway 1984).

Experimental Results

For turbulent flow the heat transfer coefficient is generally a

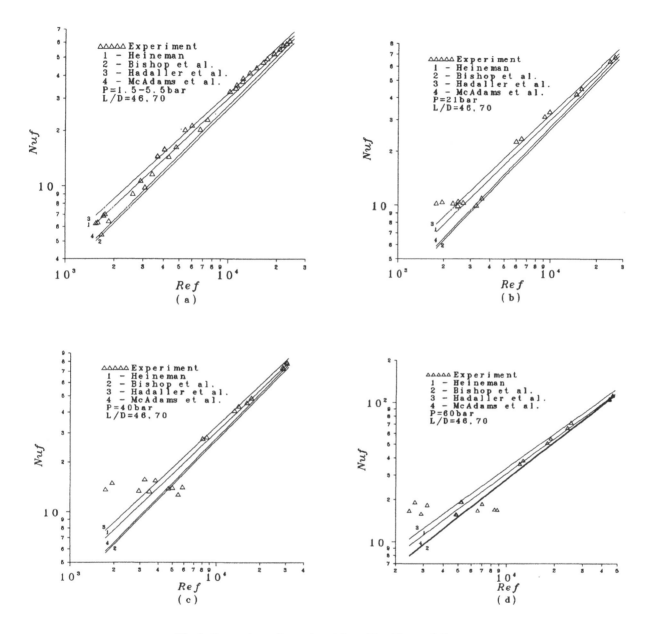

Fig. 3 Comparison of experimental results with correlations

function of Reynolds and Prandtl number. In this work the Pr is close to 1.0 and the effect of Pr is not studied. The effect of Reynolds number is illustrated in Fig.3, in which Nu_f is plotted against Re_f in logarithmic ordinate for pressure of 1.5-5.5, 21, 40 and 60 bar. The subscript f denotes the properties evaluated at film temperature $T=(T_w+T_g)/2$.

For pressure of 21, 40 and 60 bar, over higher Reynolds number region the data exhibit linear increase of Nu_f with the Re_f, suggesting a proportional increase of Nu_f with Re_f^n. The slope of the line gives the exponent n from 0.8-0.9, which is similar to that found in various turbulent correlations. At a certain Re_f a dramatic change of the trend is observed. When Re_f decreases a little below this value a sharp decrease of Nu_f is attained. This decrease seems to be more significant at high pressure. At P=60

bar, for instance, as the Reynolds number decreases from 12000 to 9000, the Nu_f decreases from 36 to 17. This distinct change of the trend appears to be the result of the change of flow from turbulent region to transition region. When the Reynolds number decreases further to around 2000 the variation of Nu_f is not appreciable. This trend leads to a crossover with the extension of the straight line at some point, and in further low Reynolds number region the data are above the line. For the pressure of 21, 40 and 60 bar, the Reynolds number Re_f for the occurance of the transition are in the range of 4000–5500, 6000–8000 and 9000–12000, respectively.

At pressure of 1.5 – 5.5 bar, the change of the trend is not distinct. As the Reynolds number decreases to a relatively low region, the Nu_f decreases a little faster, but for the whole region

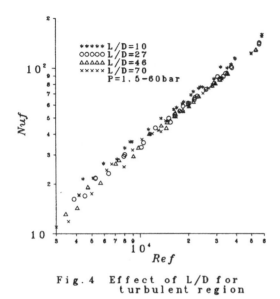

Fig.4 Effect of L/D for
turbulent region

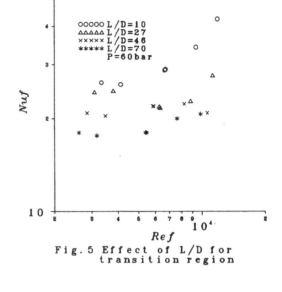

Fig.5 Effect of L/D for
transition region

the trend can be essentially viewed as a straight line with a little steeper slope.

The Reynolds number at the change of flow regime increases with the pressure increasing. This fact can be explained by the increase of the buoyant force, which tends to increase the velocity near the wall, and hence, for the same total flow rate the velocity (and gradient) in the region far from the wall is lower, compared to the flow with less buoyant force. This profile of velocity associates with less shear force in the flow. Therefore higher flow rate is needed for the inception of turbulence.

The effect of L/D for turbulent region is illustrated in Fig.4, where the data at L/D=10, 27, 46 and 70 are compared. The data for L/D of 46 well agrees with that for L/D of 70. For L/D=27 the data are a little higher. For L/D=10, at higher Re the data are about 10% above that of L/D=70, while at lower Re the effect appears to be more stronger. Generally speaking, the effect of L/D is similar to the correlation of Heineman and Hadaller, but is not so strong as the Bishop and McAdams correlations.

The present experiment indicates that for a heating section with developed velocity profile at the entrance the temperature profile is essentially developed when L/D exceeds 40. Nevertheless, for the transition region, the effect of L/D appears much stronger and complicated, as illustrated in Fig.5 for P=60bar.

COMPARISON OF THE RESULTS WITH SOME CORRELATIONS

As mentioned above there are no correlations which were derived in the range similar to the present investigation. In this paragraph six correlations are extended to the present range to compare with the data to examine their applicability. These correlations and the relevant conditions are listed in Tab.1. In Fig.3 the experimental data for different pressures are compared with the Heineman, Hadaller, Bishop and McAdams correlations, all of which were derived from steam data and based on film

temperature. In Fig.6 the Dittus-Boelter and Gnielinski correlations are compared with the data in the ordinate of Nu and Re based on bulk temperature.

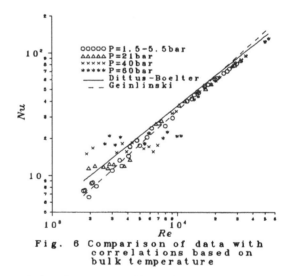

Fig. 6 Comparison of data with
correlations based on
bulk temperature

For turbulent flow region, which is identified by the distinct change of the trend as defined above, the average deviation of the McAdams and Bishop correlations from the data (at L/D=46 and 70) are -14 and -18%, respectively, with greater daviation in lower Reynolds number region and greater L/D. This is expectable, because these two correlations have stronger dependence of Nu_f on Re_f and L/D ratio, resulting in appreciable underprediction of Nu_f when extending to rather low Re_f, especially for larger L/D. The slopes of the Heineman and Hadaller correlations close to the data and their predictions are within +14 to -11% and +33 to -6% of the data, respectively. The average deviations from the data are -3.1 and +4.1%, and standard deviation 6.5 and 6.7%, respectively.

Table 1 Correlations for single phase turbulent flow heat transfer

References	Correlations	Comments
Dittus and Boelter (1930)	$Nu_b = 0.023\,Re_b^{0.8}\,Pr_b^{0.4}$	Circular geometry, Re > 10,000, L/D > 50, water
McAdams et al. (1950)	$Nu_b = 0.0126\,Re_f^{0.89}\,Pr_f^{1/3}(D_e/L)^{0.13}$	Annular 6,000<Re<400,000 1 < P < 24 MPa, steam
Heineman (1960)	$Nu_f = 0.0133\,Re_f^{0.84}\,Pr_f^{1/3}$, L/D>60 $Nu_f = 0.0157\,Re_f^{0.84}\,Pr_f^{1/3}\,(D/L)^{0.04}$, L/D<60	20,000 < Re < 370,000 2 < P < 10 MPa Steam
Bishop et al. (1961)	$Nu_f = 0.0073\,Re_f^{0.886}\,Pr_f^{0.61}(1+\dfrac{2.76}{L/D})$	High-pressure steam
Hadaller and Banerjee (1969)	$Nu_f = 0.017\,Re_f^{0.8216}\,Pr_f^{0.625}(1+0.9842\dfrac{D}{L})$	$6.10^4 < Re < 6.10^5$ $295 < T_f < 580\,°C$, 2 < P < 21.4 MPa Steam
Gnielinski (1976)	$Nu_b = \dfrac{(f/2)(Re_b-1000)\,Pr_b}{1+12.7(f/2)^{1/2}(Pr_b^{2/3}-1)}$	$2300 < Re < 5.10^6$

For the turbulent region, the predictions of Dittus-Boelter and Gnielinski correlations are within +44 to 0% and +24 to -22% of the data, respectively. The average deviations are +9.1 and -1.6%, and standard deviation 11.2 and 11.4%, respectively.

At higher Re the Gnielinski and Dittus-Boelter correlations overpredict the data by 10-15%. For lower Re under low pressure the Dittus-Boelter correlation substantially overpredicts the data. While the Gnielinski correlation has steeper slope in lower Re region, and the deviation is not so great.

For the transition region all the correlations do not predict the dramatic change of the trend. The maximum positive deviations of the Heineman, Hadaller, Gnielinski and Dittus-Boelter correlations are respectively 53, 77, 96 and 93% at pressure of 60 bar and Re of around 9000, and the maximum negative deviations are 100, 75, 56 and 91% at pressure of 60 bar and Re of around 3000.

CONCLUSIONS

An experiment of steam convective heat transfer has been performed in a round tube, covering a range of pressure 1.5-60.9bar and Re 1800-58000, the following conclusions can be drawn:

1. The change of flow from turbulent region to transition region results in dramatic change of the heat transfer trend, and the Reynolds number for the change of regime increases appreciablly with the pressure increasing.

2. The Heineman and Hadaller correlations reasonably predict the present experimental data in the turbulent region. These two correlations can be applied to lower Re region (Re>2300) only at low pressure condition, a significant error may be caused at higher pressure.

The experiment is under way for the condition of occurance of the transition and the heat transfer coefficient in this region.

ACKNOWLEDGEMENT
This work was supported by International Atomic Energy Agency under the Program of Thermohydraulic Relationships for Advanced Water-cooled Reactors.

REFERENCES
Bishop, A.A, Krambeck, F.J., and Sandberg, R.O., 1964, "Forced Convection Heat Transfer To Superheated Steam at High Pressure and High Prandtl Numbers," WCAP-2056, part III.

Chen, J.C., 1994, "Two-Phase Flow with and without Phase Change: Suspension Flows-- some Questions Answered and Unanswered," Proc. 10th Int. Heat Transfer Conf., Brighton UK, V.1, pp.369-386.

Dittus, F.W. and Boelter,L.M.K., 1930, "Heat Transfer in Automobile Radiators of the Tubular Type," University of California Publication, V.2, pp.443-461.

Gnielinski, V., 1976, "New Equations for Heat and Mass Transfer in Turbulent Pipe and Channel Flow," Int. Chem. Eng., V.16 pp.359-368.

Groeneveld, D.C. and Snoek,C.W., 1984, "A Comprehensive Examination of Heat Transfer Correlations Suitable for Reactor Safety Analysis," Multiphase Science and Technology, V.2, pp.181-274.

Hadaller, G. and Banerjee,S., 1969, "Heat Transfer To Superheated Steam in Round Tubes," Atomic Energy of Canada Limited, WDI-147.

Hartnett,J.P., 1994, "Single Phase Channel Flow Forced

Convection Heat Transfer," Proc. Tenth Int. Heat Transfer Conf., Brighton UK, V.1, pp.247-258.

Heineman, J.B., 1960, "An Experimental Investigation of Heat Transfer To Superheated Steam in Round and Rectangular Channels," ANL-6213.

Hewitt, G.F., 1984, "Dispersed Flow Heat Transfer," presented at the 1st Int. Workshop on Fundamental Aspect of Post-Dryout Heat Transfer, Salt Lake City, Ultah, NUREG/CP-0060.

Kutateladze, S.S. and Borishanskii, V.M., 1966, "A Concise Encyclopedia of Heat Transfer,"Pergamon Press, Oxford.

McAdams, W.H., Kennel, W.E., and Addoms, J.N., 1950, "Heat Transfer To Superheated Steam at High Pressures," Trans. ASME 72, pp.421-428.

Shumway,R.W. et al., 1985, "TRAC-BD1/MOD1: An Advanced Best Estimate Computer Program for Boiling Water Reactor Loss-of-Coolant Accident Analysis," NUREG/CR-3633.

MULTICOMPONENT BOILING

HEAT TRANSFER IN VERTICAL FORCED CONVECTIVE BOILING OF BINARY MIXTURES

Gian Piero Celata, Maurizio Cumo and Tommaso Setaro
Energy Department - Heat Transfer Division
ENEA
Rome, Italy

ABSTRACT

The present research focused on the heat transfer in forced convective boiling of a binary mixture of refrigerant R-12/R-114.

The test section is an industrial stainless steel (AISI 316), circular duct uniformly heated (Joule effect) over a length of 2300 mm, with an inner diameter of 7.57 mm and a wall thickness of 0.975 mm. The test section instrumentation consists of 0.5 mm, K-type insulated thermocouples distributed along the tube on the wall and inside the channel, and connected to give directly, at each location, the temperature difference between the wall and the bulk fluid. The fluid flow is upwards, with subcooled inlet conditions.

The experiments were performed using three different mixture compositions, four different pressures ranging between 1.0 and 2.5 MPa, three specific mass flow rates between 500 and 1500 kg m^{-2}s^{-1} and three heat fluxes between 18 and 40 kW/m^2. The experimental results show a relevance of the heat transfer regime identification in the evaluation of mixture heat transfer performance as referred to the pure components.

INTRODUCTION

Over the last ten years a more refined evaluation of thermophysical and thermodynamic properties of binary mixtures was performed, and in particular an evaluation of heat transfer characteristics. This increasing attention to the mixtures as working fluid in various industrial applications with moderate and/or low temperature levels is also due to a growing demand of new ecological refrigerants with low ozone depletion potential. However, most of the past research on mixtures in forced flow has been performed using horizontal channels and only in the fully developed two-phase flow. Mishra et al. (1981) made an experimental research using a binary mixture of R-12 and R-22 with various mole fractions to evaluate the local heat transfer coefficients in forced convective boiling in a horizontal channel. The experimental results showed a lower heat transfer coefficient of the binary mixture with respect to the simple linear interpolation between the values corresponding to the two pure components (ideal behaviour). This result was also confirmed by Fink et al. (1982) with a different mixture, obtained using R-11 and R-113, in the nucleate boiling region, as shown in Fig. 1. In the same figure, the negligible influence of the velocity on the heat transfer coefficient is confirmed. Recently, Jung et al. (1989) have performed tests with a nonazeotropic binary mixture of R-22 and R-114, using a horizontal test section; the results obtained by Jung et al. (1989) verify the heat transfer coefficient trends already known for nonazeotropic binary mixtures as shown in the Fig. 2, i.e., the heat transfer coefficient has a reduction for low values of mole fraction, while tends to a constant value for intermediate values of the mole fraction (0.2 < mole fraction < 0.6) and then increases sharply for mole fractions of R-22 higher than 0.8. Another interesting investigation has been performed by Kedzierski and Didion (1990) who visualized the nucleate flow boiling of a R-22/R-114 mixture in a horizontal channel. They verified a comparable amount of nucleation between the more volatile component, R-22, and the mixture, even though the mixture was mostly R-114 by mole, while R-114 exhibited much more nucleation than either R-22 or the R-22/R-114 mixture. Bennett and Chen (1980) have made one of the few studies

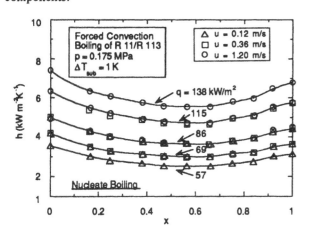

Fig. 1 Boiling heat transfer coefficient as a function of composition at different liquid velocity and heat flux, for R-11/R-113 mixtures (Fink et al. (1981))

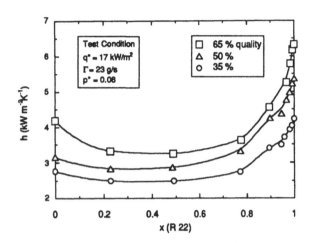

Fig. 2 Boiling heat transfer coefficient as a function of composition at different steam quality, for R-22/R-114 mixtures (Jung et al. (1989))

with vertical channels on binary mixtures in forced convective boiling. They used an aqueous mixture of ethylene/glycol and evaluated the local heat transfer coefficient with different mixture compositions and at different qualities (0 < quality < 30%). Sivagnanam and Varma (1990) also verified the presence of hysteresis at the onset of subcooled boiling using aqueous mixtures in forced convection outside a vertical heating wire provided along the center line of a glass tube, and particularly revealed an increase of the wall superheating at the onset of subcooled boiling with the increase of the mole fraction of the more volatile component. Kawano (1987) made another study with binary mixtures in forced convective boiling along tubes on the shell side of a vertical shell-and-tube heat exchanger, using R-114/R-152a mixtures, and verified the lower heat transfer coefficients of mixtures with respect to those of single-component fluids.

The above analysis of the experimental research devoted to binary mixtures shows a lack of experimental data in the vertical forced convective flow, where very few has been performed in the last two decades, as previously reported by Stephan (1982) and Shock (1982) and recently reviewed by

Steiner (1991) and Thome (1994).

Furthermore, most of the experimental research has not focused the role of the flow pattern on the evaluation of the mixtures heat transfer performance. The present paper investigates the flow characteristic and the connected heat transfer regime in a wide range of thermal-hydraulic conditions, and the relationship with the heat transfer degradation showed by the mixture.

EXPERIMENTAL APPARATUS

The experimental loop, schematically represented in Fig. 3, consists mainly of a piston pump, an electric heater, a condenser and a storage tank. The maximum operating pressure of the loop is 3.5 MPa, while the maximum specific flow rate is 1800 kg/m^{2}s; the available electrical power is 10 kW for the electric heater and 15 kW for the test section. The test section is an industrial stainless steel (AISI 316), circular duct uniformly heated (Joule effect) over a length of 2300 mm, with an inner diameter of 7.57 mm and a wall thickness of 0.975 mm. The test section is connected to the loop using fast connectors and is insulated to avoid any heat dispersion. The test section instrumentation consists of 0.5 mm, K-Type insulated thermocouples distributed according to the scheme of Fig. 4 for the wall (eleven) and the fluid (fifteen) temperature measurements. Thermocouples located at the same section of the channel have been connected to give directly the temperature difference between the outer wall and the bulk fluid, $\Delta T = T_{w,o} - T_{bulk}$. Each couple of sensors has been chosen with the same calibration curve to avoid any offset error. The temperature drop across the wall, $T_{w,o} - T_{w,i}$, calculated using the Fourier equation, is deduced by the temperature difference,

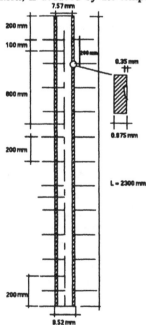

Fig. 4 Schematic of the THOM test section (ENEA)

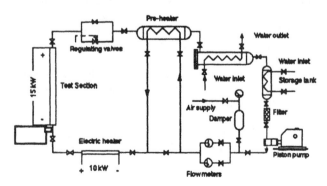

Fig. 3 Schematic of the THOM loop (ENEA)

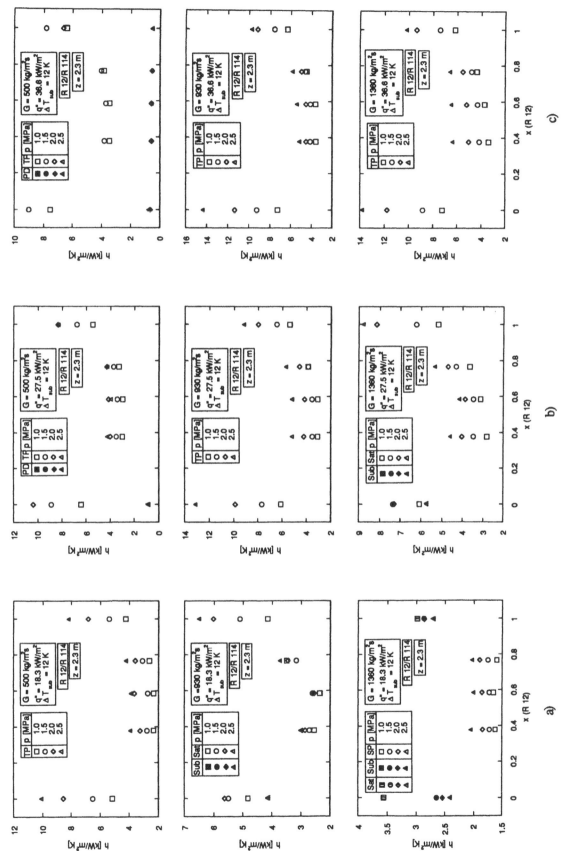

Fig. 5 Heat transfer coefficient as a function of composition at various mass fluxes and pressures, for R 12/R 114 mixtures: a) q" = 18.3 kW/m²; b) q" = 27.5 kW/m²; c) q" = 36.6 kW/m² (ENEA)

253

ΔT. From this point on, when speaking of wall temperature, T_w, we intend the inner tube wall temperature, i.e., $T_w = T_{w,i}$. The remaining four thermocouples for fluid temperature measurements provide the absolute fluid temperature trend. Two pressure transducers measure the test section inlet and outlet pressure. A turbine flowmeter measures the volumetric flow rate at the test section inlet. The heating power has been gaged with a wattmeter. The fluid flow is upwards with subcooled inlet conditions. The system was charged with the mixture of R-12 and R-114 of the required composition by introducing known masses of each refrigerant into the storage tank.

For the analysis of the data, all thermodynamics properties, such as density, composition, enthalpy, and saturation temperature, for both pure and mixed refrigerants were calculated by the Carnahan-Starling-DeSantis (CSD) equation of state, as proposed by Morrison and McLinden (1986). A detailed description of physical properties evaluation is given in the paper by Celata et al. (1993).

EXPERIMENTAL RESULTS AND DISCUSSION

A test matrix was defined to investigate the different parameters affecting the heat transfer coefficients of binary mixtures. One value of inlet subcooling and three fluid inlet flow rates were chosen, while, because of the significant dependence of heat transfer on the pressure, four pressures were investigated. Furthermore, three different heat fluxes for each mass flux have been investigated to ensure a wide range of thermal-hydraulic conditions. Globally, the range of variation of the parameters in the tests performed was as follows:

Heat Transfer Experiments	
Fluid	R-12 (CCl_2F_2), R-114 ($C_2Cl_2F_4$)
x(R-12)	0, 0.377, 0.586, 0.767, 1
p [MPa]	1.0, 1.5, 2.0, 2.5
G [kg/m^{2}s]	500, 930, 1360
q" [kW/m^2]	18.3, 27.5, 36.6
$\Delta T_{sub,in}$ [K]	12

In Fig. 5, the heat transfer coefficient at the exit level of the test section versus the more volatile component mole fraction, x(R-12), is shown for three different mass and heat fluxes with the pressure as a parameter. All the graphs show a lower heat transfer coefficient of the mixture respect to the pure refrigerants. The top graph of Fig. 5a shows the typical trend of the average heat transfer coefficient in the two-phase region (TP) for a heat flux equal to 18.3 kW/m^2 and the lower mass flux (500 kg/m^{2}s). The middle graph of Fig. 5a shows three lower heat transfer coefficients than expected at the higher pressure (p = 2.5 MPa); this different trend is due to the fact that the heat transfer lies in the subcooled flow boiling regime (sub) in the case of the pure R-114 and for the mixture with x(R-12) below 0.6 at the pressure of 2.5 MPa. A further differentiation is shown for the bottom graph of Fig. 5a, where three different heat transfer regimes are present: single phase (SP), subcooled flow boiling (sub) and saturated flow boiling

(sat). The binary mixture, for all the tested compositions, leaves the heated channel without reaching the boiling regime, neither subcooled nor saturated. On the other hand, the pure fluids at the exit of the test section are under subcooled flow boiling conditions and, at p = 1.0 MPa, also into the saturated flow boiling regime.

In the Fig. 5b similar results for the heat flux equal to 27.5 kW/m^2 have been plotted. In the top graph of Fig. 5b a very low heat transfer coefficient has been obtained in the case of pure R-114 at 2.5 MPa pressure. This low value is connected to the presence of thermal crisis on the wall at the exit of the tube; thus, the post-dryout (PD) regime of heat transfer has been reached. In fact, further increasing of the heat flux causes the onset of the post-dryout region also in the binary mixture at two different pressure: 2.0 and 2.5 MPa, as shown in the top graph of Fig. 5c, where the experimental data for the highest

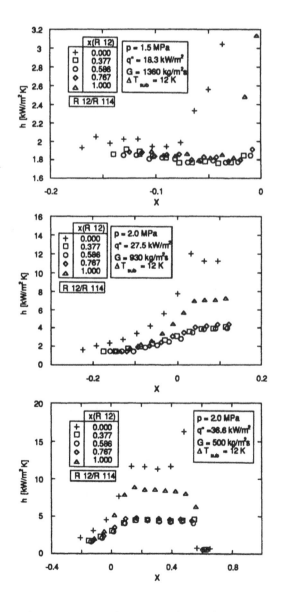

Fig. 6 Typical trends of heat transfer coefficient as a function of local quality at various compositions, for R 12/R 114 mixtures (ENEA)

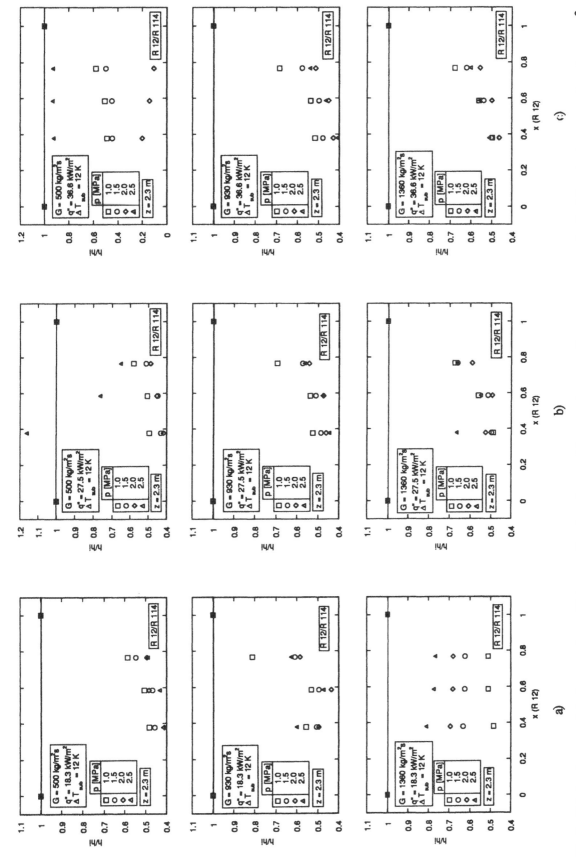

Fig. 7 Relative decrease of the heat transfer coefficient as a function of composition at various mass fluxes and pressures, for R 12/R 114 mixtures: a) q" = 18.3 kW/m²; b) q" = 27.5 kW/m²; c) q" = 36.6 kW/m² (ENEA)

255

heat flux (36.6 kW/m^2) are plotted.

Thus, the choice of the same test conditions, i.e., same pressure, inlet subcooling, mass and heat flux, does not ensure the same heat transfer regime between the pure fluids and the binary mixture. This aspect is verified in the graphs of Fig. 6, where the local heat transfer coefficient versus the calculated local quality (thermodynamic equilibrium quality) along the test section has been plotted, for three typical test conditions. Looking at the top graph of Fig. 6 (which corresponds to the same tests as the bottom graph of Fig. 5a), the heat transfer coefficient of pure R-12 and R-114 reaches higher values than the binary mixture at different compositions. In fact, the mixture, independently of the composition, continues to show a dominant single-phase heat transfer regime, while the two pure fluids behaviour is typically into the subcooled flow boiling.

In the middle graph of Fig. 6, characterized by a higher heat flux and a lower mass flux than the top graph, the mixture heat transfer coefficient at different compositions also shows an increasing value as the quality increases and approaches the zero value (i.e., saturation conditions). This corresponds to the subcooled boiling inception and the establishing of the subcooled flow boiling regime, with a comparable amount of nucleation in the different mixtures and in the pure R-12. Similarly to the case of pure fluids, the saturated flow boiling regime is also exhibited by the mixture. However, the amount of nucleation in the pure R-114 is always much more than either in the R-12/R-114 mixture or in the pure R-12, which corresponds to a higher heat transfer coefficient in both the subcooled boiling region and the saturated boiling regime.

Increasing the heat flux and further reducing the mass flux, the fully developed saturated flow boiling is reached, up to the thermal crisis condition, as shown in the bottom graph of Fig. 6. Such a behaviour is exhibited by both the pure fluids and the mixture at different compositions. As in the previous graphs, the pure R-114 shows a higher heat transfer coefficient. Both the pure fluids have a better performance than the mixture in the two-phase flow region, while the pure R-12 tends to have a trend similar to that of the mixture in the single-phase and in the subcooled flow boiling region.

Under the above premises, we have now to evaluate the heat transfer degradation in the mixture, in terms of the heat transfer coefficient, with respect to the ideal heat transfer coefficient, h_i, defined using the linear mole fraction method, as:

$$h_i = \ x\,h_1 + (1 - x)\,h_2$$

The ratio between the actual and the ideal heat transfer coefficient, h/h_i, is plotted in Fig. 7 versus the more volatile component mole fraction, $x(R-12)$, for three different heat and mass fluxes, with the pressure as a parameter, and the inlet subcooling being constant and equal to 12 K. In the two-phase flow region the heat transfer degradation shown by the mixture with respect to ideality turns out to be dependent on the saturation pressure, increasing as the saturation pressure increases. A typical trend of the h/h_i ratio in the boiling region is plotted in the top graph of Fig. 7a. On the contrary, when the thermal hydraulic conditions of the mixture are such

to lie in the single-phase region, while both pure fluids are characterized by the two-phase flow condition, an apparently opposite trend is observed, as reported in the bottom graph of Fig. 7a. In this case, the heat transfer degradation seems to be almost independent of the mixture composition, and the pressure dependence is reversed (i.e., heat transfer degradation increases as the pressure decreases).

A different trend is observed when one of the two pure fluids, for the same thermal hydraulic conditions, is characterized by a different heat transfer regime with respect to the other pure fluid, as plotted in the top graph of Fig. 7b and Fig. 7c. In particular, looking at the top graph of Fig. 7b, and keeping in mind the top graph of Fig. 5b (same tests), one can notice that the h/h_i ratio shows a value over the unity for a mole fraction of the more volatile component, $x(R-12)$, up to 0.5. This is due to the very low heat transfer coefficient of the pure R-114 (Fig. 5b) typical of the thermal crisis condition reached by the R-114 under the tested conditions. Furthermore, the top graph of Fig. 7c (which corresponds to the heat transfer coefficient plotted in the top graph of Fig. 5c) shows a very low degradation of the heat transfer in the mixture for the p = 2.5 MPa data and for all compositions. Such data are typically in the post-dryout heat transfer region for both the mixture and pure fluids. This behaviour could be expected, as evidenced by Celata et al. (1994), who found a relatively scarce heat transfer degradation in the mixture at the critical heat flux inception.

On the contrary, a consistent heat transfer degradation is observed in the mixture at p = 2.0 MPa, but this is due to the pure R-12 heat transfer coefficient which has a value typical of the annular flow regime, as clearly plotted in the top graph of Fig. 5c, while the mixture exhibits post-dryout heat transfer conditions.

Expected trends are those observed at p = 1.0 and 1.5 MPa, other conditions being equal (i.e., top graph of Fig. 7c), and for increasing mass flux (i.e., middle and bottom graphs of Fig. 7c).

CONCLUDING REMARKS

In conclusion, the experimental results confirmed the general heat transfer degradation of a binary mixture in the boiling region with respect to the ideal behaviour (linear interpolation, weighted with the mole fraction, of the pure fluids heat transfer coefficients).

The analysis of the experimental data of a R-12/R-114 binary mixture in upflow forced convective boiling has shown that the degradation of the binary mixture boiling heat transfer appears to be dependent on saturation pressure (increasing as the pressure increases), heat flux (increasing as heat flux increases) and mass flux (decreasing as mass flux increases).

The comparison of the heat transfer coefficient with the ideal one must be carefully made under thermal hydraulic conditions such as to determine different heat transfer regimes in the pure fluids and in the mixture. Therefore, the choice of constant test conditions does not necessary ensure the same flow pattern and an identical heat transfer regime.

The ideal heat transfer coefficient definition does not hold any longer when describing the heat transfer performance of a mixture under thermal hydraulic conditions that give raise to a non-homogeneous flow pattern with respect to pure components.

ACKNOWLEDGEMENTS

Authors wished to thank Mr. G. Farina for his contribution in the execution of the experiments. His sudden and unexpected death is remembered here, and the present paper is dedicated to his memory. Thanks are also due to Mrs. A. M. Moroni for the editing of the paper.

NOMENCLATURE

G specific mass flow rate, [kg/m^{2}s]
h heat transfer coefficient, [kW/m^{2}K]
p pressure, [MPa]
PD post-dryout
q" heat flux [kW/m^2]
PD post-dryout
Sat saturated flow boiling
SP single phase flow
Sub subcooled flow boiling
T temperature, [K]
TP two-phase
X quality, [-]
x liquid mole fraction of the more volatile component, [-]

Subscripts

1 pertains to the more volatile component
2 pertains to the less volatile component
bulk pertains to the bulk
i ideal
in inlet
sub subcooled conditions
w wall

REFERENCES

Bennett D.L. and Chen J.C., 1980, "Forced Convective Boiling in Vertical Tubes for Saturated Pure Components and Binary Mixtures", AIChE Journal, Vol. 26, N. 3, pp. 454-461.

Celata G.P., Cumo M. and Setaro T., 1993, "Forced Convective Boiling in Binary Mixtures", Int. J. Heat Mass Transfer, Vol. 36, N. 13, pp. 3299-3309.

Celata G.P., Cumo M. and Setaro T., 1994, "Critical Heat Flux in Upflow Convective Boiling of Refrigerant Binary Mixtures", Int. J. Heat Mass Transfer, Vol. 37, N. 7, pp. 1143-1153.

Fink J., Gaddis E.S. and Vogelpohl A., 1982, "Forced Convection Boiling of a Mixture of Freon-11 and Freon-113 Flowing Normal to a Cylinder", Proc. 7th Int. Heat Transfer Conference, Paper FB5, Vol. 4, pp. 207-212.

Jung D.S., McLinden M., Radermacher R. and Didion D.A., 1989, "Horizontal Flow Boiling Heat Transfer Experiments With a Mixture of R22/R114", Int. J. Heat Mass Transfer, Vol. 32, N. 1, pp. 131-145.

Kawano S., 1987, "Experimental Study on Heat Transfer During Evaporation and Condensation of Binary Mixtures Flowing Axially Along the Outside of Tubes", N.H. Afgan and J. Bougard Eds. - Heat and Mass Transfer in Refrigeration and Cryogenics, Hemisphere Publishing Corp., pp. 231-242.

Kedzierski M.A. and Didion D.A., 1990, "Visualization of Nucleate Flow Boiling for an R22/R114 Mixture and its Components, Experimental Heat Transfer", Vol. 3, pp. 447-463.

Mishra M.P., 1981, Varma H.K. and Sharma C.P., "Heat Transfer Coefficients in Forced Convection Evaporation of Refrigerants Mixtures", Lett. in Heat and Mass Transfer, Vol. 8, pp. 127-136.

Morrison G. and McLinden M., 1986, "Application of a Hard Sphere Equation of State to Refrigerants and Refrigerants Mixtures", NBS Technical Note 1226, NBS, Gaithersburg, Maryland.

Shock R.A.W., 1982, "Boiling in Multicomponent Fluids, Multiphase Science and Technology", Ed. G.F. Hewitt, J.M. Delhaye and N. Zuber, Vol. 1, pp. 281-386.

Sivagnanam P. and Varma Y.B.G., 1990, "Subcooled Boiling of Binary Mixtures under Conditions of Forced Convection", Experimental Thermal and Fluid Science, Vol. 3, pp. 515-522.

Steiner D., 1991, "Wärmeübertragung beim sieden gesättigter Flussigkeiten", VDI - Wärmeatlas, Hbb, pp. 1-35.

Stephan K., 1982, "Heat Transfer in Boiling Mixtures", Proc. 7th Int. Heat Transfer Conference, Paper RK14, pp. 59-81.

Thome J.R., 1994, "Two-Phase Heat Transfer to New Refrigerants", Proc. 10th Int. Heat Transfer Conference, Paper SK2, pp. 19-41.

CONVECTIVE FLOW BOILING OF BINARY MIXTURES IN A VERTICAL TUBE

Yasunobu Fujita and MasayukiTsutsui
Department of Energy and Mechanical Engineering
Kyushu University
Hakozaki, Fukuoka 812, JAPAN

ABSTRACT

Boiling heat transfer to binary mixtures of HFC-134 a(more volatile) and HCFC-123(less volatile) flowing in a uniformly heated vertical tube, 10mm i.d., and 2000mm heated length, was experimentally investigated to obtain new experimental data required for discussion about mixture effects on nucleate boiling and two-phase convection heat transfer.

In the low quality region where nucleate boiling is likely dominant, the mixtures gave lower heat transfer coefficients than either those of an equivalent single component of the same physical properties, or those predicted from a linear interpolation of heat transfer coefficients of the two pure components. This observed mixture effect is as usual in pool boiling.

In the two-phase convection region at higher quality where nucleation is supposed to be fully suppressed, heat transfer coefficients of the mixtures were neither noticeably reduced nor dependent on heat flux.

INTRODUCTION

Heat transfer of mixtures has been investigated rather extensively in pool boiling, whereas the reports on flow boiling are very limited, especially in vertical tubes. As reviewed by Thome(1994) for flow boiling tests of refrigerants mixtures conducted in the past few years, most experiments were devoted to horizontal tubes.

Toral et al. (1982) measured boiling heat transfer coefficient of ethanol- cyclohexane in a low quality region of vertical upward flow. They found that the hysteresis of boiling curves during an increase and a decrease in heat flux was notable for binary mixtures, and the boiling curves for mixtures had a gentler slope than single components. Hihara et al. (1987) experimented on horizontal flow of R12-R22 and R114-R22 mixtures. For R12-R22 mixture with small temperature difference between the dew and bubble points at constant composition, heat transfer deterioration peculiar to mixtures was not observed. On the contrary, for R114-R22 mixture having large temperature difference, significant deterioration occurred especially at low quality. Composition dependence of heat transfer coefficient was obscure, but their data were well correlated in terms of Martinelli parameter and boiling number. Singal et al. (1974) studied the effects of mass velocity, heat flux and quality on boiling heat transfer of R13- R12 mixture flowing in a horizontal tube, and found that these three parameters affect heat transfer similarly as in the case of single components. Heat transfer coefficient of mixtures was less than that of R12 for quality less than 0.3, and vice versa.

Nakanishi et al. (1987) experimented with boiling of R11-R113 mixture at high quality in vertical upward flow. In their experiment deterioration of heat transfer was observed also at high quality. Since nucleation was considerably suppressed in an annular liquid film at higher quality, the observed deterioration could not be attributable to the mass transfer resistance in forming bubbles.

Recently, using a horizontal tube motivated for heat pump applications, Ross et al. (1987) performed tests with R152a-R13B1 mixture, and Jung et al. (1989) and Takamatsu et al. (1990) made experimental researches for R 22-R114 mixture. The results obtained verify sharply lower heat transfer coefficients than either pure component. Celata et al. (1992) started the investigation on heat transfer performance of R12-R114 mixture in upflow forced convective boiling in a vertical tube. They compared the data with available correlations and confirmed the good performance of the Bennett-Chen correlation (1980) to predict heat transfer coefficient of mixtures. Ostrovski (1987) experimented four kinds of immiscible mixtures in a vertical tube arranged in a natural circulation loop. Heat transfer coefficients were the same as those for the constituent more volatile components, except one mixture which forms an emulsion.

Some theoretical studies or correlating methods of data have been presented so far. Calus et al. (1973) correlated their data of both non-azeotropic and azeotropic mixtures by using one of available correlations developed for flow boiling of single components after two modifications. One modification was the adoption of effective temperature difference between the wall and bubble interface instead of the conventional wall superheat. The other was the use of surface tension evaluated at the interfacial composition. Shock (1976) solved heat and mass balance equations for a vertical upward annular flow of mixtures, and concluded that mass transfer resistance in the vapor core might have an important role on heat transfer rather than the resistance in a liquid film. Bennet and Chen (1980) applied the Chen correlation (1966) developed for single components to mixtures by modifying the suppression factor and the temperature driving force respectively included in the boiling and convection terms in the original correlation.

This report is the first stage to investigate mixture effect on flow boiling of mixtures in a vertical tube.

EXPERIMENTAL APPARATUS AND PROCEDURE

Experimental Apparatus

Figure 1 shows a schematic view of the test loop and the vertical test section used in this study. The test section composed of a 4m long and 10mm I.D. stainless steel tube with a wall thickness of 1.5mm, the upper 2m of which serves as a heat transfer section uniformly heated by passing AC through it. The lower 1m section heated by DC works as a preheated section. Both sections are smoothly joined across a 0.5 m unheated section by flanges with Teflon sheet used for electric isolation.. The test section was wrapped with fiber glass insulation 30cm in diameter to reduce heat loss to the surroundings.

The outside temperature of the heat transfer tube was measured at twenty one stations. Chromel-Alumel thermocouples of 0.3 mm diameter were spot-welded and electrically insulated with mica sheets inserted between the tube outer wall and the exposed thermocouple wires.Thermocouples are at 50mm intervals for the first and last two stations from both ends and at 100mm intervals for the remaining central seventeen stations. At three stations, the second from both ends and the center, temperatures were measured on the opposite sides across the tube diameter to verify their equivalence.

The inner wall temperatures, T_w , were calculated from the measured outer tube wall temperatures by the one-dimensional heat conduction equation under compensation

for the heat loss to the surroundings.

Heat flux was determined from uniform heat generation by reducing the calibrated radial heat loss from the outside of the tube.

Binary mixtures of HFC-134a and HCFC-123 were selected as the test fluid because their boiling pressure and temperature are at moderate levels for experimental convenience and the temperature difference between the dew and bubble points, i.e. the flash temperature difference, is rather large among other potential mixtures.

Figure 2 shows the phase equilibrium diagram for the test mixture at an experimental pressure of 0.6 MPa.On the top graph the vapor/liquid mole fraction difference for the more volatile component, $|X_1-Y_1|$, and the flash temperature difference, ΔT_{bp} , are shown.

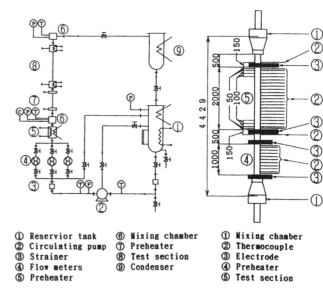

① Reservior tank	⑥ Mixing chamber
② Circulating pump	⑦ Preheater
③ Strainer	⑧ Test section
④ Flow meters	⑨ Condenser
⑤ Preheater	

① Mixing chamber
② Thermocouple
③ Electrode
④ Preheater
⑤ Test section

Figure. 1 Test Loop and Vertical Test Section

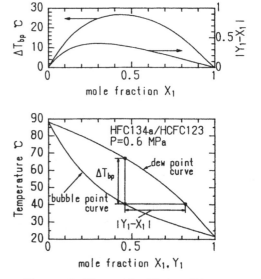

Figure. 2 Phase Equilibrium Diagram

Table 1. Experimental Conditions

pressure at inlet	: 0.6	MPa
mass velocity	: 300	kg/m²s
heat flux	: $10^3 \sim 5 \times 10^4$	W/m²
quality at inlet	: -0.05 ~ 0.0	
mole fraction at inlet	: 0.0, 0.25, 0.5, 0.75, 1.0	

The range of experimental conditions covered in this study is summarized in Table 1.

Definition of Heat Transfer Coefficient

The local heat transfer coefficient at the stations of wall temperature measurement on the heat transfer section is defined by

$$\alpha = \frac{q}{T_w - T_b} \qquad (1)$$

For single components, the thermodynamic equilibrium temperature is used as T_b in Eq.(1).

For mixtures, the thermodynamic equilibrium temperature, T_b, the vapor quality, β, and the mole fraction of the more volatile component, X_1 and Y_1, in the liquid and vapor phases respectively, are determined under an assumption of the phase and thermodynamic equilibria, so as to satisfy the following relations.

$$h = h_{in} + \frac{4qz}{Gd} \qquad (2)$$

$$h = h_l(1 - \beta) + h_v \beta \qquad (3)$$

$$h_l = h_l(P, T_b, X_1) \qquad (4)$$

$$h_v = h_v(P, T_b, Y_1) \qquad (5)$$

$$\beta = \frac{X_{1,in} - X_1}{Y_1 - X_1} \qquad (6)$$

The local pressure needed in solving the above equations is interpolated from the measured pressures at the inlet and outlet mixing chambers assuming homogeneous flow through the tube. Since an overall composition at the inlet of the heat transfer tube, $X_{1,in}$ is known and pressure, P, and specific enthalpy, h, are specified at each station, the remaining four unknown quantities, T_b, X_1, Y_1 and β are uniquely determined from Eqs.(3) to (6) by an iteration method.

Thermodynamic Properties

The BWR equation of state modified by Starling et al. (1972), and by Nishiumi and Saito (1975) was used to predict thermodynamic properties, such as density, enthalpy, composition, and dew and bubble point temperatures, for mixture and pure components.

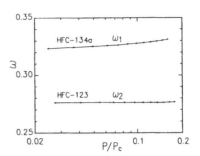

Figure. 3 Acentric Factor

The interaction parameter between HFC-134a and HCFC-123 included in the BWR equation is estimated as 0.991 from the Hudson and McCoubey equation but the value of 0.985 was used. The acentric factor was evaluated as $\omega_1 = 0.3252$ for HFC-134a and $\omega_2 = 0.2815$ for HCFC-123 from the Pitzer equation. Vapor pressure curves of pure components predicted using these values show a slight deviation from the tabulated pressure curves compiled by JAR (1990). To make them coincide, the acentric factors were revised as a function of the reduced pressure as shown in Fig. 3.

EXPERIMENTAL RESULTS AND DISCUSSION

Axial Variation of Heat Transfer Coefficient

Figure 4 shows typical axial variations of the tube wall temperature, the bulk fluid temperature and the heat transfer coefficient for single component and mixture. In the two cases boiling was initiated at the middle of the heat transfer section. Note that the bulk temperature of mixture slightly increases in the flow direction even after it has reached the saturation temperature at $\beta=0$. From the boiling initiation point downstream, heat transfer coefficient begins to rise from a nearly constant level of single phase flow and approaches a higher constant level of nucleate boiling.

Heat Transfer Coefficient in Quality Region

Measured heat transfer coefficients, α, are plotted in a conventional non-dimensional form in Fig. 5, where α_{lo} denotes the heat transfer coefficient calculated from the Dittus-Boelter equation for a total flow assumed liquid, and χ_{tt} denotes the Lockhart-Martinelli parameter.

$$\alpha_{lo} = 0.023 \left(\frac{\lambda_l}{d}\right) \left(\frac{Gd}{\mu_l}\right)^{0.8} \left(\frac{c_{pl} \mu_l}{\lambda_l}\right)^{0.4} \qquad (7)$$

$$\chi_{tt} = \left(\frac{1 - \beta}{\beta}\right)^{0.9} \left(\frac{\rho_v}{\rho_l}\right)^{0.5} \left(\frac{\mu_l}{\mu_v}\right)^{0.1} \qquad (8)$$

Here, mixture properties were evaluated at the equilibrium compositions of liquid and vapor. As known in flow boiling of single components, heat transfer coefficient becomes higher with an increase both in heat flux and quality, although heat flux effect is dominant. When heat flux is low, thus at low Boiling number, α starts to rise at lower quality from a nearly constant level and increases gradually with quality and merges into a single curve irrespective of heat flux. When heat flux is high, on the contrary, α remains nearly constant up to higher quality. These features are commonly observed in single

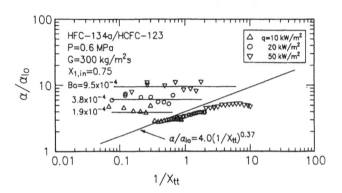

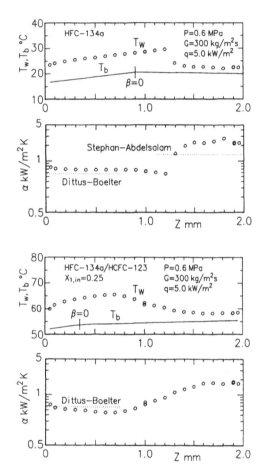

Figure. 4   Axial Variations of Tube Wall and Fluid Bulk Temperatures and Heat Transfer Coefficient

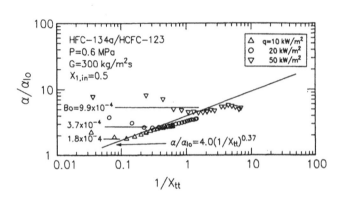

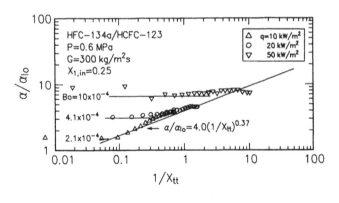

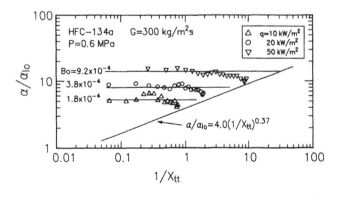

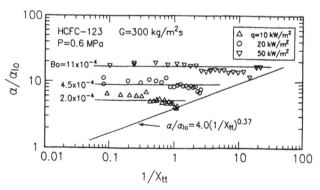

Figure. 5   Non-dimensional Heat Transfer Coefficient

components and mixtures, although heat transfer coefficients compared at the same value of Boiling number are lower for mixtures than for single components at low quality in particular. This means deterioration of nucleate boiling heat transfer in mixtures.

Comparisons with Correlations for Single Components

There have been proposed a lot of correlations for flow boiling of single component fluids by taking into account both contribution of nucleate boiling and convective evaporation. Most of them are classified into the Chen type and the Schrock-Grossman type. The present data for single components were compared with those correlations proposed by Gungor and Winterton (1986), Kandlikar (1990), Pujol and Stenning (1968), Shah (1982), Stephan and Auracher (1981) and Wadekar (1990). Among them the Stephan and Auracher correlation predicted most data well within ± 25 % error. However it was found their correlation overpredicted the present heat transfer coefficients for mixtures greatly as exemplified in Figure 6. As clearly seen in Fig. 5, nucleate boiling is predominant over two phase forced convection in the present experimental range. Thus the overprediction is attributable to deterioration of nucleate boiling in mixtures, as is the case in pool boiling.

Heat Transfer Deterioration of Mixture

Figure 7 shows the effect of mixture composition on heat transfer coefficient for three different heat fluxes in the low quality region where boiling contribution is judged predominant. In this figure, X_1 refers local mole fraction in the liquid phase, which becomes a little smaller than $X_{1,in}$ at the inlet because of evaporation. As observed in pool boiling, heat transfer coefficients of mixtures are largely reduced below the linearly interpolated coefficients between their single components. And this reduction is an increasing function of heat flux. The predicted results from the

correlations developed for pool boiling of binary mixtures by Stephan-Korner(1969) and by Schlunder(1982) are shown at a heat flux of 50 kW/m² for comparison in Fig.7. Measured heat transfer coefficients fall between two predictions.

CONCLUSION

Experimental data for boiling heat transfer of HFC-134a and HCFC-123 mixture flowing in a uniformly heated vertical tube were presented. For the range of parameters covered in the present experiments, it was revealed that there is no drastic change in overall variation of heat transfer coefficient along the heated tube. Absolute value of heat transfer coefficient for mixture, especially at low quality where nucleate boiling is considered predominant over convective evaporation, is largely reduced compared with the interpolated value between their pure components, and this reduction or heat transfer coefficient in this region is unpredictable from existing correlations developed for flow boiling of single components.

Further research is needed and under way to extend the experimental conditions and to elucidate the mixture effects on flow boiling heat transfer.

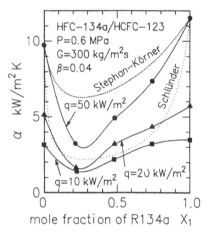

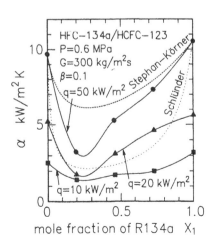

Figure. 7 Heat transfer Deterioration of Mixture

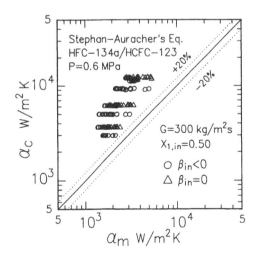

Figure. 6 Comparison of Mixture Data with Prediction
from Correlation for Single Component

NOMENCLATURE

B_o Boiling number , q/Gh_{fg}
d tube inner diameter
h specific enthalpy
h_{fg} latent heat of evaporation
P pressure
T temperature
T_w wall temperature
T_b bulk fluid temperature
X mole fraction in liquid phase
Y mole fraction in vapor phase
z axial distance from the inlet

Greek symbols

α heat transfer coefficient
β vapor quality
μ viscosity
λ thermal conductivity
χ_{tt} Lockhart-Martinelli parameter

Subscripts

in   inlet of the heat transfer section
l   liquid
v   vapor
1   more volatile component
2   less volatile component

REFERENCES

Bennet, D. L., and Chen, J. C., 1980, "Forced Convective Boiling in Vertical Tubes for Saturated Pure Components and Binary Mixtures," AIChE J.,Vol. 26, No. 3, pp. 451-461.

Calus, W. F., Montegnacco, A., and Denning, R. K., 1973, "Heat Transfer in a Natural Circulation Single Tube Reboiler , Part 2: Binary Liquid Mixtures," Chem. Engng. J., Vol. 6, No. 3, pp.251-264.

Celata, G. P., Cumo, M., and Setaro, T., 1993, "ForcedConvective Boiling in Binary Mixtures," Int. J. Heat Mass Transfer,Vol. 36, pp.3299-3309.

Chen, J. C., 1966, "Correlation for Boiling Heat Transfer to Saturated Liquids in Convective Flow," Ind. Eng. Chem. Process Design Develop., Vol. 5, No. 3, pp.3 22-329.

Gungor, K. E., and Winterton, R. H. S., 1986, "A General Correlation for Flow Boiling in Tubes and Annuli," Int. J. Heat Mass Transfer, Vol. 29, pp.351-358.

Hihara, E., Tanida, K., and Saito, T., 1987, "Forced Convective Boiling Experiments of Binary Mixtures," Proc. 2nd JSME-ASME Thermal Engng. Joint Conf.,Vol. 5, pp.119-126.

Japanese Aso. Refrigeration, 1990, "Thermophysical Properties of Environmentally Acceptable Fluorocarbons-HFC 134a and HCFC 123,"Japanese Aso. Refrigeration

Jung, D. S., McLinden, M., Radermacher, R., and Didion, D. A., 1989, "Horizontal Flow Boiling Heat Transfer Experiments with a Mixture of R22/R114," Int. J.

Heat Mass Transfer,Vol. 32, pp.131-145.

Kandlikar, S. G., 1990, "A General Correlation for Saturated Two-Phase Flow Boiling Heat Transfer Inside Horizontal and Vertical Tubes," J. Heat Transfer,Vol. 112, pp.219-228.

Nakanishi, S., Kaji, M., Matoba, H., and Kaji, N., 19 87, "Flow Boiling in Tube of Mixtures of Refrigerants R11 and R113," Trans. JSME,Vol. 53, No. 486, pp.2626-2632.

Nishiumi, H., and Saito, S., 1977, "Correlation of the Binary Interaction Parameter of the Modified Generalized BWR Equation of State," J. Chem. Eng. Japan,Vol. 10, No. 3, pp.176-180.

Ostrovski, N. YU., 1987, "Boiling of Immiscible liquids in a Natural-Circulation Loop," Heat Transfer, --Soviet Res.,Vol. 19, No. 4, pp.1-7.

Pujol, L., and Stenning, A. H., 1968, "Effect of Flow Direction on the Boiling Heat Transfer Coefficient in Vertical Tubes," Symp. Ser. Can. Soc. Chem. Eng.,No. 1, pp.401-453.

Ross, H., Radermacher, M. di Marzo, and Didion, D. A., 1987, "Horizontal Flow Boiling of Pure and Mixed Refrigerants," Int. J. Heat Mass Transfer,Vol. 30, pp.979-9 92.

Schlunder,E.M.,1982, "Uber den Warmeubergang bei der Blasenverdampfung von Gemischen," Verfahrenstechnik, Vol.16,No.9, pp.692-698.

Shah, M. M., 1982, "Chart Correlation for Saturated Boiling Heat Transfer : Equations and Further Study," ASHRAE Trans.,Vol. 88, Pt.1, pp.185-196.

Shock, R. A. W., 1976, "Evaporation of Binary Mixtures in Upward Annular Flow," Int. J. Multiphase Flow,Vol. 2, No. 4, pp.411-433.

Singal, L. C., Sharma, C. P., and Varma, H. K., 197 4, "Experimental Heat Transfer Coefficient for Binary Refrigerant Mixtures of R13 and R12," ASHRAE Trans., Vol. 80, pp.175-188.

Starling, K. E., and Han, M. S., 1972, Hydrocarbon Processing, Vol. 51, No. 5, p.129

Stephan, K., and Auracher, H., 1981, "Correlations for Nucleate Boiling Heat Transfer in Forced Convection,"Int. J. Heat Mass Transfer, Vol. 24, pp.99-107.

Stephan,K., and Korner, M., 1969, Berechnung des Warmeubergangs verdampfender binarer Flussigkeitsgemische, Chemie Ing. Techn.,Vol.41,No.7, pp.409-417.

Takamatsu, H., Miyara, A., Koyama, S., Fujii, T., and Yonemoto, K., 1990, "Forced Convective Boiling of Nonazeotropic Refrigerant Mixtures of R22 and R114 inside a horizontal tube,"Heat Transfer – Jap. Res. Vol. 19, No. 3, pp.68-82.

Thome,J.R., 1994, "Two-Phase Heat Transfer to New Refrigerants," Proc. 10th Int. Heat Transfer Conf., Vol.1,pp.19-41.

Toral, H., Kenning, D. B. R., and Shock, R. A. W., 19 82, "Flow Boiling of Ethanol/Cyclohexane Mixture,"Proc. 7th Int. Heat Transfer Conf.,Vol. 4, pp.255-260.

Wadekar, V. V., 1990, "Flow Boiling – A Simple Correlation for Convective Heat Transfer Component," Proc. 9th Int. Heat Transfer Conf., Vol. 2, pp.87-91.

HEAT TRANSFER
FROM A HORIZONTAL TUBE TO BOILING BINARY MIXTURES
WITH SUPERIMPOSED CONVECTIVE FLOW

Dieter Gorenflo, Martin Buschmeier and Paul Kaupmann

Laboratorium für Wärme- und Kältetechnik
Universität (GH) Paderborn
Paderborn, Germany

ABSTRACT

Pool boiling heat transfer at a single horizontal tube with large diameter (88.4mm) and superimposed convective flow of bubbles and liquid from below has been investigated experimentally simulating the convective effects caused by bubbles and liquid streaming upwards within a shell-and-tube evaporator. The boiling liquids are Propane, n-Butane, i-Butane and binary mixtures of these.

The additional convection of the vapour bubbles significantly increases the heat transfer coefficients at low heat fluxes, above all for the mixtures. The well known deterioration of heat transfer with mixtures below the molar average of the heat transfer coefficients of the pure components may entirely be compensated at low heat fluxes, where heat transfer is dominated by convective effects.

The data for this domain can be represented by a calculation method taken from the literature and having been modified. Local measurements of the circumferential variation of the wall superheat and calculations of the local heat transfer coefficients show, however, that some of the assumptions of the calculation should be revised.

INTRODUCTION

At single horizontal evaporator tubes the heat transfer coefficients at nucleate boiling of binary mixtures are smaller than the molar average of the heat transfer coefficients for the pure components due to the preferential evaporation of the lower boiling component and the subsequent enrichment of the higher boiling component at the heated wall. Within shell-and-tube evaporators, the additional convective flow induced by bubbles and liquid streaming upwards from the lower tubes of the tube bundle will enhance the heat transfer in comparison with single tubes, especially at small heat fluxes and low pressures, as is well known for pure substances. For binary mixtures, the enhancement will be even more pronounced and extended to higher heat fluxes because of the deterioration of heat transfer at nucleate boiling compared to pure liquids, without additional convection.

The effect of the additional convection within tube bundles has been simulated by bubble production at a heater below a horizontal test tube with big diameter (88.4mm). The big tube was chosen in order to get a pronounced effect of the superimposed convective flow along the tube surface. Former measurements with Propane/n-Butane (Buschmeier et al., 1994) have been completed for mixtures with approx. 70 mole % Propane at higher saturation pressures, see Table 1.

In the same range of normalized saturation pressures ($0.1 \leq p^* = (p_s/p_c) \leq 0.5$), new experiments have been performed with the binary systems Propane/i-Butane and i-Butane/n-Butane. The results within the region of small heat fluxes, dominated by convective effects, are the same for these systems as for Propane/n-Butane, within the experimental limits of error.

Table 1: Compositions, Critical Pressures p_c and Saturated States of the Binary System Propane/n-Butane

	Propane C_3H_8[1]	n-Butane n-C_4H_{10}[1]	Mixtures / mole % 25	50	70
$p^* = 0.1$					
x/mole%	100	0	22.1[2]	45.2	70.6
p_c/bar	42.64	37.96	40.2	42.3	42.9
p_s/bar	4.36	3.79	4.0	4.2	4.3
T_s/°C	-3.02	39.98	24.4	13.4	6.1
$p^* = 0.2$					
x/mole%	100	0	23.6	45.8	70.8
p_c/bar	42.64	37.96	40.3	42.4	42.9
p_s/bar	8.48	7.59	8.1	8.5	8.6
T_s/°C	20.46	67.27	51.3	41.9	31.8
$p^* = 0.5$					
x/mole%	100	0	27.3	48.7	71.9
p_c/bar	42.64	37.96	40.7	42.5	42.9
p_s/bar	21.21	18.90	20.35	21.2	21.45
T_s/°C	60.05	111.22	96.15	84.4	73.25

1) Impurities < 0.5 vol%; 2) x data taken at q = 20000 W/m^2

This paper being focussed on the domain of small fluxes, the Propane/n-Butane data will be used for the discussion throughout the paper, and the other binaries merely enter the final comparison between measurements and calculations.

EXPERIMENTAL RESULTS

The measurements were performed in a modified version of the Standard Apparatus for pool boiling heat transfer developed by Gorenflo et al. (1982), the main features of which have been outlined by Gorenflo (1993) and Buschmeier et al. (1994), with further details given by Sokol (1994). Cross sections of the test tube (88.4mm O.D.) are shown in Fig. 1. It consists of an inner copper tube (with stainless steel barriers at mid height) and two mild steel tubes, all of them combined by soft solder on their entire lengths (heated section: 192.4mm).

An additional (U-shaped) heater below the test tube simulates bubbles produced by a tube at a distance of roughly 2D, cf. the sketch in Fig. 2. The reference junctions of 26 thermocouples to monitor the wall temperature (circumferentially distributed at half the tube length) are placed about 40mm below the test tube within the upstreaming liquid (and bubbles, if the additional heater is active). Therefore, the heat transfer coefficient α defined by the ratio of the heat flux q and the superheat ΔT of the tube surface,

$$\alpha = q / \Delta T \qquad (1),$$

is always calculated with the temperature difference actually existing between the heated wall and the rising fluid.

The experimental uncertainties of the primary measurements are (cf. Sokol, 1994):

ΔT: ± 0.05 K; q: $\pm 0.2\%$; p: $< \pm 1\%$; x: $< \pm 0.5\%$;

α: $< \pm 5\%$ at $p^* = 0.1$; α: $< \pm 8\%$ at $p^* = 0.5$.

In Fig. 2, the double logarithmic plot of the heat transfer coefficient α_2 over the heat flux q_2 at the upper tube (No 2)

demonstrates the effect of the additional flow of bubbles on α_2 at different constant heat fluxes q_1 at the lower tube (No 1); the new results for the mixture with 70/30 mole % Propane/n-Butane at 20 per cent of the critical pressure have been taken as an example. Fig. 3 shows the new results for $q_1 = 10.7$ kW/m² together with former data (Gorenflo, 1993; Buschmeier et al., 1994) for the same heat flux q_1 and other compositions, as well as the pure components. It is seen that at small heat fluxes q_2 at the test tube (= upper tube 2), the α_2-values with additional flow of bubbles

- are significantly higher than without,
- are nearly independent of the heat flux q_2 and increase with q_1 in an asymptotic manner (Fig. 2), and
- are only sligthly smaller for the mixtures than for the pure substances (Fig. 3).

At fully developed nucleate boiling, however, the additional bubbles are no longer favourable to heat transfer, and the deterioration of the heat transfer coefficients of mixtures compared to the pure components remains unchanged. Between both domains, a transition zone exists that extends over a wider range of heat fluxes q_2 for the mixtures than for the pure substances (dashed lines in Fig. 3).

The discussion of results being restricted to the convective domain at small heat fluxes q_2 in the following, Figs. 4 and 5 summarize this domain for the binary system Propane/n-Butane and the three investigated normalized pressures $p^* = p_c/p_s$. It is evident from the vanishing differences in the α-values for Propane and n-Butane (cf. Fig. 4) that the binaries Propane/i-Butane and i-Butane/n-Butane will behave in the same way and that there is no need to discuss these separately. The comparison of the diagrams in Fig. 4 reveals

- that the very slight increase of the heat transfer coefficients for the pure components with rising heat flux q_2 at $p^* = 0.1$ gradually augments with increasing pressure, (note the much smaller scale for q_2 than for α_2),

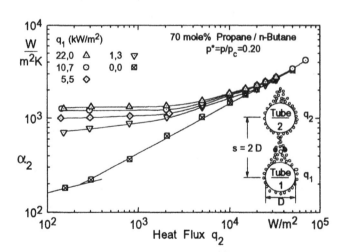

Figure 2: Heat Transfer Coefficient α_2 at the Upper Tube 2 as Function of the Heat Flux q_2 for Boiling Propane/n-Butane Mixtures at Constant Pressure $p^* = 0.20$ with or without Additional Convective Flow Induced by Bubbles at Constant Heat Inputs q_1 at the Lower Tube 1.

Figure 1: Cross Section of the Test Tube with Enlarged Detail, and Position of the Additional Heater (to scale) Producing Bubbles to Simulate Another Tube Positioned below at roughly 2D (cf. Sketch in Fig. 2).

- that this increase is always somewhat more pronounced for the mixtures than for the pure components at the same conditions,
- that the small deterioration of the α_2-values for the mixtures compared to the pure components increases with rising pressure, and
- that the convective domain extends to gradually smaller ranges of the heat flux q_2 with rising pressure at constant heat flux q_1.

The first three effects are due to the bubble formation at the test tube increasing with pressure at constant heat flux, and due to the fact that with mixtures increasingly less bubbles are formed than with pure substances when the pressure rises. The last effect is caused by fully developed nucleation shifting to smaller heat fluxes at higher pressures.

Fig. 5 shows the slight reduction of the heat transfer coefficients at intermediate compositions in linear α_2-scales for constant heat flux q_1 and rising pressure (top), or constant pressure and rising heat flux q_1 (bottom). The vertical bars indicate the variation of the α-values throughout the q_2-range dominated by convection at the three pressures investigated (full interpolating lines in Fig. 4). The *decreasing* α_2-values at rising pressure in the upper diagram demonstrates that the pressure dependence of α_2 within the convective domain is vice versa to fully developed nucleate boiling. Furthermore, it can be seen that heat transfer with 70 mole % Propane

- is the same or slightly better than with the pure components at $p^* = 0.1$, and
- decreases below the level of the pure components with rising normalized pressure.

Both results are consistent with investigations of Bednar (1993) on nucleate boiling of Propane/n-Butane and other

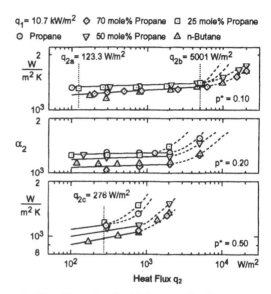

$q_1 = 10.7 \text{ kW/m}^2$ ◇ 70 mole% Propane   ☐ 25 mole% Propane
○ Propane   ▽ 50 mole% Propane   △ n-Butane

Figure 4: Heat Transfer Coefficient α_2 for Pure Components and Propane/n-Butane Mixtures at Constant Heat Input q_1 and Three Different Normalized Pressures p^*.
q_{2a}, q_{2b}: Lower or Upper Limit of Convective Domain at $p^* = 0.1$, referred to in Fig. 6.
q_{2c}: Characteristic Experimental Heat Flux for Convective Domain at $p^* = 0.5$, referred to in Fig. 7.

binaries of hydrocarbons without superimposed convection. He found the same relative dependencies of the density of active nucleation sites from composition and normalized pressure as follows from the diagram on top of Fig. 5 for α_2. This underlines that α_2 at the *upper* tube reflects the bubble formation at the *lower* (cf. also the comments on Fig. 4).

The circumferential variation of the superheat ΔT_{exp} of the tube wall, as measured by the thermocouples at different positions along the circumference (cf. Fig. 1), is given at the top of Fig. 6 for one of the experimental heat fluxes, $q_{2a,exp} = 123.3 \text{ W/m}^2$, at the lower end of the

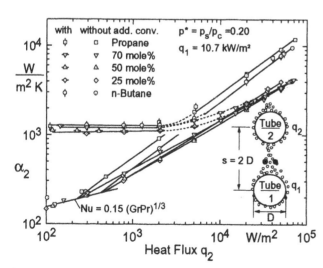

Figure 3: Heat Transfer Coefficient α_2 as a Function of the Heat Flux q_2 for Boiling Propane/n-Butane Mixtures and Pure Components with or without Additional Flow of Bubbles at Constant Pressure $p^* = 0.2$ and Constant Heat Input $q_1 = 10.7$ kW/m² (Symbols without Dashes: $q_1 = 0$).

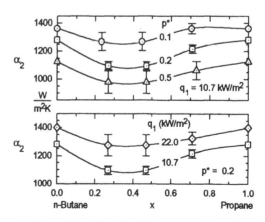

Figure 5: Heat Transfer Coefficient α_2 for the System Propane/n-Butane as Function of the Molar Fraction x of the Liquid at Constant Heat Input q_1 and Varying Pressure (Top) or Constant Pressure and Varying Heat Input q_1 (Bottom).

convective domain at $p^* = 0.1$, and for one at the upper end, $q_{2b,exp} = 5001$ W/m^2 (cf. Fig. 4). Fig. 7 contains the corresponding results for a characteristic experimental heat flux $q_{2c,exp} = 276$ W/m^2 within the convective domain at $p^* = 0.5$ (cf. Fig. 4). In both figures, the results for Propane have been chosen as typical examples for pure substances and mixtures as well. It is seen that the relative variation of the experimental superheat ΔT_{exp} over the circumferential angle φ is quite similar in all three cases and that the ratio $\bar{q}_{2a}/\bar{q}_{2b} = 40.56$ is also found in the pertaining local superheats, almost within the experimental limit of error. (Note that the ratio of the ΔT-scales in Fig. 6 is 41.0). The significant circumferential variation of ΔT implies that heat is not only transferred in *radial* direction within the tube wall, but also from the top to the bottom of the tube. This has been calculated simplifying the real structure of the wall (given in Fig. 1) by a homogeneous layer consisting of mild steel. The resulting circumferential variations of the radial heat fluxes $q_{2,local}$ or of the heat transfer coefficients $\alpha_{2,local} = q_{2,local}/\Delta T_{exp}$ are plotted in the middle or lower diagrams of Figs. 6 and 7, resp.. It can be concluded that

– the heat transfer coefficient *at the bottom* of the tube, resulting from the rising bubbles produced at constant heat flux q_1 at the lower tube, is independent of the heat flux $\bar{q}_2$ and the saturation pressure ($\alpha_{2,local} \approx 1900$ W/m^{2}K),

– the shape of the $\alpha(\varphi)$-curves is almost entirely independent of q_2 at constant pressure $p^* = 0.1$, and it is but slightly dependent on the pressure,

– the heat transfer coefficient at the top of the tube decreases somewhat with increasing pressure, from $\alpha_{2a,b} \approx 900$ to $\alpha_{2c} \approx 770$ W/m^{2}K when the saturation pressure is raised to the fivefold. This will be due to the slower motion of the bubbles because of their much smaller size at $p^* = 0.5$ than at $p^* = 0.1$ and because of the smaller density difference between vapour and liquid.

CALCULATIONS

Fujita et al. (1986) developed a calculation method for the heat transfer coefficient caused by bubbles rising to horizontal evaporator tubes from below. The method based on measurements with the former refrigerant R113 ($C_2F_3Cl_3$) at comparatively low normalized saturation pressures has already been applied successfully to our experiments with Propane and n-Hexane at pressures $p^* \leq 0.12$, cf. e.g. Buschmeier et al. (1994) and Fig. 9.

In the calculation procedure, the total area A of the tube is divided into two parts, A_U and A_F, with convection enhanced by the bubbles, and an upper part A_O for which free convective heat transfer is assumed (left hand side of Fig. 8). According to a proposal of Windisch (1988), this has been adapted to the

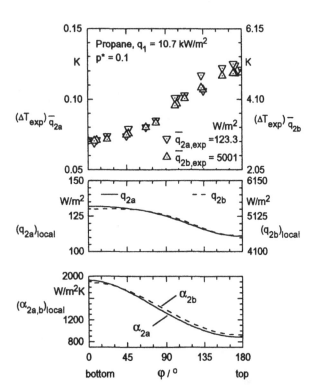

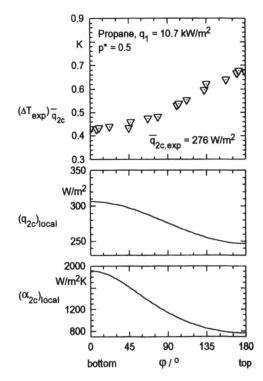

Figure 6: Local Superheat ΔT_{exp} of the Tube Surface (Top), Heat Flux $q_{2,local}$ Calculated for the Radial Direction (Middle), and Heat Transfer Coefficient $\alpha_{2,local}$ (Bottom) as Functions of the Circumferential Angle φ for Pure Propane at two Experimental Heat Fluxes $\bar{q}_{2a}$ or $\bar{q}_{2b}$ at the Lower or Upper End, resp., of the Convective Domain at $p^* = 0.1$.

Figure 7: Local Superheat ΔT_{exp} of the Tube Surface (Top), Heat Flux $q_{2,local}$ Calculated for the Radial Direction (Middle), and Heat Transfer Coefficient $\alpha_{2,local}$ (Bottom) as Functions of the Circumferential Angle φ for Pure Propane at a Characteristic Experimental Heat Fluxe $\bar{q}_{2c}$ within the Convective Domain at $p^* = 0.5$.

greater distance of the tubes our heater is simulating (Fig. 8, on the right):

$$\alpha_{conv} = \frac{A_O}{A}\alpha_{free\ conv} + \frac{A_F}{A}\alpha_{enhanced\ conv} \qquad (2),$$

with $(A_O/A) = 0.1$ and $(A_F/A) = 0.9$. Besides that, the empirical correlation of Fujita for the domain with enhanced convection has been slightly modified replacing the saturation pressure p_s by the normalized pressure p^*,

$$\alpha_{enhanced\ conv} = 7000\ (V/k)^{0.177}\ p^{*-0.18} \qquad (3),$$

with k = (enthalpy of the vapour) / (total heat input to lower tube);
V = vapour volume produced $(m^3/m^2 s)$; α $(W/m^2 K)$.

The comparison of eqs. (2) and (3) with our experimental results for the pure hydrocarbons Propane, n-Butane, i-Butane, n-Hexane at normalized pressures $0.012 \le p^* \le 0.65$ is given in the upper diagram of Fig. 9. As can be seen, the experimental heat transfer coefficients at $p^* = 0.011$ (stars) or $p^* = 0.070$ (rhombs) are represented very well by the modified calculation method (dashed lines # 6 or 5, resp.). For $p^* = 0.1$ and 0.12 (triangles on tips or dashed line # 4, resp.), the representation is fairly good, too (except the triangle at the smallest vapour volume, $(V_2/k) = 2.3 \cdot 10^{-4}$ m/s, in the upper diagram). The calculation procedure extrapolated to higher pressures does not contain, however, the vanishing pressure dependence of the experimental α_2-values at pressures $p^* > 0.10$.

The heat transfer coefficients measured with the mixtures of Propane/n-Butane at $p^* = 0.1;\ 0.2;\ 0.5$ display similar $\alpha_2(p^*,V_2)$-relationships as the pure components, with *smaller* α_2-values for 25 and 50 mole% Propane than for the pure substances, cf. the dot-dashed interpolation line in the lower diagram of Fig. 9, and with the *same* α_2-values for 70 mole% Propane as for the pure substances, see the triangles and the full line (cf. also Figs. 4 and 5.).

Taking into account that the velocity of the bubbles decreases with increasing pressure and that less bubbles are produced with boiling mixtures than with pure substances at the same conditions, the area A_F with enhanced convection should be reduced with increasing p^* and should be smaller for mixtures than for pure substances, as has been done according to Fig. 10. Introducing this into eq. (2), it can be shown that the experimental results are represented with deviations approximately within the experimental limit of error.

The extent of the variation of the area ratios with pressure, and the differences of the ratios between pure substances and mixtures seem to be significantly too high, however, if the small differences in the heat transfer coefficients for pure substances and mixtures at different pressures, following from Figs. 3 to 7 are taken into account. Obviously, these modifications also compensate further inadequate assumptions in the calculation procedure that should be modified, like the pressure dependence of eq. (3) at higher normalized pressures and the circumferential variation of the heat transfer coefficients.

The discrepancy between the $\alpha(\varphi)$-relationships evaluated from the experiments (full curves in Fig. 11) and the assumption of two constant α-values, either enhanced (upper

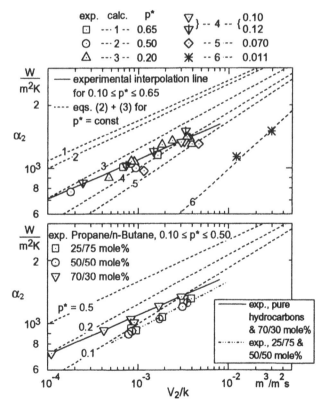

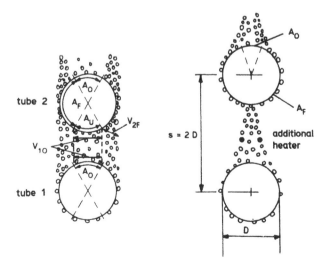

Figure 8: Left: Area Distribution by Fujita et al. (1986)
          Right: Area Distribution according to Windisch (1988)
          for Large Tube Pitch.

Figure 9: Comparison of the Modified Calculation Procedure of Fujita (Dashed Lines) with the Experimental Results (Symbols and Full or Dot-Dashed Interpolation Lines) for Various Pure Hydrocarbons in the Pressure Range $0.011 \le p^* \le 0.65$ (Upper Diagram) and for Propane/n-Butane Mixtures at $0.1 \le p^* \le 0.5$ (Lower Diagram).

dashed horizontal lines in Fig. 11) or free convective (lower dashed lines), is demonstrated in Fig. 11 for the examples already discussed in Figs. 6 and 7. In fact, the influence of pressure is very small and the shape of the curves is not met by the assumptions, as is quite obvious from the comparison of the experimental $\alpha(\varphi)$-relationships of Fig. 11 with the dashed horizontal lines representing the calculation based on the areas modified according to Fig. 10, on the left.

CONCLUSIONS

- The experiments discussed in this paper show that only slight differences exist between the heat transfer coefficients of mixtures and pure substances boiling at the outside of horizontal tubes with superimposed convective flow at small heat fluxes that means under conditions with heat transfer dominated by convective effects.

- Subsequently, the convective domain extends to higher heat fluxes with mixtures and the enhancement of heat transfer by convection is more pronounced than with pure substances, because within the domain of nucleate boiling heat transfer is significantly worse for mixtures than for pure substances.

- A calculation method for the heat transfer coefficient in the literature has been modified and can be used successfully for pressures up to approx. 10% of the critical pressure of the mixtures (and pure components).

- For higher pressures, various steps of the calculation procedure should be improved in future, especially a more realistic local variation of the heat transfer coefficient along the circumference of the tube should be introduced.

NOMENCLATURE

A	Surface area of the tubes	Subscripts	
V	Vapour volume	1	Lower tube
k	Ratio of vapour enthalpy to total heat input	2	Upper tube (= test tube)
p	Pressure	c	Critical
$p^* = p_s/p_c$	Normalized pressure	s	Saturated
q	Heat flux	O	Upper part of circumference
T	Temperature		
ΔT	Superheat of tube surface	U	Lower part of circumference
x	Molar fraction of the lower boiling component in the liquid	F	Flanks
		calc	Calculated
α	Heat transfer coefficient	conv	Convective
φ	Circumferential angle	local	Local value at circumference

REFERENCES

Bednar, W., 1993. "Wärmeübergang beim Blasensieden von binären Kohlenwasserstoffgemischen." Diss. Universität Karlsruhe (TH).

Buschmeier, M., Sokol, P., Pinto, A.D., Gorenflo, D., 1994. "Pool Boiling Heat Transfer of Propane/n-Butane Mixtures at a Single Tube with Superimposed Convective Flow of Bubbles and Liquid." *Proc. 10th Int. Heat Transfer Conf.*, Brighton, Vol. 5, pp. 69-74.

Fujita, Y., Ohta, H., Hidaka, S., Nishikawa, K., 1986. "Nucleate Boiling Heat Transfer on Horizontal Tubes in Bundles." *Proc. 8th Int. Heat Transfer Conf.*, San Francisco, Vol. 5, pp. 2131-2136.

Gorenflo, D., Goetz, J., Bier, K., 1982. "Vorschlag für eine Standard-Apparatur zur Messung des Wärmeübergangs beim Blasensieden." Wärme- und Stoffübertragung, Vol 16, pp. 69-78; cf. also: Goetz, J., 1980. Diss. Universität Karlsruhe (TH).

Gorenflo, D., 1993. "Pool Boiling Heat Transfer from Horizontal Tubes to Alternative Refrigerants." *Proc. Int. Sem. Heat Transfer, Thermophysical Properties and Cycle Performance of Alternative Refrigerants*, Jap. Ass. of Refrig., Kitakyushu, pp. 91-106.

Sokol, P., 1994. "Untersuchungen zum Wärmeübergang beim Blasensieden an Glatt- und Rippenrohren mit großem Außendurchmesser." Diss. Universität (GH) Paderborn.

Windisch, R., 1988. "Wärmeübergang beim Sieden von Kältemitteln an Einzelrohren, Zweirohranordnungen und Rohrbündeln." Diss. Universität Stuttgart.

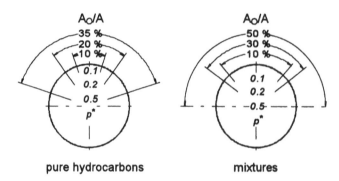

Figure 10: Modified Area Ratios A_O/A growing from 10 to 35% for Pure Hydrocarbons (left) and from 10 to 50% for Mixtures (right) at the Same Increase of the Normalized Pressure from $p^* = 0.1$ to 0.5.

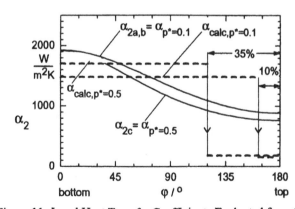

Figure 11: Local Heat Transfer Coefficients Evaluated from the Measurements at $p^* = 0.1$ or 0.5, resp. (full curves, taken from Figs. 6 and 7, bottom), and Comparison with the Modified Calculation Method according to Fig. 10, left (Dashed Horizontal Lines).

FORCED CONVECTIVE VAPORIZATION
OF SATURATED LIQUID MIXTURES

Dieter Steiner

Institut für Thermische Verfahrenstechnik
Universität Karlsruhe
Karlsruhe, Baden-Württemberg
Bundesrepublik Deutschland

ABSTRACT

Convective vaporization of liquid mixtures is characterized by the decline of heat-transfer, which depends on the mixture composition. As the total heat-transfer is a function of the convective evaporation and the nucleate flow boiling components, the effect of liquid composition and the other variables of each mechanism have been analysed in this article. The analysis has been performed to identify the fundamental changes in convective vaporization of mixtures compared to convective vaporization of the pure components. Additionally, it has been observed that a high ratio of heat flux to mass velocity reduces the heat-transfer coefficient of the nucleate flow boiling component. In this case it is possible that the convective evaporation component gives significantly higher heat-transfer coefficients. Therefore, a remarkable amount of the decline of heat-transfer at flow vaporization can be reduced by using sufficiently high mass velocities.

1. INTRODUCTION

Vaporization of liquid mixtures in the presence of forced flow is frequently used in industrial applications, such as in reboilers to supply vapours to distillation columns, evaporators to vaporize a liquid, and various types of evaporators in refrigerating plants and for heat pump systems. Although many experimental data and heat-transfer correlations exist, prediction of heat-transfer of wide-boiling mixtures results in large uncertainties up to now, as reported by Palen (1983). The reason for this is that the reduction of heat-transfer coefficient and the effect of various variables on the reduction mechanism is not correctly predicted.

Additionally, there is a further difficulty in the analysis of the experimental data because two mechanisms of heat-transfer occur as long as the critical boiling regime is not exceeded. The first one is convective evaporation. In this case, the heat-transfer coefficient depends on flow variables and the physical properties of the mixture. The second one is nucleate flow boiling, which is dependent mainly on the heat flux (wall superheat), the pressure and the composition of the mixture. If the characteristic variables are not systematically varied in a sufficiently large range, the different mechanisms cannot be detected. The derived correlations then contain an array of variables and cannot be used for other conditions than the tested ones.

2. COMPARISON OF EXPERIMENTAL RESULTS WITH MODELS

In this article only the behaviour of saturated liquid mixtures without miscibility gap and without surfactant component is considered. The results for two mixtures with different boiling point ranges are given in Fig. 1 and 2. The mixture $CFCl_3$ (1)/$C_2F_3Cl_3$ (2) has a difference in boiling points, ΔT_S, of 23.8 K whereas SF_6 (1)/CF_2Cl_2 (2) has a difference of 48 K at 0.2 MPa because SF_6 is not liquid at normal pressure. The experiments with the mixture $CFCl_3$/$C_2F_3Cl_3$ are performed with a submerged cylindrical rod in forced cross flow. Flow vaporization of the mixture SF_6/CF_2Cl_2 has been investigated in a horizontal tube. To study the circumferential variation of the heat-transfer coefficient a special and segmented test section has been used. A detailed description of the test apparatus is given by Niederkrüger (1991) and Niederkrüger et al. (1992). All the results are obtained at the isothermal boundary condition. As for a given mixture and other conditions the void fraction is the same in vertical upflow and in a horizontal tube, the average velocities of both phases are the same. Additionally, it was verified for the pure components that the pressure effect on the heat-transfer coefficient in segment S4 (the fully wetted bottom segment), i.e. the pressure function, was – within statistical uncertainties – identical with the equation for vertical tube heat-transfer. Therefore, only the data of the fully wetted bottom segment S4 has been used. It is assumed that the data determined from such a set-up should be representative for flow vaporization at vertical upflow as well.

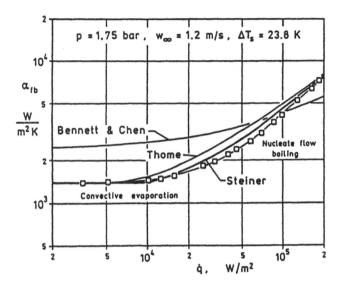

Fig. 1.   Heat-transfer coefficient as a function of
          heat flux. $CFCl_3/C_2F_3Cl_3$; $\tilde{x}_1$= 0.47.
          Fink (1982).

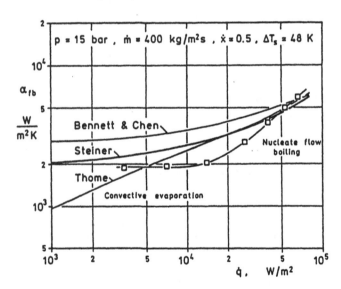

Fig. 2.   Heat-transfer coefficient as a function of heat-
          flux. SF_6/CF_2Cl_2; $\tilde{x}_1$ =0,14.
          Niederkrüger (1991).

In both experiments the regions of convective evaporation
and nucleate flow boiling are observed. Additionally to the
experimental data the calculated results of the following
methods are plotted: Bennett and Chen (1980), Thome
(1989) and Steiner (1994). The ideal heat transfer coefficient,
a_{id}, as given in this paper, has been used as basis in
Thome's (1989) and in Steiner's (1994) equation. The
contribution of the pure components in Thome's equation
has been calculated using Gorenflo's (1988) *pool* boiling
equation. However, in Steiner's equation the nucleate *flow*
boiling equations have been used as given by Steiner and
Taborek (1992).

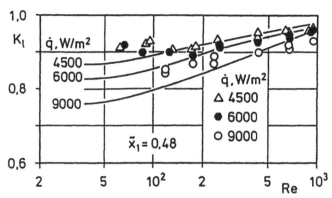

Fig. 3.   Kinetic separation factor K_ℓ (analytical results
          and experimental data) as a function of the film
          Reynolds number. Heat flux $\overset{\circ}{q}$ is parameter.
          Gropp (1985).

The comparison with the experimental data shows that
there are remarkable differences between the various methods
and the measured data which increases for mixtures with
high boiling point difference, see Fig. 2. Only the method
of Steiner (1994) predicts both regions independent from the
ratio heat-transfer affected by the convective evaporation and
heat-transfer by nucleate flow boiling. This asymptotic
superposition model will be analysed further.

2.1 Convective Evaporation

Convective evaporation occurs if the thermal resistance of
the boundary layer near the wall is lower than in nucleate
flow boiling. In this case, superheating of the liquid at the
wall no longer suffices to meet the condition for the activation of boiling sites and bubble formation. In forced flow
the heat-transfer coefficient can be predicted with well-
known equations by inserting physical properties of the
mixture. Gropp (1985) experimentally verified that the
mass-transfer resistance of the liquid side, although higher
than the one of the vapour side, has comparatively little
effect on convective evaporation of mixtures of normal
viscosity. Using the liquid-side kinetic separation factor K_ℓ
as a measure of the depletion of the more volatile
component 1 in the boundary layer, this dimensionless
number is defined as

$$K_\ell \equiv \frac{\tilde{y}_{1,Ph} - \tilde{x}_1}{\tilde{y}_{1,Ph} - \tilde{x}_{1,Ph}} \qquad (1).$$

Solving the equation of the mass flux of component 1 by
use of the film theory the following result can be obtained

$$K_\ell = \exp(-\phi_\ell) \qquad (2).$$

The ratio of vaporization flux and diffusion flux is expressed
as

$$\phi_\ell = \frac{\overset{\circ}{n}}{\beta_{\ell,1}\,\tilde{\rho}_\ell} = \frac{\overset{\circ}{q}}{\beta_{\ell,1}\,\tilde{\rho}_{\ell}\,\Delta\tilde{h}_v} \qquad (3).$$

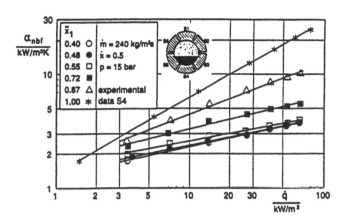

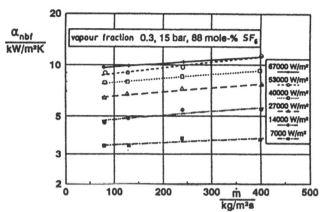

Fig. 4. Heat-transfer coefficient in nucleate flow boiling as a function of heat flux at various liquid compositions.

Fig. 5. Dependence of nucleate flow boiling heat-transfer coefficient on the mass velocity.

$\overset{\circ}{n}$ is the molar mass flux which passes through the interface, $\beta_{\ell,1}$ is the liquid-phase mass-transfer coefficient and $\tilde{\rho}_\ell$ the molar density of liquid mixture. The analytically and experimentally obtained results for a binary mixture of $CFCl_3/C_2F_3Cl_3$ with a bulk mole fraction $\tilde{x}_1 = 0.48$ are presented in Fig. 3. As the experiments are performed with falling films, K_ℓ is plotted against the film Reynolds number, $Re = \Gamma/\eta_\ell$, where Γ is the mass flow per wetted perimeter, η_ℓ the dynamic viscosity of the liquid.

Our results can be compared with the conclusion of Jung et al. (1989) who wrote: "Non-ideal variations in physical properties account for 80 % of the heat-transfer degradation seen with mixtures (R22/R114 with $\Delta T_S = 44,44$ K) and the other 20 % is believed to be caused by mass-transfer resistance in this region".

All the experimental data in segment S4 could be predicted with correlations given for vertical upflow. The heat-transfer coefficient for convective evaporation, α_{ce}, in a vertical tube may be calculated from

$$\alpha_{ce} = \alpha_{LO}\left[(1-\overset{\circ}{x})^{1.5} + 1.9\,\overset{\circ}{x}^{\,0.6}\left(\frac{\rho_\ell}{\rho_v}\right)^{0.35}\right]^{1.1} \qquad (4)$$

as long as the vapour fraction $\overset{\circ}{x} \leq 0.6$. In this equation α_{LO} is the tube-side single phase heat-transfer coefficient, based on the total flow as liquid with $Re_{LO} = \overset{\circ}{m}\,d/\eta_\ell$ using recommended equations. All the properties required for calculation are those of the mixture at bulk composition at a point z along the tube. The heat-transfer calculation for the whole range of vapour fraction and other tube orientations are given by Steiner (1988). For vertical flow see Steiner and Taborek (1992).

2.2 Nucleate Flow Boiling

As long as the wall superheating is sufficiently high to activate the cavities on the wall surface, nucleate flow boiling takes place. However, this superheating depends on the convective evaporation heat-transfer. Results of local-point heat-transfer coefficients are presented in Fig. 4 for

SF_6/CF_2Cl_2 mixtures as investigated by Niederkrüger (1991).

It can be seen from Fig. 4 that, for a given composition and other fixed variables, the heat-transfer coefficient can be approximated by the relationship $\alpha_{nbf} \sim \overset{\circ}{q}^{\,n}$. The slope n, however, depends on the composition of the mixture. The effect of mass velocity, $\overset{\circ}{m}$, and vapour fraction, $\overset{\circ}{x}$, on the heat-transfer has been systematically investigated. From the results it can be concluded that in the developed nucleate flow boiling region of mixtures, if the composition is low or high, the heat-transfer coefficient is generally, except for small deviations, not affected by mass velocity, as shown in Fig. 5.

Also the vapour fraction, $\overset{\circ}{x}_1$ shows only a slightly reduction on the heat-transfer coefficient in the developed nucleate flow boiling region, as presented in Fig. 6. In these experiments the vapour fraction was increased from 0.1 to 0.7. The reduction is not due to changes in flow conditions, it is a result of the boiling process. At a given overall composition, the vapour produced is enriched with the more volatile component, which results in depletion of this component in the remaining liquid and hence heat-transfer reduction. It has been proved experimentally and demonstrated with flash calculations by Niederkrüger et al. (1992) that with increasing vapour fraction the change in liquid composition becomes larger within the test section. The composition given in the Figures is always measured at the inlet of the test section with a heated length of 200 mm.

In nucleate flow boiling, the heat-transfer coefficient of the mixture is generally lower than that of a mixture following the linear molar mixing law of the corresponding pure components, shown as dashed-dotted line. Typical results are plotted in Fig. 7 for two heat fluxes.

The magnitude by which the heat-transfer coefficient is reduced in relation to the composition obviously depends on the heat flux. Moreover, the minimum of the heat-transfer coefficient does not always correspond to the maximum of the difference in vapour-liquid composition. The experimental data of Niederkrüger (1991) and of Celata, Cumo et al. (1993) in a vertical tube demonstrate that the degree of

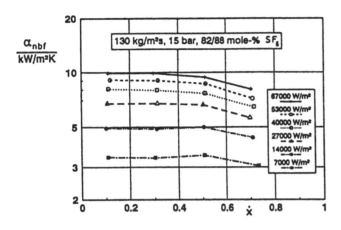

Fig. 6. Dependence of nucleate flow boiling heat-
transfer coefficient on the vapour fraction.

heat-transfer reduction also depends on the pressure in such a
way that it increases with pressure, though the driving
composition difference $\tilde{y}_1 - \tilde{x}_1$ decreases. These effects are
shown in Fig. 8 with the data of Bayer (1988).

The degradation of heat-transfer coefficients has been
calculated from the equation of Schlünder (1982) and
compared with the heat-transfer coefficient given by the ideal
molar mixing law. If a mixture without azeotropic points is
considered, this equation is given as

$$\alpha_{id} = \left(\sum_{i=1}^{n} \frac{\tilde{x}_i}{\alpha_{nbf,i}} \right)^{-1} \tag{5}.$$

The mixture heat-transfer coefficient in nucleate flow
boiling is obtained from

$$\frac{\alpha_{nbf}}{\alpha_{id}} =$$
$$\left\{ 1 + \frac{\alpha_{id}}{\overset{\circ}{q}} \left[\sum_{i=1}^{n-1} (T_{sn} - T_{si})(\tilde{y}_i - \tilde{x}_i)\left(1 - \exp\left(\frac{-B_0 \overset{\circ}{q}}{\beta_{\ell,i}\, \tilde{\rho}_\ell\, \Delta\tilde{h}_v} \right) \right) \right] \right\}^{-1} \tag{6}.$$

Comparison of nucleate flow boiling with pool boiling
data of a given mixture shows that the heat-transfer coefficients are not identical, even though all parameters (heat
flux, pressure, composition, wall material, wall roughness)
remain the same. This results mainly from the contribution
of the pure components to the mixture heat transfer as given
in α_{id}. The heat-transfer coefficient of the pure components,
$\alpha_{nbf,i}$, must only be calculated with nucleate flow boiling equations, dependent on the tube orientation, see Steiner
(1988) and Steiner and Taborek (1992). The heat-transfer coefficient in flow to pool nucleate boiling appeared to be a
function of molecular weight. At a reduced pressure of 0.1,
fluids with a molecular weight $\tilde{M}$ greater than approximately 60 kg/kmol have a higher heat-transfer coefficient in

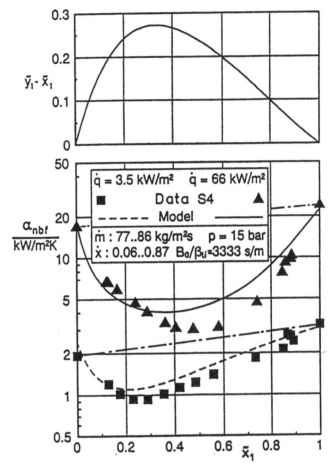

Fig. 7. Flow boiling heat-transfer coefficient as a
function of liquid composition.

nucleate flow boiling than in pool boiling. For fluids with
$\tilde{M}$ less than ca. 60 kg/kmol, the reverse is true. For
example, the mixture SF_6/CF_2Cl_2 has a ratio $\alpha_{nbf,i}/\alpha_{nb}$
of ca. 1.65 for the component CF_2Cl_2 and 2.07 for the
component SF_6.

On the basis of 668 data sets, measured heat-transfer coefficients have been compared with the ones calculated from
Eq.(6). The results showed that the reduction in heat-transfer
could be predicted quite well with an adjusted parameter
$B_0/\beta_{\ell,i} = 3333$ s/m for all the values taken by the variables.
However, an effect of mass velocity was observed. At
certain conditions, defined by the ratio of heat flux to mass
velocity, the decline of the heat-transfer is so considerable
that convective evaporation gives higher heat-transfer
coefficients than those resulting from nucleate flow boiling.
This led to a remarkable contribution of convective
evaporation to the total heat-transfer in this range of
composition where the drop is largest, as shown in Fig. 9.

Hence, the heat-transfer in convective vaporization, where
the flow may be induced by forced or by natural circulation,
has to be determined from the superposition of both
mechanisms according to

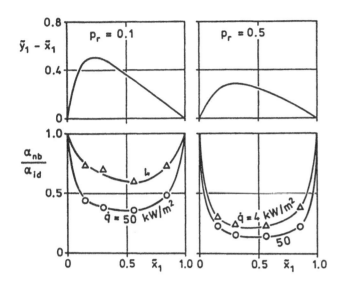

Fig. 8. Dependence of the reduced heat-transfer coefficient on heat flux and liquid composition for a CHF$_3$/C$_2$F$_5$ mixture.

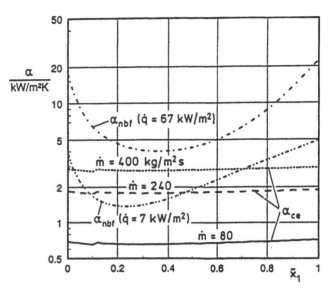

Fig. 9. Calculated convective evaporation and nucleate flow boiling heat-transfer as a function of liquid composition. SF$_6$/CF$_2$Cl$_2$ mixture, $p = 15$ bar, $\overset{\circ}{x} = 0.5$.

$$\alpha_{fb} = \sqrt[3]{\alpha_{ce}^3 + \alpha_{nbf}^3} \qquad (7).$$

The equation for the convective evaporation heat-transfer coefficient, α_{ce}, is given in section 2.1. The nucleate flow boiling contribution, α_{nbf}, needs to be calculated only if a certain wall superheat or a certain heat flux, $\overset{\circ}{q}_{onb}$, required for the onset of nucleate flow boiling, are exceeded. The following equation approximately applies where $T_s(\tilde{x},p)$ is the boiling-point temperature at $z = 0$:

$$\overset{\circ}{q}_{onb} = \frac{2 \, \sigma \, T_s \, \alpha_{LO}}{r_{cr} \, \tilde{\rho}_V \, \Delta \tilde{h}_V} \qquad (8).$$

The values to be inserted for the properties are those to be calculated for the mixture composition at the tube entrance ($z = 0$). $\tilde{\rho}_V$ is the molar density of the vapour mixture, $\Delta \tilde{h}_V$ is the molar differential latent heat of vaporization, σ is the surface tension of mixture, r_{cr} is a given parameter with $r_{cr} = 0.3 \cdot 10^{-6}$ m for all mixtures. The liquid single-phase heat-transfer coefficient α_{LO} is based on $\overset{\circ}{m}$ as total flow and calculated with mixture properties at tube entrance conditions ($z = 0$).

The results are presented in Fig. 10. The dominant contribution to the heat-transfer coefficient at the lowest mass velocity is that of nucleate flow boiling. This is the lowest curve calculated from Eq.(6). Convective evaporation has only a small additional contribution. The experimental data at higher mass velocities are close to the calculated curve, using Eq.(7), which is significantly affected by convective contribution.

The application of forced convective vaporization gives the possibility to increase the heat-transfer coefficient in the

range of mixture composition where the difference of vapour-liquid composition is a maximum if sufficiently high mass velocities are used.

CONCLUSION

Analysis of reduced heat-transfer coefficients in nucleate flow boiling, α_{nbf}/α_{id}, shows a dominant effect of mixture composition, pressure and heat flux. In view of these data, the liquid side mass-transfer controlled model of Schlünder (1982) is up to now the best available one. However, in this equation, where the linear molar mixing law is applied, the contribution of the pure components has to be calculated from nucleate flow boiling heat-transfer coefficients only.

The total heat-transfer in convective vaporization, where the flow may be induced by forced or natural circulation, has to be calculated by taking into account the contribution of convective evaporation and nucleate flow boiling by an asymptotic superposition model. As a consequence, the drop of the mixture heat-transfer can be reduced by using sufficiently high mass velocities.

NOMENCLATURE

B_0	parameter in Eq.(6)
d	inside tube diameter, m
g	acceleration due to gravity, m^2 s^{-1}
Δh_V	latent heat of vaporization, J kg^{-1}
$\Delta \tilde{h}_V$	molar differential latent heat of vaporization, J mol^{-1}
K_ℓ	kinetic separation factor, liquid side
$\overset{\circ}{m}$	mass velocity, kg m^{-2} s^{-1}
$\tilde{M}$	molecular weight, kg kmol^{-1}
$\overset{\circ}{n}$	molar flux, kmol m^{-2} s^{-1}
p	pressure, Pa
p_{cr}	pressure, Pa

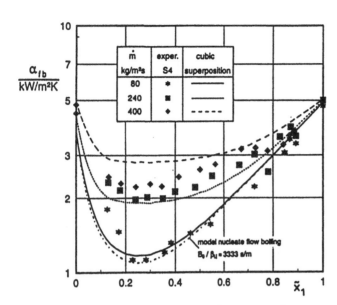

Fig. 10. Comparison of measured heat-transfer coefficients in convective vaporization with calculated results. Parameter is the mass velocity. SF_6/CF_2Cl_2 mixture with $\overset{\circ}{q} = 7 \, kW/m^2$, $p = 15 \, bar$ and $\overset{\circ}{x} = 0.5$.

p_r	reduced presure ($= p/p_{cr}$)
$\overset{\circ}{q}$	heat flux, W m^{-2}
Re	Reynolds number
T	temperature, K
T_s	boiling-point temperature, K
T_{si}	boiling-point temperature of component i, K
T_w	wall surface temperature, K
$\overset{\circ}{x}$	vapour fraction
$\tilde{x}_i$	liquid mole fraction of component i
$\tilde{x}$	liquid mole fraction vector ($\tilde{x}_1, \tilde{x}_2, ..., \tilde{x}_n$)
$\tilde{y}_i$	vapour mole fraction
z	coordinate in main flow direction, m
α	heat-transfer coefficient, $\alpha = \overset{\circ}{q}/(T_w - T_s)$, W m^{-2} K^{-1}
$\beta_{\ell,i}$	mass-transfer coefficient of component i in the liquid phase, m s^{-1}
η	dynamic viscosity, kg m^{-1} s^{-1}
ρ	density, kg m^{-3}
$\tilde{\rho}$	molar density, mol m^{-3}
σ	surface tension, N m^{-1}
ϕ_ℓ	flux ratio

Subscripts

ce	convective evaporation
cr	critical
fb	flow boiling
GO	based on $\overset{\circ}{m}$ as total flow as vapour
id	ideal (linear) mixing law
ℓ	liquid based
LO	based on $\overset{\circ}{m}$ as total flow as liquid
nb	nucleate pool boiling
nbf	nucleate flow boiling

nbf,i	nucleate flow boiling of component i
onb	onset of nucleate boiling
Ph	interphase vapour-liquid
v	vapour based

REFERENCES

Bayer, A., 1988, "Untersuchungen zum Blasensieden von binären Stoffgemischen in einem großen Druckbereich", *Dissertation*, Universität Karlsruhe.

Bennett, D.L. and Chen, J.C., 1980, "Forced convective boiling in vertical tubes for saturated pure components and binary mixtures", *AICHE J.*, Vol. 26, pp. 454-461.

Celata, G.P., Cumo, M. and Setoro, T., 1993, "Forced convective boiling in binary mixtures", *Int. J. Heat and Mass Transfer*, Vol. 36, pp. 3299-3309.

Fink, J., 1982, "Verdampfung von R11/R113-Gemischen an einem quer angeströmten waagerechten Zylinder", *Dissertation*, Techn. Universität Clausthal.

Gorenflo, D., 1988, "Behältersieden (Sieden bei freier Konvektion)", *VDI-Wärmeatlas*, Sect. Ha, VDI-Verlag, Düsseldorf.

Gropp, U., 1985, "Wärme- und Stoffübergang bei der Oberflächenverdampfung und beim Blasensieden eines binären Kältemittelgemisches am Rieselfilm", *Dissertation*, Universität Karlsruhe.

Jung, D.S., McLinden, M., Radermacher, R. and Didion, D., 1989, "Horizontal flow boiling heat transfer experiments with a mixture of R22/R114", *Int. J. Heat Mass Transfer*, Vol. 32, pp. 131-145.

Niederkrüger, M., 1991, "Strömungssieden von reinen Stoffen und binären zeotropen Gemischen im waagerechten Rohr bei mittleren und hohen Drücken", *Fortschr.-Ber. VDI-Z.*, Vol. 3, pp. 1-249.

Niederkrüger, M., Steiner, D. and Schlünder, E.-U., 1992, "Horizontal flow boiling experiments of saturated pure components and mixtures of R846/R12 at high pressures", *Int. J. Refrig.*, Vol. 15, pp. 48-58.

Palen, J.W., 1983, "Shell-and-tube reboilers", in *HEDH-Heat Exchanger Design Handbook*, Hemisphere, New York.

Schlünder, E.-U., 1982, "Über den Wärmeübergang bei der Blasenverdampfung von Gemischen. *Vt. Verfahrenstechnik*, Vol. 16, pp. 692-698.

Steiner, D., 1988, "Wärmeübertragung beim Sieden gesättigter Flüssigkeiten, *VDI-Wärmeatlas*, Sect. Hbb, VDI-Verlag, Düsseldorf.

Steiner, D., 1994, "Pool and forced convective vaporization of saturated liquid mixtures", *Chem. Eng. Proc.*, Vol. 33, pp. 337-351.

Steiner, D. and Taborek, J., 1992, "Flow boiling heat transfer in vertical tubes correlated by an asymptotic model", *Heat Transfer Eng.*, Vol. 13, pp. 43-69.

Thome, J.R., 1989, "Prediction of the mixture effect on boiling in vertical thermosyphon reboilers", *Heat Transfer Eng.*, Vol. 10, pp. 29-38.

SUBCOOLED FLOW BOILING HEAT TRANSFER TO MIXTURES AND SOLUTIONS

H. Müller-Steinhagen
Dept. of Chemical and Process Engineering
The University of Surrey, Guildford, England GU7 1YL

M. Jamialahmadi
University of Petroleum Industry
Ahwaz, Iran

Abstract

Extensive measurements of subcooled flow boiling heat transfer coefficients have been performed for pure liquids, electrolyte solutions and ternary mixtures of water, isopropanol and acetone over a wide range of concentration, flow velocity, bulk temperature and heat flux. Furthermore convective and subcooled boiling heat transfer coefficients were measured for two complex industrial liquids, namely Kraft liquor from the pulp and paper industry and Bayer liquor from Bauxite refineries. The experimental results are compared with a prediction model combining the Chen model with the Gorenflo correlation and the Schlünder model, for single and multicomponent boiling, respectively.

INTRODUCTION

Subcooled boiling occurs if the heat transfer surface temperature is above the boiling point, while the fluid bulk temperature is still subcooled. Bubbles departing from the wall then collapse in the subcooled liquid. A common type of heat transfer equipment is a vertical long tube evaporator. The evaporating liquid flows through vertical tubes heated from the outside by condensing steam. The inlet temperature of the liquid usually is below the boiling point because of the hydrostatic head. Depending on the total operating pressure, subcooled boiling will, therefore, occur over a considerable length of the evaporator. For instance, up to 50% of the total heat duty in low pressure evaporators may be transferred by subcooled boiling. A computer literature search revealed that the application of existing models for the prediction of heat transfer coefficients is restricted either to single component fluids or to fully developed saturated nucleate boiling heat transfer. The calculation procedure presented in this paper is not limited by those restrictions. It predicts heat transfer from the convective regime up to the fully developed boiling regime for single component fluids, electrolyte solutions and polynary mixtures under saturated and subcooled flow boiling conditions.

DESCRIPTION OF MODEL

The model for the prediction of local subcooled boiling heat transfer coefficients is based on an additive superposition of convective and boiling heat transfer, as suggested by Chen [1]:

$$\dot{q} = \alpha_{conv,tp}(T_W - T_b) + \alpha_{boil}S(T_W - T_{Sat}) \quad (1)$$

An enhancement factor F is incorporated in the prediction of the convective heat transfer coefficient to account for the increased heat transfer due to the co-current flow of liquid and vapour. The suppression of nucleate boiling due to forced convection is considered by the suppression factor S. Collier suggests the following empirical correlations [2]:

$$F = 1 \quad \text{for} \quad \frac{1}{X_{tt}} \leq 0.1 \quad (2)$$

$$F = 2.35\left(\frac{1}{X_{tt}} + 0.213\right)^{0.736} \quad \text{for} \quad \frac{1}{X_{tt}} \geq 0.1 \quad (3)$$

$$S = \frac{1}{1 + 2.53 \cdot 10^{-6} Re_{tp}^{1.17}} \quad (4)$$

with:

$$Re_{tp} = \frac{\dot{m}(1 - \dot{x})d_h}{\mu_l} \cdot F^{1.25} \quad (5)$$

X_{tt} is the Martinelli Parameter, which is a function of the vapour mass fraction:

$$X_{tt} = \left(\frac{1 - \dot{x}}{\dot{x}}\right)^{0.9}\left(\frac{\rho_g}{\rho_l}\right)^{0.5}\left(\frac{\mu_l}{\mu_g}\right)^{0.1} \quad (6)$$

where the vapour mass fraction $\dot{x}$ is defined as:

$$\dot{x} = \frac{\dot{m}_g}{\dot{m}_g + \dot{m}_l} \qquad (7)$$

To calculate the enhancement and suppression factors, the local vapour mass fraction has to be determined. Schröder presents a calculation method [3] for the local vapour mass fraction, which is applicable for subcooled and saturated boiling:

$$\dot{x} = Ph - Ph_n \exp\left(\frac{Ph}{Ph_n} - 1\right) \qquad (8)$$

where Ph is the phase change number and is defined as:

$$Ph = \frac{\overline{h}_f - h_{l,Sat}}{\Delta h_v} \qquad (9)$$

This number describes the local thermodynamic condition of the fluid. As long as the mean fluid stream is subcooled, Ph has negative values. Ph becomes positive and equivalent to the local vapour mass fraction, if the mean enthalpy of the fluid is higher than the saturation enthalpy of the liquid. Ph_n is the value of the phase change number which is reached, once the mean fluid temperature is high enough to permit the existence of vapour bubbles in the bulk of the liquid. Schröder [3] suggests to calculate Ph_n with a correlation valid for laminar and turbulent flow using the boiling number Bo and the Peclet number Pe:

$$Ph_n = \frac{-Bo}{\left[\left(\frac{455}{Pe}\right)^2 + 0.0065^2\right]^{\frac{1}{2}}} \qquad (10)$$

$$\text{with} \quad Bo = \frac{\dot{q}}{\dot{m}\Delta h_v} \quad \text{and} \quad Pe = \frac{\dot{m}c_p d_h}{\lambda} \qquad (11)$$

Convective Heat Transfer

Turbulent flow occurs in almost all practical applications. Therefore, only correlation of turbulent flow will be considered in this paper, for laminar flow calculations see [4]. For the same reasons, the contribution of natural convection to convective heat transfer is ignored. Instead of the somewhat outdated Dittus/Boelter correlation [5] suggested by Chen [1], a correlation by Petukhov and Popov [6] is used to predict heat transfer during turbulent flow in pipes. Experiments with all fluids discussed in this paper showed that this equation predicts the measured convective heat transfer data with better accuracy.

$$\alpha_{turb} = \frac{\lambda}{d_h} \cdot \frac{f/8 \, RePr}{1 + 12.7\sqrt{f/8}\,(Pr^{2/3} - 1)} \qquad (12)$$

The friction factor f is given by Filonenko [7]:

$$f = (1.82 \cdot \log Re - 1.64)^{-2} \qquad (13)$$

The Reynolds number used in equations (12,13) is the two-phase Reynolds number defined in equation (6). Variations in the physical properties of the fluid due to the wall superheat are accounted for by multiplying the heat transfer coefficient with the viscosity ratio at bulk and wall temperature to the power of 0.11, as suggested by Gnielinski in [8]. For annular flow, equation (12) is multiplied by a factor suggested by Petukhov and Roizen [11]:

$$\Phi = 0.86\left(\frac{d_l}{d_o}\right)^{-0.16} \qquad (14)$$

Nucleate Boiling

The calculation of the nucleate boiling heat transfer coefficient is only initiated, if the wall temperature is higher than the saturation temperature. Heat transfer under boiling conditions is a function of heat flux and fluid composition. It is often recommended to use correlations developed for pool boiling conditions even though forced convection is present. One of the most reliable pool boiling correlations was developed by Gorenflo [10]:

$$\frac{\alpha}{\alpha_o} = F_p \left(\frac{\dot{q}}{\dot{q}_o}\right)^n \left(\frac{R_p}{R_{po}}\right)^{0.113} \qquad (15)$$

The pressure function F_p and the exponent n are calculated using the reduced pressure p^*:

for organic liquids

$$F_p = 1.2 \cdot p^{*0.27} + \left(2.5 + \frac{1.0}{1 - p^*}\right) \cdot p^* \qquad (16)$$

for water and low boiling liquids

$$F_p = 1.73 \cdot p^{*0.27} + \left(6.1 + \frac{0.68}{1 - p^{*2}}\right) \cdot p^{*2} \qquad (17)$$

The exponent n is calculated from

$$n = 0.9 - 0.3 \cdot p^{*a} \qquad (18)$$

with a = 0.3 for organic liquids and a = 0.15 for water and low boiling liquids. Values of the reference heat transfer coefficient α_o, the reference heat flux $\dot{q}_o$ and the surface roughness R_{po} are listed in [10]. If no tabulated α_o-values are available, the reference heat transfer coefficient α_o is calculated for a reference heat flux q = 20,000 W/m^2 and a reference pressure of p_r = 0.1 using the correlation suggested by Stephan and Preußer [11]

$$Nu - \frac{\alpha d_B}{\lambda_l} - 0.0871 \left(\frac{\dot{q} d_B}{\lambda_l T_{sat}} \right)^{0.674} \left(\frac{\rho_v}{\rho_l} \right)^{0.156} \left(\frac{\Delta h_v d_B^2}{a^2} \right)^{0.37}$$

$$\cdot \left(\frac{a^2 \rho_v}{\sigma d_B} \right)^{0.35} \left(\frac{a}{\nu_l} \right)^{0.162} \qquad (19)$$

with the thermal diffusivity a

$$a - \frac{\lambda_l}{\rho_l c_{p,l}} \qquad (20)$$

and the bubble departure diameter d_B

$$d_B - 0.0146 \beta \sqrt{\frac{2\sigma}{g(\rho_l - \rho_v)}} \qquad (21)$$

The above calculation procedure is recommended for pure liquids and for solutions. For boiling of liquid mixtures, a reduction of the heat transfer coefficient defined with the difference between wall and bulk saturation temperature is observed which is caused by the preferential evaporation of the more volatile component. Owing to the liquid-side mass transfer resistance, the liquid at the heat transfer surface becomes depleted of the component with the higher vapour pressure, which results in an increase in the saturation temperature at the vapour/liquid interface. The local saturation temperature is used by Schlünder [12] to correlate the heat transfer coefficient of mixtures:

$$\alpha_{boil} - \frac{\alpha_{id}}{1 + (\alpha_{id}/\dot{q}_{boil})(T_{Ph} - T_{Sat})} \qquad (22)$$

The ideal heat transfer coefficient α_{id} of the mixture is estimated from the single component heat transfer coefficients of the various components of the mixture:

$$\alpha_{id} - \left(\sum_{i-1}^{n} \frac{\tilde{x}_i}{\alpha_i} \right)^{-1} \qquad (23)$$

The saturation temperature T_{Ph} at the interface is a function of the interface compositions of liquid and vapour, which depend on the diffusion of each component in each phase. The gas-side mass transfer resistance can be neglected according to Gropp and Schlünder [13] because of the high velocity of the generated vapour. The evaporation of polynary mixtures results in a multicomponent, non-equimolar diffusion process. Assumptions outlined in [4] lead to equation (24)

$$\frac{\tilde{y}_{i,Ph} - \tilde{x}_i}{\tilde{y}_{i,Ph} - \tilde{x}_{i,Ph}} - \exp \left(\frac{-\dot{q}_{boil}}{\rho_l \beta_l \Delta h_v} \right) \qquad (24)$$

which provides the relationship between the interface

composition, the boiling heat flux and the mass transfer coefficient. Using appropriate equations for phase equilibrium, equation (24) can be solved by iteration to give the liquid concentration of each component at the interface. The saturation temperature corresponding to the liquid composition at the interface is the temperature T_{Ph} required in equation (22). Different approaches to determine the mass transfer coefficients using the Stefan-Maxwell equations have been evaluated in [4], but did not provide any improvement over the use of a constant value of $0.5 \cdot 10^{-4}$ m/s.

COMPARISON BETWEEN PREDICTED AND MEASURED DATA

All experimental data used for the following comparisons have been obtained for vertical up-flow in an annular test section. The measured heat transfer coefficients are local values. A detailed description of the test apparatus is given in [4].

Pure components

The calculation procedure outlined above has been tested against experimental data for flow of water, acetone, isopropanol and heptane over a wide range of operating conditions [4,15,17]. As α_o-values for the boiling component the following values have been used: Acetone 3,700 W/m²K; Heptane 2,400 W/m²K; Isopropanol 3,000 W/m²K and Water 5,600 W/m²K.

For all fluids, the mean relative error was below ± 10%. Figure 1, which is typical for all these experiments, shows that the model predicts the influence of heat flux, flow velocity and the transition between the two heat transfer regimes with excellent accuracy.

Solutions

Salt Solutions

Subcooled flow boiling experiments have been performed with aqueous solutions of the following salts: NaCl, Na_2SO_4, $Na_2S_2O_3 \cdot 5H_2O$, KNO_3, $CaCl_2$ and $CaSO_4$ [16]. The criteria for selection of the salts were based on the solubility of the salts and valence and size of the respective ions. Care was taken to avoid any formation of deposits on the heat transfer surfaces during the experiments. The range of operating conditions and salt concentrations is given in Table 1.

The effect of electrolyte concentration on subcooled flow boiling heat transfer is shown in Figure 2. Both, forced convective heat transfer coefficients and subcooled flow boiling heat transfer coefficients for electrolyte solutions are lower than the corresponding values for distilled water. The difference is increased by increasing the electrolyte concentration and this must be attributed to the different physical properties and the changes in the mechanism of bubble formation [16]. The heat transfer coefficients of calcium sulphate solution are much lower than those of pure water and of electrolyte solutions with similar concentration but with positive solubility. Similar to pool boiling [16] significant improvement in heat transfer to the calcium sulphate solutions was observed when the ionic strength of the solution was increased by adding sodium chloride (with positive solubility) to the system. For salts with positive solubility and negative solubility, values of 5,600 W/m^2 and 4,800 W/m^{2}K were used as average reference heat transfer coefficients for the correlation of the boiling component.

All measured heat transfer coefficients are compared to the corresponding predictions in Figure 3. Considering the errors involved in the determination of physical properties and the approximate nature of the correlations, a Root Mean Square error of 8.5% is an satisfactory indication that the Chen model and associated correlations can be recommended for the prediction of subcooled flow boiling heat transfer to electrolyte solutions.

Bayer Liquor

Most aluminium oxide is produced with the Bayer process where the alumina is extracted from the bauxite using caustic soda. Spent liquor from the Bayer process is a complex mixture with physical properties which vary greatly with temperature and composition. A typical liquor contains about 240 g/l caustic soda, 95 g/l aluminium oxide, 45 g/l carbonate, 45 g/l organic sodium, 9 g/l chlorine and various minor organic and inorganic components. The major operating cost in the Bayer process is the cost of energy. The reliable prediction of heat transfer coefficients is, therefore, of great importance for the overall plant design. A large number of heat transfer coefficients for forced convective and subcooled flow boiling heat transfer have been measured and compared with the prediction of the above model [18]. A reference value of 5700 W/m^2/K was calculated using the Stephan-Preußer correlation [17]. Figure 4 shows a comparison between measured and predicted heat transfer coefficients for spent liquor as a function of the heat flux under various degrees of subcooling. Generally the agreement between measured and predicted heat transfer coefficients is good. The mean average error is ±8%, with no bias for low and high heat transfer coefficients.

Kraft Black Liquor

In the Kraft wood pulping process, wood chips are digested in a solution of dissolved sodium hydroxide, sodium carbonate and sodium sulphide. After digestion, fibres and black liquor are separated and the liquor is concentrated from 15% to above 65% solids content for recovery and recycling of the process chemicals. The major components of Kraft spent liquor solids are alkali lignin (30-45%), cellulosics (25-35%), Na_2CO_3 (7-13%), Na_2SO_4 (1-10%), $Na_2S_2O_3$ (1-6%), and 4%-8% Na_2S, NaSH and NaOH. The total sodium content of the liquor is 15-12%. The physical properties of black liquor vary greatly with temperature and concentration, with Prandtl numbers at room temperatures as high as 500 for the 65% liquor [18]. Measurements of heat transfer coefficients for forced convection in the laminar and turbulent

flow regime and for subcooled flow boiling are reported in [20]. The data have been compared with the above prediction model. The calculated reference heat transfer coefficient α_o at a reduced pressure of 0.03 is shown in Figure 5. Figure 6 shows the comparison between measured and predicted heat transfer coefficients in the developed subcooled boiling region for liquor solids concentrations between 0 and 65%. Agreement between measured and predicted data is excellent.

Mixtures

Binary and ternary mixtures of water, acetone and isopropanol have been investigated [4,20]. Prediction of heat transfer coefficients of mixtures requires knowledge of the liquid-side mass transfer coefficients and a value of $1 \cdot 10^{-4}$ m/s was suggested by Gropp and Schlünder [13]. The comparison between measured and calculated heat transfer coefficients shows that the agreement can be improved by choosing a mass transfer coefficient of $0.5 \cdot 10^{-4}$ m/s. The use of mass transfer coefficients smaller than $0.5 \cdot 10^{-4}$ m/s did not result in a further significant improvement of the model's accuracy. Figure 7 illustrates predictions and measurements for binary systems. The largest discrepancies are found for 75.0 and 87.5 mol % of acetone in the acetone/water system and for 50.0 and 62.5 mol % of isopropanol in the isopropanol/water system. Figure 8 shows measured and predicted heat transfer coefficients of an acetone/water mixture as a function of the heat flux. The predicted heat transfer coefficients compare favourably with the experimental data covering convective, transition and nucleate boiling heat transfer regimes. In total, 2,400 measurements of subcooled flow boiling heat transfer to mixtures have been compared with the predicted values. The agreement between predictions and measurements is good with a mean error of 7.6% and a root mean square error of 11.3 %.

CONCLUSIONS

A model to predict local heat transfer coefficients for single component fluids, solutions, binary and ternary mixtures is presented. It is applicable for subcooled and saturated conditions, covering the regimes of convective, transition and fully developed nucleate boiling heat transfer. The model uses correlations for convective and pool boiling heat transfer, which are superimposed if the local wall temperature is higher than the saturation temperature of the fluid.

Comparison between a large number of experimental data and the predictions of the model shows good agreement for all heat transfer regimes, independent of heat flux, sub-cooling, fluid composition and flow velocity. For all fluids investigated, the average relative error was less than ± 10%.

NOMENCLATURE

Bo	boiling number	-
c_p	specific heat	J/(kg K)
d	diameter	m

f	friction factor	-
F	enhancement factor	-
F_p	pressure function	-
h	enthalpy	J/kg
Δh_v	latent heat of evaporation	J/kg
$\dot{m}$	mass flux	$kg/(m^2\ s)$
n	exponent	-
Nu	Nusselt number	-
p^*	reduced pressure $(=p/p_o)$	-
Pe	Peclet number	-
Ph	phase change number	-
Pr	Prandtl number	-
$\dot{q}$	heat flux	$W/(m^2)$
R_p	surface roughness	μm
Re	Reynolds number	-
S	suppression factor	
T	temperature	K
x	mole fraction in liquid	-
$\dot{x}$	vapour mass fraction	-
X_{tt}	Martinelli parameter	-
y	mole fraction in vapour	-

Greek symbols

α	heat transfer coefficient	$W/(m^2\ K)$
β_l	mass transfer coefficient	m/s
β	contact angle	rad
λ	thermal conductivity	$W/(m\ K)$
μ	dynamic viscosity	$kg/(m\ s)$
ρ	density	kg/m^3
ν	kinematic viscosity	m^2/s

Indices

b	bulk	B	bubble
boil	boiling	conv	convective
f	fluid	g	gas
h	hydraulic	i	inner
id	ideal	l	liquid
o	outer	Ph	interface
Sat	saturation	tp	two phase
turb	turbulent	v	vapour
W	wall	0	reference conditions

REFERENCES

1 J.C. Chen, Correlation for boiling heat transfer to saturated fluids in convective flow, Ind. Eng. Chem. Process Design and Development 5, 322-329 (1966)

2 J.G. Collier, Boiling within vertical tubes, Heat Exchanger Design Handbook, Sect. 2.7.3-12, (1983)

3 J.J. Schröder, Wärmeübertragung beim unterkühlten Sieden, VDI-Wärmeatlas, Sect. Hba 4th ed. VDI-Verlag (1984)

4 U. Wenzel and H. Müller-Steinhagen, Heat Transfer to Ternary Mixtures of Acetone, Isopropanol and Water Under Subcooled Flow Boiling Conditions. Part I: Experimental Results. Int. Journal Heat Mass Heat Transfer, Vol. 37, No. 2, pp.175-183, pp.185-194, 1994

5 F.W. Dittus and L.M.K. Boelter, Heat transfer in automobile radiators of the tubular type, University of California Press 2, 13 (1930)

6 B.S. Petukhov and V.N. Popov, Theoretical calculation of heat exchange and frictional resistance in turbulent flow in tubes of an incompressible fluid with variable physical properties, High Temp. (USSR) 1, 69-83 (1963)

7 G.K. Filonenko, Hydraulic resistance in pipes, Teploenergetica 1, 40-44 (1954)

8 V. Gnielinski, Wärmeübertragung bei der Strömung durch Rohre, VDI-Wärmeatlas, Sect. Gb 4th ed. VDI-Verlag (1984)

9 B.S. Petukhov and L.J. Roizen, Generalized relationship for heat transfer in turbulent flow of gas in tubes of annular section, High Temp. (USSR) 2, 65-68 (1964)

10 D. Gorenflo, Behältersieden, VDI-Wärmeatlas, Sect. Ha 4th ed. VDI-Verlag (1984)

11 K. Stephan and P.Preußer, Wärmeübergang und maximale Wärmestromdichte beim Behältersieden binärer und ternärer Flüssigkeitsgemische, Chem. Ing. Techn. MS 649/79 (1979)

12 E.U. Schlünder, Über den Wärmeübergang bei der Blasenverdampfung von Gemischen. Verfahrenstechnik. Vol. 16, pp. 692-698 (1982).

13 U. Gropp and E.U. Schlünder, The influence of liquid side mass transfer on heat transfer and selectivity during surface and nucleate boiling of liquid mixtures in a falling film, Chem. Eng. Process 20, pp. 103-114 (1986)

14 U. Wenzel, Saturated pool boiling and subcooled flow boiling of mixtures at atmospheric pressure. Ph.D. thesis University of Auckland (1993).

15 H. Müller-Steinhagen, A.P. Watkinson and N. Epstein, Subcooled-Boiling and Convective Heat Transfer to Heptane Flowing Inside an Annulus and Past a Coiled Wire. Part I: Experimental Results. Part II: Correlation of Data. Journal of Heat Transfer, Vol. 108, No. 4, pp. 922-927. pp. 928-933. (1986)

16 S.H. Najibi, H. Müller-Steinhagen and M. Jamialahmadi, Boiling and Non-Boiling Heat Transfer to Electrolyte Solutions. submit. for publ. in Heat Transfer Engineering (1994).

17 M. Jamialahmadi and H. Müller-Steinhagen, Convective and subcooled boiling heat transfer to BAYER process liquor, Light Metals, pp. 141-150 (1991)

18 C.A. Branch and H. Müller-Steinhagen, Physical Properties Of Kraft Black Liquor. APPITA J. Vol.44, No. 5, pp. 339-341 (1991).

19 C.A. Branch and H. Müller-Steinhagen, Convective and subcooled flow boiling heat transfer to Kraft pulp black liquor. APPITA Journal (1993)

20 U. Wenzel and H. Müller-Steinhagen, Wärmeübergang beim Strömungssieden unterkühlter Flüssigkeitsgemische. Wärme- und Stoffübertragung, 26, pp. 265-271 (1991).

Table 1   Range of Investigation

Velocity, m/s	Heat flux kW/m²	Bulk tempe-rature, °C	Solutions g/l
0.4 - 3	10 to 350	50 to 98	$CaSO_4$ 0.08 - 2.5 $NaCl$ 1.0 - 70 $CaCl_2$ 0.05 - 10 $Na_2S_2O_3.5H_2O$ 1.0 - 32 KNO_3 1.0 - 95

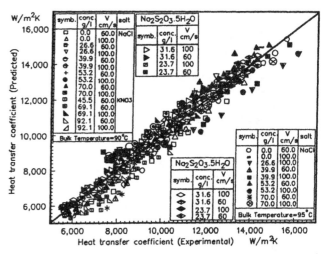

Fig. 3 Measured and predicted heat transfer coefficients for electrolyte solutions

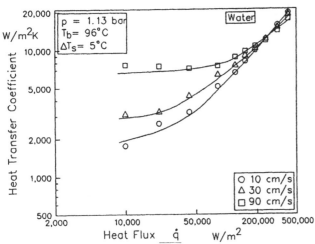

Fig. 1 Influence of heat flux and flow velocity on heat transfer to pure water

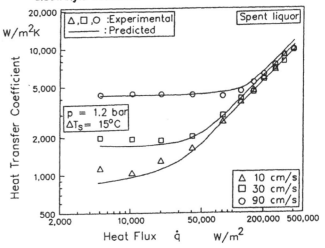

Fig. 4 Influence of heat flux and flow velocity on heat transfer to Bayer liquor

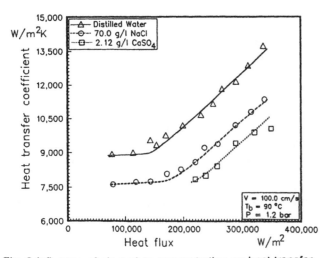

Fig. 2 Influence of electrolyte concentration on heat transfer

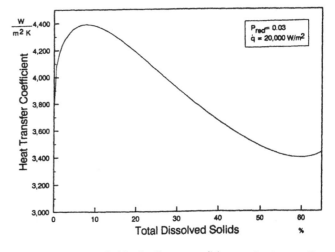

Fig. 5 Influence of black liquor solids content on the calculated reference heat transfer coefficient

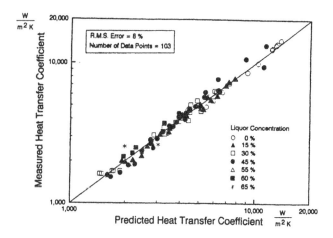

Fig. 6 Measured and predicted heat transfer coefficients for Kraft black liquor

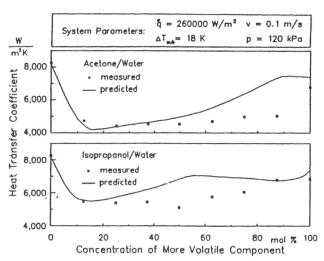

Fig. 7 Heat transfer coefficient as a function of liquid composition

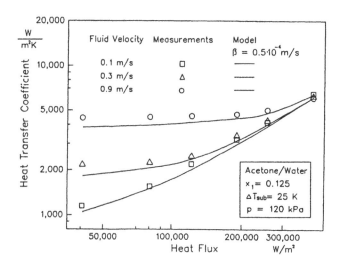

Fig. 8 Heat transfer coefficient as a function of heat flux and flow velocity

FLOW BOILING OF REFRIGERANT-OIL MIXTURES :
A CURRENT REVIEW

John R. Thome

Consulting Engineer
Rome, Italy
and
Industrial Energetics Laboratory
Swiss Federal Institute of Technology - Lausanne
CH-1015 Lausanne, Switzerland

ABSTRACT

Recent advances in refrigerant-oil flow boiling research are reviewed together with identification of the important trends in published experimental data, while shortcomings in data reduction methods used in most of these previous tests are also noted. New explanations for these seemingly unusual trends in refrigerant-oil experimental data are presented and discussed.

INTRODUCTION

Flow boiling research on evaporation of refrigerant-oil mixtures inside horizontal tubes has now been underway for several decades, but without development of an accurate design method to model the effect of oil on boiling heat transfer. In fact, the effect of oil is probably one of the most industrially significant, unresolved problems in flow boiling heat transfer. In refrigeration, air-conditioning and heat pump systems, concentrations usually range from 0.5-5.0 wt.% oil in the refrigerant charge, depending on the type of compressor and whether or not an oil separator is used; for example, reciprocating compressors tend to have from 0.5-1.0 wt.% oil while screw compressors have from 3-5 wt.% oil, referring to the concentration in the subcooled liquid prior to the expansion device. The local wt.% oil in the liquid-phase rises during passage through the expansion device and the evaporator as the refrigerant flashes and evaporates into the vapor-phase, respectively. At the evaporator's exit, the local concentration can reach 50 wt.% oil or more in the remaining liquid on the tube wall or entrained mist. Hence the oil's effect on heat transfer must be analyzed locally to model its local influence on the evaporation process.

The objective of the present paper is not to provide a comprehensive state-of-the-art review but rather a critical look at what has been done, what are the problems, and what can be drawn from existing test results.

TRENDS IN REFRIGERANT-OIL FLOW BOILING DATA

Figure 1 from Hambraeus (1991) depicts many of the trends found in local heat transfer coefficients measured for refrigerant-oil mixtures, plotted as local coefficients with oil, h(oil), normalized by the pure refrigerant values, h(no oil). The oil concentrations cited are the nominal values in the subcooled liquid prior to the test section. The tests were run with a 0.10 vapor quality change in each test section. Thus tube numbers in Figure 1 correspond to local vapor qualities of 0.1, 0.2, to 0.9. The trends observed are:

1. For $x < 0.15$-0.20, the oil coefficients are below those of pure R-134a;
2. For 0.15-$0.20 < x < 0.50$-0.60, the oil coefficients are above those of pure R-134a;
3. For $x > 0.5$-0.6, the oil coefficients are below those of pure R134a, decreasing rapidly with increasing oil concentration, by as much as 50% for the 2 wt.% oil mixture.
4. The peaks in the ratio h(oil)/h(no oil) occur at about 0.4-0.5 vapor quality, depending on the oil concentration.

Figure 2 from Ha and Bergles (1993) depicts the ratio of h(oil)/h(no oil) for mean heat transfer coefficients measured for R-22 and R-12 for two different grades of oil, determined for inlet vapor qualities of about 0.15 and outlet qualities of 0.80-0.85. The trends observed are:

1. For R-22/150 SUS oil at a mass velocity of 200 kg/m²s, the mean coefficient rises as the inlet oil concentration increases to about 2.3% and then falls off. For the same combination at 400 kg/m²s, very little variation occurs up to 5.0 wt.% oil.
2. For R-22/300 SUS oil at mass velocities of 200 and 400 kg/m²s, the mean coefficient with oil decreases

monotonically as inlet oil concentration increases, dropping off by 25% at 5 wt.% oil.

3. At lower mass velocities of 50 and 100 kg/m²s for R-12/150 SUS oil, the degradation attributable to oil is very severe, with h(oil)/h(no oil) falling off by 25% with only 1 wt.% oil present and by 70% for 5 wt.% oil.

4. The higher viscosity of 300 SUS oil compared to 150 SUS oil causes a decrease in the flow boiling heat transfer coefficient.

Explanations for the above trends can be sought in: (i) thermodynamic effects on heat transfer coefficients and measurements, (ii) local physical properties of refrigerant/oil mixtures, (iii) influence of oil on the nucleate boiling process in flow boiling, (iv) influence of oil on the convective contribution to flow boiling, (v) mass transfer effects on the evaporation process, (vi) oil holdup, and other effects. Each of these will be discussed below.

THERMODYNAMICS OF REFRIGERANT-OIL MIXTURES

The integrity of existing test data must be addressed as a starting point...is there some type of systematic error in previous refrigerant-oil flow boiling tests?

Oil Contamination Approach

Up to now, the saturation temperatures of the pure refrigerants have been used to reduce raw experimental measurements to local and mean flow boiling heat transfer coefficients. This method can be termed the "Oil Contamination Approach" since the oil is treated as an impurity rather than a component of the mixture. From a thermodynamic standpoint, the bubble point temperature T_{bub} of the actual fluid must be used to determine the heat transfer coefficient,

$$h \ (mixture) = q/(T_w - T_{bub}) \qquad (1)$$

and not the saturation temperature of the pure refrigerant T_{sat} at the same pressure

$$h = q/(T_w - T_{sat}) \qquad (2)$$

For a pure fluid, $T_{bub} = T_{sat}$ and Eq. (1) is consistent with Eq. (2). Hence, the widely known effect of oil increasing the bubble point temperature with increasing concentration (i.e. "solubility curves") has been ignored when reducing raw test data. In addition, local vapor qualities have been derived using the following pure fluid expression:

$$Q = (x_2 - x_1) h_{LV} \qquad (3)$$

where the subscripts 1 and 2 refer to the initial known vapor quality and local quality, respectively, of the evaporating mixture, h_{LV} is the latent heat of the pure

refrigerant, and Q is the heat added per kg of flow. This equation neglects the heat added to the oil flowing in the tube with the refrigerant and the temperature rise.

Binary Mixture Approach

Thome (1995) has developed a "Comprehensive Thermodynamic Approach" for modelling refrigerant-oil mixtures as binary zeotropic mixtures, essentially applying chemical engineering methods to these mixture systems. The approach includes general methods for determining: bubble point temperatures, specific heats, local oil concentrations, and enthalpy changes for evaporating miscible refrigerant-oil mixtures.

As an example of this approach, Table 1 shows a temperature-enthalpy curve in tabular form calculated for R-134a mixed with 5 wt.% oil (Mobil Arctic EAL 68: specific gravity at 15°C of 0.971, viscosities of 62.5 and 8.1 mm²/s at 40°C and 100°C, respectively, and an assumed molecular weight of 590) at the isobar of 2.93 bar (T_{sat} = 0.0°C for pure R-134a at this pressure). Column one gives the local vapor qualities evaluated in the calculation while columns 2 and 3 list the calculated bubble point temperatures and local oil concentrations, w_{oil}. The total heat absorbed by the fluid (in kJ/kg of flowing fluid) is given in column 4 while the next two columns show the individual contributions of latent and sensible heat.

Referring to the table, the bubble point temperature first increases slowly relative to T_{sat} of pure R-134a and then rises rapidly at vapor qualities above 0.80. For $x <$ 0.90, sensible heat represents less than 1% of the total but its influence rises rapidly afterwards; thus for test data for $x >$ 0.90, the reported vapor qualities become ever too high. For example, if 181.41 kJ/kg of heat is added to the refrigerant-oil mixture, its local vapor quality should be 0.902 according to Table 1; using Eq. (3) it would be 0.9114. Thus, the error in reported vapor qualities is not very significant except at vapor qualities approaching 1.0. However, many data reported in the literature use x_R, which is the local mass of refrigerant vapor divided by the total mass of refrigerant (not the thermodynamic vapor quality x, i.e. local mass of vapor divided by the total mass of refrigerant and oil). Also, for the highest vapor quality shown in the table, x = 0.93, T_{bub} is predicted to be 8.015°C above the pure refrigerant value, such that a temperature cross in the evaporator may be predicted to occur, invalidating design calculations or evaporator bench test data.

In the last two columns in Table 1, the effect of erroneously using T_{sat} in place of T_{bub} to convert "raw" data to experimental heat transfer coefficients is shown for wall superheats of 5°C (i.e. heat transfer coefficient of 2000 W/m² K at a heat flux of 10,000 W/m², typical of plain tubes at medium mass velocities) and 2°C (i.e. heat transfer coefficient of 5000 W/m² K, typical of microfin tubes at medium mass velocities). Consequently, existing published data that utilized T_{sat} are always too low compared to their more correct values. For instance, at x = 0.837 and a wall superheat of 5°C, published values [based on Eq. (2)] would be 13.3% lower than the correct values [based on Eq. (1)].

For a plain tube this systematic error in h becomes significant (> 5%) at local oil concentrations greater than 15% while for a microfin tube it becomes significant at w_{oil} > 6.5%. At high vapor qualities and oil concentrations this error becomes very significant, substantially changing the trends reported in the literature. Thus, according the Table 1, the ratio of h(oil)/h(no oil) does not really drop off as much as indicated in Figures 1 and 2, diagrams which were obtained using the old definition, i.e. Eq. (2). Hence previous published test data that fall in these ranges should be used with caution or better yet, be recalculated using Eq. (1) with the method for predicting T_{bub} given in Thome (1995). Table 1 represents a worst case (5 wt.% oil) and oil effects would be less significant for smaller concentrations.

PROPERTIES OF REFRIGERANT-OIL MIXTURES

Typically only the liquid viscosities and densities of refrigerant-oil mixtures are reported by manufacturers but not other important properties required for predicting boiling heat transfer coefficients. Adopting some petroleum engineering methods for crude oil cuts to model lubricating oils and comparing mixture property methods to test data where available, Thome (1993) put together a comprehensive set of prediction methods that may be the best currently available (but would benefit from further research).

Table 2 shows physical properties calculated for representative points of the enthalpy curve in Table 1. Since oil vapor pressures are about one-ten millionth those of refrigerants, effectively no oil enters the vapor-phase and hence vapor properties are determined for the pure refrigerant at a saturation temperature corresponding to the local bubble point temperature. For the present conditions, the physical properties that demonstrate a significant change from inlet to outlet (> 5%) are: liquid density (-17.2%), liquid viscosity (+9,003%), liquid specific heat (+23.6%), liquid thermal conductivity (+11.7%), critical pressure (-58.1%), critical temperature (+103%) and liquid Prandtl number (+10,021%). Also shown is the ratio of the liquid Reynolds number relative to that for the inlet at x = 0.0. Thus, the variation in local physical properties can potentially have a significant effect on the local heat transfer coefficients, both for nucleate pool boiling and for flow boiling.

OIL EFFECTS ON THE NUCLEATE POOL BOILING PROCESS

No explanation for the maximum in h(oil)/h(no oil) found in many experimental studies, typical of that observed in Figure 1, is mentioned in the literature.

However, various nucleate pool boiling studies on plain tubes have found maxima in h(oil)/h(no oil) at about 3-6 wt.% oil; hence, a convincing hypothesis proposed here is that the oil augments the nucleate boiling contribution to the flow boiling coefficient, producing the maximum in h(oil)/h(no oil).

For example, Figure 3 depicts the ratio of h(oil)/h(no oil) obtained by Memory, Bertsch and Marto (1993) in pool boiling tests for R-124 with an alkylbenzene oil, boiling on plain and enhanced tubes at a temperature of 2.2°C. For the plain tube, boiling performance increased by up to 22% at 6 wt.% oil relative to the pure R-124 value. This could thus also occur for nucleate boiling's contribution in flow boiling. On the other hand, sometimes plain tubes demonstrate only a monotonic decrease in nucleate pool boiling coefficients with the addition of oil, such as the results for R-123 and R-11 obtained by Webb and McQuade (1993) using a mineral oil. This monotonic decline is also found in some flow boiling experiments for plain tubes, i.e. without a maximum in h(oil)/h(no oil), and thus the relative importance of this nucleate boiling oil effect has to be established, and why it affects some pool and flow boiling tests and not others.

OIL EFFECTS ON THE CONVECTIVE CONTRIBUTION TO FLOW BOILING

The influence of oil on liquid-phase convection in flow boiling of refrigerants may also be significant. Referring to the liquid Prandtl numbers and relative liquid Reynolds numbers in Table 2, one quickly deduces a strong effect on the liquid convective heat transfer coefficient used in flow boiling correlations. Obtaining h_L from the Dittus-Boelter correlation, it is directly proportional to the liquid Prandtl number ($Pr_L^{0.4}$) and the liquid Reynolds number ($Re_L^{0.8}$). Consequently, h_L drops off rapidly as the local liquid viscosity increases with oil concentration along an evaporator tube [h_L proportional to the -0.4 power of viscosity].

Figure 4 depicts the Gungor-Winterton (1987) flow boiling correlation for pure fluids evaluated with local mixture properties for R-134a/EAL 32 and R-134a/EAL 68 mixtures without any modification to the correlation (Mobil Arctic oil EAL 32: specific gravity at 15°C of 0.993, viscosities of 31.8 and 5.5 mm^2/s at 40°C and 100°C, respectively, and assumed molecular weight of 590). A standard operating condition of DX-evaporators was assumed (4.44°C or 40°F) where P_{sat} is 3.43 bar for pure R-134a. A tube diameter of 12 mm and a uniform heat flux of 10 kW/m^2 were assumed. The values of h(oil) for four inlet oil concentrations show a distinct drop off at high vapor qualities and with increasing oil concentration relative to R134a. This physical property effect on the convective contribution to flow boiling thus is one of the important keys to explaining and predicting the trend observed at high vapor quality in Figure 1. The higher viscosity grade oil (68 compared to 32) decreases the flow boiling coefficient, qualitatively predicting the trend shown in Figure 2 for SUS 150 and 300 oils.

MASS TRANSFER EFFECTS OF OIL ON FLOW BOILING

For boiling in mixtures, it is common to correlate the degradation due to mass transfer effects with an expression proposed by Stephan and Korner (1969):

$$\frac{h}{h_I} = \frac{(T_w - T_{bub})}{(T_w - T_{bub}) + dT_{bub}} \qquad (4)$$

where the ideal coefficient h_I is for an equivalent pure fluid with physical properties identical to the mixture and the mixture coefficient is defined as in Eq. (1). The rise in the local bubble point temperature at the interface of a growing bubble, dT_{bub} is caused by depletion of the volatile component (i.e. refrigerant) across the diffusion shell formed around the bubble, and it reduces the effective temperature driving force, such that h is less than or equal to h_I.

Referring to Table 1, the rise in T_{bub} with local oil concentration is relatively small for $w_{oil} < 20$ wt.%. For example, the bubble point temperature only rises by 0.27°C for a change in w_{oil} from 5 to 19.5 wt.%. Instead, for $w_{oil} > 20$ wt.% the slope becomes increasingly higher. Hence at low and medium vapor qualities where $w_{oil} < 20$ wt.%, the mass diffusion effect is minimal while at higher local oil concentrations it may be significant, further reducing h(oil) values in Figure 4 relative to the pure R-134a, particularly at high vapor qualities where the local oil concentrations are higher. [For a comprehensive review of mixture boiling see: Thome and Shock (1984), Thome (1990) or Collier and Thome (1994)].

OTHER EFFECTS OF OIL ON FLOW BOILING

Oil Holdup. In Figure 2, the large degradation in h(oil)/h(no oil) at low mass velocities almost certainly represents oil holdup in the evaporator tube, i.e. the flow of the oil rich liquid is impeded by the larger viscous and surface tension forces. Thus, the local oil concentration is higher than that predicted by equilibrium thermodynamics and has to be modelled hydrodynamically.

Foaming/Surfactant Effect. Some oils tend to foam and this may influence the local flow pattern; foaming in stratified flow or stratified wavy flow may cause the entire perimeter of a horizontal evaporator tube to be wetted rather than only partially wet. The oil probably also increases the contact angle, perhaps making nucleation sites more stable such that larger sites are activated at lower wall superheats.

CONCLUDING REMARKS

Explanations were proposed for many of the trends in existing refrigerant-oil experimental data as part of the review. Some of these trends can be predicted with existing flow boiling correlations utilizing the new thermodynamic approach for modelling vapor-liquid equilibria of refrigerant-oil mixtures together with locally-calculated mixture properties. In flow boiling the maximum often found in the ratio of h(oil)/h(no oil) at vapor qualities of about 0.4-0.5 appears to be caused by the maximum in the nucleate boiling contribution at this local oil concentration. The drastic decline in h(oil)/h(no oil) at high vapor quality is primarily caused by (i) the rapid rise in the local liquid viscosity and subsequent reduction in the convective boiling contribution via h_L, (ii) the adverse effect of mass diffusion on evaporation of mixtures and (iii) oil holdup at low mass velocities.

ACKNOWLEDGEMENT

The present research was funded by the Swiss Federal Office of Energy (OFEN).

NOMENCLATURE

G	mass velocity of total flow (kg/m^{2}s)
h	heat transfer coefficient (W/m^2 K)
h_I	ideal heat transfer coefficient (W/m^2 K)
h_L	liquid convective heat transfer coefficient from Dittus-Boelter correlation (W/m^2)
h_{LV}	latent heat of vaporization (J/kg)
Q	heat duty per unit mass (J/kg)
Pr_L	liquid Prandtl number
q	heat flux (W/m^2)
Re_L	liquid Reynolds number
T_{bub}	bubble point temperature of mixture (°C)
T_{sat}	saturation temperature (°C)
T_w	wall temperature (°C)
w_{oil}	local oil concentration in liquid (wt.%)
x_1	vapor quality at start of zone
x_2	vapor quality at end of zone

REFERENCES

Collier, J.G. and Thome, J.R. 1994. Convective Boiling and Condensation. 3rd Edition, Oxford University Press, Oxford.

Gungor, K.E. and Winterton, R.H.S. 1987. Simplified General Correlation for Saturated Flow Boiling and Comparisons of Correlations with Data, Chem. Eng. Res. Des., Vol.65, pp. 148-156.

Ha, S. and Bergles, A.E. 1993. The Influence of Oil on Local Evaporation Heat Transfer inside a Horizontal Microfin Tube. ASHRAE Trans., Vol. 99, Part 1, pp. 1244-1255.

Hambraeus, K. 1991. Heat Transfer Coefficient during Two-Phase Flow Boiling of HFC-134a, Int. J. Refrig., Vol. 14, pp. 357-362 (November).

Memory, S.B., Bertsch, G. & Marto, P.J. 1993. Pool Boiling of HCFC-124/Oil Mixtures from Smooth and Enhanced Tubes, ASHRAE Trans., Vol. 99.

Schlager, L.M., Pate, M.P. and Bergles, A.E. 1989. A Comparison of 150 and 300 SUS Oil Effects on Refrigerant Evaporation and Condensation in a Smooth Tube and a Micro-Fin Tube, ASHRAE Trans., Vol. 95(1), pp. 387-397.

Stephan, K. and Korner, M. (1969). Calculation of Heat Transfer in Evaporating Binary Liquid Mixtures, Chemie-Ingenieur Technik, Vol. 41(7), pp. 409-417.

Thome, J.R. 1986. Prediction of In-Tube Boiling of Mixtures in Vertical Thermosyphon Reboilers, 16th HTFS Research Symp., Heriot-Watt University, Paper RS677.

Thome, J.R. 1989. Prediction of the Mixture Effect on Boiling in Vertical Thermosyphon Reboilers, Heat Transfer Engineering, Vol.10, No. 2, pp. 29-38.

Thome, J.R. 1990. Enhanced Boiling Heat Transfer. Hemisphere, New York.

Thome, J.R. 1993. Thermodynamic and Transport Properties of Refrigerant and Lubricating Oil Mixtures, LENI Internal Report, Dept. of Mech. Engng., Swiss Federal Inst. of Technology - Lausanne (Jan. 28).

Thome, J.R. 1995. Comprehensive Thermodynamic Approach to Modelling Refrigerant-Lubricating Oil Mixtures, Int. J. HVAC&R Research, ASHRAE, Vol. 1, No. 2., pp. 107-123.

Thome, J.R. and Shock. R.A.W. 1984. Boiling of Multicomponent Liquid Mixtures, Advances in Heat Transfer, Academic Press, Orlando, Vol. 16, pp. 59-156.

Webb, R.L. & McQuade. W.F. 1993. Pool Boiling of R-11 and R-123 Oil-Refrigerant Mixtures on Plain and Enhanced Tube Geometries, ASHRAE Trans., Vol. 99, Part 1, pp. 1225-1236.

Table 1. R134a-oil temperature-enthalpy curve at 2.93 bar for 5 wt.% oil inlet concentration.

Local Vapor Quality	T_{bub} °C	w_{oil} wt.%	Total Q Absorbed kJ/kg	Latent Heat kJ/kg	Sensible Heat kJ/kg	% Error (5°C)	% Error (2°C)
0.000	0.076	5.00	0.00	0.00	0.00	-1.5	-3.8
0.093	0.084	5.51	18.53	18.52	0.01	-1.7	-4.2
0.186	0.094	6.14	37.06	37.04	0.02	-1.9	-4.7
0.279	0.106	6.93	55.60	55.56	0.04	-2.1	-5.3
0.372	0.123	7.96	74.14	74.08	0.06	-2.5	-6.2
0.465	0.146	9.35	92.68	92.59	0.09	-2.9	-7.3
0.558	0.180	11.31	111.23	111.10	0.13	-3.6	-9.0
0.651	0.236	14.33	129.80	129.61	0.19	-4.7	-11.8
0.744	0.346	19.53	148.42	148.11	0.31	-6.9	-17.3
0.837	0.663	30.67	167.22	166.59	0.63	-13.3	-33.2
0.846	0.729	32.47	169.14	168.44	0.70	-14.6	-36.5
0.856	0.819	34.72	171.08	170.29	0.79	-16.4	-41.0
0.865	0.922	37.04	173.03	172.14	0.89	-18.4	-46.1
0.874	1.056	39.68	175.00	173.98	1.02	-21.1	-52.8
0.883	1.240	42.74	177.02	175.82	1.20	-24.8	-62.0
0.893	1.541	46.73	179.15	177.66	1.49	-30.8	*
0.902	1.974	51.02	181.41	179.50	1.91	-39.5	*
0.911	2.726	56.18	183.96	181.33	2.63	-54.5	*
0.921	4.435	63.29	187.43	183.15	4.28	*	*
0.930	8.015	71.43	192.70	184.95	7.75	*	*

(% error determined assuming a heat flux of 10,000 W/m^2).

Table 2. R134a-oil properties at 2.93 bar for 5 wt.% oil inlet concentration.

Local Vapor Quality	0.000	0.279	0.558	0.744	0.930
Oil Concentration	5.00	6.93	11.31	19.53	71.43
Liq. Molecular Weight	126.4	135.9	157.2	197.3	450.6
Bubble Pt. Temp. (°C)	0.076	0.106	0.180	0.346	8.015
Total Q Absorbed (kJ/kg)	0.00	55.6	37.1	148.4	192.6
Liquid Density (kg/m^3)	1276	1270	1256	1230	1057
Vapor Density (kg/m^3)	14.5	14.5	14.5	14.6	16.8
Liquid Viscosity (cp)	0.398	0.458	0.628	1.137	35.83
Vapor Viscosity (cp)	0.011	0.011	0.011	0.011	0.011
Liq. Sp. Heat (kJ/kg K)	1.363	1.372	1.391	1.428	1.684
Vap. Sp. Heat (kJ/kg K)	0.889	0.889	0.889	0.889	0.906
Liq. Th. Cond. (W/m K)	0.094	0.095	0.095	0.096	0.105
Vap. Th. Cond. (W/m K)	0.012	0.012	0.012	0.012	0.013
Surf. Tension (dyne/cm)	11.7	11.7	11.7	11.8	17.2
Latent Heat (kJ/kg)	199.1	199.1	199.1	198.9	195.5
Critical Pressure (bar)	38.9	38.2	36.7	33.9	16.3
Critical Temperature (K)	405.7	417.9	445.5	497.4	824.7
Liquid Prandtl Number	5.75	6.64	9.20	16.9	576.2
Relative Liquid Re No.	1.000	0.869	0.634	0.350	0.011

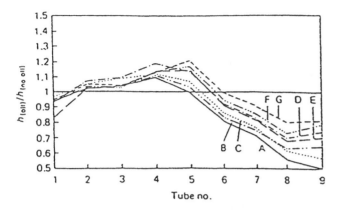

Figure 1. R-134a/oil mixtures evaporating at 2.3 bar and $q = 6$ kW/m^2 from Hambraeus (1991) in 12.0 mm bore tube. w_{oil}: A(2%), B(1.8%), C(1.7%), D(1.3%), E(1.1%), F(0.8%) and G(0.3%).

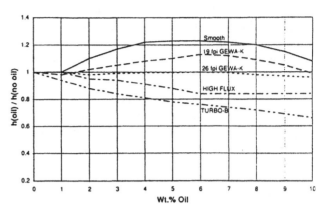

Figure 3. Oil Effect for Boiling of R-124 at $q = 25$ kW/m^2 [Memory, Bertsch and Marto (1993)].

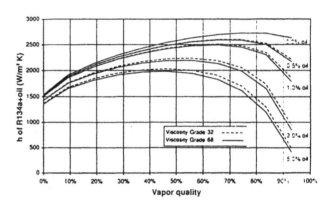

Figure 4. Predicted heat transfer for R-134a/oil mixtures for $G = 200$ kg/m^{2}s and $q = 10$ kW/m^2.

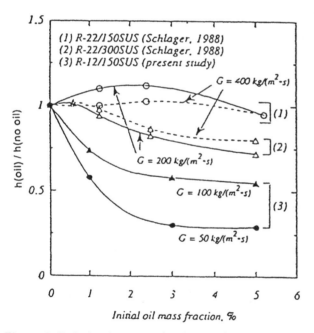

Figure 2. Relative heat transfer degradation for R-12 and R-22 oil mixtures from Ha and Bergles (1993).

SPECIAL SYSTEMS
AND EFFECTS

FLOW BOILING IN BUBBLY FLOW

Youyou Yan and D.B.R. Kenning
Department of Engineering Science
University of Oxford
Oxford, UK

ABSTRACT

High-speed video photography combined with liquid crystal thermography has been used to measure heat transfer near isolated steam or air bubbles sliding along the underside of a sloping, heated plate in water, in simulation of bubbly flow boiling on the lower surface of a tube in a horizontal bundle. The high rate of evaporative cooling under a vapour bubble masks the convective heat transfer that may occur in its wake. The experiments with gas bubbles, in which the evaporative cooling is negligible, show that there is a high rate of convective heat transfer in a very limited region just at the back of a sliding bubble but convection in its wake is only about 30% higher than in single-phase natural convection. There is limited evidence that convective cooling may be much more effective in a swarm of bubbles.

1. INTRODUCTION

In a kettle reboiler bubbly flow rises through the bundle of horizontal tubes. Flow boiling heat transfer may occur by a combination of nucleate boiling, evaporation of liquid microlayers under sliding bubbles and convection to the bulk flow driven by the rising bubbles, Cornwell (1990). Heat transfer correlations could be improved if the relative importance of these mechanisms were known.

In simulation of the conditions on the lower surface of a tube, experiments were performed on the heat transfer from the underside of an electrically-heated flat plate, sloping at an angle of 15^0 and immersed in a pool of water subcooled by 4K, Figure 1. The heat flux and the wall superheat were kept low enough to eliminate nucleate boiling. Steam bubbles were generated by a source outside the pool and injected so that they slid up the plate one at a time. The motion of the bubbles and the colourplay of thermochromic liquid crystal applied to the upper, dry surface of the 0.075 mm thick stainless steel plate were recorded simultaneously at 200 frames per second by a NAC high-speed colour video system. The experimental methods and preliminary

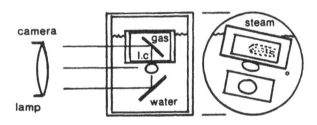

Fig.1 Experimental arrangement

results were described by Yan and Kenning (1994) and Kenning and Yan (1994). Since then the methods of processing the video recordings have been further developed. The recordings are first translated to Hue, Saturation and Intensity variables. The Hue distribution in each frame is then converted to temperature distribution, using a calibration determined for the liquid crystal on an isothermal block. The wall heat flux distribution is calculated from local heat balances using the measured rate of change of temperature and allowing for lateral conduction. The calculation is sensitive to signal noise, which is reduced by filtering, Kenning and Yan (1995). In Section 2 of this paper we present the wall temperature, heat flux and heat transfer coefficient distributions under a large sliding vapour bubble. The wall is strongly cooled by evaporation of the thin liquid layer between it and the bubble. Because the wall has to be thin (in order to use the liquid crystal on its rear surface) and the input heat flux is low (to avoid nucleation), the wall temperature falls nearly to the saturation temperature and remains depressed in the wake behind the bubble. The resulting small temperature difference between the wall and the bulk liquid, even with some subcooling, makes it difficult to measure accurately any further cooling by convection in the wake. In Section 3 we describe an experiment in which the steam bubble is replaced by an air bubble in subcooled water. This eliminates the evaporative cooling and makes it possible to measure the convective cooling more accurately.

2. SLIDING STEAM BUBBLE

Figure 2 shows the wall temperature and heat flux contours at five times as a vapour bubble slides at a velocity of 0.13 ms^{-1} along the underside of a plate inclined at an angle of 15^0 in water at atmosphere pressure and a bulk temperature of 96^{0}C. The input heat flux to the wall is 5.1 kWm^{-2}. Despite the subcooling, the bubble increases in size so there must be a net evaporative heat input. The bubble leaves a cooled trail behind it. Most of the heat removal is concentrated under the bubble itself. The cooling effect is better displayed by the temperature-time and heat transfer coefficient-time graphs in Figure 3 for the six points marked on Figure 2(a). The cooling starts just after the front of the bubble passes over a point. The two points on the centreline are cooled to temperatures slightly below the saturation temperature before the arrival of the back of the bubble and there is a small further fall in temperature behind the bubble, Figure 3(b),(c). The points off the centreline are subjected to less cooling, Figure 3 (a),(b),(e),(f). The strong cooling under the bubble is consistent with evaporation of a liquid microlayer, for which the driving temperature difference is the wall superheat. Cooling below the saturation temperature can only be caused by convective cooling by the subcooled bulk liquid so some bulk liquid must be pulled in very close to the wall at the back of the bubble. In the wake region the driving temperature difference is the wall superheat plus the bulk subcooling. The presentation of the data as heat transfer coefficient plots, in order to make some allowance for the local changes in wall temperature, is therefore complicated by the choice of temperature difference to define the heat transfer coefficient. The plots extending beyond the bubble in Figure 3(g-l) are based on the wall to bulk temperature difference. An additional line based on the wall superheat is plotted for the region covered by the bubble but it tends to infinity near the back of the bubble on the centreline, because of the convective cooling. In the wake the heat transfer coefficient based on wall to bulk temperature difference is about three times the value for single-phase natural convection ahead of the bubble near the centreline. The increase of the heat transfer coefficient at points off the centreline is less than that on the centreline.

3. SLIDING AIR BUBBLE

In order to eliminate evaporative cooling, a single air bubble, of similar size to the steam bubble in Section 2, was injected into water at a bulk temperature of 61^{0}C. The bubble moved at a velocity accelerating from 0.07 ms^{-1} to 0.13 ms^{-1}. The wall temperature distributions during the passage of the bubble are shown in Figure 4(a), together with the temperature changes from the first frame in Figure 4(b). The changes occur in a somewhat fluctuating region just at the back of the bubble and in a narrow wake. An encapsulated liquid crystal was used in these experiments, instead of the unencapsulated liquid crystal used in

the experiments on the steam bubble. There is more signal noise, which affects the heat transfer coefficient-time graphs for particular positions shown in Figure 5. Nevertheless the graphs confirm that there is a small region of greatly enhanced heat transfer at the rear edge of the bubble. The increase further away in the wake is less than it was for the steam bubble and is about 1.3 times the coefficient ahead of the bubble. The reason for this is still under investigation.

4. MULTIPLE BUBBLES

A mechanism for controlled injection of arrays of steam or air bubbles is still under development. Experiments in which a sequence of air bubbles was injected from a single source indicated that there was a large increase in the heat transfer coefficient over most of the plate. When air bubbles were generated at 9 Hz, (with bubble velocity of 0.17 ms^{-1} and bubble projected area of 4% of the plate surface area), the heat transfer coefficient at the centreline was 3 times that when there were no bubbles.

5. CONCLUSIONS

It has been shown that evaporative cooling is very effective under isolated sliding vapour bubbles, cooling a thin heated plate almost to the saturation temperature. This indicates that the properties of the wall will have an effect on the heat transfer. The evaporative cooling is augmented by convection to the subcooled bulk liquid in a small region at the back of the bubble and in the bubble wake. The effectiveness of the wake cooling is still unresolved and requires further work at higher concentrations of sliding bubbles.

ACKNOWLEDGEMENT

This work was supported by EPSRC Grant No. GR/H45377 and was performed in collaboration with Heriot-Watt University.

REFERENCES

Cornwell, K., 1990, "The Influence of Bubbly Flow on Boiling from a Tube in a Bundle", *Int. J. Heat Mass Transfer*, 33, pp.2579-2584.

Kenning, D.B.R., and Yan, Y., 1994,"Heat Transfer Near Sliding Vapour Bubbles in Boiling", *Proc. 10th Int. Heat Transfer Conference*, Brighton 14-18 August 1994, Vol.5, pp.195-200.

Kenning, D.B.R. and Yan, Y., 1995, "Pool Boiling Heat Transfer on a Thin Plate: Features Revealed by Liquid Crystal thermograph", Oxford University, Department of Engineering Science, Report No. OUEL 2055/95.

Yan, Y. and Kenning, D.B.R., 1994, "Liquid Crystal Measurements of Wall Temperatures in Boiling Heat transfer", *Optical Methods and Data Processing in Heat and Fluid Flow*, I.Mech.E. Seminar, City University, 14-15 April 1994, pp.37-42.

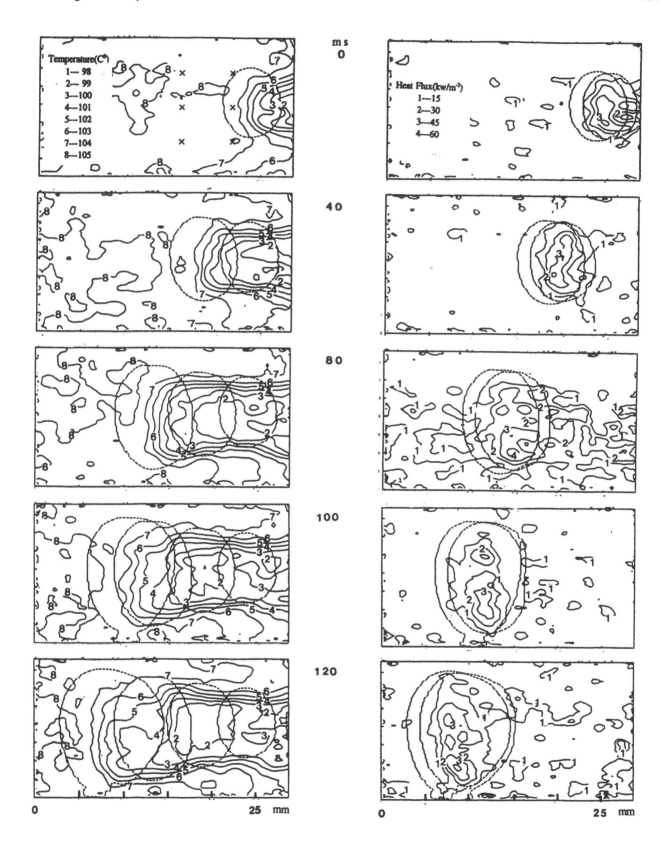

Fig. 2(a) Temperature distributions (Steam bubble)   Fig. 2(b) Heat flux distributions (steam bubble)

T (C⁰)

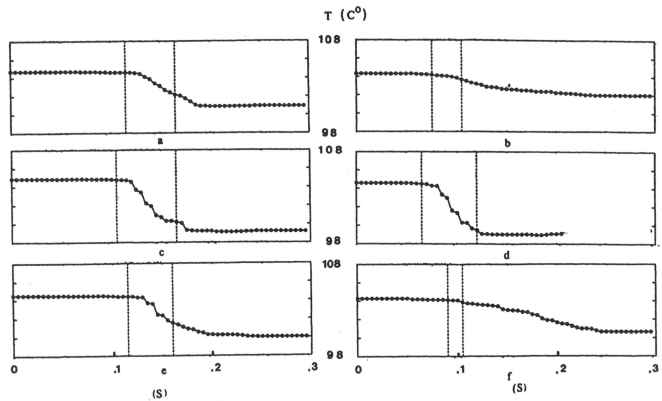

Fig. 3 (a)-(f) Temperature-time (steam bubble)

Heat Flux(kW/m⁻²)

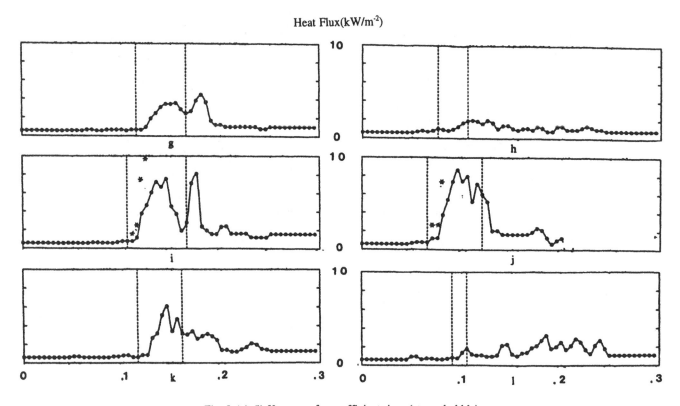

Fig. 3 (g)-(l) Heat transfer coefficient-time (steam bubble)

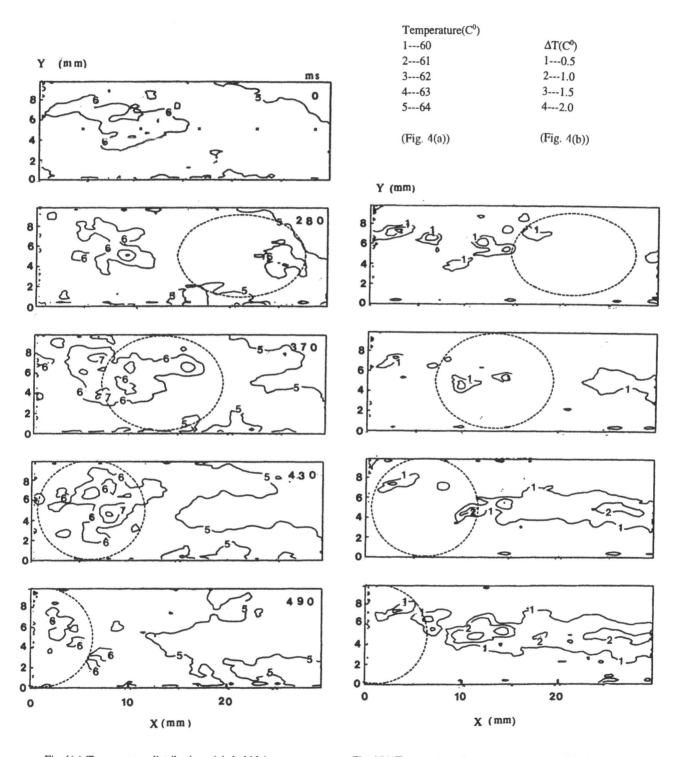

Fig 4(a) Temperature distributions (air bubble)   Fig 4(b) Temperature change contours (air bubble)

Heat Flux(kW/m^{-2})

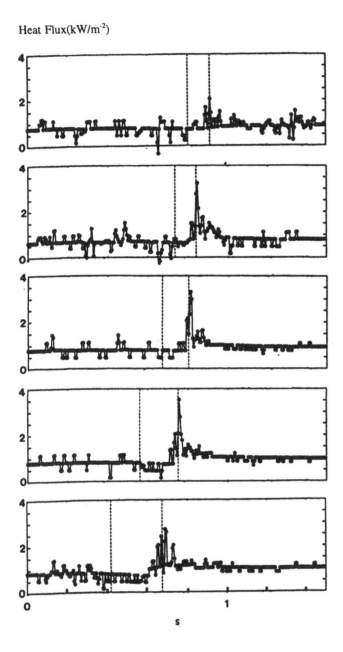

Fig 5. Heat transfer coefficient-time (air bubble)

FREE AND FORCED CONVECTIVE BOILING HEAT TRANSFER FROM A SMALL HEATING ELEMENT

Kunio Hijikata *,   Takao Nagasaki **   and   Masayuki Hinokuma *
* Department of Mechano-Aerospace Engineering
Tokyo Institute of Technology, Meguro-ku, Tokyo, Japan
** Department of Energy Sciences
Tokyo Institute of Technology, Midori-ku, Yokohama, Japan

ABSTRACT

Forced convective and pool boiling heat transfer from a small thin film square heaters on a glass plate has been investigated using various sizes of heaters ($20\times20\mu$m to 5×5mm), which are cooled by FX3250 saturated liquid in a narrow channel or in a pool. In addition to the measurement of heater temperature, boiling behavior was observed by a high-speed video camera which was synchronized with the measurement of temperature fluctuation. In the case of larger heaters, heat transfer characteristics in nucleate boiling region are similar to those for conventional wide heating surface, and transition to film boiling occurs with moderate temperature increase. The transition occurs at a larger heat flux for larger flow rate. With the decrease of heater size the heat flux increases due to the effect of substrate heat conduction, and the heat transfer characteristics approaches those for pure substrate conduction. Further, in the case of smaller heaters, only one bubble is nucleated periodically, and the heater temperature fluctuates corresponding to the bubble growth and departure. The heater temperature reaches minima just at the moments of bubble detachment from the surface. In addition, under certain conditions, a large bubble is formed which fills the channel and reduces the heat transfer. The effect of flow velocity on the boiling behavior was also observed.

INTRODUCTION

With the increase of heat dissipation of microelectronic devices the development of effective cooling methods for small heat sources with high heat flux becomes important. Direct liquid cooling with boiling is a promising method among other modes of cooling such as natural or forced convection, and many experiments have been reported on the boiling heat transfer characteristics from microelectronic chips or simulated flush-mounted heaters, including temperature overshoot at the incipient boiling and critical heat flux (Lee, et al., 1992, You, et al.,1992). Although, in many cases, the chip size is regarded as the characteristic dimension of heat source, ultimate heat source is the semiconductor junction in the chip, and the boiling heat transfer characteristics from such a small heater would also be important when the chip surface is directly cooled by the liquid.

Baker (1973) reported boiling curves for various sizes of rectangular thin-film heaters, ranging from 0.01cm^2 to 2cm^2, flush mounted on glass and sapphire substrates. The results show increasing heat flux with the decrease of the heater size, and a significant effect of the substrate conduction is observed in the case of the smallest heater. Samant and Simon (1989) investigated forced convective boiling from a small heating element (0.25mm long in the flow direction and 2mm wide in the spanwise direction) sputtered on a quartz substrate. In their study the substrate conduction was numerically estimated, and the effect of flow velocity on the wall heat flux were reported. Although these studies revealed heat transfer characteristics peculiar to small heaters, knowledge on fundamental aspects of boiling heat transfer from small heating elements on a substrate is still limited, especially when the heater size is smaller than the bubble.

From this view point the authors made experiments previously on pool boiling using various sizes of square thin film heaters ($20\times20\mu$m to 5×5mm) on a glass plate (Nagasaki et al., 1993). It was shown in these experiments that in the case of smaller heaters only one bubble is nucleated on the heater, and the heater temperature fluctuates due to the bubble growth and departure. Further the interaction of bubbles, which are nucleated at two or three heaters neighboring in various distance, was investigated (Fushinobu, et al., 1994). In this paper forced convective boiling has been investigated using the same heating elements, and the results are compared with those of pool boiling.

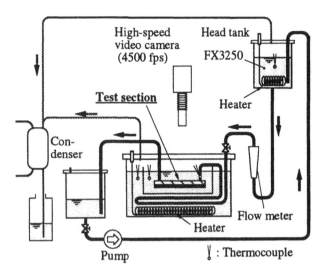

Figure 1. Experimental Apparatus

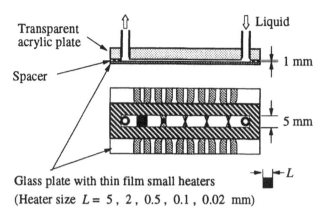

Figure 2. Details of Test Section

EXPERIMENTAL APPARATUS

A schematic diagram of the experimental apparatus is shown in Fig.1. The working fluid is FX3250 (a perfluorinated fluorocabon of the 3M Company, nearly the same coolant as FC72), and is boiled in the head tank to eliminate the dissolved gas. The liquid is supplied to the test section which is immersed in a liquid pool of FX3250. The liquid pool is also heated to maintain the test section at the saturated temperature. The liquid temperature is measured at the inlet of the test section, and the inlet subcooling is kept less than 1.5 °C.

The detail of the test section is shown in Fig.2. Heating elements are square thin films of In_2O_3 (thickness 20nm) deposited on a glass plate (thickness 1mm). The length, L, of the square heaters are 0.02mm, 0.1mm, 0.5mm, 2mm and 5mm. Each heating element has two pairs of leads (Ni 500nm and Au 50nm), one of which is used for resistive heating by DC current and another is used to measure the voltage drop across the heater. The values of the DC current and voltage drop are digitized by digital multimeters (5 1/2 digit), which is connected to a personal computer via GPIB to calculate the electric resistance for the measurement of heater temperature based on a precalibrated temperature dependency of the resistance. The data acquisition rate is 10 Hz and is not sufficient to measure the temporal change of the heater temperature caused by the bubble growth and departure in the case of smaller heaters. Therefore the above method is used only to measure the averaged temperature. In order to measure the temperature fluctuation a conventional bridge method is used with one pare of leads to detect the change of electric resistance, and the bridge voltage is recorded in a 8 bits digital storage oscilloscope.

A small channel, 5mm in width and 1mm in height, is formed on the glass plate, and the heater is cooled by the fluid flow. The top wall of the channel is transparent for the visual observation by a high-speed video camera, which stores images in IC memory (maximum 4500 fps). The

video recording and the measurement of temperature fluctuation are started simultaneously by a trigger in order to clarify the relationship between bubble behavior and temperature fluctuation.

In addition to the convective boiling, pool boiling experiments were also made. The glass plate is placed in the liquid pool with the heater surface vertically upward. In this case, side view is taken by the high-speed video to observe the bubble departure from the surface clearly.

EXPERIMENTAL RESULTS

Liquid Convection

As the most simple case of convective heat transfer, the relationship between the Nusselt number, Nu, and the Reynolds number, Re, for liquid single-phase convection is shown in Fig.3. The characteristic length of Nu and Re is the heater length, L. Nu_l and Nu_s are based on the thermal conductivity of liquid and glass, respectively. The chain-dotted line in Fig.3 denotes the heat transfer of laminar

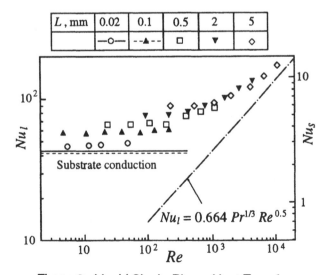

Figure 3. Liquid Single-Phase Heat Transfer

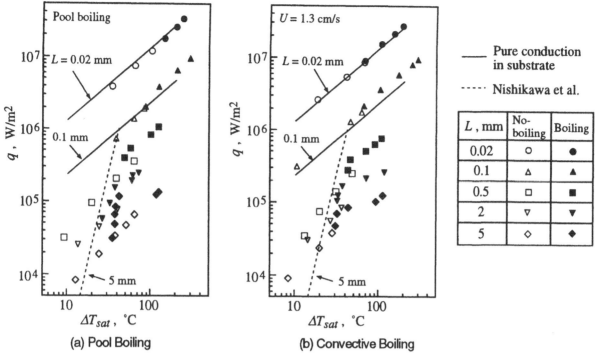

Figure 4. Boiling Curves

boundary layer on a flat plate with uniform temperature. Although the present flow does not correspond to the boundary layer flow rigorously, the data approaches the boundary-layer solution with the increase of the Re number. On the other hand, with the decrease of the Re number, the Nu number approaches a constant value. The solid and broken lines are theoretical values of the pure three-dimensional thermal conduction for a thin film heater on a semi-infinite substrate (Nagasaki, 1993). In the case the smallest heater, the experimental data are well explained by the substrate conduction. The asymptotic value of the experimental data with decreasing the Re number increases with the heater size possibly due to the effect of side walls.

Boiling Curves

Boiling curves for pool and convective boiling are shown in Figs.4(a) and (b), respectively. In the case of Fig.4(b) the inlet flow velocity, U, is 1.3 cm/s, which is the smallest one tested. Open symbols denote the data when boiling does not occur, and closed symbols denote boiling data. The ordinate, q, is a nominal heat flux, i.e. the heating rate per unit area of the heater. The broken line denotes the correlation by Nishikawa and Fujita (1977) applied for L=5mm. In the case of larger heaters ($L \geq$ 0.5mm), the slope of the present data in the nucleate boiling region is nearly the same as that of conventional correlation. The data points sift towards a larger heat flux with decreasing heater size. Such a tendency is consistent with the results of Baker (1973). Further, in the case of smaller heaters ($L \leq$ 0.1mm), the slope of the boiling data is the same as that for non-boiling data, and no difference between boiling and non-boiling data is observed in the

case of the smallest heater. The solid lines are the theoretical values of temperature increase determined by pure thermal conduction in the substrate (Nagasaki, 1993), and it is clear that the temperature increase of very small heaters is controlled by the substrate conduction.

Transition of boiling mode is illustrated in Fig.5. A temperature overshoot occurs at the onset of nucleate boiling, and further increase of heat flux brings the transition to film boiling. Figs.6(a) and (b) show snap shots of high-speed video corresponding to nucleate and film boiling, respectively, for L=2mm. In Fig.6(b) film boiling occurs in the down stream portion of the heater surface. Fig.7 shows boiling behaviors in the case of the largest flow velocity tested (U=58cm/s). In Fig.7(b) film boiling is occurring on the whole heater surface. The heat flux of Fig.7(a) is the same as that of Fig.6(b), however, the boiling mode is nucleate one due to the increased flow velocity. The temperature increase in the transition from nucleate to film boiling is moderate and does not cause damages (burn out) to the heater due to the effect of substrate conduction. Nearly the same tendency of transitions is observed for larger heaters (L= 5, 2, 0.5mm) both in pool and convective boiling as shown in Fig.4.

Detailed comparisons of results for different flow velocities and pool boiling are shown in Fig.8 for each heater size. In some regions of heat flux there appear boiling and non-boiling data for the same heat flux due to the hysteresis when heat input is increased and decreased due to the temperature overshoot at the onset of nucleate boiling. As shown in Figs.8(a) and (b) for L=2 and 5mm, the effect of flow velocity on boiling data is relatively small compared with that on no-boiling data, however,

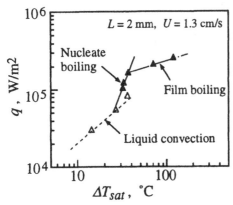

Figure 5. Boiling Modes

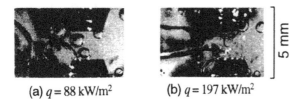

(a) $q = 88$ kW/m^2 (b) $q = 197$ kW/m^2

Figure 6. Boiling Behavior ($L = 2$ mm, $U = 1.3$ cm/s)

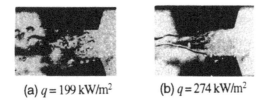

(a) $q = 199$ kW/m^2 (b) $q = 274$ kW/m^2

Figure 7. Boiling Behavior ($L = 2$ mm, $U = 58$ cm/s)

nucleate boiling is maintained at higher heat flux for larger flow velocity. On the other hand, in the case of L=0.5mm in Fig.8(c), the existence of nucleate boiling region is not so distinct as Figs.(a) and (b), due to the increasing effect of substrate conduction. Further, the boiling data for larger flow velocity (■) shows larger heat transfer than others. The reason is explained by the difference of boiling behavior as shown in the next section. In the case of L=0.02mm and 0.1mm, the heat transfer characteristics as shown in Fig.4 were not affected by the flow velocity due to the predominant effect of substrate conduction.

Boiling Behaviors and Temperature Fluctuation

Figure 9 shows an example of boiling behavior and the temporal change of heater temperature for L=0.1mm, in which case only one bubble is nucleated periodically, and the heater temperature fluctuates corresponding to the bubble growth and departure. The numbers noted in the temperature fluctuation correspond to the photograph number, and the heater temperature reaches minima at the moment of bubble departure from the heater surface. In the case of L=0.1mm the heater temperature is mainly determined by the substrate conduction, and heat flux is

	Pool boiling	U, cm/s	
		1.3	58
No-boiling	○	△	□
Boiling	●	▲	■

---- Nishikawa et al. ($L = 5$ mm)

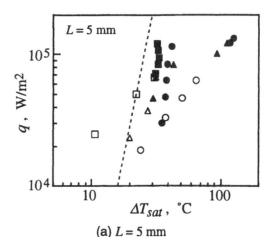

(a) $L = 5$ mm

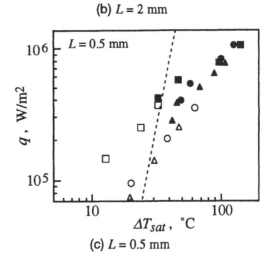

(b) $L = 2$ mm

(c) $L = 0.5$ mm

Figure 8. Effect of Flow Velocity

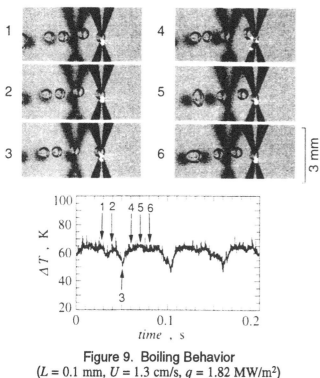

Figure 9. Boiling Behavior
(L = 0.1 mm, U = 1.3 cm/s, q = 1.82 MW/m²)

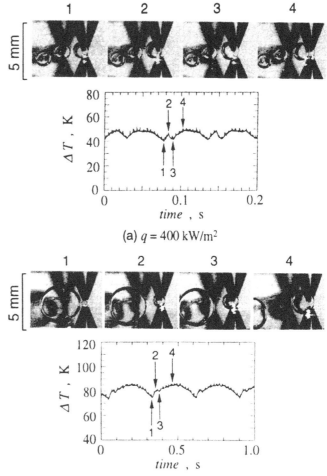

(a) q = 400 kW/m²

(b) q = 709 kW/m²

Figure 10. Boiling Behavior (L = 0.5mm, U = 1.3 cm/s)

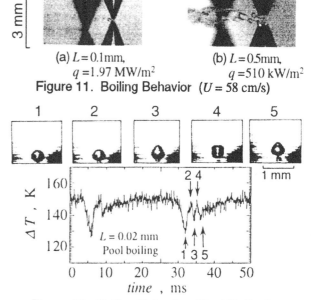

(a) L = 0.1mm,   (b) L = 0.5mm,
q = 1.97 MW/m² q = 510 kW/m²

Figure 11. Boiling Behavior (U = 58 cm/s)

Figure 12. Boiling Behavior (Pool Boiling)
(L = 0.02 mm, q = 22.5 MW/m²)

very large compared with larger heaters as was shown in Fig.4, still, the heater temperatures is affected by boiling heat transfer at the moment of bubble departure, which indicates extraordinary large surface heat flux at the bubble departure.

Nearly the same behavior is observed in the case of small heat flux for L=0.5mm as shown in Fig.10(a). In this case two minima are observed in the heater temperature at the bubble departure, because a new bubble, which is created just after the bubble departure, coalesces into the leaving large bubble. With the increase of heat flux the bubble size increases as shown in Fig.10(b). The bubble size is much lager than the channel height (1mm), then it is expected that the bubble fills the channel cross section and covers the heater surface resulting in the reduction of surface heat transfer as was shown in 8(c).

Figs.11(a) and (b) show results in the case of large flow velocity for L=0.1 and 0.5mm, respectively. Remarkable decrease of bubble size is observed compared with Figs.9 and 10 for small flow velocity. Further, in the case of L=0.5mm, several bubbles are nucleated on the heater surface which will cause large surface heat transfer evidenced in Fig.8(c). Such a effect of flow velocity is important in the case of the intermediate heater size, in which case the size of departing bubble relative to the heater size changes. In these cases no significant temperature fluctuation was observed even in the case of single-bubble formation for L=0.1mm, which might be caused by high frequency of bubble departure and the thermal inertia of the substrate.

Fig.12 shows a result in the case of pool boiling for the smallest heater, $L= 0.02$mm. In the case of pool boiling a side view was taken, and the formation and coalescence of small bubble immediately after the departure of the main bubble are clearly observed. Such a temperature fluctuation of a small heater reflects the temporal change of microscopic surface heat transfer during the bubble cycle, therefore, a quantitative estimation of the fluctuating surface heat transfer rate incorporating the unsteady heat conduction in the substrate would be important to clarify the detailed mechanism of boiling heat transfer.

Finally, in the case of the single-bubble mode (L=0.02, 0.1 and 0.5mm), the bubble departing diameter and frequency were measured from high speed video recording as shown in Fig.13. The bubble size decreases with the decrease of heater size, with the decrease of heat flux and with the increase of flow velocity. The bubble departing frequency seems to be mainly affected by flow velocity. In addition, it is shown that, in the case of L=0.5mm, extraordinary large bubbles are formed at low frequency due to the constraint by the channel, as mentioned previously. Although detailed analysis has not yet been done, several informations could be obtained by these measurements such as the contribution of vaporization on the heat transfer and the dynamics of bubble detachment.

CONCLUSIONS

The characteristics of convective and pool boiling heat transfer from small heaters on a substrate have been investigated using various sizes of heaters ranging from 5×5mm to 0.02×0.02mm, and the following conclusions were obtained.

1. The heat flux plotted against the heater temperature increases with a decrease of heater size mainly due to the effect of substrate conduction, and approaches the theoretical value for the pure conduction in the substrate.

2. The heat transfer in the nucleate boiling is not greatly affected by the flow velocity. Transition to film boiling is moderate without accompanying burn out. The transition to the film boiling occurs at larger heat flux with increasing the flow velocity.

3. In the case of intermediate size of heater at small flow velocity, the vapor bubble fills the channel and reduces the heat transfer.

4. The bubble departing diameter is greatly affected by the flow velocity, which has a significant effect on the heat transfer in the case of intermediate heater size (L= 0.5mm).

5. In the case of small heaters, only one bubble is nucleated, and the heat transfer increases at the moment of bubble departure.

REFERENCES

Baker, E., 1973, "Liquid Immersion Cooling of Small Electronic Devices," *Microelectronics and Reliability*,

Vol.12, pp.163-173.

Fushinobu,K., Nagasaki,T., Saitoh,T., UI,A. and Hijikata,K., 1994, "Boiling Heat Transfer Characteristics from very small heaters on a substrate," *Proc. 10th Int. Heat Transfer Conf.*, Vol.5, pp.51-56.

Lee,T.Y.T., Mahalingam,M and Normington,P.J.C., 1992, "Subcooled Pool Boiling Critical Heat Flux in Dielectric Liquid Mixture," *Topics in Heat Transf.*, ASME HTD Vol.206-2, pp.63-73.

Nagasaki,T., Hijikata,K., Fushinobu,K. and Saitoh,T., 1993, "Boiling Heat Transfer from a Small Heating Element," ASME HTD Vol.262, pp.15-22.

Nishikawa,K. and Fujita,Y., 1977, "Correlation of Nucleate Boiling Heat Transfer Based on Bubble Population Density," *Int.J. Heat Mass Transf.*, Vol.20, pp.233-245.

Samant,K.R. and Simon,T.W., 1989, "Heat Transfer from a Small Heated Region to R113 and FC-72," *Trans.ASME J.Heat Transf.*, Vol.111, pp.1053-1059.

You, S.M., Simon, T.W. and Bar-Cohen, A., 1992, "Pool Boiling Heat Transfer with an Array of Flush-Mounted Square Heaters," *Topics in Heat Transf.*, ASME HTD Vol.206-2, pp.55-62.

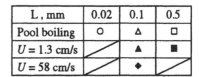

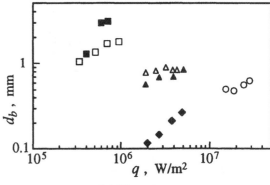

(a) Diameter

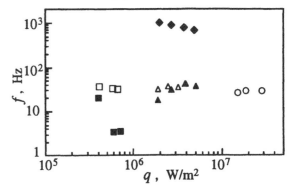

(b) Frequency

Figure 13. Bubble Departure Characteristics

CONVECTIVE FLOW BOILING IN MICROGEOMETRIES:
A REVIEW AND APPLICATIONS

R. S. Stanley, T. A. Ameel, and R. O. Warrington
Institute for Micromanufacturing
Louisiana Tech University
Ruston, Louisiana

ABSTRACT

As the field of microfluids and micro heat transfer continues to grow, it becomes increasingly important to understand the mechanisms and fundamental differences involved with heat transfer in two-phase flow in microgeometries. This paper presents a summary of the research in the area of two-phase convective heat transfer in open and closed microchannels with applications in devices such as micro heat pipes and micro heat pumps which may be used for electronic equipment cooling and potentially for macroscale cooling systems. Maximum heat fluxes of 350 W/cm², 130 W/cm², and 80 W/cm² have been found for open and closed microchannel convection flow boiling, and micro heat pipes, respectively. Some of the most promising areas for future development are the areas of liquid metal heat pipes and micro vapor compression cycles.

INTRODUCTION

The area of two-phase forced convection heat transfer in open and closed microchannels is a burgeoning field. Though extensive work has been done in the area of single phase liquid and gas flow through microgeometries (Choi, 1991; Yu,1995; Harley et al., 1989a and 1989b; Pfahler et al., 1991, 1990a and 1990b), the only area where detailed analysis of two-phase heat transfer in microgeometries has taken place is in the specialized area of micro heat pipes. As the field of microfluidics continues to grow, it is becoming increasingly important to understand the mechanisms involved with heat transfer in two-phase flow thorough microgeometries. This importance is reiterated by the increasing number of applications which use phase change as the principle mechanism to conduct or remove heat. These fields include bioengineering and biotechnology, aerospace, mini heaters, and mini heat exchangers, electronics and microelectronics, material processing and thin film deposition technology, etc.(Peng and Wang, 1993). Other areas of research include cooling of laser diode arrays, micro Joule-Thompson cooling devices, and the evaporator and condenser sections of micro vapor compression cycles. A good deal of work has been done by Little (1990a & b, 1984) in developing a commercially available mini Joule-Thompson cooling device capable of dissipating 0.25 W over a few square centimeters at cryogenic temperatures. In addition, the development of a microscale Joule-Thompson cooling device is currently underway at the Institute for Micromanufacturing. This program seeks to design, fabricate using LIGA technology, package, and test a complete microscale device which may have applicability toward electronic chip cooling. A micro heat pump concept has been proposed by researchers from Battelle Pacific Northwest Laboratories (Drost et al., 1994). The PNL concept incorporates a sheet architecture in which the various processes of a vapor compression refrigeration system are divided into individual unit operations and sheets of material with embedded micromechanical systems are developed that can perform each unit operation. The sheet unit operations could then be combined together to form more complex systems. The major advantage of two-phase flow is that the temperature gradient along the length of the channel is much smaller than in single phase flow, due to the phase change phenomena. This results in larger values of the heat transfer coefficient and reduces the requirement of large liquid flow rates.

By far, the majority of the reported research in this field are empirical studies in the area of cooling of electronic devices or systems. This is achieved either by pool boiling, open or closed channel forced convection flow boiling, or with micro heat pipes. Since pool boiling is not generally considered a forced convection process it will not be considered here. Maximum heat fluxes of 320 W/cm², 130 W/cm², and 80 W/cm² were found for open and closed channel forced convection flow boiling, and micro heat pipes, respectively. The magnitude of these heat flux data show why two-phase heat transfer is such an important area of research. Liquid metal heat pipes, because they operate at very high temperatures, have the potential of dissipating or removing the same or greater amount of heat as in open channel flow boiling, while the micro vapor compression cycles could provide more efficient operation and better energy management. The development of micro vapor compression cycles will not only provide for better thermal control on the microscale, but the macroscale as well.

It is the purpose of this paper is to present the work that has been reported in the area of forced convection two-phase heat transfer in open and closed microchannels, compare the maximum or critical heat fluxes (CHF) achieved, and to show the trends that are occurring on the microscale. The microscale is defined, here, to be geometries with hydraulic diameters of 500 μm or less.

OPEN CHANNEL MICRO FLOW BOILING

Typically, open channel micro flow boiling is defined as the boiling that occurs on protrusions or etched channels with heights or depths on the order of 500 μm or less. Only the work reported by Mudawar and Maddox (1990, 1989) and Maddox and Mudawar (1989) considered protrusions on this scale. All others had protrusions (fins) or channels on the order of 1 mm or larger, far outside the microscale. In the experiments by Mudawar and Maddox (1990, 1989) and Maddox and Mudawar (1989), heat transfer from a single flush-mounted 12.7 mm x 12.7 mm simulated microelectronic chip in vertical upflow of FC-72 was considered. Mudawar and Maddox (1989) first examined this configuration with the chip surface being smooth (no measurable protrusion). It was found that the CHF increases with increased fluid velocity or subcooling. A semi-empirical CHF model which correlated their data in a nondimensional equation based on the Weber number was developed. The maximum CHF achieved during these experiments was on the order of 100 W/cm². Mudawar and Maddox (1990) and Maddox and Mudawar (1989) extended these flow boiling experiments to include microgroove, microstud and pin surfaces extending from the smooth chip surface. Of importance to this paper were the microgrooves and microstuds which had dimensions near or in the microscale regime. The pin fin surfaces had dimensions far greater than the microscale regime, but are offered here for comparison. The dimensions for the extended surfaces were as follows: 1) microgrooves 305 μm wide and 1.02 mm deep, 2) microstuds 305 μm square with their height varying from 250 μm to 1.02 mm, and 3) pin fins with a diameter of 5.75 mm and 11.5 mm tall. A combination of moderate coolant velocity and a high degree of subcooling produced CHF values of 262, 317 and 361 W/cm² for the microstud, microgroove, and pin fin surfaces, respectively.

CLOSED CHANNEL MICRO FLOW BOILING

As with open channel micro flow boiling, the amount of theoretical development in closed channel micro flow boiling is very limited. In fact, the only general analytical development in this area are equations for the pressure drop in flow boiling given by Bowers and Mudawar (1994). These pressure drop equations include the contributions of single phase flow, boiling flow, and outlet conditions of the channel, where the boiling pressure drop consisted of acceleration and friction effects. Also, like open channel micro flow boiling, the studies to date only deal with working dimensions at or above 500 μm. This lack of data for geometries of less than 500 μm reiterates the need for extensive research in these areas.

Zhukov and Yarmak (1990) presented a strictly empirical look at the boiling behavior of superconducting magnet windings. The experimental apparatus consisted of an 800 μm diameter tube, 1m long, with the first 150 mm heated. The working fluids were helium and nitrogen. Though the data given was not conducive to the CHF comparisons that will be made later, several important observations were made. Zhukov and Yarmak (1990) observed that boiling under heat pulse resulted in an abrupt pressure rise in the heated section which can cause the flow to stop quickly and, in some cases, may result in reverse flow. They also reported that the vapor phase growth rate in channels with pulse heat is much lower than that for the steady state boiling regime and that the helium heat transfer crisis is apparently due to the pressure jump over the critical point at the onset of boiling.

Peng and Wang (1993) experimentally investigated the transition between the regions of forced convection and nucleate boiling of deionized water in microchannels that were 600 μm wide, 700 μm high, and 60 mm long. The top surface of the test section was Pyrex glass so that the flow could be visualized. Experimental observation yielded two important events; 1) the boiling regime was fully developed nucleate boiling, and 2) bubbles did not grow in microchannel boiling. Peng and Wang (1993) mentioned "that the bubble growth in liquid might be concerned with the scale of the liquid bulk. If the scale of liquid bulk is large enough, bubbles grow, otherwise, no bubbles grow and exist in liquid" flowing in microchannels. This statement suggests that in microchannel flow, the fluid properties vary enough from the bulk properties to provide a noticeable change in flow behavior. The experimental results show a dramatic increase in the slope of the heat flux when the transition to boiling is reached. Peng and Wang (1993) also compared existing macroscale Nusselt number correlations, along with experimental data of flow boiling in a 9 mm tube (He, 1988), to their experimental data. From these comparisons, they concluded that the heat flux was about the same or lower for the microchannel than for the 9 mm tube before transition; however, after transition the heat flux was remarkably higher for the flow in microchannels. No partial (nucleate) boiling was observed by Peng and Wang (1993) for the transition from single-phase liquid convection to nucleate boiling nor was there an obvious effect on the flow nucleate boiling by the velocity and liquid subcooling. The transition phenomena was probably an optical effect or the experimental configuration was such that the flow transitioned past the regime of partial nucleate boiling. A final observation made by Peng and Wang (1993) was that no bubbles grow for flow boiling in microchannels. This, also, was probably an optical effect caused by the combination of the tendency of the bubble size to decrease at it condenses back to a liquid and the tendency of the bubble size to increase with the decreasing pressure in the channel.

Bowers and Mudawar (1994) presented an experimental comparison of mini-(2.54 mm) and micro-(510 μm) channel heat sinks with a 1 cm heated length using R-113 as the working fluid. Besides the analytical pressure drop model previously mentioned, a CHF correlation was given and both were shown to be in good agreement with the experimental data. The effect on heat transfer of several variables was discussed, including pressure drop, inlet subcooling, superheated exit conditions, and channel size. One problem reported for this boiling heat transfer configuration was that the production of vapor bubbles, which give rise to the increased heat transfer rates, also increased the pressure drop. This problem was shown to be magnified, as would be expected, as the size of the channel decreases. The inlet subcooling was shown to have little or no effect on the CHF which is in agreement with the findings of Peng and Wang (1993). Some other features highlighted by the experimental results were: 1) "The two heat sinks demanded minimal flow rates..." (to allow complete evaporation), 2) "The small diameters of mini- and microchannels suggest an increased frequency and effectiveness of droplet impact with the channel wall in regions of high x_L (quality) values", and 3) "The small overall size of the heat sink seems to greatly contribute to delaying CHF by conducting heat away from the downstream region undergoing either partial or total dryout to the boiling region of the channel."

The CHF data and the resulting heat transfer coefficient (h) data as a function of the temperature difference ($T_{wall} - T_{inlet}$) reported by Peng and Wang (1993), He (1988), and Bowers and Mudawar (1994) are shown in Figs. 1 and 2, respectively. From the data given by Peng and Wang (1993), the CHF showed little or no dependency on the liquid velocity. The CHF value of 130 W/cm² corresponds to a liquid velocity of 2.32 m/s whereas the CHF value of 90 W/cm² has the highest velocity of 3.6 m/s. Comparing the 600 µm x 700 µm channel from Peng and Wang (1993) to the 9 mm tube shows the inefficiency of the 9 mm tube. A much larger temperature difference is required to achieve the same heat flux. It should be pointed out that the liquid velocity in the 9 mm tube was much lower than in the rectangular channels. The data for the micro- and minichannels indicates that, though there is more heat transfer per channel for the minichannels, the increased density of the microchannels make them a more effective system. Again, the one drawback to microchannels is the high pressure drop. Comparing the CHF values for open and closed channels, it is evident that the open channels achieve a much higher CHF. The open microchannels achieve a maximum CHF on the order of 320 W/cm² whereas the closed microchannels achieve a maximum CHF on the order of 130 W/cm². In Fig. 2, the same trends are seen and the effectiveness of the different configurations is more pronounced.

HEAT PIPES

Since the introduction of micro heat pipes by Cotter (1984), the range of application has expanded from the thermal control of laser diodes, photovoltaic cells (Peterson, 1992), and infrared detectors (Wu and Peterson, 1991a, 1991b), to the removal of heat from the leading edges of stator vanes in turbines (Longtin et al., 1992), leading edges of hypersonic aircraft, and nonsurgical treatment of cancerous tissue (Peterson, 1992). Micro heat pipes are also being developed for use in radiator fin material to remove the heat generation in space vehicles (Badran et al., 1993). However, most of the research being conducted today is in the area of cooling of semiconductor chips. Micro heat pipes are used to eliminate hot spots, reduce temperature gradients, and improve chip reliability (Wu and Peterson, 1991a, 1991b). Working fluids range from water and methanol to liquid metals such as sodium, potassium, and mercury. As will be seen, power levels on the order of one hundred W/cm² are currently possible and, along with the fact that heat pipes are self contained and self starting, make micro heat pipes an ideal heat transfer device.

While much has been written on this subject, only a few of the most important references will be discussed. The maximum heat flux data from these references will also be

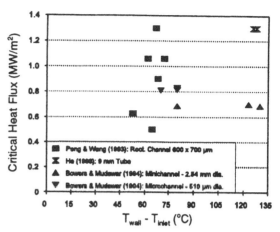

Figure 1: Critical heat flux comparison for mini- and microgeometries.

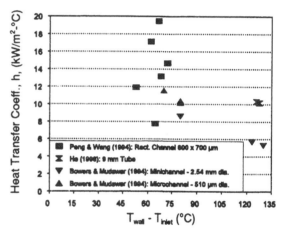

Figure 2: Heat Transfer coefficient comparison for mini- and microgeometries.

compared to that of the previous section. For a detailed review of this area, and microscale heat transfer in general, the reader is directed to Peterson (1992) and Duncan and Peterson (1994).

Longtin et al. (1992) presented a one dimensional mathematical model of the evaporator and adiabatic sections of steady-state micro heat pipes including interfacial and vapor shear stress terms. The model was solved numerically for pressure, velocity, and film thickness along the length of a heat pipe with a triangular cross-section. These results were compared with the experimental results of Babin et al. (1990) and showed good agreement. The most important feature of the Longtin et al. (1992) model is that it contains no empirical constants. An important finding from the scaling analysis and the model was that the maximum heat capacity of the pipe depends on the cube of the hydraulic diameter.

Gerner et al. (1992) developed the capillary and interfacial (flooding) limits for heat transfer in micro heat pipes. The capillary limit occurs when the capillary pressure does not exceed the sum of the vapor pressure and the liquid pressure. The interfacial limit occurs when the vapor velocity becomes large enough to entrain some liquid. This causes some of the counter flowing liquid to be carried toward the condenser. They concluded that the interfacial instability will determine the operating limit for large hydraulic diameters and that the capillary limitation will occur first for tubes that are very long.

Peterson et al. (1993) reported on an experimental investigation conducted to determine the thermal behavior of arrays of micro heat pipes fabricated in silicon wafers. This was an extension of previous work, Peterson (1992) and Peterson et al. (1991). Here, rectangular and triangular heat pipes were considered. Each configuration consisted of 39 parallel heat pipes. One array was made of 45 μm wide by 80 μm deep rectangular channels and the other was 120 μm wide by 80 μm deep triangular channels. Based on the experimental data, the triangular heat pipe outperformed the rectangular heat pipe, which, in turn, was better than a wafer without a heat pipe array. Due to the etching process, the rectangular heat pipe had slightly rounded corners, whereas the triangular heat pipe had fairly sharp corners. These differences, along with a reduction in the amount of residue in the triangular micro channels, were cited as the reason for the increased performance of the triangular heat pipe array. For further information on using heat pipes in semiconductors, the reader is directed to Mallik et al. (1992).

Swanson and Peterson (1993) developed a thermodynamic model of capillary structures in micro heat pipes. Their model included Marangoni effects, axial temperature differences, changes in local interfacial curvature, and the disjoining pressure. Expressions were developed for the heat transfer rate across the meniscus, interfacial mass flux, and nonevaporating superheated liquid film thickness. The authors presented a very detailed look at the processes and mechanisms that occur across the liquid-vapor interface in the thin film and meniscus region. The development and explanations are truly on a molecular level. The specific details of how heat pipes work and the effect of various mechanisms was presented in a clear and concise manner. The authors recently updated this paper (Swanson and Peterson, 1995).

Badran et al. (1993) addressed the feasibility of using liquid-metal heat pipes in radiator fin material. The advantages of liquid-metal heat pipes are their high operating temperatures which result in much higher heat fluxes. The need and applications for this type of heat pipe (in space systems) was presented along with a discussion of the radiator panel configuration. The maximum heat loads and specific conductivities for the three working fluids (sodium, potassium, and mercury) as a function of hydraulic diameter were presented. Using the analytical results, it was predicted that the effective thermal conductivity for a liquid-metal micro heat pipe may be as high as 200 times that of copper. Very high effective thermal conductivity, with respect to the weight ratio, was obtained by integrating micro heat pipes in radiator panels.

Cao et al. (1993) reviewed the literature for heat pipes of a larger size than considered here (hydraulic diameter > 1 mm), and discussed three limiting mechanisms for micro heat

pipes: the vapor continuum limit, the boiling limit, and the capillary limit. The vapor continuum limit occurs when the size of the heat pipe decreases to the point where the continuum characteristics are lost. They suggest that the micro heat pipe will lose its advantage for effectively cooling microelectronic devices due to the limited heat transport capability when the heat pipe is operating in a rarefied or free molecular vapor condition. The boiling limit occurs at a temperature difference in which small vapor or noncondensible bubbles are trapped on the inner wall causing degradation of the heat pipe performance. The capillary limit occurs when the vapor and liquid pressure cannot overcome the capillary pressure (surface tension) and a thin nonevaporating layer forms. Even though this thin film is superheated, it will exist in equilibrium with the saturated vapor as a consequence of the reduced vapor pressure of the film (a result of the disjoining pressure). Cao et al. (1993) also found that the disjoining pressure helped micro heat pipes have a higher maximum heat flux than macro heat pipes. This was due to the disjoining pressure pulling the liquid toward the center of the wall where it can be evaporated more readily.

Khrustalev et al. (1994) developed a detailed mathematical model which was used to examine the heat and mass transfer processes in a micro heat pipe. The model was developed for a triangular cross section, but can be expanded to configurations with a multiple of triangular corners. Expressions for the maximum heat transfer capacity and thermal resistance were developed for the condenser and evaporator sections. The numerical results were compared with the experimental data of Wu and Peterson (1991a, 1991b) and were shown to be in good agreement. One important feature included in this analysis was the shear stress at the free liquid surface. The authors indicate that failure to include this shear stress may lead to over-estimation of the maximum heat transfer capacity. Along with the shear stress, minimum contact angle and amount of liquid charge were also shown to have a pronounced effect on the maximum heat transfer capacity.

Figures 3 and 4 compare thermal conductivity (k) and maximum heat flux (q''_{max}) versus operating temperature (T_o), respectively, as reported in some of the previously mentioned references. As shown in Fig. 3, the Peterson et al. (1993) data show the higher efficiency of the triangular heat pipe compared to the rectangular heat pipe, as previously discussed. The thermal conductivity data for Longtin et al. (1992) and Wu and Peterson (1991a, 1991b) show trends similar to that of Peterson et al. (1993), above an operating temperature of 40°C. The data reported by Longtin et al. (1992) and Wu and Peterson (1991a, 1991b) follow similar trends as the operating temperature decreases. Though it looks unusual for the thermal conductivity values of Wu and Peterson (1991a, 1991b) to increase so dramatically as the operating temperature decreases, these results are in good agreement with their analytical model. The results from Babin et al. (1990) show that their "improved" heat pipe cross section is indeed a more efficient design as the operating temperature increases. This is probably due to the convex trapezoidal shape having sharper and/or smaller corner

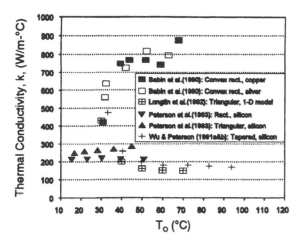

Figure 3: Thermal conductivity comparison for micro heat pipes.

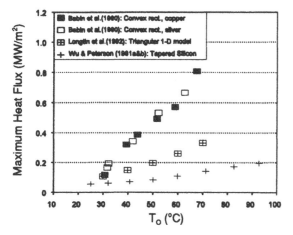

Figure 4: Maximum heat flux comparisons for micro heat pipes.

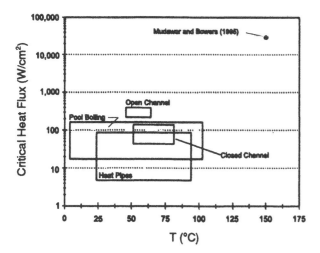

Figure 5: Critical Heat Flux Ranges.

angles than the other micro heat pipes. Note the maximum heat flux found in heat pipes was about 80 W/cm² as compared to a maximum of 320 W/cm² for open channel boiling and 130 W/cm² for closed channel flow boiling.

In Fig. 4, the most profound difference again occurs between the convex trapezoidal channels of Babin et al. (1990) and the other channels. As the operating temperature increases, the difference in maximum heat flux achieved by the convex trapezoidal channels of Babin et al. (1990), as compared to the other channels, increases tremendously. This factor is approximately 2.5 times at an operating temperature of 70°C.

As a final comparison, Fig. 5 presents the range of CHF achieved for each of the flow conditions discussed previously along with that for pool boiling in microgeometries and the ultra-high CHF achieved by Mudawar and Bowers (1995). The pool boiling range was developed using data from Anderson and Mudawar (1989), Marto and Lepere (1982), Mudawar and Anderson (1993, 1990), and Wright and Gebhart (1982). Several interesting conclusions are suggested from the ranges shown in Fig. 5. Open channel micro flow boiling has produced the highest practical CHF and is one of the most practical of convective flow boiling processes; however, this area has been the least studied. Micro heat pipes, which appear to be the most practical of the processes and for which there is considerable data, have the lowest range of CHF. Pool boiling can produce reasonably high CHF values over a wide range of temperatures. The Mudawar and Bowers (1995) data point is included to indicate possible CHF values for closed channel micro flow boiling. Unfortunately, the high pressure and flow rate requirements make this ultra-high CHF value impractical at the present time; however, this value could become achievable and possibly even exceeded as new and more powerful micropumps are developed.

CONCLUSIONS

Convective flow boiling in microgeometries appears to be a technology that can satisfy the demand for the dissipation of high heat fluxes associated with electronic and laser devices. However, at this time, the amount of work available in the open literature related to open and closed microchannels is quite limited. Only the groups at Purdue and Tsinghua Universities have reported empirical results for convective flow with phase change, while the theoretical work in this area is practically nonexistent. The minimum channel size for which data is available is approximately 500 μm. With the apparent increase in heat transfer that has been reported for single phase flow in channels of smaller size, it is recommended that these microscale effects also be investigated for phase change flows.

On the other hand, the work on microscale heat pipes is quite developed, with significant contributions coming from the groups at the Texas A&M, Cincinnati, and Wright State. The relatively high heat fluxes possible with this exciting technology suggest that microscale convective flow boiling is an area that should be receiving more attention in the near future.

REFERENCES

Anderson, T.M. and I. Mudawar,1989, "Microelectronic Cooling by Enhanced Pool Boiling of a Dielectric Fluorocarbon Liquid," *J. Heat Transfer*, v111, pp. 752-759.

Babin, B.R., Peterson, G.P., and D. Wu, 1990, "Steady-State Modeling and Testing of a Micro Heat Pipe," *J. Heat Transfer*, v112, pp. 595-601.

Badran, B., Albayyari, J.M., Gerner, F.M., Ramadas, P., Henderson, T., and K.W. Baker, 1993, "Liquid-Metal Micro Heat Pipes," *Heat Pipes and Capillary Pumped Loops*, ASME HTD-v236, pp. 71-85.

Bowers, M.B. and I. Mudawar, 1994, "High Flux Boiling Low Flow Rate, Low Pressure Drop Mini-Channel and Micro-Channel Heat Sinks," *Int. J. Heat Mass Transfer*, v37, n2, pp. 321-332.

Cao, Y., Faghri, A., and E.T. Mahefkey, 1993, "Micro/Miniature Heat Pipes and Operating Limitations," *Heat Pipes and Capillary Pumped Loops*, ASME HTD-v236, pp. 55-62.

Choi, S.B., 1991, "Friction Factors and Heat Transfer in Microtubes," Doctor of Engineering Dissertation, Louisiana Tech University.

Cotter, T.P., 1984, "Principles and Prospects for Micro Heat Pipes," *Proc 5th Int Heat Pipe Conf*, Tsukuba, Japan, pp. 328-335.

Drost, M.K., Beckette, M.R., and R.S. Wegeng, 1994, "Thermodynamic Evaluation of a Microscale Heat Pump," *Microscale Heat Transfer*, ASME HTD-v291, pp. 35-43.

Duncan, A.B. and G.P. Peterson, 1994, "Review of Microscale Heat Transfer," *Appl Mech Rev*, v47, n9, pp. 397-428.

Gerner, F.M., Longtin, J.P., Henderson, H.T., Hsieh, W.M., Ramadas, P., and S. Chang, 1992, "Flow and Heat Transfer Limitations in Micro Heat Pipes," *Topics in Heat Transfer - v3*, ASME HTD-v206-3, pp. 99-104.

Harley, J., Pfahler, J., Bau, H., and J. Zemel, 1989a, "Transport Processes in Micron and Submicron Channels," *Convection Heat Transfer and Transport Processes*, ASME HTD - v116, pp. 1-5.

Harley, J. and H. Bau, 1989b, "Fluid Flow in Micron and Submicron Size Channels," *Proceedings of the 1989 Micro Electro Mechanical Systems Workshop*, Salt Lake City, Utah, Febuary 20-22, pp. 25-28.

He, J.G., 1988, "Experimental Investigation for Enhancing Effect of Polymeric Additives on Subcooled Nucleate Flow Boiling with Higher Velocity," Master of Science Thesis, Tsinghua University, Beijing, China.

Khrustalev, D., and A. Faghri, 1994, "Thermal Analysis of a Micro Heat Pipe," *J. Heat Transfer*, v116, pp. 189-198.

Little, W.A., 1990a, "Advances in Joule-Thompson Cooling," *Adv. Cryo. Engr.*, v35, Plenum Press, New York, pp. 1305 - 1314.

Little, W.A., 1990b, "Microminiature Refrigerators for Joule-Thompson Cooling of Electronic Chips and Devices," *Adv. Cryo. Engr.*, v35, Plenum Press, New York, pp. 1325 - 1333.

Little, W.A., 1984, "Microminature Refrigeration," *Rev. Sci. Instrum.*, v55, n5, pp. 661-680.

Longtin, J.P., Badran, B., and F.M. Gerner, 1992, "A One-Dimensional Model of a Micro Heat Pipe During Steady-State Operation," *Heat Transfer on the Microscale*, ASME HTD - v 200, pp. 23-33.

Maddox, D.E. and I. Mudawar, 1989, "Single- and Two-Phase Convective Heat Transfer from Smooth and Enhanced Microelectronic Heat Source in a Rectangular Channel," *J. Heat Transfer*, v111, pp. 1045-1052.

Mallik, A.K., Peterson, G.P., and M.H. Weichold, 1992, "On the Use of Micro Heat Pipes as Integral Part of Semiconductor Devices," *J. Elec. Packaging*, v114, pp. 436-442.

Marto, P.J. and Lt. V.J. Lepere, USN, 1982, "Pool Boiling Heat Transfer From Enhanced Surfaces to Dielectric Fluids," *J. Heat Transfer*, v104, pp. 292-299.

Mudawar, I. and T.M. Anderson, 1993, "Optimization of Enhanced Surfaces for High Flux Chip Cooling by Pool Boiling," *J. Electronic Packaging*, v115, pp. 89-100.

Mudawar, I. and T.M. Anderson, 1990, "Parametric Investigation Into the Effects of Pressure, Subcooling, Surface Augmentation and Choice of Coolant on Pool Boiling in the Design of Cooling Systems for High-Power-Density Electronic Chips," *J. Electronic Packaging*, v112, pp. 375-382.

Mudawar, I. and M.B. Bowers, 1995, "Parametric Study of Ultra-High CHF in Highly Subcooled Water Flow Inside Small Diameter Tubes," Engineering Foundation Conference on Convective Flow Boiling, Banff, Alberta, April 30 - May 5, Paper II-9.

Mudawar, I. and D.E. Maddox, 1990, "Enhancement of Critical Heat Flux from High Power Microelectronic Heat Sources in a Flow Channel," *J. Electronic Packaging*, v112, pp. 241-248.

Mudawar, I. and D.E. Maddox, 1989, "Critical Heat Flux in Subcooled Flow Boiling of Fluorocarbon Liquid on a Simulated Electronic Chip in a Vertical Rectangular Channel," *Int. J. Heat Mass Transfer*, v32, pp. 379-394.

Peng, X.F. and B.-X. Wang, 1993, "Forced Convection and Flow Boiling Heat Transfer for Liquid Flowing through Microchannels," *Int. J. Heat Mass Transfer*, v36, n14, pp. 3421-3427.

Peterson, G.P., 1992, "Overview of Micro Heat Pipe Research and Development," *Appl. Mech. Rev.*, v45, n5, pp. 175-189.

Peterson, G.P., Duncan, A.B., and M.H. Weichold, 1993, "Experimental Investigation of Micro Heat Pipes Fabricated in Silicon Wafers," *J. Heat Transfer*, v115, pp. 751-756.

Peterson, G.P., Duncan, A.B., Ahmed, A.S., Mallik, A.K., and M.H. Weichold, 1991, "Experimental Investigation of Micro Heat Pipes in Silicon Wafers," *Micromechanical Sensors, Actuators, and Systems*, ASME DSC-v32, pp. 341-348.

Pfahler, J., Harley, J., Bau, H., and J.N. Zemel, 1991, "Gas and Liquid Flow in Small Channels," *Micromechanical Sensors, Actuators, and Systems*, ASME DSC - v 32, pp. 49-60.

Pfahler, J., Harley, J., Bau, H., and J.N. Zemel, 1990a, "Liquid and Gas Transport in Small Channels," *Microstructures, Sensors and Actuators*, ASME DSC-v19, Dallas, Texas, Nov. 25-30, pp. 149-157.

Pfahler, J., Harley, J., Bau, H., and J.N. Zemel, 1990b, "Liquid Transport in Micron and Submicron Channels," *Sensors and Actuators*, A21-23, pp. 431-434.

Swanson, L.W. and G.P. Peterson, 1995, "The Interfacial Thermodynamics of Micro Heat Pipes," *J. Heat Transfer*, v117, February, pp. 195-201.

Swanson, L.W. and G.P. Peterson, 1993, "The Interfacial Thermodynamics of the Capillary Structures in Micro Heat Pipes," *Heat Transfer on the Microscale*, ASME HTD - v253, pp.45-51.

Wright, N. and B. Gebhart, 1989, "Enhanced Boiling on Microconfigured Surfaces," *J. Electronic Packaging*, v111, pp. 112-120.

Wu, D. and G.P. Peterson, 1991a, "Transient Experimental Investigation of Micro Heat Pipes," *AIAA J. ThermoPhysics*, v5, n4, pp. 539-544.

Wu, D. and G.P. Peterson, 1991b, "Investigation of the Transient Characteristics of a Micro Heat Pipe," *AIAA J. Thermophysics*, v5, n2, pp. 129-134.

Yu, D., Warrington, R.O., Barron, R.F., and T.A. Ameel, 1995, "An Experimental and Theoretical Investigation of Fluid Flow and Heat Transfer in Microtubes," 4th ASME/JSME Thermal Engineering Joint Conference, Maui, Hawaii, March.

Zhukov, V.M. and I.L. Yarmak, 1990, "Transient Heat Transfer in Two-Phase Cryogenic Liquid Forced Flows Under Step Heat Flux in Narrow Channels," *Cryogenics*, v30, pp. 82-286.

FLOW BOILING IN A NARROW CHANNEL

M.R. Aligoodarz and D.B.R. Kenning
Department of Engineering Science
University Of Oxford
Oxford, U.K.

ABSTRACT

Flow boiling heat transfer in a single vertical channel of cross-section 1 mm x 2 mm has been studied experimentally using high speed video photography of bubble behaviour and simultaneous recording of the wall temperature by liquid crystal thermography. The conditions for the recordings so far obtained are flow rate 85 kg m^{-2} s^{-1} of water at 1 atm., inlet temperature 98 °C, heat flux 29 to 109 kW m^{-2} in a channel of heated length 140 mm. The bubble boundaries, contours of wall temperature, heat flux and heat transfer coefficient as well as local temperature-time curves can be obtained by this technique. Some preliminary results are given to illustrate the capabilities of the technique for investigating mechanisms of heat transfer.

1. INTRODUCTION

Compact evaporative heat exchangers offer advantages in reduction of size, cost and fluid inventory with potential improvements in safety and environmental acceptability. In channels that are so narrow that a growing bubble can bridge the cross-section, the flow boiling behaviour deviates from that in large channels . Kew and Cornwell (1993) proposed a general criterion for deviations when the confinement Number exceeds 0.5,

$$Co = [\sigma / g (\rho_f - \rho_g)]^{(0.5)} / d_e > 0.5 \qquad (1)$$

but there are relatively few heat transfer measurements for saturated flow boiling in such narrow channels . For water at atmospheric pressure their criterion predicts deviations in channels with cross-sectional dimensions below 4 mm and even smaller dimensions for liquids of lower surface tension such as refrigerants and organics . This study has been undertaken to improve the understanding of the physics of boiling in a single narrow channel, in the absence of the flow oscillations that can occur with parallel

channels, with the eventual objective of improving design methods for compact heat exchangers with boiling.

2. EXPERIMENTAL METHODS

A rig has been built to provide short-duration (typically 10 s) flow of water, refrigerant or organic liquid through a single heated channel of length 140 mm with rectangular cross- section of 1 mm x 2 mm, Figure 1.

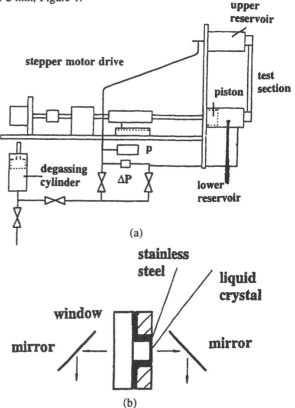

Figure 1. Schematic figure of (a) rig (b) test section

Three sides of the channel are formed from a single piece of 0.075 mm thick stainless steel, heated by the passage of ripple-free direct current from a rotary generator. The fourth (2 mm) side is a glass window sealed by silicone rubber. Considerable difficulty was encountered in making such a small channel leak-free. Liquid in a lower reservoir at a controlled, slightly subcooled temperature is pushed up the channel by a piston driven at controlled speed by a stepper motor. Liquid and vapour leaving the channel are collected in a upper reservoir that, in these preliminary experiments with water, is vented to the atmosphere. The outlet pressure and the differential pressure between the reservoirs are measured by transducers.

The back of the central metallic side of the channel is painted black and coated with a layer of unencapsulated thermochromic liquid crystal with a colourplay range of 99 °C to 107 °C. By using a pair of mirrors, Figure 1(b), the colourplay of the liquid crystal and the motion of bubbles inside the channel are recorded simultaneously by a NAC high-speed colour video camera at 200 frames s^{-1}. The NTSC video signal is replayed through a RGB convertor and a frame-grabber with on-board conversion to HSI (Hue, Saturation, Intensity) variables. The frame-grabber is controlled by a pc that stores selected sequences of frames for analysis. The Hue signal is converted to temperature by a previously-determined calibration and the Intensity signal is used to identify the boundaries of the larger bubbles. Local values of the heat flux from the wall to the liquid are calculated from the input heat flux, the rate of change of wall temperature and lateral conduction in the wall. The video signal is affected by noise during recording and replay, which affects the accuracy of the heat

flux contours. Kenning and Yan (1995) describe the techniques of high-speed video thermography applied to pool boiling and the reduction of noise by filtering. In the present experiments the changes in wall temperature of about 3K can be measured to ± 0.1 K. The spatial resolution depends on the spacing of the pixels after digitizing. When recording events in a 30 mm length of the channel, there are 20 pixels across the 2 mm width of the channel. This is about the minimum number required to resolve transverse variations in temperature, since spatial filtering is usually based on a 3 x 5 array. The difficulties of sealing the channel made it impossible, in these preliminary experiments, to apply a uniformly thin layer of encapsulated liquid crystal right up to the edge of the channel so the temperature changes within 0.2 mm of the corners of the rectangular section have not been measured. The choice of recording conditions is a compromise between spatial resolution and the size of the field. Recording events over a longer length of the channel, which would be of considerable interest, would require more sophisticated optics to lay images of 30 mm lengths side-by-side on the video screen.

3. OBSERVATIONS

The temperature of a group of pixels at a particular axial position can be measured and averaged over a long period of time. Examples are shown in Figure 2 for two heat fluxes of 61 kW m^{-2} and 109 kW m^{-2} at the same position 65 mm from the channel inlet and the same mass flux of 85 kg m^{-2} s^{-1} of water .

At the lower heat flux bubbles nucleate and grow locally, some filling the cross-section of the channel; occasionally long bubbles slide up the channel.

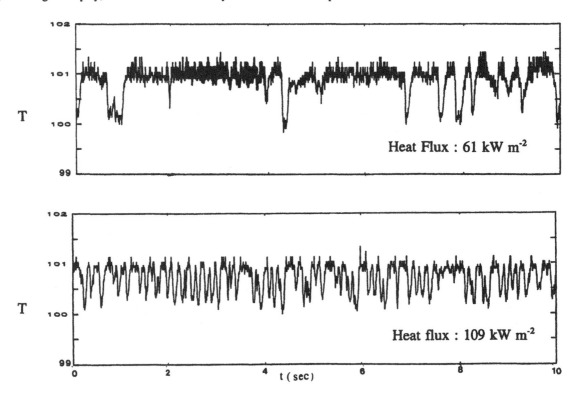

Figure 2. wall temperature variation

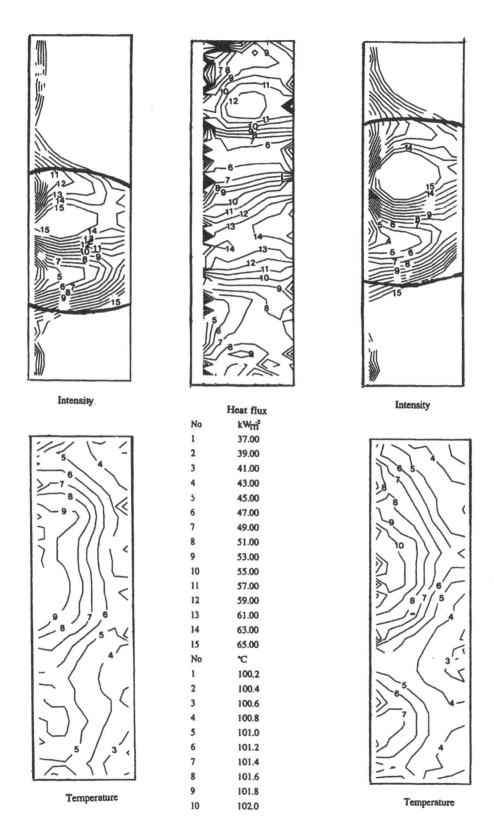

No	Heat flux kW/m²
1	37.00
2	39.00
3	41.00
4	43.00
5	45.00
6	47.00
7	49.00
8	51.00
9	53.00
10	55.00
11	57.00
12	59.00
13	61.00
14	63.00
15	65.00
No	°C
1	100.2
2	100.4
3	100.6
4	100.8
5	101.0
6	101.2
7	101.4
8	101.6
9	101.8
10	102.0

Figure 3. Growing bubble

The wall temperature fluctuates about 101 °C with an amplitude of about ± 0.1 K with occasional sharp dips towards 100 °C. The time-averaged wall temperature is 100.9 °C and the corresponding heat transfer coefficient is 68 kW m^{-2}K^{-1}, assuming a constant local saturation temperature of 100 °C. (The local fluctuations in pressure and saturation temperature cannot be measured but the measured time-averaged pressure drop along the channel is negligible at this low flow rate). At the higher heat flux the flow pattern consists mainly of long bubbles sliding up the channel. There is a marked change in the temperature-time graph, Figure 2. The maximum wall temperature is nearly the same as at the lower heat flux but there are much more frequent dips to temperatures in the range 100.2 - 100.5 °C. The time-averaged wall temperature is 100.7 °C and the heat transfer coefficient is 156 kW m^{-2} K^{-1}. The wall superheats in confined-flow boiling are nearly 15 times smaller than those measured at similar heat fluxes in pool boiling on a thin stainless steel plate, Kenning and Yan (1995) .

The large percentage changes in wall superheat indicate that the low thermal capacity of the wall restricts the supply of heat during bubble growth and motion, so wall thickness and thermal properties are likely to have significant effects on the heat transfer process. Kenning and Yan (1993) have drawn attention to the problems that may arise when applying boiling data from thin electrically-heated test sections to thin-walled fluid-fluid heat exchangers. Allowances for the effects of wall properties and the nature of the heat source must be made through modelling based on an understanding of the mechanisms of heat transfer . Measurements of the distributions of temperature and heat flux during the growth and passage of individual bubbles should provide this understanding .

Two examples are now given of the local information obtainable from liquid crystal measurements, for which detailed interpretation has not yet been completed. Both refer to a heat flux of 53 KW m^{-2}.

Figure 3 portrays events on a wall region 1.6 mm wide, 5 mm long, over an interval of 10 ms during the growth of a bubble that has nucleated locally and then grown to fill the channel. The contours of intensity should identify the boundaries of the bubble. The outermost contours correspond approximately with the boundaries as judged by eye (marked by a thick line) but the variations of intensity between the boundaries are artifacts of the lighting system . They do not provide information about the thickness of the liquid layer between the bubble and the back, heated wall and this is a disadvantage of these observations . The corresponding wall temperature contours show that there are significant variations in temperature even before the wall is covered by the bubble, making it difficult to identify the influence of the bubble. This becomes clearer from the contours of heat flux, which depend primarily on dT/dt. There are two regions of increased heat removal under the bubble, one near its rapidly-advancing upper boundary and the other ahead of the slowly-moving lower boundary. Between these regions there is a small area in which the heat removal rate is slightly less than the heat supply rate. All the changes in temperature and heat flux

are relatively modest. It is not possible to tell what contribution the heated side walls and the corner regions are making to the growth of the bubble under what may be asymmetrical conditions. Faster recording is required to resolve the rapid initial growth of the bubble and to identify its nucleation site .

In Figure 4 the propagation of the temperature disturbance caused by a long sliding bubble is shown by the temperature-time plots for three points on the centerline of the wall at axial intervals of 1.6 mm, 65 mm from the channel inlet . Shortly after the arrival of the front of the bubble, the wall is cooled rapidly to a very low superheat of about 0.2 K. This is consistent with the formation of a thin liquid microlayer . The following period of nearly steady heat transfer corresponds to conduction through a liquid layer about 3 microns thick. Well before the arrival of the rear of the bubble there is a recovery of wall superheat that is too slow to be simply due to dryout of the liquid layer under the bubble: if the wall were completely uncooled its temperature would rise at 200 Ks^{-1}. Further information is required about the thickness of the liquid layer .

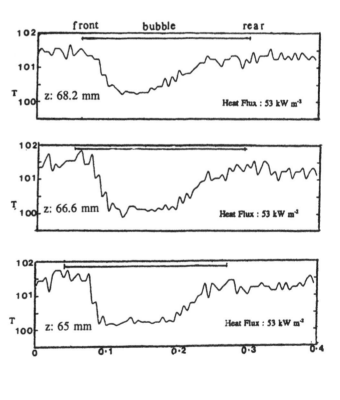

Figure 4. Passage of a long sliding bubble

4. CONCLUSION

This paper has demonstrated the feasibility of using high speed video/liquid crystal thermography to investigate boiling in channels with cross-sectional dimensions as small as 2 mm. The technique requires further improvement . The requirement for high spatial resolution conflicts with the desirability of recording events over a long length of the channel. A higher video speed of say 1000 frames s^{-1} is required to resolve the formation of confined bubbles.

Flow boiling of water in the 1 mm x 2 mm channel can be maintained at a superheat an order of magnitude smaller than that required in pool boiling on a similar surface . The heat transfer coefficients are so high that the thermal properties of the wall have a significant influence on the processes of heat transfer.

The data described in this paper were obtained very recently so their interpretation is tentative and subject to further work.

ACKNOWLEDGEMENTS

This work formed part of CEC Joule2 contract No. J002-CT 92-0045. The development of the liquid crystal analysis system was supported by EPSRC contract No. GR/H45377. M.R. Aligoodarz received a scholarship from the government of the Islamic Republic of Iran.

REFERENCES

Kenning, D.B.R. and Yan, Y., 1993, "Boiling on Thin Walls", Eurotherm 26: Compact Two-Phase Heat Exchangers, Heriot-Watt University, 16-17 September 1993.

Kenning, D.B.R. and Yan, Y., 1995, "Pool boiling on a thin plate: Features Revealed by Liquid Crystal Thermography", Oxford University Engineering Laboratory Report No. 2055/95.

Kew, P.A. and Cornwell, K., 1993 "Confined Bubble Flow and Boiling in Narrow Spaces", Eurotherm 26: Compact Two-Phase Heat Exchangers, Heriot-Watt University, 16-17 September 1993.

THE EFFECTS OF INTERNAL RIBS AND FINS ON ANNULAR TWO-PHASE TRANSPORT IN COMPACT EVAPORATOR PASSAGES

Nancy E. Hawks, Kim A. Shollenberger and Van P. Carey
Mechanical Engineering Department
University of California at Berkeley
Berkeley, California 94720
USA

ABSTRACT

This paper summarizes results of recent experimental studies of two-phase flow in passages containing idealized single fin and constriction elements. These studies used a phase-Doppler particle analyzer system to determine how droplet entrainment is affected by these types of internal structures. The relevance of the results of these experimental studies to annular flow vaporization processes is explored in detail. These studies imply that constrictions and expansions in plate-fin evaporators may enhance entrainment, thereby reducing the liquid film thickness on the surface. This may enhance heat transfer when the liquid inventory is high and may hurt heat transfer performance when the liquid inventory is low by inducing localized dryout of the film on the passage wall. Shedding of liquid from fin structures can produce a non-uniform entrainment field which facilitates redeposition of droplets on fins immediately downstream. This helps keep fin surfaces wetted when liquid inventory is low, improving heat transfer performance.

INTRODUCTION

Compact high-performance evaporators used in a variety of applications frequently incorporate flow passages with internal fin or rib structures. In applications such as automotive air-conditioning, cryogenic processing, and evaporator cold plates for electronics, most, if not all of the vaporization process occurs under annular flow conditions. In such cases, the interaction of the two-phase flow with fin or rib structures in the passage strongly affects the behavior of the two-phase flow and the associated heat transfer.

This paper will focus on offset-fin (plate-fin) surfaces (Fig. 1). In such geometries, the two-phase flow behavior is typically quite complex, resulting in distinctive features of the two-phase transport which are not present in round-tube evaporators. This paper will specifically explore the nature of such complexities and how they affect the

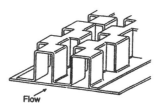

Figure 1: Example of a Plate-Fin Surface of the type that is Sometimes Used in Compact Evaporators.

transport. Previous studies of these issues will be briefly reviewed, and the results of some very recent studies of annular flow in finned passage geometries will be presented. The implications of the new findings with respect to convective vaporization processes will be discussed in detail.

There are two features of the operating characteristics of matrix-type surfaces in vaporization applications that strongly affect how they will perform. First, the length scales of the passage geometry, particularly the hydraulic diameter, are usually much smaller than the larger round tube passages most often used in power systems, refrigeration, and process heat transfer applications. The second important operating feature relates to the complexity of the passage and its impact on pressure drop. To operate heat exchangers containing torturous finned passages without a heavy pressure drop penalty, the flow through the exchanger is often kept at a relatively low mass flux level. While power and process evaporators with large round tube passages typically operate at mass flux (G) levels above 1000 kg/m^{2}s, matrix surfaces of the type considered here may have hydraulic diameters between 0.4 mm and 3 mm and operate at mass flux levels between 50 and 1000 kg/m^{2}s.

For a given working fluid, the ranges of G and hydraulic diameter d_h, together with the details of the passage geometry and the imposed wall conditions, will largely dictate the performance of the heat exchanger. To make this more concrete, we consider the example of saturated convective boiling of nitrogen at atmospheric pressure in a passage with $d_h = 1.0$ mm. For mass flux values between 50 and 300 kg/m^{2}s, the liquid and vapor Reynolds numbers lie in the ranges $0 < Re_l < 3100$ and $0 < Re_g < 90,000$. These Reynolds number ranges span the laminar, transition and turbulent ranges. In general, we expect that viscous and inertial effects may play important roles in the two-phase transport. The Weber numbers We_l and We_g are generally large compared to one, suggesting that the inertial effects associated with vapor motion are usually large compared with surface tension effects at moderate to high quality.

Unlike vertical round tubes, the two-phase transport in offset fin passages is usually non-uniform over the perimeter of the passage at any downstream location because of the irregularity of the passage geometry and because the thermal boundary condition is usually non-uniform over the passage wall. Also, at high quality, the liquid which flows in a film along fin structures may be shed from the downstream edges of these structures into the core flow.

In most common vaporization or condensation applications, a large portion of the phase change process is accomplished under conditions which result in annular flow. For annular flow, transport of heat and momentum across the liquid film on the prime surface and fins largely dictates the heat transfer and pressure loss performance of the offset fin surface. The spatial non-uniformity of the core flow which weaves its way between the fins causes the liquid film to experience substantial changes in local pressure gradient and in shear stress at the film interface (see Mandrusiak and Carey (1990)). Also, the non-uniformity of the core flow alters the entrainment and deposition, which, in turn, changes the liquid distribution over the perimeter of the passage. At low to moderate qualities, this may simply result in a variation of the film thickness over the fin and prime wall surfaces. During vaporization at moderate to high qualities, this may cause portions of the passage wall to dryout, while other areas remain wet.

For vaporization processes in offset fin geometries, the transport of liquid onto the fin surfaces, which typically oomprise 80% or more of the total surface area in the offset fin passage, is especially important. Portions of the fin surface that dry-out contribute little, if anything, to the total heat transfer, resulting in a reduced overall effectiveness of the surface. In offset fin surfaces, liquid is delivered to the fins by two mechanisms: (1) entrained liquid droplets may be deposited on the fin surfaces, and (2) liquid from the prime surface walls may be dragged onto the fins by lateral Reynolds stresses exerted by the turbulent core flow on the prime-surface liquid film. Liquid on the fin flows downstream due to the combined effects of pressure gradient, interfacial shear, and body forces. When the film reaches the trailing edge of the fin, the liquid may be shed as droplets off the trailing edge of the fin and become entrained in the core flow or, near the lateral prime surface walls, surface tension may force the liquid to merge into the liquid films on the lateral walls. Although the discussion here has

focused on offset strip fins, liquid shedding may also occur in other enhanced passage geometries having interrupted surfaces. The efficiency of convective vaporization processes in such passages will depend directly on how effectively liquid can be delivered to surfaces of fin structures downstream of the location where shedding occurs. It is desirable to keep the fins covered with liquid, but too much liquid on the fins generally increases the resistance to heat flow across the film and degrades performance. A surface that sheds liquid effectively from its fins may be prone to early partial dryout and inferior performance in flow boiling applications.

At low to moderate wall superheat levels, where nucleate boiling may persist into the slug or annular flow regimes, the non-uniformity of convective transport during two-phase flow within a finned passage can cause strong convective transport of heat from the surfaces of fins which may result in strong temperature gradients within these structures. Even with relatively uniform convection, the fin-efficiency effect will produce non-uniform surface temperatures on the fins. The non-uniformity of the convective effect can further enhance the non-uniformity in the wall temperatures. This, in turn, can lead to partial suppression of nucleation at locations where the superheat is lower, and vigorous nucleation at locations where the superheat is higher. The wake regions immediately downstream of fins are usually regions of weak convection, resulting in locally higher wall temperatures. Not surprisingly, such regions are often, but not always, observed to be regions of exceptionally vigorous nucleate boiling activity (see Carey and Mandrusiak (1986) and Cohen and Carey (1989)).

In addition, photographs of upward vaporizing annular two-phase flow in a passage with offset strip fins (Carey and Mandrusiak (1986)) indicate that the heated fin and prime surfaces in the passage become progressively drier with increasing downstream distance until the liquid inventory is so low that the walls are completely dry. The observed gradual dryout of surfaces in the fin matrix is a direct consequence of the non-uniformity of the core and film flows in the passage. This progressive dryout, which produces a gradual decrease in the localized heat transfer coefficient with downstream distance, is distinctly different from dryout in round tubes under comparable conditions.

The results of the studies discussed above indicate that the presence of fin structures in a matrix surface may strongly impact entrainment, which, in turn, may affect heat transfer and dryout during convective boiling at moderate to high quality. Two effects of these structures on entrainment are associated with (1) shedding of the liquid film from the downstream end of fins and (2) enhanced shedding due to localized reductions in the flow cross-sectional area. To explore these effects, experimental studies of the droplet entrainment field downstream of fins and a localized constriction have been studied experimentally for high void fraction conditions typical of the latter stages of vaporization processes in fin matrix surfaces. The experimental methodology used and the results of these studies are discussed in the following sections.

EFFECTS OF FLOW PASSAGE CONSTRICTION

To explore the effects of flow passage constriction on annular flow with entrainment, experimental studies were conducted with a round tube test section. A two-phase flow of liquid water and air was supplied to the test section by a recirculating flow loop. The two-phase flow was generated in the flow system by directly depositing the liquid on the tube walls through six holes around the circumference at a location 50 diameters upstream of the test section. This provided an essentially equilibrium entrainment distribution to the test section containing an orifice-plate restriction in a round tube passage.

The experiments conducted with this setup were done at a single gas flow rate corresponding to a vapor Reynolds number of approximately 3.2×10^4. This value was chosen to be representative of the weakly turbulent flow conditions typical of compact evaporators in automotive air-conditioning systems. Liquid flow rates were chosen to span the liquid Reynolds number range $170 < Re_l < 800$, which is also typical of compact evaporators.

Two configurations were investigated: a 2.54cm diameter unobstructed straight round tube and a 2.54cm diameter straight round tube with an orifice plate constriction (Fig. 2a) located 50 diameters downstream of the inlet. The orifice plate constriction had an inside diameter of 2.22cm corresponding to a diameter ratio of 0.75 and was 0.32cm long. The gas flow and liquid flow were measured using calibrated meters to accuracies of 4% and 1%, respectively. The liquid was separated from the gas downstream of the test section. Laser measurements of droplet velocities, diameters, number density, and volume flow rate were made using a phase Doppler particle analyzer (PDPA) system operating in a 30° off forward scatter configuration.

A laser access test section provided an undisturbed optical path for the laser beams to travel into and out of the gas core (Fig. 2b). Two steps were required to ensure that the optical path remained clear. First, the liquid film was removed from the walls through a slot into a film removal chamber. Second, a recess was provided so that droplet deposition did not result in immediate rewetting of the test section walls. This recess also allowed the laser beams to enter through planar container walls, permitting the test section to be traversed without unfocusing the probe volume. Tests were performed to insure that the access section did not disturb the flow (Shollenberger (1994)).

To assess the effects of a constriction and expansion in a round tube on entrainment, experiments were first conducted using a plain round tube for the conditions of interest to determine base entrainment features. Having established a baseline for plain round tube flows, PDPA measurements were then made at three axial locations downstream of the orifice plate constriction, corresponding to L_c/d values of 5, 15 and 20. We used the straight tube data to help us interpret the constriction data. We found that right above the constriction, the liquid film thickens, and then, after it passes through the constriction, it forms large droplets that eventually reattach to the tube walls, as indicated schematically in Fig. 3. Downstream, the droplet

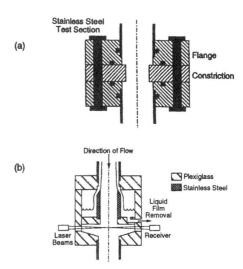

Figure 2: (a) Schematic of the Constriction Test Section and (b) the Optical Access Section.

entrainment distributions return to the equilibrium distributions found in the straight tube measurements.

The main features of this process are illustrated in Fig. 4. This figure is a plot of measured dimensionless entrained liquid flux versus radial position in the tube. The $L/d = 70$ data were obtained for the plain round tube with no constriction. The flow approaches this distribution asymptotically downstream of the restriction. Of particular note is that entrainment remains substantially above the equilibrium value at 10 diameters downstream of the orifice. Note that this is much longer than the distance required for a single phase gas flow to reattach at the same gas flow conditions (nominally about two diameters).

The results in Fig. 4 thus indicated that the restriction causes most, if not all of the liquid film to be ejected into the core flow. Much of the liquid reattaches to the walls in less than 5 diameters downstream of the restriction. However, this process causes entrainment levels significantly higher than equilibrium values to persist for 10 to 20 diameters downstream. These conclusions are also consistent with similar data obtained by McQuillan and Whalley (1984) for upward flow at higher flow velocities.

SHEDDING OF LIQUID FROM FIN STRUCTURES

In recent experiments, we have also used a PDPA system to determine the characteristics of the droplet field that results from liquid shedding from fin structures under conditions similar to annular flow in an evaporator. These experiments used a 4.47cm square passage containing two in-line fins (Fig. 5) and a large optical access section downstream of the test section to allow photographic as well as laser access. Air flowed into the test section through a long square passage with an L/d ratio of 70 to ensure fully developed flow at the leading edge of the first fin in the test section. A liquid flow loop delivered a metered flow of liquid (a 15% solution of isopropyl alcohol

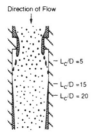

Figure 3: Schematic of Phase Distribution for Annular Mist Two-Phase Flow Past an Orifice Plate Constriction (Not to Scale).

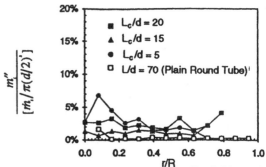

Figure 4: Entrainment versus Radial Location Downstream of a Constriction; $G = 22.7$ kg/m^{2}s, $\dot{m}_1 = 800$ g/min.

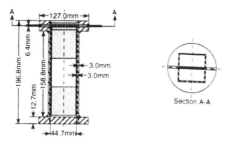

Figure 5: Fin Section.

in water) onto the leading edge of the upper fin. The fin was observed prior to the experiments to ensure that the liquid wet the fin surface.

Experiments were done with a double fin arrangement made by cutting the middle third section out of a single plate to create two fins, one upstream of the other (Fig. 5). Photographs and PDPA data were taken 4.6 cm downstream of the second fin. The gas flow was held nominally fixed at a value corresponding to $Re_g = 2.6 \times 10^4$. The liquid flow delivered to the fin was varied from 75 to 175 cc/min, corresponding to liquid film Reynolds number between 12 and 18. Based on the considerations discussed by Dressler and Kraemer (1990), the uncertainty in the PDPA diameter data obtained in these experiments is estimated to be about 2%. Typically, we experienced 70-80% signal validation which we believe is misleadingly low because often large droplets in the near wake blocked the laser beams, causing data to be rejected. Thus we believe that the number count of droplets is about 15-20% uncertain. From these uncertainties, we can conclude that the volume flux measurements are about 20-25% uncertain. We could photographically measure the larger droplet diameters to ±0.5mm, corresponding to about 12% uncertainty.

In these experiments, we studied droplet size and volume flux distributions downstream of a fin in order to better quantify the liquid distribution in an offset fin matrix like that shown in Fig. 1a. From our experiments, we concluded that the droplet field downstream of the fin consisted of two regions: (1) a near wake region within a few fin thicknesses of the fin centerline which contained a mixture of small droplets and larger droplets shed from the fin, and (2) a far wake region at lateral distances larger than a few fin thickness, which contained only smaller droplets. These regions are indicated schematically in Fig. 6. In our system, the near wake region is located downstream, within 6mm perpendicular to the plane of the fin. The far wake region is located downstream, farther than 6mm perpendicular to the plane of the fin.

The near wake region primarily contained large droplets which were shed from the trailing edge of the fin which would then strike the leading edge of the downstream fin and reattach. From photographs, the average droplet diameters downstream of the second fin were measured to be 3.5±0.5mm. As shown in Fig. 7, the measured mean droplet size in the near wake varies only weakly with liquid flow rate.

The far wake region consisted of a mist flow of small droplets which decreased in number density with increasing distance measured perpendicular to the plane of the fin. Using the PDPA system, droplet diameters were measured in the range 49 to 1740µm. The total volume flow of droplets determined from the PDPA measurements was subtracted from the total liquid supplied to the fins to determine the fraction of liquid in the near wake region. This fraction ranged from 0.62±0.02 to 0.93±0.02. Figure 7b shows the variation of this fraction of liquid in the near wake region. This fraction ranged from 0.62±0.02 to 0.93±0.02. Figure 7b shows the variation of this fraction with volume flow rate of liquid. It can be seen that the liquid fraction in the near wake decreases as the volume flow rate of liquid increases. This increased dispersion of droplets is apparently a consequence of more vigorous collisions of droplets with the fin leading edge at higher liquid flow rates.

The diameter probability distributions obtained from the PDPA measurements were found to exhibit peaks. Figure 8 shows an example of such a distribution at one lateral location. This distribution has peaks centered at 50mm, 400mm, 780mm, 1150mm, and 1530mm for the flow conditions 175cc/min, 9mm from the fin. Similar distributions were found for other locations and other flow rates. It is noteworthy that diameter probability histograms exhibiting multiple peaks were also found for the constricted round tube experiments.

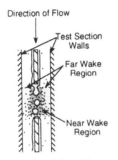

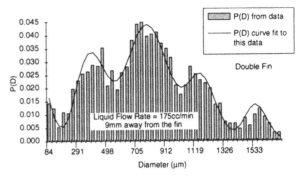

Figure 6: Schematic of the Near and Far Wake Regions (Not to Scale).

Figure 8: Probability versus Diameter Histogram in the Far Wake Region for the Double Fin 9mm off Center (i.e., 9mm Perpendicular to the Plane of the Fin) with a Liquid Flow Rate of 2.92 ml/s. In All Cases, the Gas Reynolds Number Re_g = 26,000.

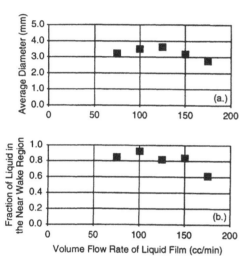

Figure 7: Plots of (a) Average Droplet Diameter and (b) Liquid Fraction versus Volume Flow Rate of the Liquid Film for the Large Droplets Found in the Near Wake Region Downstream of the Double Fin. In All Cases, the Gas Reynolds Number Re_g = 26,000.

We calculated and curve fit our fin diameter distribution data for the far wake with one log-normal distribution per peak, insuring that the total probability under the superposition of these curves remained one (see Hawks (1995) for details). The fact that the distributions used for sprays (log-normal distributions) fit the data well suggests that the peaks in the diameter probability histograms correspond to different droplet sources associated with different generation or breakup mechanisms. The most important mode of droplet formation appeared to be from satellite droplets. Each time a large droplet was shed from the fin, a cascade of small satellite droplets was shed as well. Van der Geld, et al. (1994) and Shi, et al. (1994) also documented this behavior in their studies. Van der Geld, et al. (1994) found that satellite droplets formed when a liquid filament breaks up, as in a jet or spray. Shi, et al. (1994) found that the surface tension of the fluid affected the satellite droplet cascade when a droplet is shed from a nozzle. In our experiments, we found that as the droplets

left the trailing edge of the fin, they cascaded in a sequence of progressively larger droplets with increasing distance from the trailing edge as shown in Fig. 9, similar to the behavior Shi, et al. (1994) found for lower surface tension fluids. The peaks in the diameter probability distributions could be explained by the presence of these satellite droplets. From the photographs, we found that for the double fin geometry, the ratios of the droplet diameters in the cascade to the initial (downstream) droplet were D_2/D_1 = 0.45±0.05 D_3/D_1 = 0.33±0.05. The ratios of the peaks in the distributions to the diameter of the initial droplet were also in range of D_2/D_1 = 0.40 to 0.50 for the first droplet and D_3/D_1 = 0.30 to 0.35 for the second droplet in the cascade.

ENTRAINMENT EFFECTS ON CONVECTIVE VAPORIZATION

The experimental studies described above indicate that localized constrictions and fin structures can have a strong impact on entrainment in annular flow in matrix-type flow passages. Each constriction-expansion increases the level of entrainment for a number of hydraulic diameters downstream above that which it would be in a passage of uniform cross section with the same mean hydraulic diameter. Thus, passages with repeated constriction-expansions intrinsically maintain an enhanced level of droplet entrainment in annular flow. This is likely to enhance evaporator performance at low qualities because an increase in entrainment thins the liquid film on the passage walls, reducing the resistance to heat transfer across the liquid film. At moderate to high qualities, however, increased entrainment may cause portions of the evaporator wall to dry out, reducing the heat transfer performance. For refrigerant-oil mixtures, increased entrainment may reduce the amount of liquid in the film on the passage wall, which may result in high oil concentration in the film, which in turn raises the dew point temperature of the liquid, reducing heat transfer performance.

The fin shedding experiments described above suggest that for conditions characteristic of many evaporators, the droplet entrainment field downstream of a fin typically has a high concentration of large droplets in the near wake region close to the centerline of the fin. In the far wake region, at larger lateral distances, the volume concentration of droplets is lower as is the mean droplet size. This implies that a substantial portion of the liquid shed from a fin is likely to impact the leading edge of the fin immediately downstream and that lateral transport of liquid droplets due to turbulence is relatively small for the moderate gas Reynolds number conditions in these experiments. The larger the separation between successive fins, the larger the distance that shed droplets will likely be carried by the gas. While these results were obtained in a scaled-up fin section, the essential physics behind the droplet shedding and transport processes have been preserved by the fact that important dimensionless groups (Re_g, We, Fr) were insured to be within the operating range of typical compact evaporators. This shedding model is applicable when there is enough liquid in the liquid film on the fin surfaces for droplet shedding to occur (see Mandrusiak and Carey (1991) for this limit).

The experiments summarized above reveal entrainment characteristics that may strongly affect the heat transfer performance of a matrix surface in a high performance evaporator. Accurate computational models of transport in offset strip fin evaporators must properly account for the entrainment behavior indicated in these experiments.

NOMENCLATURE
D Droplet diameter
d Test section (tube) diameter
Fr   Froude number = $3\rho_v U_v^2 D/2\rho_l gD$
G Mass flux
L Test section length
$\dot{m}$ Mass flow rate
m'' Local liquid droplet mass flux
P Probability
R Tube Radius
r Radial distance along the tube diameter
Re   Reynolds number = Ud_h/η
Re_g Gas Reynolds number = $Ud_h x/\eta$
Re_l Liquid Reynolds number = $Ud_h(1-x)/\eta$
U Velocity
We   Weber number = $G^2 d_h/\sigma\rho_l$
x Quality

Greek Symbols:
δ Film thickness
η Kinematic viscosity

Subscripts:
c   From the contraction
g   Gas
l   Liquid
h   Hydraulic
1   First droplet in a cascade
2   Second droplet in a cascade
3   Third droplet in a cascade

Figure 9: Droplet Cascade Downstream of a Fin

REFERENCES
Carey, V.P. and Mandrusiak, G.D., 1986, "Annular Film-Flow Boiling of Liquids in a Partially Heated Vertical Channel with Offset Strip Fins," *Int J Heat Mass Trans*, Vol 29, pp 927-939.

Cohen, M. and Carey, V.P., 1989, "A Comparison of the Flow Boiling Performance Characteristics of Partially-Heated Cross-Ribbed Channels with Different Rib Geometries," *Int J Heat Mass Trans*, Vol 32, pp 2459-2474.

Dressler, J.L. and Kraemer, G.O., 1990, "A Multiple Drop-Size Generator for Calibration of a Phase-Doppler Particle Analyzer," In *Liquid Particle Size Measurement Techniques*, (E.D. Hirleman, W.D. Bachalo and P.G. Felton, ed.), Vol 2, ASTM STP 1083, pp 30-44.

Hawks, N.H., 1995, "Experimental and Computational Study of Droplet Shedding Downstream of Fins for Application to Compact Heat Exchangers," PhD Dissertation, UC Berkeley, Berkeley, CA.

Mandrusiak, G.D. and Carey, V.P., 1990, "A Finite Difference Computational Model of Annular Film-Flow Boiling and Two-Phase Flow in Vertical Channels with Offset Strip Fins," *Int J Mult Flow*, Vol 16, pp 1071-1096.

Mandrusiak, G. D. and Carey, V. P., 1991, "A Method for Detecting Liquid Shedding Phenomena During Annular Flow in Channels with Offset Strip Fins," *Experimental Thermal and Fluid Science* Vol. 4, pp. 239-245.

McQuillan, K.W. and Whalley, P.B., 1984, "The Effect of Orifices on the Liquid Distribution in Annular Two-Phase Flow," *Int J Mult Flow*, Vol. 10, pp 721-729.

Shi, X.D., Brenner, M.P., and Nagel, S.R., 1994, "A Cascade of Structure in a Drop Falling from a Faucet," *Science* Vol 265, pp 219-222.

Shollenberger, K.A., 1994, "An Experimental and Analytical Investigation of Entrainment Rates for Downward Annular Mist Two-Phase Flow in Pipes and Downstream of a Localized Contraction," PhD Dissertation, UC Berkeley, Berkeley, CA.

van der Geld, C.W.M. and Vermeer, H, 1994, "Prediction of Drop Size Distributions in Sprays Using the Maximum Entropy Formalism: The Effect of Satellite Formation," *Int J Mult Flow*, Vol 20, pp 363-381.

GAS-ASSISTED EVAPORATIVE COOLING IN A
NARROW CHANNEL

Avram Bar-Cohen and Gary L. Solbreken
Department of Mechanical Engineering
University of Minnesota
Minneapolis, Minnesota

ABSTRACT

A high velocity flow of a liquid-gas mixture, through a narrow channel between parallel plates, has been recently found to provide very formidable heat flux and heat density removable capability. This technique is of particular interest for thermal management of densely-packaged electronic components and has prompted the study of FC-72 and nitrogen flowing together in a 0.5 mm channel. The paper will describe the experimental apparatus used and the thermofluid data obtained for a range of inlet mixture conditions. Flow regime maps and a theoretical model for such gas-assisted evaporative cooling will be presented and discussed.

INTRODUCTION

Recent years have witnessed increasing industrial interest in the development of very high heat flux thermal management systems. In applications ranging from high power laser optics to high power synchrotron x-ray beam line components, and from plasma processing hardware to fast-switching semiconductor transistors, peak heat fluxes are already well in excess of 1 kW/cm^2 and, in some cases approach 50 kW/cm^2 (Khounsary, 1992). Moreover, the imposed volumetric constraints can result in heat densities from 20 W/cm^3 in electronic applications to nearly 1 kW/cm^3 in x-ray beam line applications (Khounsary, 1992). Material and reliability considerations limit the allowable temperature rise of the heat dissipating component surfaces and necessitate heat transfer coefficients of 1 W/cm^2 K to 100 W/cm^2 K in order to achieve effective thermal control. Recognition of the difficulty inherent in providing such high performance cooling, has led to the development of new thermal management techniques, capable of removing both high heat fluxes and high heat densities.

A recently patented electronics thermal management technique, relying on the evaporation of a high velocity flow of a liquid-gas mixture in the narrow channels between populated substrates, appears to address this need (Sherwood and Cray, 1992).

This paper will describe the experimental apparatus at the University of Minnesota used to study GAEC behavior, and the thermofluid data obtained using FC-72 and nitrogen. Flow regime maps and a theoretical model for GAEC will also be presented and discussed.

U OF MN EXPERIMENTAL APPARATUS

In the Laboratory for the Thermal Management of Electronics, at the University of Minnesota, a flow loop was constructed to conduct detailed FC-72/Nitrogen GAEC experiments. The flow loop, which operates in a "blowdown" mode, is shown schematically in Figure 1 and is described in detail in Hodes (1994) and Bar-Cohen 1995). FC-72 is pumped from the reservoir up to the accumulator, where it is pressurized to about 1 atmosphere. The pressurized liquid flows through a spray nozzle, into a mixing chamber. The gas flows directly from a regulated nitrogen cylinder into the mixing chamber at a 90 degree angle relative to the liquid nozzle. The accuracy and repeatability of the flow meters was estimated to equal 1% of full-scale and 0.2% of full scale, respectively.

Upon exiting the mixing chamber, the FC-72/Nitrogen two-phase flow enters the asymmetrically-heated, parallel plate channel, shown schematically in Figure 2. One side of this channel contains a 20.3 x 3.81 cm Watlow MI strip heater soldered to a 1.27 cm thick, 26 x 5.6 cm copper heat spreader, with 12 thermocouples soldered 1.3 mm below the wetted surface and the thermocouples are placed at the locations shown in Figure 2. The other side of the test section contains a transparent acrylic window, with the test section itself being 5.1 cm wide by 25.4 cm long, comparable in configuration to a single channel in the

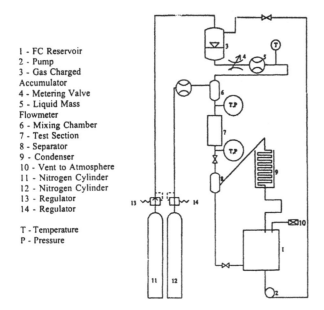

1 - FC Reservoir
2 - Pump
3 - Gas Charged
Accumulator
4 - Metering Valve
5 - Liquid Mass
Flowmeter
6 - Mixing Chamber
7 - Test Section
8 - Separator
9 - Condenser
10 - Vent to Atmosphere
11 - Nitrogen Cylinder
12 - Nitrogen Cylinder
13 - Regulator
14 - Regulator

T - Temperature
P - Pressure

Figure 1. GAEC Experimental Apparatus

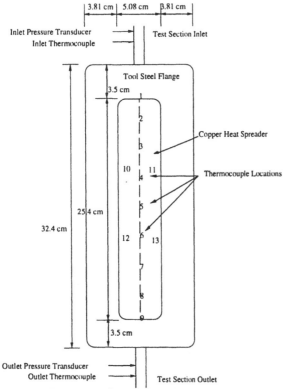

Figure 2. GAEC Test Section Detail

Cray-3 module. The gap between the heat spreader and acrylic window was 0.508 mm (+/-0.025 mm). Because the heat spreader is wider than the channel, it is assumed, for heat flux calculations, that the channel walls have the

same heat flux as the heat spreader. This yields a flux area of 132.1 cm^2

The pressure drop across the test section was monitored with transducers, calibrated to an accuracy of 1 kPa. Bulk temperature rise through the test section is measured with thermocouples at the inlet and exit of the test section.

EXPERIMENTAL DATA

The thermal tests reported herein complement the results reported in Bar-Cohen, et. al. (1995) and were carried out at liquid superficial velocities of 0.16 m/s and 0.09 m/s, with the gas superficial velocity varying from 2.74 to 4.38 m/s, and the heat flux in the test channel set at 0.91, 1.89, and 3.79 W/cm^2. The liquid/gas velocities were chosen such that annular flow was established the channel inlet.

The experimental data for the range of conditions stated above can be found in Table 1, under the headings of 'operating conditions' and 'measured values'. The two comparison columns will also be addressed later. In the Table, T_i, T_o, T_{wall}, and ΔT correspond to the mixing bulb inlet temperature, the test section bulk outlet temperature, the wall temperature at the outlet, and T_o - T_i respectively. Δp is the pressure drop through the channel and the heat transfer coefficient is based on $(T_{wall} - T_i)$, the rise in temperature from the inlet to the wall at the outlet.

From the Table, it can be seen that the bulk outlet temperature and the wall outlet temperature increase with increasing heat flux at the same fluid flow rate. Also, for a given heat flux and liquid velocity the temperature decreases with increasing gas velocity. This results from the more effective evaporation at the higher gas velocities, as can be seen clearly in figure 3.

One of the dangers of reducing the liquid velocity, however, is the increased possibility of dryout. This phenomena is illustrated in Figure 4 where the transient response of the wall temperature rise is plotted for the three heat fluxes at a constant gas/liquid velocity pair of 4.38/0.09 m/s. For the two lower heat fluxes a steady state value is attained. However, for the high heat flux case, the wall outlet temperature rises continuously, suggesting dryout.

As the fluid flow rates for a particular heat flux are increased, the pressure drop through the channel increases as expected. Likewise, as the heat flux is increased, the pressure drop for corresponding fluid flow rates increases, reflecting the evaporation (increase in the gas velocity) taking place along the channel.

Flow Regimes

In studying two-phase flow, it is useful to examine the flow regime developed by the flow. The flow regime is most conveniently defined by the superficial velocities of the gas and liquid phases. The superficial velocity is defined as the velocity a single phase would have if it occupied the entire channel. The superficial velocity, expressed in terms of flow rate, is:

$$U^s = \dot{m} / \rho A_x \qquad (1)$$

Taitel (1990) discusses and derives mechanistic relations for the flow regime boundaries using superficial velocities.

A flow regime map based on Taitel's relations and developed for FC-72 and nitrogen in a 0.508 mm gap, vertical, parallel plate channel, can be seen in Figure 5 (Hodes, 1994). On the map, the three, two-phase flow regimes associated with a vertical channel, i.e., bubble, intermittent, and annular, can be seen. At high liquid velocities (and low gas velocities) the bubble flow regime is encountered, while higher gas velocities yield annular flow. So, if a heated channel was entered by a fluid in bubble flow, considerable additional evaporation (gas generation) would have to take place over a substantial channel length before the gas velocity would e sufficient to develop annular flow. The addition of a non-condensable gas at the channel entrance (i.e., raising the gas velocity) makes it possible to enter annular flow immediately. The flow map in Figure 5 is for FC-72 without any vapor. However, flow maps generated with vapor included have the same general characteristics as the map shown (Hodes, 1994).

Annular flow provides a thin film of liquid on the channel walls, and hence, a low conduction resistance to the gas-vapor interface. Since the driving force for evaporation is the vapor pressure, the addition of gas also reduces the core vapor pressure for a given void fraction, and enhances evaporation. Reliance on annular flow may also prevent nucleate flow boiling critical heat flux. The drawbacks of gas addition include the increased possibility of dryout, the introduction of an evaporation diffusion resistance, and an inherently larger pressure drop through the channel. The inclusion of the non-condensable gas may also require a fluid separation device (condenser) if the system is to operate in a closed loop.

The flow regime map in Figure 5 can be used to find the superficial gas velocity needed to assure annular flow in the presence of a specified liquid superficial velocity. In order to keep pressure drop minimal, the smallest gas velocity possible should be used at the inlet. Continued annular flow is assured since liquid evaporation will cause the gas velocity to increase, and the flow to become "more" annular. It must be noted that while the liquid mass flow rate can be readily found from the superficial velocity using eqn. (1), the gas mass flow rate is pressure dependent.

Energy Balance

A theoretical model for calculating two phase temperature rise, pressure drop, and outlet quality in a vertically oriented, narrow, heated, parallel plate channel has been developed. For purposes of this analysis, the mixing bulb is integrated into the channel and it is assumed that the non-condensable gas enters the channel dry. In addition, diffusion resistance in the gas-vapor core is assumed to be negligible. This assumption implies that the gas/vapor core is well mixed and saturated. Furthermore, if both the gas and vapor operate far from

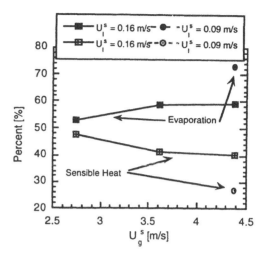

Figure 3. Distribution Of Heat Dissipation

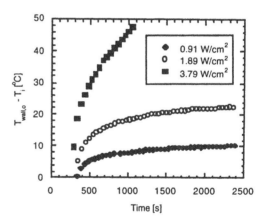

Figure 4. Transient Response For
$U_l^S = 0.09$ m/s And $U_g^S = 4.38$ m/s

their thermodynamic critical point, it may be assumed that the gas/vapor core is an ideal gas mixture.

A general energy balance for the channel, ignoring kinetic and potential energy terms, can be written:

$$\dot{Q} = \dot{m}_{L,o} i_{L,o} + \dot{m}_{v,o} i_{v,o} + \dot{m}_{g,o} i_{g,o} -$$

$$\dot{m}_{L,i} i_{L,i} - \dot{m}_{v,i} i_{v,i} - \dot{m}_{g,i} i_{g,i} \qquad (2)$$

By applying the mass balance, it is found that

$$\dot{m}_{L,o} = \dot{m}_{L,i} - \dot{m}_{v,gen} \qquad (3)$$

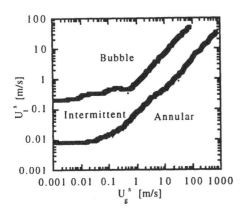

Figure 5. Flow Regime Map For FC-72 And Nitrogen In a 1 mm, Parallel Plate Channel At Atmospheric Conditions

Assuming that the vapor is saturated at the exit, i.e.

$$i_v = i_L + h_{fg} \qquad (4)$$

and that the sensible heat capacity of the gas is relatively negligible, the liquid specific heat is constant, there is no vapor at the inlet, the energy balance reduces to:

$$\dot{Q} = \dot{m}_{L,i} C_L (T_o - T_i) + \dot{m}_{v,gen} h_{fg,o} \qquad (5)$$

In order to solve for the outlet temperature, use must also be made of the ideal gas ratio for the mixture flowing in the core:

$$\frac{\dot{m}_v}{\dot{m}_g} = \left[\frac{M_v}{M_g}\right]\left[\frac{p_v}{p_g}\right] \qquad (6)$$

This equation is subject to the constraint that the sum of the vapor and gas partial pressures is equal to the outlet pressure. To determine the vapor flow rate, it is necessary to know the gas mass flow rate. This could be found using inlet conditions since there is no change in the gas flow rate through the channel. Unfortunately, only the gas superficial velocity is directly known at the inlet (from the flow regime map). To find the mass flow rate the gas density, and hence pressure, must be known (eqn. (1)). Solving iteratively for the inlet pressure, the outlet temperature can be found by using Eqns. (1), (5), and (6).

Pressure Drop

With the outlet pressure specified, the inlet pressure must obey both continuity considerations and established relations for two-phase pressure drop. Both require an

outlet temperature to be known, and solving eqn. (5) for the vapor flow rate.

The continuity relation uses eqn. (6) to solve for the gas flow at the exit (the vapor pressure, and hence gas pressure, are just functions of the outlet temperature). By continuity, the gas flow at the exit is the same as the gas flow at the inlet. Using eqn. (1), the density, and hence the pressure, can be determined.

One of the most common two-phase pressure drop approaches relies on the use of the Chisholm correlation. This relation was successfully used to predict the adiabatic pressure drop in the test channel (Hodes, 1994). The Chisholm correlation uses two-phase multipliers to relate the two phase pressure drop to the pressure drop of each individual phase, flowing alone in the channel, using the Lockhart-Martinelli parameter, X^2. This frequently used two-phase parameter is defined as:

$$X^2 = \frac{\Delta p_L}{\Delta p_g} \qquad (7)$$

The pressure drop for each phase alone is defined as:

$$\Delta p = (y) * (f)\left[\frac{L}{D_h}\right]\left[\frac{\rho U^{s^2}}{2}\right] \qquad (8)$$

where the coefficient y will be discussed later.

In all the GAEC experiments reported herein, the inlet liquid Reynolds numbers were below 1000, and thus considered to be in the laminar regime (Lockhart and Martinelli, 1949). The calculated channel exit Reynolds numbers for the gas core varied from laminar to turbulent, using 2000 as a transition value (Lockhart and Martinelli, 1949). However, since the gas flow is laminar for nearly the entire channel length, it was found appropriate to base the calculation of pressure drop on the laminar/laminar relation. In recognition of the turbulent transition experienced by the gas stream near the exit of the channel, and the desire to obtain a conservative prediction, a coefficient, y, of 1.5 is used in Eqn. (8) in place of the usual coefficient of unity.

The second Lockhart-Martinelli parameter used is the two-phase multiplier, ϕ_L^2, defined as:

$$\phi_L^2 = \left(\frac{\Delta p_{tp}}{\Delta p_L}\right) \qquad (9)$$

The Chisholm correlation (Chisholm, 1967) for ϕ_L^2 in terms of the Lockhart-Martinelli parameter is:

$$\phi_L^2 = 1 + \frac{c}{X} + \frac{1}{X^2} \qquad (10)$$

To obtain an upper-bound estimate of the pressure drop in the channel it was assumed that the outlet conditions prevail over the entire channel, while neglecting the pressure drop in the mixing bulb.

The density for the gas/vapor core, was determined by assuming the mixture to be an ideal gas, while the outlet mixture viscosity was taken as the weighted average of the gas and vapor values. The liquid properties were assumed to be constant along the channel, with the liquid mass flow progressively reduced, in the axial direction, by evaporation (see eqn. (3)).

Using the foregoing relations, the channel inlet pressures can be found in terms of both continuity and pressure drop considerations. In the present study, the outlet temperature was varied iteratively until the two pressure relations converged to identical values.

The temperature which results from the model is the bulk fluid temperature at the channel outlet. For cooling applications, the temperature of interest is generally along the channel wall. For a thin film, conduction through the film provides a good approximation of the temperature difference. For the present apparatus, Hodes (1994) found film thicknesses on the order of 10 μm for outlet qualities of 50% and comparable flow velocities. It is of interest to note that the surface roughness of the milled copper heater plate is estimated to be in the range of 2 μm. A conduction approximation gives a temperature difference of between 1 and 5 °C across the thickness of the liquid film. This is in generally good agreement with experimental data at the wall near the exit of the channel. The exit wall temperature can then be used to define a global heat transfer coefficient, based on the average surface heat flux and the difference between the exit wall temperature and the inlet liquid temperature, as $h = q''/(T_{wall,o} - T_i)$

MODEL VALIDATION

The measured channel inlet and outlet pressures and temperatures, obtained in a series of experiments performed on the GAEC facility at the University of Minnesota, are shown in Table 1, along with predicted values for the exit conditions, calculated exit quality, and heat transfer coefficients. In the Table, columns under the heading "measured values" are experimental results, columns under "predicted values" are model calculated results, and columns under "operating conditions" are specified experimental parameters which are also used as inputs for model calculations. The temperature and pressure drop errors, under the heading 'comparison', are defined under to Table.

The experimental uncertainty in the measured values was estimated to be +/- 1 kPa (yielding an average uncertainty of approximately 13%) for the pressure drop and an estimated uncertainty of +/- 1K (yielding an average uncertainty of 7%) for the axial temperature rise in the channel. It should be noted that the predicted pressure drop was based on fluid properties associated with the measured outlet pressure of each run (approximately 125 kPa).

The predicted outlet fluid temperatures are seen to be relatively close to, though consistently above, the

experimental values, with an average discrepancy of 3 °C, vs. an experimental error of +/- 1 K, with an axial fluid temperature rise that varies from 5.5 °C to 26 °C. It might have been expected that ignoring the resistance to the diffusion of vapor from the liquid/vapor interface to the gas/vapor core, might have led to an underestimate of the liquid temperature. The results obtained suggest that the diffusion resistance is negligible in the geometry and conditions tested at the University of Minnesota facility, and that ignoring the sensible heat absorption in the gas/vapor, along with the other assumptions discussed above, yields a conservative prediction of the outlet temperature.

Examining this Table, it may be seen that, as expected, all the predicted pressure drops for the 25.4 cm long channel - which are based on the outlet vapor/gas velocities and a 50% increase in the single-phase pressure drop coefficient - exceed the experimental values. The predicted values are, relatively close to the measured pressure drops and the discrepancy is always within the experimental error for the heat flux runs of 3.79 W/cm^2. However, for the 2 low heat flux data sets, the pressure drop error is between 45% and 75%, when the expected error is between 13% and 23%. For as yet undetermined reasons, case O21 (low liquid flow/low heat flux) has an inordinate pressure drop error of 135%, when the error is expected to be just 30%. These errors are largely due to the choice of the coefficient y, and the conservative nature of other assumptions made in the pressure drop calculation, as discussed earlier.

The last column in Table 1 presents the experimental heat transfer coefficient, h, defined in terms of the axial temperature rise between the inlet fluid and the wall at the exit. It may be seen that the GAEC heat transfer coefficients, in the range of 1000 W/m^{2}K (0.1 W/cm^{2}K), are extremely high for such a low conductivity liquid and are approximately six times the fully-developed, single-phase convective heat transfer coefficients in a 0.5 mm channel. Neglecting the channel wall thickness, these results point to volumetric heat transfer coefficients in the range of 1.9 MW/m^{3}K (1.9 W/cm^{3}K). While these parameters are at the low end of the spectrum of high heat removal requirements, chemical and electrical constraints dictate the use of FC liquids in the immersion cooling of electronic components (Bar-Cohen, 1993).

CONCLUSIONS

Examination of the GAEC behavior of nitrogen-assisted, evaporating FC liquid flow has served to establish the potential of this technique for the thermal management of electronic components. No less importantly, this study provides a test-bed for the development of a GAEC modeling procedure. The generally good agreement obtained for higher heat flux in comparing the experimental and predicted thermofluid behavior of FC-72/nitrogen flow through the 0.5 mm University of Minnesota GAEC channel, sets the stage for exploring the performance of this unique heat transfer technique with other, perhaps more thermally-robust, fluid/gas combinations.

NOMENCLATURE

A = area [m^2]
c = constant for Chisholm correlation
C = specific heat [kJ/kg K]
D_h = effective hydraulic diameter [m]
f = friction factor
h = heat transfer coefficient [kW/m^2 K]
h_{fg} = latent heat of vaporization [kJ/kg]
i = enthalpy [kJ/kg]
L = channel length [m]
k = thermal conductivity [W/mK]
m = mass flow rate [kg/s]
M = molar mass [kg/kmol]
p = pressure [kPa]
Q = heat flow [kW]
q" = heat flux [kW/m^2K]
t = channel gap [m]
T = temperature [^{o}C]
U^s = superficial velocity [m/s]
V = volumetric flow rate [m^3/s]
w = channel width [m]
x = quality
X^2 = Martinelli parameter
y = coefficient for single phase pressure drop
Δ = change in value
ϕ^2 = Martinelli parameter
μ = viscosity [kg/m s]
ρ = density [kg/m^3]

Subscripts

exp = experiment
g = gas phase
gen = generated vapor
i = inlet conditions
L = liquid phase
mod = model
o = outlet conditions
tot = equivalent or mixture
tp = two phase
v = vapor phase
x = channel cross-section

REFERENCES

Bar-Cohen, A., Sherwood, G., Hodes, M., and Solbreken, G.L., 1995, "Gas-Assisted Evaporative Cooling of High Density Electronic Modules," *IEEE Transactions on CPMT: Part A*, Vol. 18, No. 3, pp. 502-509.

Bar-Cohen, A, 1993, "Thermal Management of Electronic Components with Dielectric Liquids," *JSME International Journal*, Vol. 36, No.1, pp . 1-25.

Chisholm, D., 1967, "A Theoretical Basis for the Lockhart-Martinelli Correlation for Two-Phase Flow," *Int. J. Heat Mass Transfer*, Vol. 10, pp. 1767-1778.

Hodes, M., 1994, "Gas-Assisted Evaporative Cooling in Downflow Through Vertical Channels," Master's Thesis, Department of Mechanical Engineering, University of Minnesota.

Khounsary, A., editor, 1992, "High Heat Flux Engineering," SPIE Proceedings Series, Vol. 1739, Washington D.C.

Lockhart, R. W., and Martinelli, R. C., 1949, "Proposed Correlation of Data for Isothermal Two-Phase, Two-Component Flow in Pipes," *Chemical Engineering Progress*, Vol. 45, No. 1, pp. 39-48.

Sherwood, G., and S. Cray, 1992, "Gas-Liquid Forced Turbulence Cooling," United States Patent # 5,131,233.

Taitel, Y., 1990, "Flow Pattern Transition in Two-Phase Flow," *Proceedings of the Ninth International Heat Transfer Conference*, Hemisphere Publishing Corporation, Vol. 1, pp. 237-254.

Table 1: Comparison Between Experimental And Predicted Values For FC-72/Nitrogen

Run	Operating Conditions				Measured Values						Predicted Values			Comparison	
	q"	U_L^s	U_g^s	T_i	T_{wall}	T_o	ΔT	P_o	ΔP	h	x_o	T_o	ΔP	ΔT_{err}	ΔP_{err}
	[W/cm^2]	[m/s]	[m/s]	[^{o}C]	[^{o}C]	[^{o}C]	[^{o}C]	[kPa]	[kPa]	[kW/m^2K]	[%]	[^{o}C]	[kPa]	[%]	[%]
O11	3.79	0.16	2.74	24.4	57.6	49.2	24.8	125	9.9	1.14	48	53.7	10.7	18.3	8.1
O12	3.79	0.16	3.62	19.6	51.7	43.2	23.6	133	10.2	1.18	45	51.5	11.1	35.1	8.5
O8	3.79	0.16	4.38	22.8	57.4	48.8	26.0	142	11.0	1.09	49	51.6	11.3	10.6	2.3
O13	1.89	0.16	2.74	22.9	44.1	39.1	16.2	119	6.2	0.89	18	42.2	9.9	19.3	59.0
O14	1.89	0.16	3.62	22.8	42.2	37.0	14.2	123	7.2	0.98	20	40.3	10.7	23.0	48.5
O15	1.89	0.16	4.38	22.1	41.1	36.0	13.9	130	7.8	1.00	21	38.7	11.3	19.2	44.4
O20	1.89	0.09	4.38	24.3	46.8	40.9	16.6	122	4.4	0.84	50	42.8	7.6	11.7	72.7
O22	0.91	0.16	2.74	21.6	32.2	29.5	7.9	115	5.3	0.86	9	31.2	9.2	21.8	73.0
O17	0.91	0.16	3.62	22.7	31.8	29.1	6.4	125	6.4	1.00	11	30.6	10.1	23.8	58.1
O18	0.91	0.16	4.38	22.5	30.7	28.0	5.5	131	6.9	1.11	12	29.4	10.9	25.1	57.5
O21	0.91	0.09	4.38	23.8	34.1	30.8	7.0	123	3.3	0.88	25	32.5	7.8	23.9	135.5

$$\Delta T_{err} = \frac{\left(T_{o,exp} - T_i\right) - \left(T_{o,mod} - T_i\right)}{\left(T_{o,exp} - T_i\right)} \qquad \Delta p_{err} = \frac{\Delta p_{exp} - \Delta p_{mod}}{\Delta p_{exp}}$$

HEAT TRANSFER IN A SUBCOOLED FALLING LIQUID FILM

Erol Ulucakli

Department of Mechanical Engineering
Lafayette College
Easton, Pennsylvania USA

ABSTRACT

A continuous two-phase flow apparatus was built to study the convective heat transfer in subcooled falling liquid films of water. Film Reynolds numbers in the range of 800 to 1560 and heat fluxes up to 35 kW/m² were employed. An unheated calming section (82 cm long) preceded the heating surface (105 cm long, 2.2 cm OD). Heating surface temperature was measured at eight locations along the heater by internal thermocouples with an uncertainty of 0.05°C. The calming section allowed the hydrodynamic boundary layer to develop before the thermal boundary layer and the onset of the evaporative processes. The variation of the heat transfer coefficient along the heater was obtained as a function of the heat flux and the film Reynolds number. The separation of the influences of the hydrodynamic and the thermal entrance regions contributed to a better understanding of these effects on the heat transfer coefficient.

INTRODUCTION

Falling liquid films are frequently employed in a variety of technological processes. Further, new areas of applications have been recognized in recent years, prompting a revival of research interest in falling films. These areas include vaporization processes at very low temperature differences, the vaporization of solvents from viscous liquids, and the heating of heat sensitive fluids. Recent reviews of heat transfer in falling films have been presented by Seban (1978), Ganic (1983), Yih (1986) and Palen (1988).

The characteristics of nucleate boiling in thin liquid films are observed to be different from those of pool boiling (Cerza and Sernas, 1989). In falling films, the bubble moves with the fluid while it grows and may soon burst or escape very rapidly. Drops are formed when a bubble is ejected from the film. The process of bubble growth in the film offers a low thermal resistance path to the heat flow which eventually causes lower temperature differences. The quick escape of a bubble and replacement of the film improve the heat flux.

The investigations reported in the heat transfer literature concerning the boiling processes in liquid falling films are limited (Palen, 1989). The nucleate boiling might be the desired effect in certain evaporators where increased heat transfer coefficients are needed. In certain other situations, the nucleation might not be desired since it will increase the local rates of heat transfer and cause dry-out, usually at the lower part of the heating surface. This might cause the destruction of the heater or the damage of the product. Very little information is available on the determination of the incipience and effect of nucleate boiling in thin falling liquid films (Palen, 1988).

In the investigation reported here a liquid film is heated and evaporated as it flows on the outer surface of a single cylindrical heater under the influence of gravity. The independent variables of the problem were the heat flux from the heating surface, the liquid subcooling (the difference between the saturation and liquid inlet temperatures), the mass flow rate of the liquid, and the system pressure. The physical condition of the surface, as inadequately described by its surface roughness, was not modified during experiments. The thickness of the liquid film and the temperature of the heating surface were the dependent variables. The surface temperature of the heater at a given location was calculated from the internal temperature measurements. The characteristics and the behavior of the falling films depended on the film Reynolds number.

Continuous flow falling film evaporators employ a convenient device (distributor) to distribute the fluid evenly over the solid, heated surface. As the fluid is discharged from the distributor, the depth of the film changes and reaches a constant value after a short distance. This distance is called the hydrodynamic entrance length. The entrance length in falling films is found to be larger than those of pipe and rectangular channel flows as noted by Pierson and Whitaker (1977). They presented an estimate of the hydrodynamic entrance length for water as $L_e = 0.006\, Re^{3/4}$ cm where the Reynolds number is based on laminar film thickness and free surface velocity for the uniform flow.

By the time the film thickness reaches its constant thickness, the axial velocity profile is supposed to reach its fully developed parabolic shape calculated by Nusselt (Brauer, 1956). However, it is well known that vertical falling films are unconditionally unstable. Consequently the film surface becomes unstable after a short distance from the distributor and the interface is no longer smooth but wavy. This condition occurs when the Reynolds number is in laminar range (less than 1600). The waves may appear even before the entrance length is reached. Pierson and Whitaker (1977) state that for Reynolds numbers 400 and larger, the small disturbances grow to a finite amplitude entirely within the entrance region. Brauer (1956) obtained data for aqueous ethylene glycol solutions and for water at the onset of waves as $Re_i = 1.22\, Ka^{-0.1}$ where R_i is the Reynolds number at the onset of the waves. Due to the complexity of the processes in this region there is no agreement among many investigators. However, Kapitza's (1982) two dimensional stability analysis yields $Re_i = 2.44\, Ka^{-1/11}$ which is close to that of Brauer's (1956).

As the film flows down in a wavy laminar regime a transition to turbulent flow occurs at a Reynolds number range 1000 to 1600. Takahama and Kato (1980) discovered that the critical Reynolds number at which the transition may occur depends on the distance from the distributor. After the transition, the flow continues to have its wavy interface which is observed to be random and three dimensional by Salazar and Marschall (1978).

The objective of the work described here was to obtain the much needed variation of the heat transfer coefficient in a falling film in the axial flow direction by measuring the surface temperatures locally along the heating surface as functions of Reynolds number and heat flux. An unheated section preceding the heating surface allowed the hydrodynamic boundary layer to develop before the thermal boundary layer. Specifically, enhancement of the heat transfer coefficient beyond the developing thermal boundary layer was investigated.

EXPERIMENTAL FACILITY

The schematic of the experimental facility used in this

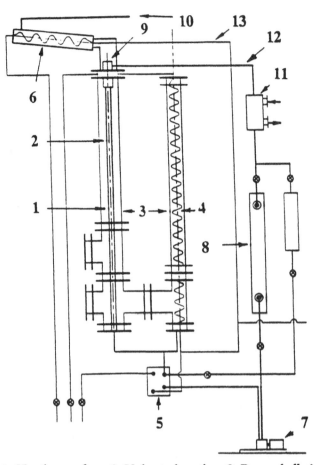

1. Heating surface, 2. Unheated section, 3. Pyrex shell, 4. Main condenser, 5. Liquid cooler, 6. Knockback condenser, 7. Pump, 8. Flowmeters, 9. Liquid distributor, 10. Vacuum line, 11. Preheater, 12. Test fluid line, 13. Coolant line.

Figure 1. Experimental Facility

study is shown in Figure 1. The continuous flow test apparatus consists of a vertical test section enclosed in a Pyrex shell, a knock-back condenser, a subcooling heat exchanger and a magnetic-seal circulation pump. The test section consists of the liquid film distributor followed by an unheated inlet section, and the heated section. The unheated "calming" section helps to establish a fully developed hydrodynamic flow regime before the fluid reaches the heated section. The test section is enclosed in a series of a pipe, a tee and a cross, all made of Pyrex. The inside diameter of the Pyrex shell is 10.2 cm and the diameter of the main heater is 2.2 cm. The main heater shown in Figure 2(a) was fabricated from a cartridge heater. A 99 cm long cartridge heater of 1.6 cm diameter, rated 7 kW at 208 volt AC was used. A copper tube was placed over the cartridge heater and the assembly was passed through a die, forming a "sinked" product. Eight sheathed, ungrounded thermocouples were

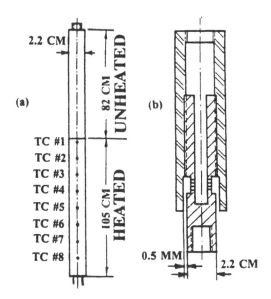

Figure 2. Heating Surface and Film Distributor

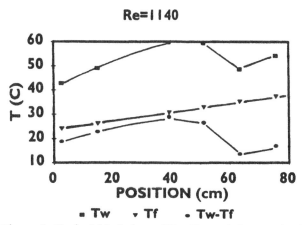

Figure 3. Typical Variation of T_W, T_F, and T_W - T_F in the Axial Direction.

inserted into the longitudinal grooves cut on the copper tube. A stainless steel 316 tube was "sinked" over the copper tube.

A liquid film distributor shown in Figure 2(b) was made of stainless steel 316 and was used to form a liquid film of uniform thickness. The distributor design was based on Palen's (1994). The top part had a small reservoir for the test liquid. The width of the annular gap in Figure 2(b) was 0.5 mm. The test section was carefully aligned with vertical direction to insure symmetrical liquid film at the exit from the distributor. A vacuum pump was used to purge the inert gases from the system. A small water cooled condenser was attached to the top of the upper flange of the test section to remove and condense vapor. The electrical outputs of the thermocouples were measured with a precision digital multimeter with 6 1/2 digit reading and 100 nV resolution. The accuracy of the voltage readings at this range was 0.0001 mV. A data acquisition system (a personal computer, a 16-bit board, and a screw terminal panel) was used to monitor the temperatures. The data acquisition system and the thermocouples were calibrated together.

The heat losses from the lower end of the heater exposed to the ambient were estimated and found to be insignificant. The upper end of the heater was assumed to be insulated. The heating surface temperature, T_W, was calculated using the Fourier equation, $T_W = T_I - q'' \, d/k_s$, for the heat conduction through wall of the stainless steel tube. T_I was the temperature read by the thermocouples embedded between the copper and stainless steel layers over the cartridge heater. The liquid temperature was measured at the inlet and exit of the test section. The pressure of the system was atmospheric. The mass flow rate and the inlet temperature of the test fluid were kept

constant during the experiments. The thermocouple readings for the surface temperatures along the heater were recorded after the steady-state was reached.

RESULTS

The film Reynolds number during all experiments was below 1600 which is commonly accepted transition number to the turbulent flow regime from the laminar. Thus the film Reynolds numbers used in the experiments covered the wavy laminar flow regime and the films were indeed wavy and assumed to be laminar. No attempt was made to measure the turbulence levels. The films were allowed to develop fully in the hydrodynamic sense before they reached the heated section. At lower Reynolds numbers the liquid films were thin and developed dry spots due to the evaporation without nucleation. At higher Reynolds numbers, the films were thicker and it was possible to reach higher heat fluxes without damaging the heater.

In Figure 3, variations of the wall, liquid temperatures and the difference between the wall and liquid temperatures with position in the axial direction are shown for a fixed Re = 1140 and q" = 21.95 kW/m². The abscissa in Figure 3 represents the temperature measurement locations on the heating surface and shown in Figure 2. Figure 3 shows the almost linear increase of the wall temperature along the heating surface followed by a decrease and a gradual increase closer to the end of the heating surface. The reason for this behavior will be discussed below. The inlet and exit temperatures were measured and the intermediate liquid temperatures were calculated from energy balances using the measured heat flux and the mass flow rate. A reference temperature, defined as the average of the local wall and liquid temperatures, was used in determining the liquid properties. The increase of the liquid temperature in Figure 3 along the heating surface is linear as expected

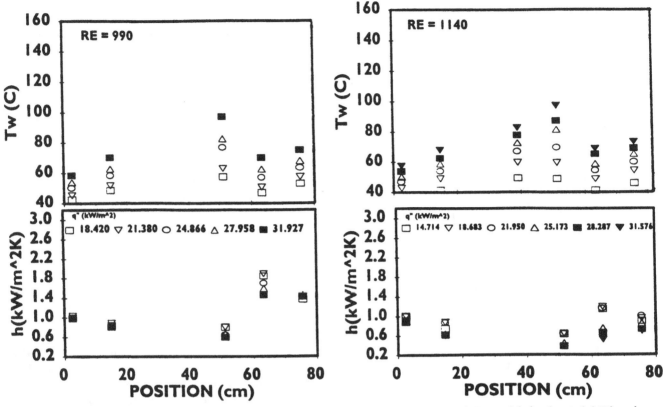

Figure 4. Variations of T_w and h in the Axial Direction, Re = 990.

Figure 5. Variations of T_w and h in the Axial Direction, Re = 1140.

and the calculated exit temperature agrees well with measured exit temperature.

The influence of the heat flux on the variations of the local surface temperature and the local heat transfer coefficient with position along the heating surface is shown in Figure 4 for Re = 990. In the upper part of Figure 4, the variation of T_w with heat flux is shown. It is observed that, the surface temperature increase in the axial directionwas insignificant at low heat fluxes. As the heat flux was increased gradually, the average surface temperatures increased. At heat fluxes 20 kW/m^2 and above, the axial gradient of T_w became pronounced. In the lower part of Figure 4, the variation of the heat transfer coefficient, h, corresponding to the variation of T_w on the upper part of the figure, is shown. The local heat transfer coefficient was calculated from the expression $h (T_w - T_p) = q''$. The development of the boundary layer is indicated by the gradual decrease in h observed at the upstream positions on the heater. It is noted that the decrease in h slowed near the middle of the heater and then began a gradual increase.

Figures 5 and 6 show the influence of the mass flow rate or the Reynolds number. In Figures 5 and 6, the variations in T_w and h with heat flux and axial position are shown for the Reynolds numbers of 1140 and 1543,

respectively. Re = 1543 was the highest Reynolds number employed in the experiments. The trends in T_w and h are similar to those of Figure 4. The thermal entrance lengths are also similar. The upper part of Figure 5 shows the gradual increase in the magnitude of T_w as the heat flux is increased. At constant heat flux, a gradual increase in T_w and then a decrease was observed in all high heat flux experiments. The local heat transfer coefficients varied along the heater in a manner similar to those of Figure 4 for Re = 990.

The examination of the behavior of the local heat transfer coefficients in Figures 4, 5, and 6 indicates a common trend that occurred at all heat fluxes. The data show that the local heat transfer coefficient decreased within the entrance length and after passing through a minimum, it experienced an increase over a relatively short distance. After that h either decreased by a small amount or reached an asymptotic value. Many other investigators, Wilke (1962), Struve (1969), Ganchev et al. (1972), Oosthuizen and Cheung (1977), Gimbutis et al. (1978), Shmerler and Mudawwar (1988), Tuzla et al. (1995), also observed similar film behavior over long heating surfaces. Shmerler and Mudawwar (1988) described the increase in h as the "downstream enhancement" of the heat transfer coefficient.

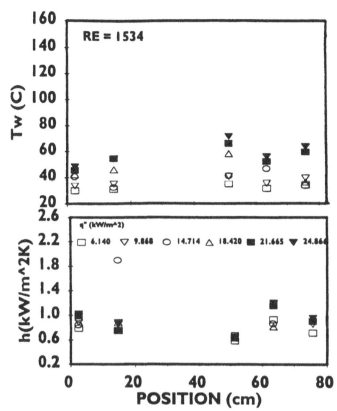

Figure 6. Variations of T_W and h in the Axial Direction, Re = 1545.

The key to understanding this behavior is provided by Salazar and Marschall (1978) and Takahama and Kato (1980) who performed experiments on surface waves in unheated falling films. They measured the maximum and minimum thicknesses of the film and found that thicknesses varied along the heater as a function of the axial location. They showed that as the distance from the distributor increased, the minimum thickness decreased, and eventually reached an asymptotic value. The maximum film thickness increased almost linearly with the axial distance as the mass of the fluid in the film was accumulated into faster and larger turbulent waves. The groth of large waves in turn resulted in an increase in the characteristic wavelength which eventually caused the time-averaged film thickness to decrease in the axial direction. It is reasonable to assume that the thinner film subjectd to increased turbulent wave activity is the cause of the enhanced heat transfer at the end region of the heating surface.

CONCLUSIONS

Variations of the surface temperature and the heat transfer coefficient in the axial direction in a wavy laminar falling water film at atmospheric pressure were measured for heat fluxes up to 35 kW/m², and Reynolds numbers in

the range of 800 to 1560. The influences of the hydrodynamic and thermal boundary layers were separated by employing an unheated section preceding the heated section. The local heat transfer coefficients were found to be influenced by the Reynolds number and the level of heat flux. The gradual decrease of the heat transfer coefficient along the upper half of the heated length - due to the developing thermal boundary layer - was observed. Instead of reaching an asymptotic value similar to the pipe flow, the heat transfer coefficient showed an enhancement after approximately the midpoint of the heated section. This phenomenon is explained by the thinning of the falling film caused by a comlex interaction of wave motion and turbulence. It is also concluded that the characteristics of the interfacial waves such as maximum and minimum film thicknesses significantly influence the heat transfer in falling liquid films.

NOMENCLATURE

d	Thickness of the sheath
g	Gravitational acceleration
h	Local heat transfer coefficient
k_s	Thermal conductivity of stainless steel
Ka	Kapitza number, $g \, \mu^4 / \rho \, \sigma^3$
L_e	Hydrodynamic entrance length
Pr	Prandtl number
q"	Heat flux
Re	Reynolds number, $4 \, \Gamma / \mu$
Re_i	Reynolds number for the onset of waves
T_F	Fluid temperature
T_I	Measured inside heater temperature
T_R	Reference temperature, $(T_W + T_F)/2$
T_{sat}	Saturation temperature
T_W	Heating surface temperature
Γ	Mass flow rate of liquid per unit circumference of the heater
μ	Dynamic viscosity of liquid
ρ	Density of liquid
σ	Surface tension of liquid

ACKNOWLEDGEMENTS

The author gratefully acknowledges the assistance of his students, Anand Gupta and Rasika Abeysinghe, and his colleagues, Dr. Kemal Tuzla of Lehigh University, and Professor Arthur Gorman of Lafayette College, for valuable discussions.

REFERENCES

Brauer, H., 1956, "Strömung und Wärmeübergang bei Rieselfilmen," *VDI-Forschungsheft 457*.

Cerza, M. and Sernas, V., 1989, "A Comparison of

Boiling in Thin, Falling Films and Pool Boiling from a Vertical Cylinder," *Heat Transfer with Phase Change,* ed. Habib, I. S., ASME HTD-Vol. 114, pp. 63-67.

Chun, K. R. and Seban, R. A., 1971, "Heat Transfer to Evaporating Water Films," *J. Heat Transfer,* Vol. 93, pp. 391-396.

Fujita, T. and Ueda, T., 1978, "Heat Transfer to Falling Liquid Films and Film Breakdown-I: Subcooled Liquid Films," *Int. J. Heat Mass Transfer,* Vol. 21, pp. 97-108.

Ganchev, B. G., Kozlov, V. M. and Lozovetsky, V. V., 1972, "A Study of Heat Transfer to a Falling Fluid Film at a Vertical Surface," *Heat Transfer-Soviet Research,* Vol. 4(2), pp. 102-110.

Ganic, E. N. and Mastanaiah, K., 1983, "Hydrodynamics and Heat Transfer in Falling Film Flow," *Low Reynolds Number Flow Heat Exchangers,* ed. Kakac, S. et al., pp. 487-527, Hemisphere.

Gimbutis, G. J., Drobavicius, A. J. and Sinkunas, S. S., 1978, "Heat Transfer of a Turbulent Water Film at Different Initial Flow Conditions and High Temperature Gradients," *Heat Transfer-1978,* Paper FB-9, pp. 321-326.

Kapitza, P., 1982, "Wave Flow of Thin Liquid Layers of a Viscous Fluid," *Collected Papers of P. Kapitza,* pp. 662-709.

Oosthuizen, P. H. and Cheung, T., 1977, "An Experimental Study of Heat Transfer to Developing Water Film over Cylinders," *J. Heat Transfer,* Vol. 99, pp. 152-155.

Palen, J. W., Wang, Q., and Chen, J. C., 1994, "Falling Film Evaporation of Binary Mixtures," *AIChE J.,* Vol. 40, pp. 207-214.

Palen, J. W., 1988, "Falling Film Evaporation in Vertical Tubes," *Heat Exchanger Technology,* ed. Chisholm, D., pp. 189-218, Elsevier.

Pierson, F. W. and Whitaker, S., 1977, "Some Theoretical and Experimental Observations of the Wave Structure of Falling Liquid Films," *Ind. Eng. Chem. Fundam.,* Vol. 16, pp. 401-408.

Salazar, R. P. and Marschall, E., 1978, "Time-Average Local Thickness Measurement in Falling Liquid Flow," *Int. J. Multiphase Flow,* Vol. 4, pp. 405-412.

Seban, R. A., 1976, "Transport to Falling Films," *Heat Transfer-1976,* Vol. 6, pp. 417-428.

Shmerler, J. A. and Mudawwar, I., 1988, "Local Heat Transfer Coefficient in Wavy Free-Falling Turbulent Liquid Films Undergoing Sensible Heating," *Int. J. Heat Mass Transfer,* Vol. 31, pp. 67-77.

Takahama, H. and Kato, S., 1980, "Longitudinal Flow Characteristics of Vertically Falling Liquid Films without Concurrent Gas Flow," *Int. J. Multiphase Flow,* Vol. 6, pp. 203-215.

Tuzla, K., Warn, K. and Chen, J. C., 1995, "Boiling Inception in a Binary Liquid Film," *Proc. 1995 National Heat Transfer Conference,* HTD-Vol. 314, pp. 32-35.

Yih, S-M, 1986, "Modeling Heat and Mass Transport in Falling Liquid Films," *Handbook of Heat and Mass Transfer-Volume 2,* ed. Cheremisinoff, N. P., pp. 111-210, Gulf.

FALLING FILM EVAPORATION OF SOLUBLE MIXTURES

Reiner Numrich
Chemical Engeneering
University of Paderborn
Germany

ABSTRACT

The liquid phase in falling film evaporators represents the main transfer resistance, even if there are several components in the gas phase. For optimal dimensioning of such evaporators it is necessary to describe correctly transfer processes in the falling film which often shows a turbulent flow pattern.

Investigations show that a correct calculation for soluble mixtures depends on the knowledge of heat transfer and also on mass transfer. Great deviations occur in comparison with conventional dimensioning methods which are based on heat transfer controlled processes. Therefore the falling evaporators are often underdimensioned especially for mixtures with high Prandtl numbers using these conventional methods.

Our investigations yield that the discrepancy cited is not caused by mass transfer as proposed above. In fact the heat transfer coefficient at high Prandtl numbers for turbulent falling films is much lower then mentioned previously in literature. Based on a "Three–Zone–Model", a new Nusselt equation for turbulent film flow is derived and compared with measurements in a wide range of Prandtl numbers. Using this equation, it is shown that mass transfer in the liquid mixture is mostly negligible. Additionally, significant flow of the gas phase increases mass transfer which is very good anyhow. Therefore the mass transfer resistance can usually be disregarded for practical dimensioning of falling film evaporators.

INTRODUCTION

The industrial application of distillation processes with a large number of products (e.g. fatty alcohols, glycerine) requires high boiling temperatures, even at low system pressures, while the products mentioned are characterised by particularly high thermal sensivity. As a rule, in order to reduce the temperature stress and the dwell time in the separator, falling film evaporators are used. These are vertical tube bundle arrays which are heated by condensing steam in the shell. The liquid product is fed into the tubes at the top and flows downwards by gravitation. The heat input from the hot steam causes evaporation in the tubes of those components with lower boiling points which, at the bottom of the evaporator, may then be passed in vapour form into a rectification column, for example.

The remaining liquid phase, mostly still about 80 – 90 % of the initial quantity, is fed in again at the top together with the fresh product so that, as a result of the relatively strong circulatory current, a turbulent flow pattern exists in the falling film.

The evaporation of the components takes place on the surface of the falling film. Evaporation on the surface of the tube leads to the thermal decomposition mentioned and is prevented by a limitation of the heat flux. In addition to the low pressure loss, a characteristic feature of the falling film evaporator is the low temperature difference between condensates and vapours. Therefore, in addition to the applications already mentioned, this type of evaporator is also found in combination heating circuits or in multiple stage plants.

DIMENSIONING

In order to calculate the heat exchange area required, we usually start from the enthalpy balance. The difference between the enthalpy flow on entry into the evaporator $\dot{H}_E$ and on exit $\dot{H}_A$, with the mean heat flux $\dot{q}$, gives the required exchange area A_q:

$$A_q = \frac{\dot{H}_A - \dot{H}_E}{\dot{q}} \qquad (1)$$

The mean heat flux $\dot{q}$ is calculated from the mean coefficients of heat transfer k and the mean driving temperature drops $\Delta\vartheta = \vartheta_S - \vartheta_V$ between the hot steam side (Index S) and the vapour side (Index V) (see Figure 1).

$$\dot{q} = k \cdot (\vartheta_S - \vartheta_V) \qquad (2)$$

While the temperature ϑ_S is regarded as constant over the whole area, corresponding with the saturation temperature dependent on system pressure, ϑ_V must be used as an arithmetical aid between the entry and exit of the evaporator. Analogously, the arithmetical averaging of characteristic values and coefficients is regarded as sufficiently exact for the design of falling film evaporators. Accordingly, the following descriptions always refer to mean values.

The mean heat transfer coefficient k can be determined by the serial calculation of the individual heat transfer resistances, nearly flat surfaces being assumed for simplicity. Figure 1 shows qualitatively the radial temperature profile during the evaporation of soluble mixtures.

In accordance with Equation (2), the law for the heat transfer coefficient k is

$$\frac{1}{k} = \frac{1}{\alpha_S} + \frac{1}{(\frac{\lambda}{s})_W} + \frac{1}{\alpha_L} + \frac{1}{\alpha_V} \qquad (3)$$

α_S -   heat transfer coefficient in steam condensate
$(\frac{\lambda}{s})_W$ - heat conductivity in tube wall
α_L -   heat transfer coefficient in falling film
α_V -   heat transfer coefficient in vapour

If the individual terms are analysed, it is found, especially in the case of the above-mentioned substances because of their poor thermal conductivity, that the main transport resistance occurs to about 90 % in the falling film [1]. An approximation for design in practice is therefore

$$\dot{q} \approx \alpha_L (\vartheta_S - \vartheta_V) \qquad (4)$$

Thus the determination of the heat transfer coefficient α_L in the falling film gains an elementary significance.

HEAT TRANSFER IN A TURBULENT FALLING FILM

The calculation of the heat transfer coefficient α_L in a turbulent flowing film is normally carried out by means of Nusselt realtionships as a function of the Reynolds number Re_L and the Prandtl number Pr_L in the form

$$Nu_L = \frac{\alpha_L \cdot L}{\lambda} = const \ Re_L^a \ Pr_L^b \qquad (5)$$

where the characteristic length is defined as $L = (\frac{\nu_L^2}{g})^{1/3}$ with ν_L as the cinematic viscosity and g as the gravitional acceleration. In falling film evaporation, the equation from Chun and Seban [2] is often indicated or recommended.

$$Nu_L = 0.0035 \ \cdot \ Re_L^{0.4} \ Pr_L^{0.65} \qquad (6)$$

Here in the case of a vertical tube with diameter d and the falling film mass flow $\dot{M}_L$, the Reynolds number Re_L is with the dynamic viscosity η_L given by:

$$Re_L = \frac{4\dot{M}_L}{\pi d \eta_L} \qquad (7)$$

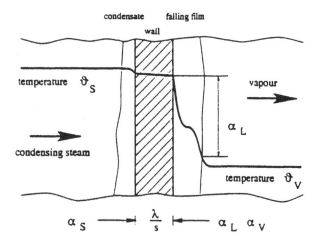

Figure 1. Radial temperature profile during falling film evaporation

The Prandtl number is defined as $Pr_L = \frac{\nu_L}{a_L}$ where a_L is the thermal conductivity number. Equation (6) shows good correlation with measured data acquired with water as pure substance [2,3]. In its application to soluble binary mixtures, Palen et al. [4] have shown experimentally on the substance pairs water/ethylene glycol and water/propylene glycol that this equation does not adequately describe the transport processes. Figure 2 shows a comparison of measured heat transfer coefficients with the values calculated according to Equation (6).

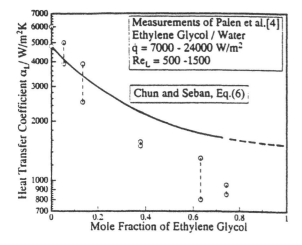

Fig 2.: Measured and calculated heat transfer coefficients in relation to the mole fraction of ethylene glycol.

It can be seen that the measured heat transfer coefficient is up to 80 % smaller than the calculated one. As the difference increases with an increasing mole fraction of the glycols used, Palen et al. attribute this observed effect to the additionally occuring mass transfer in the liquid phase. This mass transfer, as a result of the equilibrium condition, produces an increase in the phase interface temperature and so to a deterioration in the effective heat transfer in the falling film. This explains the experimental findings. Schutte et al. [5] came to similar conclusions in their comparison of different theoretical procedures.

From their measurements, Palen et al. have derived a Sherwood relationship to describe the mass transfer with which their own measured results can be satisfactorily reproduced. However, the comparison, shown in Figure 3, of the reported Sherwood relationship with the data of other authors [6,7] measured in connection with absorption processes shows considerable differences. Herein the Sherwood number Sh_L is defined analogously to the Nusselt number in Equation (5) with the mass transfer coefficient β_L and also the same characteristic length L.

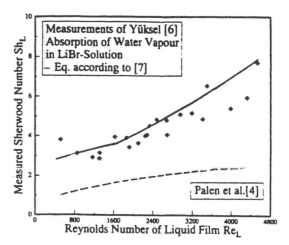

Figure 3. Calculated and measured Sherwood numbers in relation to the Reynolds number of the falling film

The equation from Palen et al. produces larger Sherwood numbers than the values determined by pure substance transport measurements. The question arises how the facts described can be explained without contradiction. Measurements by Müller [8] made on the turbulent film condensation of vapour showed that, at high Prandtl numbers of the liquid phase, considerable discrepancies arise between values measured and those calculated in accordance with Chun and Seban. In a study of our own on the basis of the model conceptions of Yüksel [6] and the extensive experiments ($Pr_L \leq 50$) of Müller, a simple Nusselt relationship was derived by modification of the Prandtl analogy which reproduces well the known measured values including water[9].

$$Nu_L = 0.003 \cdot Re_L^{0.44} Pr_L^{0.4} \qquad (8)$$

Crucial in the comparison with the equation from Chun and Seban is the difference in the dependence of the Nusselt number on the Prandtl number and thus, in the end, on the thermal conductivity in the falling film.

Figure 4 shows measured Nusselt numbers from various authors as a function of the Prandtl number at a Reynolds number of $Re_L = 4000$.

There is clearly a good correlation between the measured values and the graph as calculated with Equation (8). For comparison, the results for Equation (6) are also shown, with deviations of up to 100 % at high Prandtl numbers.

When Equation (8) is applied to the measured values for falling film evaporation already shown in Figure 2, the results are as shown in Figure 5.

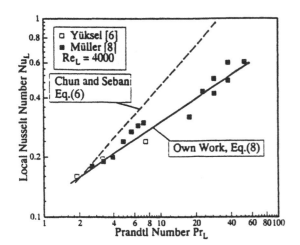

Figure 4. Dependence of the Nusselt number on the Prandtl number in a turbulent falling film

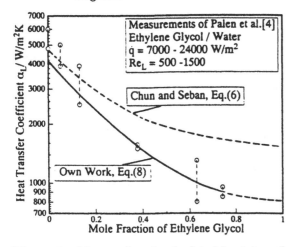

Figure 5. Measured and calculated heat transfer coefficients in relation to the mole fraction of ethylene glycol.

The reported Nusselt relationship reproduces the measured values for the present mixture very well in the whole experimental range without consideration of mass transfer processes. This was also found during experimental investigations in condensation of binary mixtures [1] and absorption [6], where the Schmidt numbers reached values of $Sc_L \approx 600$. The initially described increase in the difference between calculation and measurement with an increase in the mole fraction of glycol in the falling film should thus not be attributed to the mass transfer, but to the Prandtl number which rises from the value for water of $Pr_L \approx 2$ to a maximum of the value for pure ethylene glycol $Pr_L = 21.5$. This Prandtl number results in an actual heat transfer coefficient α_L for ethylene glycol which is only 46 % of the value calculated according to Chun and Seban. The comparatively low heat transfer coefficient α_L for substances with higher

Prandtl number underlines the assumption of the negligibility of all other heat transfer resistances upon which Equation (4) is based. This results of the present work must be verified in further experiments.

CONCLUDING NOTES

This study clearly shows that falling film evaporation, like the condensation of binary vapour mixtures, is a heat transfer controlled process in which the main transport resistance is usually to be found in the falling film. That does not mean, as often proposed, that the concentration of the two components should be regarded as constant over the falling film. Especially in the case of mixtures with a broad boiling range, a more or less distinct concentration profile, dependent on the phase equilibrium, will indeed be formed in the layer near the surface of the falling film. However, as this profile depends on the planned evaporation capacity, a thermally based design, as described initially, is entirely adequate for the dimensioning of a falling film evaporator. The assumption of complete mixing is certainly accurate enough for mixtures with narrow boiling ranges and where a significant gas flow exists.

REFERENCES

[1] Hadley, M.; Numrich, R: Kondensation eines im flüssigen Zustand löslichen binären Dampfgemisches, Chemie-Ingenieur-Technik, 66(1994)No.3, pp 387 - 390.

[2] Chun,K. R.; Seban, R. A.; Heat Transfer to Evaporating Liquid Films, Journal of Heat Transfer 93(1971), pp 391 - 396.

[3] Papendieck, H.: Strömung und Wärmeübergang in welligen Rieselfilmen, Dissertation, TH Braunschweig 1984.

[4] Palen, J. W.; Wang, Q.; Chen, J. C.: Falling Film Evaporation of Binary Mixtures, AIChE–Journal 1994 (40) No. 2, pp 207 - 214.

[5] Schutte, P.; Gorak, A.; Wozny, G.; Jeromin, L.: Stationäre und instationäre Simulation von Fallfilmverdampfern, Chemie–Ingenieur–Technik, MS 2043/92.

[6] Yüksel, L. M.; Schlünder, E. U.: Heat and mass transfer in non isothermal absorption of gases in falling liquid films, Chemical Engineering Processing 22(1987) Part I pp 193 - 202, Part II pp 203 - 213.

[7] Brauer, H.: Stoffaustausch, Sauerländer–Verlag, Aarau 1971.

[8] Müller, J.: Heat transfer during film condensation Chemical Engineering Processing 33(1994) pp. 277–283.

[9] Numrich, R.: Heat Transfer in Turbulent Falling Films, Preprints Eurotherm Seminar No. 26, 16./17. September 1993, Edinburgh, UK.

Critical Heat Flux in Falling Films Undergoing Nucleate Boiling

Abdulmalik A. Alhusseini, Bryan C. Hoke, John C. Chen
Department of Chemical Engineering
Lehigh University
Bethlehem, PA 18015 USA

Abstract

Falling film evaporation is used in the chemical industry for vaporization of both single component and multi-component liquids. This process is a special case of single component boiling, wherein bulk flow is induced by gravitational body force and the liquid flows as a thin film on vertical heat transfer surfaces. With appropriate wall superheats, bubble nucleation and boiling occurs on the heat transfer surface within the flowing film. The critical heat flux (CHF) is governed by quite different mechanisms than those commonly found for either pool boiling or convective boiling in channels.

An experimental investigation was conducted to measure the critical heat flux at approximately atmospheric pressure using three test fluids: F-113, water, and aqueous isopropanol. The new experimental data and observations supported Ueda et al. (1981) findings regarding the existence of more than one breakdown region in films undergoing nucleate boiling. The dry patch model failed to predict CHF data because it required unrealistic contact angles.

INTRODUCTION

Falling film evaporators find increasing applications due to their high thermal efficiency and suitability to heat sensitive fluids, viscous fluids, and subatmospheric processes. Under certain conditions, falling films fail to completely wet the evaporator surface which reduces the effective heat transfer area and, as a result, adversely affects the overall efficiency of the evaporator. This condition is often termed film breakdown.

Film breakdown is a complicated phenomenon that can be triggered by several mechanisms which are film thinning due to insufficient flow, thermocapillary effect, film disruption and thinning due to nucleate boiling, the Leidenfrost effect at high wall temperatures, and flooding by countercurrent vapor flow. The first three mechanisms are the most common in the

applications of falling film evaporators. Breakdown due to film thinning takes place when a stable dry patch is formed at sufficiently small flow rates. This phenomenon is governed by surface tension and the contact angle of the liquid on the evaporator surface. Hartley and Murgatroyd (1964) proposed the dry patch model to predict film breakdown by this mechanism.

Film breakdown due to the thermocapillary effect is a condition where gradients in surface tension cause lateral diversion of the flow and thin the film beyond a critical thickness in local regions. Surface tension gradients can be caused by temperature or concentration gradients. The thermocapillary effect is usually the triggering mechanism when heating subcooled films or evaporating mixtures where the more volatile component has a higher surface tension than the less volatile component. Lateral diversion models were proposed by Simon and Hsu (1970), Fujita and Ueda (1978b), and more recently by Hoke and Chen (1992).

Falling films undergoing nucleate boiling breakdown by more than one mechanism. Ueda et al. (1981) studied the breakdown of F-11, F-113, and water falling films undergoing nucleate boiling on a test tube that was 18 cm long. Ueda et al. identified three breakdown regions depending on the exit film flow rate. Breakdown in the first region occurred at relatively low heat fluxes and was caused by the progressive reduction of flow rate along the evaporator surface due to evaporation and droplet entrainment by the bubbles bursting through the film. When the exit flow rate was smaller than the minimum wetting flow rate (about 0.01-0.02 kg/m/s for water films), stable dry patches were formed near the bottom of the test section. Breakdown in the second region occurred at higher heat fluxes and was caused by disruption of the film by the bursting vapor bubbles. In this region, the critical heat flux (CHF) increased with the exit film flow rate. Breakdown in the third region occurred at even higher heat fluxes where high vapor effusion separated most of the film from the heater surface in spite of considerable film flow rates at the exit of the test section.

Breakdown of falling films undergoing nucleate boiling has been studied by several investigators in connection with the cooling of high temperature surfaces, such as nuclear fuel rods, where heat fluxes are much higher than that encountered in the typical applications of falling film evaporators. Data at lower heat fluxes are limited to that reported by Fujita and Ueda (1978a), and Ueda et al. (1981). This is somewhat surprising considering the wide applications of falling film evaporators. Therefore, more research is needed in this area.

In this paper, new data for breakdown of falling films undergoing nucleate boiling are reported. The data were collected for three test fluids: F-113, water, and a binary mixture of isopropanol and water. The latter fluid did not breakdown by the thermocapillary mechanism because the more volatile component (isopropanol) has a lower surface tension than the less volatile component (water).

EXPERIMENT

A schematic of the test loop is shown in Figure 1. The test loop consisted of a metering pump, preheater, instrumented test section, knock-back condenser, condenser, and subcooler. Test fluids were F-113, water, and a binary mixture of water and isopropanol. Parameters not adjusted in these experiments were tube diameter and heated length. Flow rate was controlled as the independent variable, and minimum heat flux causing permanent film breakdown was determined as the dependent parameter. Both flow rate and heat flux were directly determined from measurements. Temperatures were measured at various locations in the loop by type K thermocouples as indicated in Figure 1. The temperatures of the stream entering the test section, the vapor surrounding the test section, the inner wall of the evaporator, and the inlet stream to the pump were all measured.

The test section was a 316 stainless steel tube that had been hand sanded with 240-weight sandpaper. The test section was 107 cm long and the outside diameter was 2.1 cm. The surface was prepared for each fluid by "washing" with clean test fluid for several hours, and the loop was discharged and then recharged before experimental measurements were taken.

The film was distributed on the outer surface of the test tube by a conical distributor whose length was 5.7 cm. The conical distributor was found experimentally to provide nearly uniform film thickness at the test section inlet. For isothermal films, uniform waves traversed the test section without distortion. This provided confirmation that the test section was vertical and the film distribution was uniform. Fluid was introduced to the conical distributor of the test section by a feed tube mounted in the top flange above the test section. After a 1.7-cm-long cylindrical unheated region, the test section had a 75-cm heated length. A maximum heat flux of 120 kw/m^2 was provided by a cartridge heater imbedded within the test section. Three thermocouples were imbedded in the inner wall of the test section and were located 15.2 cm

from the bottom into the heated length and spaced 120° around the circumference. While not sensitive enough to detect film thickness variation, wall temperature measurements assisted in steady-state determination and the calculation of wall superheats. A 2-inch-diameter beaded glass pipe enclosed the test section allowing visual observation of the test section.

Vapor generated in the test section flowed to the condenser where it condensed on copper cooling coils and dripped to the subcooler section. The balance of the fluid introduced to the test section formed a pool at the base of the test section and overflowed into the subcooler.

The test procedure was to set the flow rate and gradually increase heat flux until permanent breakdown was observed visually. After each increase in heat flux, the system was allowed to come to steady state before the heat flux was increased again. Inlet temperatures were maintained constant and close to the saturation temperature by adjusting the subcooler and preheater as necessary.

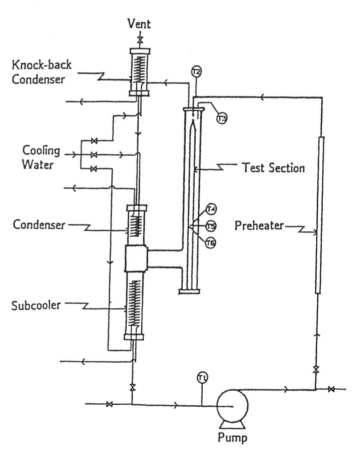

Figure 1. Test Loop Schematic.

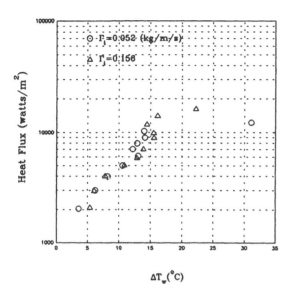

Figure 2. Boiling Curves for F-113 at the Lowest
And Highest Inlet Flow Rates

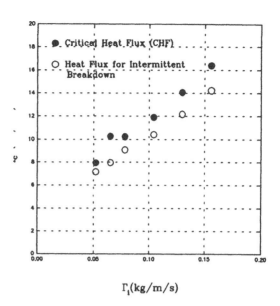

Figure 3. Critical Heat Fluxes for the Formation of
Intermittent and Permanent Dry Patches
Versus Inlet Flow Rate

DISCUSSION

Experimental Results and Observations

Sample boiling curves for F-113 at atmospheric pressure are presented in Figure 2 for two runs at the lowest and highest inlet flow rates. The heat flux was calculated by dividing the total electrical power input to the cartridge heater by the surface area of the heated length of the test tube. The wall temperature, which was measured on the inner surface of the test tube, was corrected for heat conduction across the thickness of the test tube. The saturation temperature was measured by the vapor thermocouple.

As seen in Figure 2, the onset point for nucleate boiling, which is marked by a sharp increase in the slope of the boiling curve, seemed to occur at higher wall superheats as flow rate was increased which is consistent with the observations of Mudawwar et al. (1987) for F-72 films. At the onset point, nucleation sites were seen along the test section especially near the exit end. As heat flux was increased beyond the onset point, nucleation sites became more vigorous and covered a larger portion of the test tube. This was especially true for F-113 at heat fluxes larger than about 10 kw/m² where more vigorous nucleate boiling resulted in noticeable droplet entrainment by vapor bubbles bursting through the film. Further increase in heat flux resulted in the formation of intermittent dry patches which seemed to occur near the bottom of the test tube. The formation of intermittent dry patches marks the beginning of the flattening boiling curves because of the slightly higher wall temperatures. At even higher heat fluxes, the duration and size of the intermittent dry patches increased which eventually led to the formation of stable dry patches near the bottom of the test section. This was

accompanied by a sharp rise in the measured wall temperature near the bottom of the test section and marked the critical heat flux (CHF) for a given inlet flow rate. The critical heat flux was larger at higher inlet flow rates which is expected and consistent with the observations made by Fujita and Ueda (1978a) and Ueda et al. (1981). The nucleate boiling regime seemed to extend over a relatively narrow heat flux range (about 5 kw/m²) due to the small inlet flow rates covered in this experiment.

The intermittent and permanent critical heat fluxes for F-113 are plotted versus the inlet flow rate in Figure 3. The permanent critical heat flux (CHF) is larger than that at the onset of the intermittent dry patches by about 10-20% for the conditions covered in this experiment. The effect of the inlet flow rate is clear; both the CHF and the onset point for intermittent dry patches can be delayed by increasing the inlet flow rate which is in accordance with findings in previous studies.

The CHF for three fluids (water, F-113, and aqueous isopropanol) is plotted versus the film inlet flow rate in Figure 4. The CHF for water films is higher than that for aqueous isopropanol and much higher than that for F-113. In addition, the CHF is much more sensitive to flow rate variations for water than for F-113 films. The above two observations suggest that the fluid's latent heat affects the CHF characteristics in saturated films more than any other thermodynamic or physical property.

Dry Patch Model

As a first attempt to correlate the obtained data, we

examined the dry patch model, which was proposed by Hartley and Murgatroyd (1964) to predict the minimum wetting flow rate in isothermal and smooth falling films. In its simplest form, the dry patch model accounts for the stagnation force and the resultant of surface tension forces. The stagnation force is a stabilizing downward force resulting from liquid deceleration around the dry patch. For laminar films, the stagnation force is given by,

$$F_s = \frac{\rho_{1}^3 g^2 \delta}{15\mu_{1}^2} \qquad (1)$$

where δ is the film thickness, g is gravitational acceleration, and ρ_1 and μ_1 are the liquid density and viscosity. The resultant of the surface tension forces is a destabilizing upward force and is given by,

$$F_\sigma = \sigma \left(1 - \cos\left(\Theta\right)\right) \qquad (2)$$

where σ is surface tension and θ is the contact angle of the liquid on the evaporator surface. The stability condition is obtained by equating the stagnation to surface tension forces as given by,

$$\frac{\rho_{1}^3 g^2 \delta^5}{15\,\mu_{1}^2} = \sigma\left(1 - \cos(\Theta)\right) \qquad (3)$$

Equation (3) can be used to calculate the critical film thickness which in turn can be used to calculate the minimum wetting flow rate if the contact angle is known.

The applicability of the dry patch model can be assessed by backcalculating the contact angle from (3) using exit flow rates calculated at CHF conditions. The exit flow rate in films undergoing surface evaporation can easily be calculated as the difference between the inlet liquid flow rate and that lost due to evaporation as given by,

$$\Gamma_o = \Gamma_i - \frac{qL}{\lambda} \qquad (4)$$

where q is heat flux, L is the heated length of the test section, λ is the latent heat of evaporation, and Γ_i and Γ_o are the inlet and outlet liquid mass flow rates per unit flow width. The situation is more complex in films undergoing nucleate boiling because droplet entrainment by bursting vapor bubbles is another mechanism for the reduction of film flow rate along the test section. To account for droplet entrainment, equation (4) can be modified as,

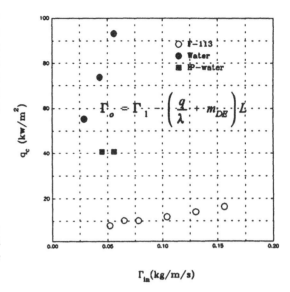

Figure 4. Critical Heat Flux for F-113, Water, and Aqueous Isopropanol versus Inlet Flow Rate

$$\Gamma_o = \Gamma_i - \left(\frac{q}{\lambda} + m_{DE}\right) L \qquad (5)$$

where m_{DE} is the droplet entrainment rate per unit evaporator area. In this study, the exit flow rates for water and F-113 were calculated with the aid of the droplet entrainment correlation proposed by Ueda et al. (1981) which is given by,

$$\frac{m_{DE}}{q/\lambda} = C\left[\frac{(q/\lambda)^2\,\delta_m}{\sigma\rho_g}\right]^{3/4} \qquad (6)$$

where δ_m is the film thickness calculated at the mean flow rate along the evaporator, $(\Gamma_i + \Gamma_o)/2$, and C is a constant that depends on the fluid. For water, $C=477$, and for F-113, $C=663$. It must be noted that the above droplet entrainment correlation was extrapolated to heat fluxes smaller than that in Ueda et al. database. Droplet entrainment for aqueous isopropanol films was ignored due the lack of entrainment correlations. However, droplet entrainment were only important for F-113 films for the conditions covered in this study, as will be discussed subsequently.

Table 1 lists the measured inlet flow rates, measured CHF, calculated exit flow rates, ratio of droplet entrainment rate to inlet flow rate, and backcalculated contact angles. The backcalculated contact angles for water are smaller than that expected for water on stainless steel surfaces, which is about 45°. On the other hand, the backcalculated contact angles for

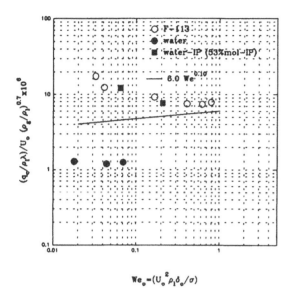

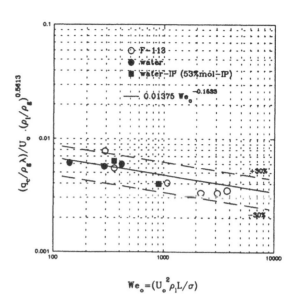

Figure 5. Comparison of the Obtained Data and the CHF Correlation Proposed by Ueda et al. (1981) for the Second Region

Figure 6. Proposed Correlation of the New Data

F-113 are larger than that expected for F-113 on stainless steel surfaces which is smaller than 10°. It is also noted that the backcalculated contact angles seem to increase with flow rate. This discrepancy suggests that the dry patch model, which is intended for smooth films, is inadequate for films undergoing nucleate boiling due to the vigorous disruption of the free interface by vapor bubbles. The ratio of droplet entrainment

Table 1.   Backcalculated Contact Angles Based on the Dry Patch Model

Fluid	Γ_1 (kg/m/s)	q_c (kw/m²)	Γ_o (kg/m/s)	$m_{EB}L/\Gamma_1$	θ (Degrees)
water	0.029	55.4	0.0101	0.017	8.7
water	0.043	73.8	0.0174	0.025	13.7
water	0.056	93.2	0.0229	0.038	17.3
IP-water	0.045	40.7	0.0109	--	16.5
IP-water	0.056	40.7	0.0219	--	29.7
F-113	0.052	7.94	0.0090	0.043	13.2
F-113	0.065	10.26	0.0078	0.068	11.7
F-113	0.078	10.23	0.0207	0.061	26.6
F-113	0.104	11.90	0.0353	0.073	42.2
F-113	0.130	14.07	0.0455	0.094	52..8
F-113	0.156	16.42	0.0529	0.120	60.5

rate to inlet flow rate was less than 5% for water films and less than 12% for F-113 which indicates that film flow rate reduction by droplet entrainment was not significant for most of the conditions covered in this study. This is consistent with our observations during the experiments where few liquid

droplets impinged on the glass pipe for all conditions except for F-113 at heat fluxes larger than about 10 kw/m².

Comparison With Ueda et al. (1981) CHF Correlation

It is interesting to note that the data presented in Table 1 seem to support the existence of at least two film breakdown regions based on the relationship between the film exit flow rate and the CHF. Data for water, aqueous isopropanol, and F-113 at inlet flow rates smaller than 0.08 kg/m/s are probably associated with the first film breakdown region reported Ueda et al. (1981), where breakdown is due to film thinning. The exit flow rates for these data were roughly between 0.01-0.02 kg/m/s, which is the minimum wetting flow rate as reported by Fujita and Ueda (1978a) for water films. For F-113 films at inlet flow rates larger than about 0.1 kg/m/s, the exit flow rates seemed to increase more steadily with the CHF which suggests that these data may fall in the second breakdown region reported by Ueda et al., where breakdown is due to disruption of the film by the bursting vapor bubbles.

To verify the existence of more than one breakdown region, all data are compared with the CHF correlation proposed by Ueda et al. for the second region in Figure 5. The correlation is given by,

$$\left(\frac{\frac{q_c}{\rho_1 \lambda}}{U_o} \right) = 60 \times 10^{-6} \left(\frac{\rho_1}{\rho_g} \right)^{07} \left(\frac{\rho_1 U_o^2 \delta_o}{\sigma} \right)^{0.10} \quad (7)$$

where U_o and δ_o are the film mean velocity and thickness at the exit end of the test section. As can be seen in Figure 5, the data seem to merge with Ueda et al. correlation as Weber number based on exit conditions increases. This supports the existence of the second region and establishes more confidence in Ueda et al. correlation. Agreement with the correlation is poor at low Weber numbers which supports the existence of another breakdown region and is consistent with Ueda et al. findings.

Correlation

A tentative correlation of the new data is given by,

$$\frac{\left(\dfrac{q_c}{\rho_1 \lambda}\right)}{U_o} = 0.01375 \left(\frac{\rho_1}{\rho_g}\right)^{0.5613} \left(\frac{\rho_1 U_o^2 L}{\sigma}\right)^{-0.1533} \quad (8)$$

This correlation form was originally proposed by Monde and Katto (1978) for the CHF in saturated impinging jets and wall jets. As shown in Figure 6, the correlation is in very good agreement with data. However, the correlation should be used with caution because the data is admittedly limited.

CONCLUSIONS

The following conclusions can be withdrawn from this study,

1. The new data support Ueda et al. (1981) findings regarding the possible existence of more than one breakdown region in falling films undergoing nucleate boiling.

2. The dry patch model failed to predict breakdown in falling films undergoing nucleate boiling because unrealistic contact were required to fit data. The disruption of the film surface by vapor bubbles and wave activity is a likely cause for the failure of the dry patch model.

3. At relatively small film flow rates, the critical heat flux is only slightly larger than that at the onset point of nucleate boiling. Therefore, it is not recommended to operate falling film evaporators in the nucleate boiling regime at small film flow rates.

NOMENCLATURE

C constant;
F_s stagnation force;
F_σ surface tension forces resultant;
g gravitational acceleration;
L heated length of the test section;
m_{DE} droplet entrainment rate per unit evaporator area;
q heat flux;
U mean film velocity;
We Weber number, $We = \rho_1 U^2 \delta / \sigma$;

Greek Symbols

ρ_g vapor density;
ρ_1 liquid density;
μ_1 liquid dynamic viscosity;

Γ mass flow rate per unit flow width;
δ mean film thickness;
δ_m mean film thickness based on the average flow rate;
θ contact angle;
λ latent heat of evaporation;
σ surface tension;

Subscripts
c   critical heat flux conditions;
i   inlet of test section;
o   outlet of test section;

REFERENCES

Fujita, T, and Ueda T., 1978a, "Heat Transfer to Falling Liquid Films and Film Breakdown-II. Saturated Liquid Films with Nucleate Boiling," *International Journal of Heat and Mass Transfer*, Vol. 21, pp. 109-118.

Fujita, T, and Ueda T., 1978b, "Heat Transfer to Falling Liquid Films and Film Breakdown-I. Subcooled Liquid Films," *International Journal of Heat and Mass Transfer*, Vol. 21, pp. 97-108.

Hartley, D. E., and Murgatroyd, W., 1964, "Criteria for the Break-up of Thin Liquid Layers Flowing Isothermally over Solid Surfaces," *International Journal of Heat and Mass Transfer*, Vol. 7, pp. 1003-1015.

Hoke, B. C., and Chen, J. C., 1992, "Thermocapillary Breakdown of Subcooled Falling Liquid Films," *Ind. Eng. Chem. Res.*, Vol. 31, pp. 688-69.

Monde, M, and Katto, Y., 1978, "Burnout in A High Heat Flux Boiling System with An Impinging Jet," *International Journal of Heat and Mass Transfer*, Vol. 21, pp. 295-305.

Mudawwar, I. A., Incropera, T. A., and Incropera, F. P., 1987, "Boiling Heat Transfer and Critical Heat Flux in Liquid Films Falling on Vertically-Mounted Heat Sources *International Journal of Heat and Mass Transfer*, Vol. 30, pp. 2083-2094.

Simon, F. F., and Hsu, Y. Y., 1970, "Thermocapillary Induced Breakdown of a Falling Liquid Film," NASA TN D-5624, Lewis Research Center.

Ueda, T., Inoue, M. M., and Nagatome, S., 1981, "Critical Heat Flux and Droplet Entrainment Rate in Boiling of Falling Liquid Films," *International Journal of Heat and Mass Transfer*, Vol. 24, pp. 1257-1266.

MODELING OF HYDRODYNAMIC PARAMETERS AND CHF IN SPRAY COOLING

Issam Mudawar[†] and Kurt A. Estes[††]
Boiling and Two-Phase Flow Laboratory
School of Mechanical Engineering
Purdue University
West Lafayette, Indiana, U.S.A.

ABSTRACT

Experiments were performed to better understand both the hydrodynamic structure of pressure sprays and cooling characteristics in the nucleate boiling regime. Sprays with large volumetric fluxes (dense sprays) were found to greatly inhibit vapor production from the heated surface, producing an unusually small boiling curve slope in the nucleate boiling regime. Experimental data demonstrated critical heat flux (CHF) is a complex function of the spray's mean volumetric flux, spray angle, nozzle-to-surface distance, Sauter mean diameter (SMD), and subcooling. By using a model for the spatial distribution of the spray's volumetric flux across the impact area and a correlation for SMD, a new method was developed to determine CHF without having to conduct costly and laborious drop sizing measurements. This method was successful at correlating CHF data for three fluids (FC-72, FC-87, and water) and both circular and square heaters over a broad range of Weber numbers, including both light and dense sprays.

1. INTRODUCTION

Sprays can be formed either by supplying liquid at high pressure through a small orifice (pressure or plain-orifice spray) or by breaking the liquid with the aid of a high pressure supply of air (atomized spray). The present study is concerned with plain-orifice sprays which simplify cooling in many applications by requiring a single (pure) working fluid. Spray cooling has long been employed in the quenching of metals upon exit from a high temperature furnace, and many studies exist in the literature for predicting spray heat transfer for those situations (*e.g.*, Bolle and Moureau, 1976). Since furnace temperatures typically correspond to the film boiling regime, these studies are not applicable to systems involving removal of high heat fluxes at relatively low surface temperatures such as

cooling of supercomputer electronics, avionics, x-ray medical devices, and lasers; devices for which the findings of the present study are intended.

Sprays utilize the momentum of liquid entering the spray nozzle to induce breakup into fine drops which impinge individually upon the heated surface. Not only does the breakup increase the surface area to volume ratio of the liquid, but it also helps produce a fairly uniform spatial distribution of heat removal compared to jet cooling. Another key feature of spray cooling is resisting liquid separation from the surface during vigorous boiling by the momentum of the individual drops.

Despite these advantages, sprays remain less popular for cooling of low temperature, high heat flux devices for several reasons. Breakup of the liquid flow to fine drops demands relatively high pressures which may not be desirable in certain situations, and the small size of the spray orifice increases the likelihood of clogging (Hall and Mudawar, 1995). However, the single most important reason for the limited use of spray cooling in low temperature, high heat flux situations is undoubtedly the lack of understanding of the mechanisms of spray boiling, especially CHF, and the absence of useful heat transfer correlations which can be employed over broad ranges of spray parameters.

Spatial Variations in Spray Cooling

Mudawar and Deiters (1994) investigated the spatial variation of the heat transfer coefficient across the impact area in water sprays. They also measured the spatial variations of the key hydrodynamic parameters of the spray just prior to impact. These are volumetric flux, Q'', mean drop velocity, U_m, and Sauter mean diameter, d_{32}. Both Q'' and U_m possess the units of velocity, the former being the volume flow rate of liquid impacting an infinitesimal portion of the impact area divided by the area of the same portion. They observed that, while U_m was virtually constant for the entire spray impact area, the heat transfer coefficient peaked at the center of the impact area and decayed gradually away from the center, becoming negligible just outside of the impact area as shown in Fig. 1. This spatial

---

[†] Professor and Director of the Purdue University Boiling and Two-Phase Flow Laboratory; author to whom correspondence should be addressed

[††] Graduate Student

distribution followed essentially the distribution of volumetric flux, Q'', across the impact area.

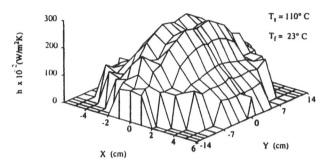

Fig. 1 Variation of heat transfer coefficient across spray impact area.

The dimensionless correlations developed by Mudawar and Valentine (1989) reveal what was suggested earlier by Toda (1972): the proper scaling velocity for heat transfer correlations near CHF is Q'' not U_m, and the scaling length is d_{32}. Increasing Q'' for water sprays improved heat transfer, while increasing drop diameter adversely affected CHF.

Unlike the studies by Mudawar and Valentine and by Mudawar and Deiters, which resulted in correlations for a small heater positioned in a much larger spray impact area, the present study is also concerned with spray boiling and CHF from a small square heater almost entirely impacted by the spray. Key objectives of the present study are to (1) develop a better understanding of how volumetric flux influences cooling performance for the full cone sprays commonly employed in low temperature, high heat flux applications; (2) optimize nozzle-to-surface distance in order to maximize CHF; and (3) develop a CHF correlation which precludes the need to perform costly drop sizing experiments.

2. EXPERIMENTAL METHODS
Spray chamber

The working fluid (FC-72 or FC-87) was deaerated and conditioned to the desired operating pressure and temperature using a closed two-phase flow loop. Inside the loop's test chamber, Fig. 2, liquid exiting the nozzle impinged upon the test heater and drained directly into the loop reservoir. Vapor produced within the chamber was recovered by a condensing coil situated within the chamber itself. A micrometer translation platform mounted atop the test chamber enabled precise positioning of the spray nozzle relative the test heater's surface.

The polished impingement surface, which measured 12.7 × 12.7 mm², was fabricated from oxygen free copper. Heat was dissipated to the impingement surface by a thick film electrical resistor which was soldered to the underside of the copper block. The heater assembly was insulated on all surfaces save the impingement surface with fiberglass plastic.

Spray Sampling Techniques

The spray heat transfer data measured in the present study were correlated with respect to the spray Weber number, ratio of spray momentum to surface tension forces exerted upon the spray drops, as proposed by Mudawar and Valentine (1989).

$$We = \frac{\rho \, Q''^2 d_{32}}{\sigma} . \qquad (1)$$

The Weber number can be based on either local or mean values of volumetric flux. The local volumetric flux, Q'', is the volumetric flow rate for an infinitesimal portion of the impact surface divided by the area of the same portion. Another convenient measure of spray flux is the average volumetric flux, $\overline{Q''}$, which can be readily determined by dividing the total volumetric flow rate of the spray by the portion of the surface directly impacted by the spray, the so-called spray impact area.

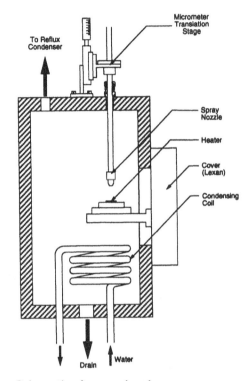

Fig. 2   Schematic of spray chamber.

The Sauter mean diameter (SMD), d_{32}, used to calculate the spray Weber number is defined as the diameter of the drop with a volume to surface area ratio that equals the volume to surface area ratio of the entire spray.

$$d_{32} = \frac{\Sigma \, n_i \, d_i^3}{\Sigma \, n_i \, d_i^2}, \qquad (2)$$

where n_i represents the number of drops in a size class i and d_i is the middle diameter of this ith class. SMD was measured in the present study for FC-72 and water using a Phase Doppler Particle Analyzer (PDPA).

When the nozzle is remote from the heater surface, the fraction of liquid leaving the nozzle that impacts the surface is a function of spray angle, θ, nozzle-to-surface distance, H, and heater length, L. To measure this fraction experimentally, a spray sampler fitted with a graduated cylinder (collector) was fabricated. The sampler possessed a knife-edged inlet with a sampling area equal to the heater surface area, 12.7 × 12.7 mm². The sampler was mounted in the heater flange itself with its inlet positioned at the exact location of the heater surface.

Since the spray's volumetric flux is not uniform across the heater surface area, the spray sampler could not measure the spatial distribution of volumetric flux. This distribution was inferred with the aid of three different sized inserts which were slid into the square opening of the original sampler. The circular openings of these inserts enabled measurement of flow rate over sampling areas which were smaller than the heater surface area.

3. NUCLEATE BOILING AND CHF TRENDS

Boiling curves were measured for three different nozzles and broad ranges of flow rate and subcooling. Shown in Fig. 3a are FC-72 boiling curves for the largest of the three nozzles and three flow rates. A mild increase in slope marks the commencement of nucleate boiling, and CHF seems to occur following only a short nucleate boiling heat flux span. These trends were characteristic of most FC-72 and FC-87 data except for the smallest flow rates and lowest subcooling, 13 °C.

In sharp contrast, the boiling curves for water sprays shown in Fig. 3b depict a broad nucleate boiling heat flux span. Also evident are the convergence of nucleate boiling data onto a single line and the relatively large slope in the nucleate boiling regime. CHF increased with increasing Q'' and/or decreasing d_{32}. Comparing these water boiling curves to the curves for FC-72, Fig. 3a, clearly points to the sensitivity of spray boiling to working fluid, volumetric flux, drop diameter, and subcooling.

A parameter characterizing the efficiency of liquid evaporation was employed to explain these differences in the shape of boiling curve. This efficiency was defined as the percentage of the total heat that could be removed by the spray, sensible and latent, that was actually removed at CHF.

$$\eta = \frac{q''_m}{q''_{dryout}} \times 100\% = \frac{q''_m}{\rho_f \overline{Q''} h_{fg} \left[1 + c_{p,f} \Delta T_{sub} / h_{fg} \right]} \times 100\%$$

(3)

Figure 4 shows efficiency versus spray Weber number for FC-72 and FC-87 at all volumetric flow rates and subcoolings and all three nozzles. Also shown are water data for full cone sprays measured by Mudawar and Valentine (1989). Figure 4 shows evaporation efficiency decreases with increasing Weber number, from a nearly 100 % efficiency for $We < 10^{-5}$ to less than 10 % for $We > 0.1$. Water sprays had much smaller Weber numbers than the FC-72 and FC-87 sprays, resulting in more of the liquid impinging upon the heated surface evaporating. A relatively small Weber number (water data) corresponds to a smaller volumetric flux and, to a lesser extent, greater surface tension. A small volumetric flux reduces liquid buildup, exposing the surface to direct impingement by the drops. Evaporation of a large percentage of the spray liquid is manifest both in a greater heat flux span in the nucleate boiling regime and a large boiling curve slope in the nucleate boiling regime.

4. SPRAY HYDRODYNAMIC PARAMETERS
SMD Correlation

Many correlations have been developed in the past for the purpose of determining SMD for various spray types, however, none are available for full cone sprays. Therefore, a new correlation is developed in the present study using SMD data for the three full cone spray nozzles and two fluids, FC-72 and water.

Lefebvre (1989) proposed a model for the breakup of a sheet of liquid into spray droplets that was successful at correlating SMD data for hollow cone sprays. He suggested breakup is the net result of both turbulent fluctuations in the liquid upstream of the orifice as well as aerodynamic forces downstream of the orifice. He recommended a general correlation for dimensionless SMD, ratio of SMD to thickness of the liquid sheet, as a function of a

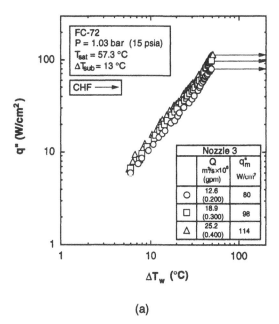

(a)

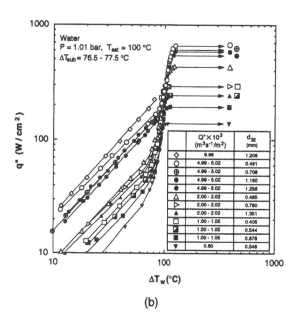

(b)

Fig. 3   Boiling curves for (a) FC-72 and (b) water (Mudawar and Valentine, 1989).

Weber number and a Reynolds number, both based on liquid velocity and thickness of the liquid sheet.

Since full cone spray nozzles do not produce a well defined liquid sheet prior to exiting the orifice, the characteristic length and velocity chosen for correlating SMD in the present study were orifice diameter, d_0, and liquid velocity at the orifice defined as $(2\Delta P/\rho_f)^{0.5}$. The Weber and Reynolds numbers based on the orifice conditions were defined, respectively, as

$$We_{d_0} = \frac{\rho_a (2\Delta P / \rho_f) d_0}{\sigma}$$

(4)

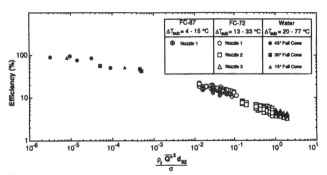

Fig. 4   Evaporation efficiency versus spray Weber number (Estes and Mudawar, 1995).

and

$$\text{Re}_{d_0} = \frac{\rho_f (2 \Delta P / \rho_f)^{1/2} d_0}{\mu_f}, \qquad (5)$$

where ρ_a is the density of ambient fluid (air or vapor).

The resulting correlation,

$$\frac{d_{32}}{d_0} = 3.67 \left[We_{d_0}^{1/2} \text{Re}_{d_0} \right]^{-0.259}, \qquad (6)$$

fits the SMD data for both FC-72 and water with a mean absolute error of 12.4% (Estes and Mudawar, 1995).

Volumetric Flux Distribution Model

As shown in Fig. 1, volumetric flux is not uniform across the spray impact area. Mudawar and Estes (1995) developed a model to predict how the spray flow rate is distributed across the heated surface. This model is based on the assumption that the total spray flow rate, Q, is uniformly distributed over any spherical surface centered at the spray orifice and bound by the spray cone angle, θ.

As the spray impacts a flat surface located a distance H from the orifice, the uniform volumetric spray flux along the spherical surface of radius H produces a volumetric flux on the impact surface which decreases radially from the center of the impact surface. As shown in Fig. 5, a differential area, dA', along the spherical surface is given by

$$dA' = 2 \pi H^2 \sin \gamma \, d\gamma. \qquad (7)$$

Integrating γ between 0 and $\theta/2$ yields the area of the spherical surface bound by the spray angle,

$$A' = 2 \pi H^2 [1 - \cos(\theta/2)]. \qquad (8)$$

Since a uniform volumetric spray flux is assumed over the spherical surface area A', this flux can be defined as

$$Q''_{sp} = \frac{Q}{2 \pi H^2 [1 - \cos(\theta/2)]}. \qquad (9)$$

The projection of dA' upon the flat impact surface is another differential area, dA, which can be defined as

$$dA = \pi (H \tan(\gamma + d\gamma))^2 - \pi (H \tan \gamma)^2 = 2 \pi H^2 \frac{\sin \gamma}{\cos^3 \gamma} d\gamma. \qquad (10)$$

Thus, the spray volumetric flux along the impact surface is given by

$$Q'' = Q''_{sp} \frac{dA'}{dA} = \frac{Q}{[1 - \cos(\theta/2)]} \cdot \frac{1}{2 \pi H^2 \cos^3 \gamma}$$
$$= \frac{1}{2} \left[\frac{Q}{\pi (H \tan(\theta/2))^2} \right] \left[\frac{\tan^2(\theta/2)}{1 - \cos(\theta/2)} \right] \frac{1}{\left[1 + \left(\frac{r}{H} \right)^2 \right]^{3/2}}. \qquad (11)$$

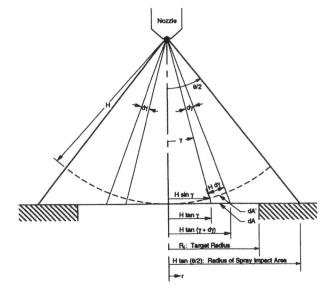

Fig. 5   Geometry used in development of volumetric flux distribution model (Mudawar and Estes, 1995).

Equation (11) indicates volumetric flux decreases from a maximum along the spray axis to roughly one half of the maximum toward the outer edge of the impact area.

These results prove the fraction of the spray flow rate received by the heater surface can be determined purely from geometrical arguments.

CHF optimization is further complicated with square heaters which possess corner regions which are less adequately cooled than the center. CHF measurements were, therefore, repeated at different nozzle-to-surface distances while maintaining the same flow rate. Figure 6 shows the ratio of CHF measured at a given nozzle-to-surface distance to the maximum CHF measured over the entire range of distances, plotted against $H \tan(\theta/2) / (L/2)$.

Also shown in the same figure is the ratio of flow rate captured by the sampler (i.e., for a sampling area equal to the heater surface area) to the total spray flow rate. The largest CHF seems to occur near the point where the spray just inscribes the square heated area, i.e., $H = (L/2) / \tan(\theta/2)$.

5. CHF CORRELATION

Mudawar and Valentine (1989) demonstrated local CHF for a small heated portion of a large impact area can be correlated according to the equation

$$\frac{q''_{m,p}}{\rho_g Q'' h_{fg}} = f \left[\frac{\rho_f}{\rho_g}, \frac{\rho_f Q''^2 d_{32}}{\sigma}, \frac{\rho_f c_{p,f} \Delta T_{sub}}{\rho_g h_{fg}} \right], \qquad (12)$$

where $q''_{m,p}$, Q'' and d_{32} are all based on *local* conditions within the spray impact area. Lowest CHF values were attained at the outer edge of the impact area, where Q'' is lowest; CHF was negligible outside the impact area.

Unlike the study by Mudawar and Valentine, the present study is also concerned with cooling a small square heater. It is postulated CHF would commence at the outer edge of the impact area (which just inscribes the square heater), where volumetric flux is a minimum, and propagate inward in an unstable manner. Thus, at the outer edge, critical heat flux is given by

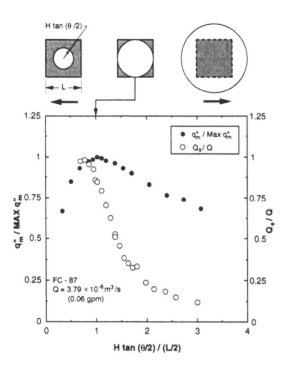

Fig. 6   Optimization of nozzle-to-surface distance for maximum CHF (Mudawar and Estes, 1995).

$$q''_{m,p} = \frac{L^2}{\frac{\pi}{4}L^2} q''_m \ . \tag{13}$$

The volumetric flux at the outer edge of the impact area can be related to the impact area's mean volumetric flux $(\overline{Q}'' = Q / (\pi L^2 / 4))$ by setting $H = (L/2) / \tan(\theta/2)$ and $r = L/2$ in equation (11),

$$\frac{Q''}{\overline{Q}''} = \tfrac{1}{2}\left(1 + \cos(\theta/2)\right)\cos(\theta/2) \ . \tag{14}$$

Figure 7 shows the CHF data for water, FC-72 and FC-87 and different full cone spray nozzles fitted over broad ranges of flow rate and subcooling according to the equation

$$q''_* \equiv \frac{\dfrac{q''_{m,p}}{\rho_g h_{fg} Q''}}{\left(\dfrac{\rho_f}{\rho_g}\right)^{0.3}\left[1 + 0.0019\dfrac{\rho_f c_{p,f} \Delta T_{sub}}{\rho_g h_{fg}}\right]} = 2.3\left[\frac{\rho_f Q''^2 d_{32}}{\sigma}\right]^{-0.35} \tag{15}$$

with a mean absolute error of 11.5 %.

Combining equations (13) - (15) yields another more convenient form of the CHF correlation relating q''_m to $\overline{Q}''$.

$$\frac{q''_m}{\rho_g h_{fg} \overline{Q}''} = 1.467\left[\left[(1 + \cos(\theta/2))\cos(\theta/2)\right]^{0.3} \times \right.$$
$$\left(\frac{\rho_f}{\rho_g}\right)^{0.3}\left[\frac{\rho_f \overline{Q}''^2 d_{32}}{\sigma}\right]^{-0.35}\left[1 + 0.0019\frac{\rho_f c_{p,f} \Delta T_{sub}}{\rho_g h_{fg}}\right]. \tag{16}$$

Close inspection of equation (16) reveals which spray parameters actually influence CHF. These parameters include thermophysical properties (ρ_f, ρ_g, σ, h_{fg}, $c_{p,f}$), flow parameters (ΔT_{sub}, ΔP, Q), orifice parameters (d_0, θ), and heater length (L). The important effect of nozzle-to-surface distance (H) is implicitly a function of both heater size and spray cone angle

for a spray which has been configured to optimize CHF, *i.e.*, whose impact area just inscribes the square heated surface.

6. CONCLUSIONS

Experiments were performed to better understand nucleate boiling and CHF in sprays, and to develop correlations for the spray's SMD and CHF. The key findings are as follows:

(1) The shape of the boiling curve of sprays is markedly different from that of other boiling systems. Sprays with high volumetric flux (high Weber number) show little increase in slope of the boiling curve between the single phase and nucleate boiling regimes due to a suppression of nucleation and reduced evaporation efficiency.

(2) SMD for full cone sprays is dependent upon orifice diameter and the Weber and Reynolds numbers based on the orifice flow conditions prior to liquid breakup. A dimensionless correlation was developed which gives good predictions for fluids with vastly different surface tensions.

(3) Volumetric flux is fairly uniform along any spherical surface centered at the spray orifice and confined by the spray angle. A theoretical model based on this assumption shows volumetric flux is greatest at the center of the spray and decays radially toward the outer edge of the impact area.

(4) Sprays produce drastically different cooling performances for the same nozzle and same flow rate depending upon the nozzle-to-surface distance. CHF is a maximum when the spray impact area just inscribes the square surface of the heater.

(5) A new correlation was developed for full cone sprays which accurately predicts CHF for water, FC-72, and FC-87 and many different full cone nozzles over broad ranges of flow rate and subcooling. This correlation indicates CHF is influenced by thermophysical properties (ρ_f, ρ_g, σ, h_{fg}, $c_{p,f}$), flow parameters (ΔT_{sub}, ΔP, Q), orifice parameters (d_0, θ), and heater length (L).

ACKNOWLEDGMENTS

Financial support for this work by IBM is greatly appreciated. The authors also thank 3M Company for donating Fluorinert samples, and Spraying Systems Company for both donating spray nozzles and assisting with the spray droplet sizing.

NOMENCLATURE

A	area defined along heater surface
A'	area defined along a spherical surface
c_p	specific heat at constant pressure
d_i	middle diameter of ith class of drops
d_0	nozzle orifice diameter
d_{32}	Sauter mean diameter (SMD)
H	nozzle-to-surface distance
h_{fg}	latent heat of vaporization
L	heater length (12.7 mm)
n_i	number of drops in ith class
P	test chamber pressure
ΔP	pressure drop across spray nozzle
q	electrical power input to heater
q''	heater power divided by heater surface area
q''_{dryout}	heat flux associated with complete evaporation of liquid (dryout)
q''_m	critical heat flux for square heater, q / L^2
$q''_{m,p}$	local critical heat flux
q''_*	dimensionless CHF
Q	total volumetric flow rate of spray

Q'' local volumetric flux
$\overline{Q''}$ average volumetric flux over spray impact area, $Q/(\pi R^2)$ (or $Q/(\pi L^2/4)$ for an optimized spray)
Q_s flow rate incident on heater surface
Q''_{sp} uniform volumetric flux along a spherical surface
r radial coordinate
R radius of spray impact area
Re_{d0} Reynolds number based on orifice flow parameters
T temperature
ΔT_{sub} liquid subcooling at nozzle inlet, $T_{sat} - T_f$
T_f spray inlet temperature
U_m mean drop velocity
We spray Weber number
We_{d0} Weber number based on orifice flow conditions

Greek symbols

γ integrating angle in volumetric flux distribution model
θ spray cone angle
η evaporation efficiency
μ viscosity
ρ density
σ surface tension

Subscripts

a ambient (air or vapor)
f liquid
g vapor
sat saturation
sub subcooling.

REFERENCES

Bolle, L. and Moureau, J. C., 1976, "Spray Cooling of Hot Surfaces: A Description of the Dispersed Phase and a Parametric Study of Heat Transfer Results," in *Two Phase Flow Heat Transfer, Proc. NATO Advanced Study Inst.*, Vol. 3, Washington, DC, pp. 1327-1346.

Estes, K. A. and Mudawar, I., 1995, "Correlation of Sauter Mean Diameter and CHF for Spray Cooling of Small Surfaces," *Int. J. Heat Mass Transfer*, in press.

Hall, D. D. and Mudawar, I., "Experimental and Numerical Study of Quenching Complex-Shaped Metallic Alloys with Multiple, Overlapping Sprays," *Int. J. Heat Mass Transfer*, in press.

Lefebvre, A. H., 1989, *Atomization and Sprays*, Hemisphere Publishing Corporation, New York, NY.

Mudawar, I. and Deiters, T. A., 1994, "A Universal Approach to Predicting Temperature Response of Metallic Parts to Spray Quenching," *Int. J. Heat Mass Transfer*, Vol. 37, pp. 347-362.

Mudawar, I. and Estes, K. A., "Optimizing and Predicting CHF in Spray Cooling of a Square Surface," *J. Heat Transfer*, in review.

Mudawar, I. and Valentine, W. S., 1989, "Determination of the Local Quench Curve for Spray-Cooled Metallic Surfaces," *J. Heat Treating*, Vol. 7, pp. 107-121.

Toda, S., 1972, "A Study in Mist Cooling," *Trans. JSME*, Vol. 38, pp. 581-588.

	FC-72, P = 1.03 bar, T_{sat} = 57.3 °C			**FC-87**, P = 1.03 bar, T_{sat} = 30.7 °C	**Water**, P = 1.01 bar, T_{sat} = 100 °C
	Nozzle 1	Nozzle 2	Nozzle 3	Nozzle 1	Several full cone spray nozzles
	○	□	△	⊕	●
$Q'' \times 10^3$ $m^3 s^{-1}/m^2$	16.6 - 52.4	55.6 - 145	86.3 - 216	16.6 - 49.7	0.600 - 0.216
$d_{32} \times 10^6$ m	140 - 196	181 - 225	182 - 214	110 - 195	405 - 1350
ΔT_{sub} °C	13 - 33	13 - 33	13 - 33	4 - 15	20 - 77

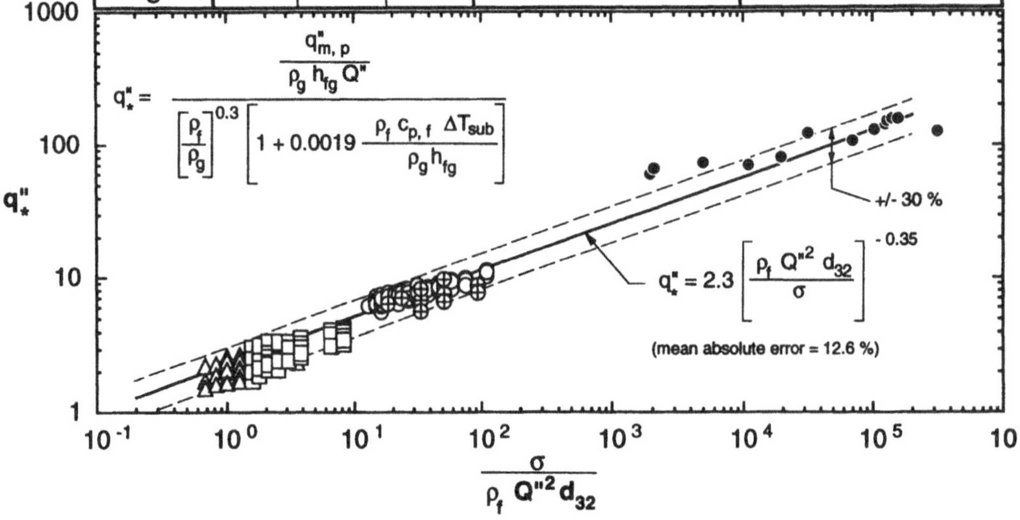

Fig. 7 CHF Correlation.

EFFECTS OF AN ELECTRIC FIELD ON NUCLEATE POOL BOILING AND BUBBLE BEHAVIOR ON A HORIZONTAL WIRE

Y. C. Kweon[1], M. H. Kim[1], H. J. Cho[2] and I. S. Kang[2]
[1] Department of Mechanical Engineering, POSTECH, Pohang, Korea
[2] Department of Chemical Engineering, POSTECH, Pohang, Korea

S. J. Kim
Korea Institute of Machinery and Metals, Daejeon, Korea

ABSTRACT

To study nucleate boiling heat transfer enhancement by DC and AC applied voltages, this electrohydrodynamic nucleate boiling experiment was carried out under subcooled and saturated boiling. The boiling curve, the onset of nucleate boiling(ONB), the critical heat flux(CHF) and the bubble behavior on a horizontal wire were investigated. In the present study, boiling heat transfer enhancement is more effective under AC applied voltage than DC applied voltage and under subcooled boiling rather than saturated boiling. By increasing the applied voltage, the boiling heat transfer rate, ONB and CHF increased linearly. The behavior of bubbles under DC and AC applied voltages was consistent with a trend of flow visualizations and nucleate boiling heat transfer enhancement.

INTRODUCTION

The world wide interest in environmental problems and energy conservation has recently increased. One of the active efforts to develop new heat transfer equipment with better performance has been made by applying an electric field to a heat transfer system. This technique using an electric field is based on the principle of electrohydrodynamics. The advantages of the electrohydrodynamic technique are active control of the heat transfer rate, negligible electric power consumption and a higher heat transfer coefficient. For more energy transfer per unit area of a heat exchanger, many researchers have studied heat transfer enhancement techniques to overcome the barrier of existing technologies. The effects of an electric field on boiling heat transfer are as follows : destabilization of a thermal layer on a heating surface, mixing or turbulence of a boiling fluid and promotion of bubble departure. These electrohydrodynamic effects are the main causes of nucleate boiling heat transfer enhancement compared with conventional methods. Boiling experiments

using an electric field(Cooper 1990, Ogata and Yabe 1993) have shown that nucleate boiling is related to bubble behavior on the heating surface. The change in bubble dynamic behavior results from the mutual interaction between the boiling fluid and the electric field.

Bonjour et al.(1962) found that the electric field could enhance natural convection and nucleate boiling heat transfer and suppress film boiling. Poulter and Allen(1986) reported that the heat transfer performance of a small shell and tube exchanger under an electric field was nearly equivalent to that of 60-tube conventional one. Cooper(1990) developed a model for electrohydrodynamic enhanced nucleate boiling with a relationship between the applied voltage and the bubble departure diameter. He observed that the electric field eliminated boiling hysteresis, enhanced nucleate boiling heat transfer about ten times for a pure fluid and reduced the foaming of an oil mixture. Ogata et al.(1992) found that the boiling heat transfer enhancement of a tube bundle is as effective as a single tube.

Although many studies on nucleate boiling heat transfer have been performed experimentally and theoretically, the mechanisms of electrohydrodynamic boiling heat transfer have not been clarified yet. However, it is clear that the electric field changes the heat transfer processes of nucleate boiling which relates to the dynamic behavior of bubbles. To understand electrohydrodynamic boiling heat transfer enhancement, the present study is focused on nucleate boiling under DC and AC applied voltages. The pool boiling curve, the onset of nucleate boiling, the critical heat flux and bubble behavior were investigated. Flow visualizations were also carried out to facilitate the interpretation of nucleate boiling heat transfer under DC and AC applied voltages.

GENERAL STATEMENTS

The electrical body force($F^{(e)}$) which acts on a dielectric

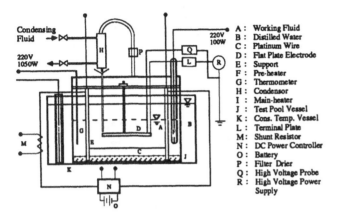

Figure 1   Experimental Boiling Apparatus

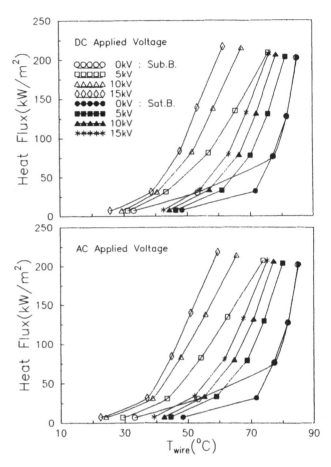

Figure 2   Pool Boiling Curves (Sub.B. : Subcooled Boiling, Sat.B. : Saturated Boiling)

fluid from the concept of electrostatic energy can be derived as follows

$$\overline{F}^{(e)} = \rho_e \overline{E} - \frac{1}{2} E^2 \nabla \varepsilon + \frac{1}{2} \nabla \left(E^2 \rho \frac{\partial \varepsilon}{\partial \rho} \right) \qquad (1)$$

where ρ_e, ε_i, $\overline{E}$ and ρ are the free charge density, the permittivity, the electric field strength and the density of a dielectric fluid.

The first term is the electrophoretic force due to a free charge. In convective heat transfer, free charges in the dielectric fluid are redistributed due to the thermal gradient of the electrical conductivity. When the dielectric fluid is heated, the electrophoretic force caused by redistributed free charges can move the dielectric fluid and affect the stability of a thermal layer near the heating surface. This electroconvective effects may enhance convective heat transfer(Grosu and Bologa, 1968; Turnbull, 1968; Takashim and Aldridge, 1976).

The second term is the dielectrophoretic force due to the inhomogeneous permittivity of the dielectric fluid. This force moves the dielectric fluid with higher permittivity toward the region of greater electric field. Also, when bubbles exist in a nonuniform electric field, bubbles with lower permittivity than the dielectric fluid move toward the region of the lower electric field strength. The dielectrophoretic force can have no effect on the dielectric fluid unless there is inhomogeneity in the dielectric fluid. Since the permittivity is a function of temperature, its distribution in the thermal gradient system becomes inhomogenous. The variation of the permittivity with temperature is due to the change in density. The change in the permittivity with density is given by the Clausis-Mossotti relation.

The third term is the electrostrictive force produced by the inhomogenous distribution of the electric field strength and the change of the permittivity with density. Hakim and Higham (1962) showed that the electrostrictive force could be

expressed as the excess pressure produced in the dielectric fluid by the electric field, using a schlieren optical system.

When boiling heat transfer occurs, vapor bubbles can be affected by the electrophoretic force due to the polarized charges on their surface and the dielectrophoretic force due to the difference of the permittivity between bubbles and the dielectric fluid. The relative magnitude of two electric forces depends on the electric field strength, the electrical conductivity and the permittivity. Induced polarized charge which is built up on the bubble surface is related to the dielectric relaxation time and the bubble rising time. If the bubble rising time is much less than the dielectric relaxation time, there are no polarized charges on the bubble surface. For vapor bubbles rising at a finite depth in the boiling fluid, the rising time of vapor bubbles is much less than the charge relaxation time. Therefore, the polarized charges on the surface of vapor bubbles may be negligible. Also, since the permittivity of vapor bubbles is less than that of the boiling fluid, vapor bubbles will be driven toward the region of the lower electric field. The faster departure of vapor bubbles will promote the instability of the thermal layer on

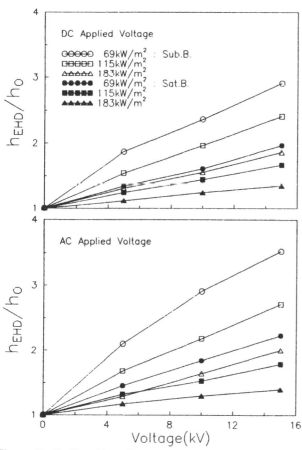

Figure 3 Boiling Heat Transfer Enhancement Ratios (Sub.B. : Subcooled Boiling, Sat.B. : Saturated Boiling)

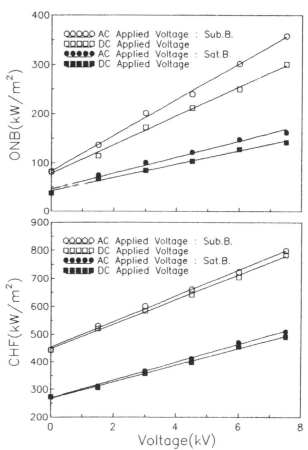

Figure 4 Onset of Nucleate Boiling and Critical Heat Fluxes (Sub.B. : Subcooled Boiling, Sat.B. : Saturated Boiling)

the heating surface and disturb the bulk fluid more effectively, thus the thermal resistance on the heating surface is reduced and nucleate boiling heat transfer will be enhanced.

EXPERIMENTAL APPARATUS AND PROCEDURE

Fig. 1 shows a schematic diagram of the boiling apparatus. It consists of DC and AC high voltage power supplies with low current, an electrode system, a boiling chamber, a shunt resistor, a condenser coil and a DC variable controller. The electrode system consists of a copper flat plate electrode and a platinum wire. DC and AC high voltages are applied to the copper flat plate electrode. The platinum wire which is to be used as an electric resistance heater is electrically heated and grounded. The DC variable controller precisely adjusts the heat flux of the platinum wire. A high-voltage probe was used to measure the applied voltage accurately. The average temperature of the bulk fluid was measured by four K-type thermocouples, which were fixed at the height of the wire. Special care was taken to avoid disturbance of the electric field near the wire. The working fluid was Freon-113.

Before each experiment, the fluid was preheated for two hours to degas it. The condenser was used to condense vapor generated by evaporation during boiling. The temperature of the bulk fluid was maintained at steady-state conditions. The resistance of the wire was obtained from the current and the voltage drop measured at two ends of the wire. It was used to determine the average temperature of the wire. Heat flux was obtained by the product of the voltage drop and the current across the wire and the average temperature of the wire was estimated from a temperature-resistance calibration curve which was obtained from the distilled water in a constant temperature bath. Errors of the calibration curve were estimated to be within ±0.5%. The experimental data were obtained after reaching the steady-state boiling conditions. Each boiling experiment was repeated six times to assure the validity of the data.

The experiments were carried out with changing DC and AC applied voltages(0~15kV), the heat fluxes of the wire (6.3~183kW/m^2) under subcooled (T_{sub} = 27.6°C) and saturated (T_{sat} = 48.7°C) temperatures and ambient pressure. For qualitative analyses, the dynamic behavior of bubbles was

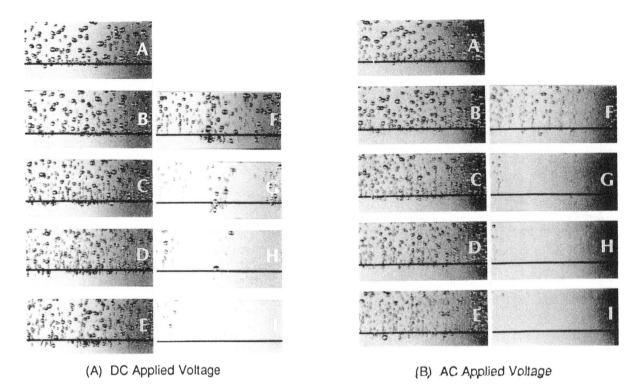

(A) DC Applied Voltage   (B) AC Applied Voltage

Figure 5   Flow Visualizations
(A:0kV, B:1.3kV, C:2.6kV, D:3.9kV, E:5.2kV, F:6.5kV, G:7.8kV, H:9.1kV, I:10.4kV)

visualized.

RESULTS AND DISCUSSIONS

Heat Transfer Enhancement

The rate of heat transfer is determined by the voltage drop and the current across the heating wire as follows

$$Q = \Delta V_{wire} \cdot I \tag{2}$$

where ΔV_{wire} and I are the voltage drop and the current of the wire. To investigate the effects of the electric field on subcooled and saturated pool boiling, pool boiling curves under DC and AC applied voltages were compared with the pool boiling curve under zero voltage. Fig. 2 shows the pool boiling curve between the surface temperature and the heat flux of the wire under subcooled and saturated boiling. With increasing the applied voltage, boiling curves shift toward the left side of the boiling curve under zero voltage. The degree of shift is more pronounced under AC applied voltage than under DC applied voltage and under subcooled boiling than saturated boiling.

The heat transfer coefficient is determined from the heat flux and the temperature difference between the wire surface (T_{wire}) and the bulk fluid (T_{liq}). If the thermal conductivity is assumed to be constant, the heat transfer enhancement ratio

is expressed as follows

$$\frac{h_{EHD}}{h_O} = \frac{q''_{EHD}}{q''_O} \cdot \frac{(T_{wire} - T_{liq})_O}{(T_{wire} - T_{liq})_{EHD}} \tag{3}$$

where h_{EHD} and h_O are the heat transfer coefficient under an applied voltage and zero voltage, respectively. Fig. 3 shows the boiling heat transfer enhancement ratio (h_{EHD}/h_O) in subcooled and saturated boiling under DC and AC applied voltages. The experiment was carried out by changing heat fluxes of the wire(69, 115 and 183kW/m^2). The boiling heat transfer ratio increased linearly along with increasing the applied voltage. Under DC applied voltage, the maximum heat transfer enhancement ratio was 180% in subcooled boiling and 90% in saturated boiling. Under AC applied voltage, it was 250% in subcooled boiling and 120% in saturated boiling. Heat transfer enhancement was more effective in subcooled boiling under AC applied voltage. The magnitude of the heat transfer enhancement ratio decreased by increasing the heat flux of the wire.

Onset of Nucleate Boiling and Critical Heat Flux

Fig. 4 shows the onset of nucleate boiling(ONB) and the critical heat flux(CHF) under DC and AC applied voltages in subcooled and saturated boiling. ONB can be calculated by the current and the voltage drop measured at the moment that

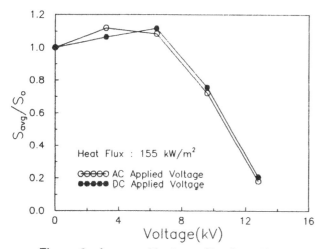

Figure 6   Average Nucleate Site Densities

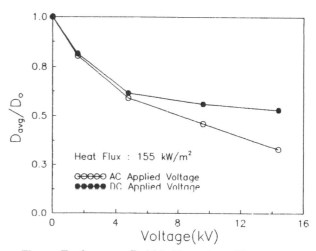

Figure 7   Average Bubble Departure Diameters

bubbles generate on the heating wire and CHF can be also calculated by the current and the voltage drop measured at the moment that a vapor film appears partially on the heating wire. CHF points were validated by the heat flux and the surface temperature of the heating wire. Above the CHF point, the heat flux of the wire decreased slightly, whereas the surface temperature increased. At the higher heat flux, the heating wire melted. ONB increased linearly by increasing the applied voltage. ONB in saturated boiling increased 2.4 times under DC applied voltage and 2.8 times under AC applied voltage. ONB in subcooled boiling increased 3.7 times under DC applied voltage and 4.4 times under AC applied voltage. From this, it is believed that the heat transfer process of ONB differs under DC and AC applied voltages. In addition, near the ONB point, it was observed that the motion of the hot liquid flow was induced around the heating wire within the thermal layer under DC applied voltage. The flow moved from the inside to the outside along the heating wire. The liquid flow within the thermal layer was also propagated by the wavelike flow which moved upward from the heating wire under AC applied voltage. By increasing the applied voltage, the speed of induced liquid flows were faster. CHF also increased linearly by increasing the applied voltage. CHF under DC and AC applied voltages increased about 1.8 times both in saturated and in subcooled boiling. By increasing the applied voltage, the behavior of bubbles became very active. From this, it is found that the increase in ONB and CHF in subcooled boiling under AC applied voltage is larger than that in saturated boiling under DC applied voltage. The electrohydrodynamic secondary flows and the active behavior of bubbles observed in experiments are believed to be one of the main causes of boiling heat transfer enhancement.

Flow Visualizations

Flow visualizations of nucleate pool boiling in saturated

condition are shown in Fig. 5. Visualization experiments were carried out by varying the DC and AC applied voltages(0kV~10.4kV) at the heat flux of 73 kW/m^2. Fig. 5(A) and 5(B) represent visualizations under DC and AC applied voltages, respectively. The number of bubbles apparently increases by increasing the applied voltage(up to 5.2kV), whereas bubbles begin to disappear above 6.5kV. Disappearance of bubbles is more remarkable under AC applied voltage than DC applied voltage. It is supposed that this may be due to the bubble oscillation by AC applied voltage, as shown by Kweon et al. (1994). Also, by increasing the applied voltage, the bubble departure diameter decreased and the bubble departure velocity increased. It was observed that the departure velocity of bubbles under DC applied voltage was larger than that under AC applied voltage. Since bubbles under the electric field were repelled from the heating wire by the dielectrophoretic force, the motion of vapor bubbles became active and faster, therefore, electrohydrodynamic boiling heat transfer is supposed to be enhanced more effectively than zero voltage. When the electric field is applied, the departure volume of bubbles depends on the balance among the surface tension force, the buoyancy force and the electric force. If the electric force acting on bubbles is very strong, the bubble size decreases with increasing the applied voltage. At above 10.4kV, the heat transfer process changed from nucleate boiling to convective heat transfer. Moreover, bubbles in the flow field disappeared and the flow field became turbulent. From these results, it is supposed that the heat transfer mechanism under the strong, nonuniform electric field mainly contributes to the turbulent and violent convection.

Bubble Behaviors

Electrohydrodynamic nucleate pool boiling can be controlled by the dynamic behavior of bubbles near the heat transfer surface. Therefore, the study on bubble behavior was

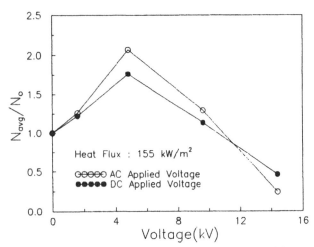

Figure 8   Average Number Densities of Bubbles

CHF increase linearly. Also, by increasing the heat flux of the wire, the magnitude of the heat transfer enhancement ratio decreases. Moreover, it was shown from flow visualizations that the electrohydrodynamic secondary flows and the modified bubble behavior. These results were consistent with a trend of the enhancement of boiling heat transfer. More elaborate experiments are needed to understand the fundamental mechanisms of nucleate boiling heat transfer enhancement and the dynamic behavior of bubbles under an electric field.

ACKNOWLEDGMENT

This work was performed with the support of the Korea Institute of Machinery and Metals and the Advanced Fluids Engineering Research Center at the Pohang University of Science and Technology.

examined under saturated condition (heat flux of 155kW/m^2). Fig. 6 shows the average nucleation site densities(S_{avg}) under DC and AC applied voltages, where S_0 is the average nucleation site density under zero voltage. The nucleate site was considered to be active when bubbles attached to the heating wire. It was observed that some parts of the wire were almost always active whereas other parts were sporadically active. The nucleation site density under AC applied voltage was similar to DC applied voltage. It increased a little up to 3.2kV(AC applied voltage) and 6.4kV(DC applied voltage) but decreased sharply by increasing the applied voltage. Fig. 7 shows the average bubble departure diameters(D_{avg}) under DC and AC applied voltages, where D_0 is the average bubble diameter under zero voltage. With increase of the applied voltage, the bubble departure diameter decreases. In Fig. 8, the average number densities of bubbles(N_{avg}) in the boiling fluid are shown, where N_0 is the average number density under zero voltage. The number of bubbles increases up to about 4.8kV but decreases with increasing the applied voltage. In addition, the number of bubbles is richer under AC applied voltage than under DC applied voltage(under 11.6kV). These results were consistent with a trend of flow visualizations of nucleate boiling heat transfer in the presence of an electric field.

CONCLUSIONS

In this study, the effects of an electric field on nucleate pool boiling under DC and AC applied voltages were investigated to understand electrohydrodynamic nucleate boiling heat transfer enhancement. Experiments were performed by changing the applied voltage and the current across the wire under saturated and subcooled boiling, respectively. The following conclusions were obtained. The application of an electric field has an important role in the enhancement of nucleate boiling heat transfer. By increasing the applied voltage, the boiling heat transfer rate, ONB and

REFERENCES

Bonjour, E., Verdier J. and Weil, L., 1962, "Electroconvection Effects on Heat Transfer," *Chem. Eng. Prog.* Vol. 58, pp.63-66.

Cooper, P., 1990, "EHD Enhancement of Nucleate Boiling," *Transactions of the ASME*, Vol. 112, pp. 458-464.

Grosu, F. P. and Bologa, M. K., 1968, "Similarity Criteria for Convective Heat Exchange in an Electric Field," *Appl. Electr. Phenom. (USSR)*, Vol. 20, pp. 50-56.

Hakim, S. S. and Higham, J. B., 1962, "An Experimental determination of the Excess Pressure produced in a Liquid Gielectric by an Electric Field," Proc. Phys. Soc., Vol. 80, pp. 190-198.

Kweon, Y. C., Kim, M. H., Cho, H. J., Kang, I. S. and Kim, S. J., 1994, "Study on the Deformation and Departure Processes of a Bubble under an Electric Field," *Proc. 3rd JSME-KSME Fluids Engineering Conf.*, Japan, pp. 33-38.

Ogata, J., Iwafuji, Y., Shimada, Y. and Yamazaki, T., 1992, "Boiling Heat Transfer Enhancement in Tube-Bundle Evaporators Utilizing Electric Field Effects," *ASHARE Transaction: Symposia* (BA-92-5-2), pp. 435-444.

Ogata, J and Yabe, A., 1993, "Augmentation of Boiling Heat Transfer by Utilizing the EHD effect - EHD behavior of boiling bubbles and heat transfer characteristics," *Int. J. Heat Mass Transfer* Vol. 36(3), pp. 783-791.

Poulter, R. and Allen, P. H. G., 1986, "Electrohydrodynamically Augmented Heat and Mass Transfer in Shell/Tube Heat Exchanger," *Proc. 8th International Heat Transfer Conf.*, Vol. 6, pp. 2963-2968.

Turnbull, R. J., 1968, "Electroconvective Instability with a Stabilizing Temperature Gradient(I. Theory)," *The Physics of Fluids*, Vol. 11(12), pp. 2588-2596.

Takashima, M. and Aldridge, K. D., 1976, "The Instability of a Horizontal Layer of Dielectric Fluid under the Simultaneous Actions of a Vertical DC Electric Field and a Vertical Temperature Gradient," *Q. j. Appl. Math.*, Vol. 29, pp. 71-87.

INDUSTRIAL APPLICATIONS
AND DEVELOPMENTS

CROSSFLOW BOILING OVER A VERTICAL COLUMN OF PLAIN TUBES

N.F.Shire and B.M.Burnside
Department of Mechanical and Chemical Engineering
Heriot Watt University
Edinburgh, United Kingdom

ABSTRACT

An experimental investigation is described in which *iso*-octane at 1 atm, was boiled while flowing vertically up over a bundle of electrically heated in-line tubes 5 columns wide by 17 rows high. Heat transfer coefficient data is presented for the 10th and 11th rows of the column at uniform heat fluxes of 20, 40 50 and 60 kW/m² with mass velocities and qualities in the ranges, 150 - 370 kg/m²s and 0.01 - 0.1 respectively. At q = 60 kW/m² nucleate boiling dominates with little effect of Reynolds number or quality on heat transfer coefficient. At 20 kW/m² there is a strong effect of both Reynolds number and quality and the heat transfer coefficient is greater than the isolated tube pool boiling value in all the tests. Transition between these phenomena was observed at the intermediate heat fluxes.

Results are compared to the Chen superposition model using the Jensen and Hsu (1988) approach modified by the addition of a heat flux dependent multiplying factor in the Chen (1966) expression for the flow factor. This model predicted the transition from convective evaporation dominated to nucleate boiling dominated vaporisation.

1. INTRODUCTION

Crossflow boiling occurs in a variety of heat exchangers. The authors approach the subject from the point of view of kettle reboilers where, due to substantial recirculation rates, the qualities are lower than in once-through configurations. The review of Webb and Gupte (1992) illustrates the variety of methods which have been used to correlate flow boiling experiments. Most of these are for tube side heat transfer and are based on a combination of nucleate boiling and convective evaporation components. The Chen (1966) superposition model and the Kutateladze (1961) asymptotic models are used currently in correlating crossflow boiling experimental data. Jung et al (1989) observed complete suppression of nucleate boiling in flow of refrigerant mixtures at high quality in tubes, but whether nucleate boiling is suppressed by crossflow at the quality levels in kettle reboilers is a matter of some controversy.

Although there is now a substantial amount of crossflow data available on refrigerants, particularly that of Jensen and his colleagues, correlated by Jensen and Hsu (1988), Cornwell and Scoones (1988) and Polley et al (1980), very little non-proprietary data is available on hydrocarbons. Data on *iso*-octane is reported here and the extent to which it can be correlated using methods developed from refrigerant crossflow boiling data is tested.

2. APPARATUS

Figure 1 shows the apparatus used in the tests. The column is configured from a short shell and tube reboiler slice. The front flange of the shell is a borosilicate glass window. Only the five centre columns of tubes are installed, with brass walls, C, machined to accommodate half tubes at each side. The 19 mm o.d., 56 mm long, 90Cu:10Ni tubes were arranged in a square in-line array of 25.4 mm pitch. A 10 mm blind pocket machined in each tube housed a 250 W cartridge heater. The tube was sealed to the shell by means of an integral flange and viton washer tensioned by a locking collar on the back face of the tube plate. Insulation between this collar and the tubeplate, and at the outside end of the tube, together with the thin, 1 mm annular wall thickness of the tube enclosed by the tubeplate, ensured a very low heat loss between the heater and the heated fluid.

Tube wall temperature was measured by a 1mm overall diameter K type thermocouple in the wall of each tube, with the measuring junction positioned at the top and halfway between the tubeplate and the closed 'window' end. The tube surface temperature was estimated by extrapolating the measured value to the surface. Shellside, eight K-type thermocouples positioned at the centre of squares, formed by vertices at the tube centres, recorded the fluid temperature, figure 1.

The power supply to groups of tube heaters connected in parallel was measured by electronic wattmeter. The power to individual heaters was determined using comparative volt drops measured across standard resistances in series with each heater. The power to each heater group was controlled by variable transformer and the groups were arranged in such a way that the 4 rows below row 10 were left unheated during the tests. This provided a calming section before rows 10 and 11 at entry to which the vertically upward flow quality was controlled. All the temperature measurements reported here were taken from the tube thermocouples in rows 10 and 11 of the column and the liquid temperature from the thermocouple between the rows, figure 1. Power to the 'half tubes' at the sides of the column was separately controlled and was set at half the power supplied to the centre tubes on the basis that very little heat leaked through the walls, which were insulated at the outside. The results were found to be very little affected by the power supply to the wall tubes in the range between half the power and the full power supplied to the centre tubes, provided correct account was taken of the effect on the fluid quality.

Liquid is pumped from hotwell, I, through rotameter G to electrically heated feedheater, F, where it is raised to its boiling point before passing into electrically heated pre-boiler, E. Vaporisation is accomplished here and using rows 1 to 5 of the column itself. Power to the pre-boiler is controlled and measured in the same way as used in the column. Fluid passing out of the shell was separated in glassware M, N, the vapour rising through section P to a vented water cooled condenser, and the liquid flowing into the hotwell to mix with the condensate from the condenser. The flowrate of the condensate was measured using rotameter K. All parts of the liquid circuit were heavily insulated. An energy balance in the column between electrical power input and the enthalpy flux to the liquid during the convection tests near the boiling point indicated negligible heat loss within the limits of experimental error.

3. EXPERIMENTS AND DATA HANDLING

Table 1 shows the range of conditions set in the tests. Heat flux density, q is based on the tube outside wall area and G_{min} on the minimum cross sectional flow area.

To avoid hysteresis, the heat flux was raised well above the nominal value and then reduced. After setting the desired conditions, the rig was run under steady conditions for about 2 hours before data collection started. The controlling computer program carried out datalog scans of tubewall and liquid thermocouples and called for readings of power to heater banks and volt drops across individual heaters, system pressure and rotameter float levels. Tube surface temperatures were calculated on the assumption of radial heat flow. From these and the measured local liquid temperature, individual tube heat transfer coefficients were calculated from the tube heat flux. Arithmetic average values of h_{fb} for the six tubes in rows 10 and 11 are

TABLE 1. RANGE OF EXPERIMENTAL CONDITIONS

q (kW/m²)	G_{min} (kg/m²s)	Quality, x
20 - 60	148 - 368	0 - 0.1

presented here normalised as the ratio to the isolated pool boiling heat transfer coefficient at q = 60 kW/m². *Iso*-octane isolated tube pool boiling data obtained by Milloy (1993) on identical tubes was employed in this reduction. By operating at heat fluxes below onset of nucleate boiling, single phase forced convection data was obtained with the liquid temperature very close to the boiling

point. The Reynolds number, Re_{lo} used is $(G_{min} D/\mu_l)$ and the quoted quality is the mean value calculated for row 10.

4. RESULTS AND DISCUSSION

Typical results are presented in figures 2, 3 and 4. Figure 2 shows (h/h_{pb60}) plotted versus Re_{lo} for values of q = 20, 40 50 and 60 kW/m², at quality, x = 0.05. The normalisation of the ordinate chosen emphasises the difference in slope of the curves at different q and at the same time avoids confusing overlap which occurs if h is normalised by h_{ipb} at each value of q. The curves shown are predictions which will be referred to below. Also shown are the single phase forced convection data. Within experimental error this data agreed with ESDU 73031 (1973) for tube bundles.

As expected, h increases with heat flux. For comparison the isolated pool boiling values of Milloy (1993), h_{ipb}, are shown in the figure. At 60 kW/m² there is no significant difference between h in the range 12,000 < Re_{lo} < 30,000 and h_{ipb}. At q = 20kW/m² h is strongly dependent on Re_{lo}, rising by over 40% between Re_{lo} = 12,000 and 30,000. At the highest value of Re_{lo}, h at 20 kW/m² is only 14% lower than that at 60 kW/m². Further, it is greater than h_{ipb}, over the whole range of Re_{lo}. This phenomena was observed also by Cornwell and Scoones (1988) at all heat fluxes in the range of their measurements, 6 < q < 36 kW/m², boiling R113 at 1 atm. At intermediate heat fluxes, transition occurs between this strong bundle effect at 20 kW/m² and nucleate boiling domination at 60 kW/m². At the lower values of Re_{lo}, h was equal to or a little lower than h_{ipb}, rising above it at higher Re_{lo}.

It appears that the 2-phase flow is very effective in stripping superheated liquid from the heated wall and increasing the phase interface area available for convective evaporation thereby reducing heat transfer resistance at the lower heat fluxes. Indeed, the effect increases with Reynolds number and leads to heat transfer coefficients much above nucleate boiling values. However, as nucleate boiling at the surface becomes well developed, nucleation and growth of bubbles shields the surface from convective evaporation completely. Figure 3 shows the influence of quality on h at fixed Re_{lo}. Again the effect is negligible at high q but h increases with quality at low q. As x increases, agitation and phase interface area increase, both favouring convective evaporation.

Finally, figure 4 shows the effect of heat flux on h when quality is held constant at two different values of Re_{lo}. h increases with q, but not as fast as the isolated nucleate pool boiling value. An increase in Re_{lo} increases h at low heat fluxes but has no affect when boiling is well developed.

5. PREDICTION

Prediction and correlation of cross-flow boiling of hydrocarbons is handicapped by lack of correlations of two phase friction multiplier, $(\phi^2)_l$ and void fraction, ε with 2-phase flow parameters. However, the authors felt that it would be useful to use correlations tested for refrigerants to compare their data with the superposition flow boiling model.

The procedure of Jensen and Hsu (1988) is based on the superposition method

$$h = Fh_c + Sh_{ipb} \qquad (1)$$

where the flow factor, F, is correlated with $(\phi^2)_l$, following Chen (1966)

$$F = A\left(\phi^2\right)_l^{m/(2-n)} \qquad (2)$$

Here, $A = 1$ and m and n are respectively the Reynolds number exponents for turbulent heat transfer and friction in tubes. For crossflow, it is common to determine m and n from data for the flow configuration, although, clearly, there is a form drag effect. Webb and Gupte (1992) suggest that this may be accommodated by allowing a Re variation of m and n. However, following Jensen and Hsu the authors have determined m (=0.63) from their forced convection data, figure 2, and n (=0.132) from ESDU 79034 (1979). $(\phi^2)_l$ was calculated using the Ishihara et al (1979) Martinelli parameter correlation, successful for R113, equation 3, as a starting point

$$\left(\phi^2\right)_l = 1 + \frac{8}{X_{tt}} + \frac{1}{X_{tt}^2} \qquad (3)$$

The Bennett et al (1980) correlation, equation 4, which postulates an exponential temperature profile to approximate the thermal boundary layer in the bubble growth region, was used to calculate suppression factor S.

$$S = \frac{k_l}{Fh_c X_o}\left[1 - \exp\left(\frac{-Fh_c X_o}{k_l}\right)\right] \qquad (4)$$

$$X_o = C_x\left[\frac{\sigma}{g(\rho_l - \rho_v)}\right]^{0.5}$$

X_o is the distance from the heated wall over which the average superheat is the driving force for nucleate boiling. Bennett et al suggested a value of the constant $C_x = 0.041$, (S.I. units) based on in-tube data for water, nitrogen, methane and ethylene glycol. They found that this value gave good correlations for shellside boiling of R11 also and it is used in the present correlations.

Making use of equations 1-4, flow boiling heat transfer coefficients were calculated for the experimental conditions. The tendency was for underprediction of h at the highest Re and quality at 20 and 40 kW/m² and overprediction at 60 kW/m². In no case was the transition observed, from the strong effect of Re_{lo} and x on h at 20 to the almost constant isolated tube pool boiling value at 60 kW/m² predicted, and could not be predicted because of the form of equations 1-4. Introducing the same flow regime dependence of $(\phi^2)_l$ as used by Schrage et al (1988) for R113 improved the prediction but the rate of increase of h with x and particularly with Re was not great enough.

The transition with heat flux was modelled by introducing a heat flux dependent factor A in equation 2.

$$A = (1 - q/60) \qquad (5)$$

$A \to 0$ as $q \to 60$ kW/m² reducing F to zero and removing suppression, $S \to 1$, equation 4. Thus, the observed condition that $h = h_{ipb}$ is imposed at 60 kW/m² and A decreases linearly with increase in q, decreasing the convective evaporation effect on h.

Figures 2 and 3 show the results of this modelling. The slope in equation 5 was chosen to optimise the fit to the data. $(\phi^2)_l$ was calculated by the method of Schrage et al (1988) based on the Grant and Chisholm (1979) flow map. Increase in h with Re is still underpredicted at 20 and 40 kW/m², and h increases too slowly with x at these heat fluxes, but the transition to nucleate boiling control as q increases is there. Table 2 shows values of F and S corresponding to A determined from equation 5 for two values of Re_{lo} at $x = 0.05$. Flow factors $F < 1$ at 40 and 50 kW/m² with suppression factors $S < 1$, indicate that the action of nucleate boiling reduces convective enhancement below that which would accompany forced convection to the liquid only. Nevertheless, nucleate boiling is suppressed by the presence of convection. F and S should be viewed as average values for the heated surface. The rate of transition to nucleate boiling control is likely to be different at different positions on the surface, depending on their inclination to the flow direction. There are areas where $S \approx 1$ with no convective enhancement and areas where $S < 1$ and $F > 1$, making up the average value for the surface, $S < 1$, $F < 1$. $(\phi^2)_l$ is a very weak function of Re_{lo} and factor A, equation 5 is independent of it. F is insensitive to Re.

If the model is to predict the variation of h with quality, far less the change observed by Cornwell and Scoones (1988) from nucleate boiling back to convective evaporation domination at higher qualities, factor A would have to have quality and flow regime as well as heat flux, dependence.

The asymptotic model, equation 6,

$$h = \{(Fh_S)^n + (Sh_{ipb})^n\}^{1/n} \qquad (6)$$

has the potential to predict the transition from nucleate boiling to convective evaporation control as quality rises. If index n is high enough, the nucleate boiling term is suppressed. However, at lower qualities, as observed in this work, the dependence of flow boiling heat transfer coefficient on Reynolds number and quality changes gradually as heat flux rises. At moderate heat fluxes, where the magnitude of the nucleate boiling and convective evaporation terms are nearly the same, the same problem arises as addressed above using the superposition model. Further, it is unlikely that a single value of index n can predict both the

TABLE 2. FLOW AND SUPPRESSION FACTORS

| | $Re_{lo} = 12,000$ | | $Re_{lo} = 30,000$ | |
| $x = 0.05$ | | | $x = 0.05$ | |
q	S	F	S	F
20	0.69	1.35	0.52	1.42
40	0.84	0.63	0.73	0.63
60	1	0	1	0

transition from nucleate boiling to convective evaporation control at high qualities and from co-existing nucleate boiling and convective evaporation to nucleate boiling domination at low qualities. In other words, a flow regime and heat flux dependence of index n is likely to be required just as for factor A in the superposition model.

CONCLUSIONS

Iso-octane flow boiling data at low quality and at 1 atm pressure showed the following trends:-

1. At low heat fluxes heat transfer coefficients were well above isolated tube pool boiling values at all flowrates and qualities tested. They increased strongly with Re_{lo} and x.

2. As heat flux was increased nucleate boiling dominated with the flow boiling heat transfer coefficient equal to the isolated tube pool boiling value at all values of Re_{lo} and x.

3. The transition observed between mixed boiling and nucleate boiling dominated modes could be correlated by a superposition model only if a heat flux dependent multiplying factor was included in equation 2.

ACKNOWLEDGEMENT

This work was carried out under the research programme of HTFS, AECL Contract No. CRD/43312, and supported by the Canadian Ministry of Energy, Mines and Resources through the Inter-Governmental Panel on Energy Research and Development - PERD.

NOMENCLATURE

A	factor in equation 5
D	tube diameter [m]
F	flow factor (equation 1)
G	mass velocity, [kg/m²s]
g	gravitational acceleration [m²/s]
h	heat transfer coefficient [kW/m²K]
k	thermal conductivity [W/mK]
q	heat flux density [kW/m²]
Re	Reynolds number
S	suppression factor
x	quality
X_{tt}	Martinelli parameter

Greek Symbols

ε	void fraction
$(\phi^2)_l$	2-phase friction multiplier
μ	viscosity [kg/ms]
ρ	density [kg/m³]
σ	surface tension [N/m]

Subscripts

c	convective
ipb	isolated tube pool boiling
l	liquid phase
min	based on minimum flow area
pb	pool boiling
pb60	pool boiling at 60 kW/m²

REFERENCES

Bennett, D.L., Davis, M.W. and Hertzler, B.L., 1980, "The Suppression of Saturated Nucleate Boiling by Forced Convective Flow", *AIChE Symp. Ser.* Vol. 76, pp.91-103.

Chen, J.C., 1966, "A Correlation for Boiling Heat Transfer to Saturated Fluids in Convective Flow", *Ind. Eng. Chem., Process Design and Development,* Vol. 5, pp.322-329.

Cornwell, K.C. and Scoones, D.S., 1988, "Low Quality Boiling on Plain and Low Finned Tube Bundles", *Proceedings 2nd UK Heat Transfer Conf.,* Vol. 1, pp.21-32, IMechE, London.

ESDU, 1973, "Convective Heat Transfer during Crossflow of Fluids over Plain Tube Banks", Item No. 73031, Engineering Sciences Data Unit, London.

ESDU, 1979, "Crossflow Pressure Loss over Banks of Plain Tubes in Square and Triangular Arrays including Effects of Flow Direction", Item No. 79034, ibid..

Grant, I.D.R and Chisholm, D., 1979, "2-Phase Flow on the Shell-Side of a Segmentally Baffled Shell-and-Tube Heat Exchanger", *Jl. Heat Transfer,* Vol. 101, pp. 38-42.

Ishihara, K., Palen, J.W. and Taborek, J., 1980, "Critical Review of Correlations for Predicting 2-Phase Pressure Drops across Tube Banks", *Heat Transfer Engineering,* Vol. 1, pp. 23-32.

Jensen, M.K. and Hsu, J.T., 1988, "A Parametric Study of Boiling Heat Transfer in a Horizontal Tube Bundle", ASME *Journal of Heat Transfer,* Vol. 110, pp. 976-981.

Kutateladze, S.S., 1961, "Boiling Heat Transfer", *Int. J. Heat Mass Transfer,* Vol. 4, pp. 31-45.

Milloy, P., 1993, "Pool Boiling Pentane and *Iso*-octane over Plain and Enhanced Tubes', *B. Eng Honours Project Dissertation,* Mech. Eng. Dept., Heriot-Watt University.

Polley, G.T., Ralston, T. and Grant, I.D.R., 1980, ASME Paper No. 80-HT-46.

Schrage, D.S., Jensen, M.K. and Hsu, J.T., 1988, "Two-Phase Pressure Drop in Vertical Crossflow across a Horizontal Tube Bundle", *AIChE Jl.,* Vol. 34, pp. 107-115.

Webb, R.L. and Gupte, N.S, 1992, "A Critical Review of Correlations for Convective Vaporisation in Tubes and Tube Banks", *Heat Transfer Engineering,* Vol. 13, pp. 58-81.

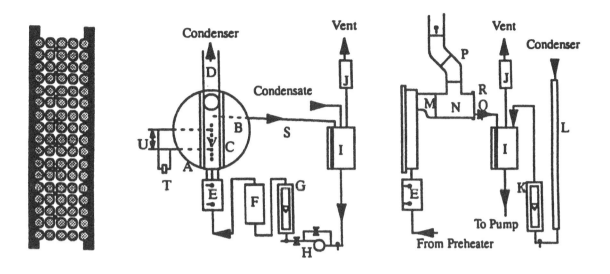

Fig.1. Flow Boiling Rig and Column Configuration

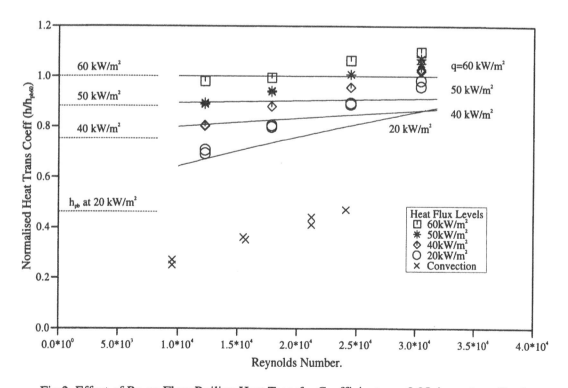

Fig.2. Effect of Re on Flow Boiling Heat Transfer Coefficient; x = 0.05, iso-octane, P = 1 atm
(Prediction curves; eqns.1,2,3 & 4 with A = 0.9{1-q/60})

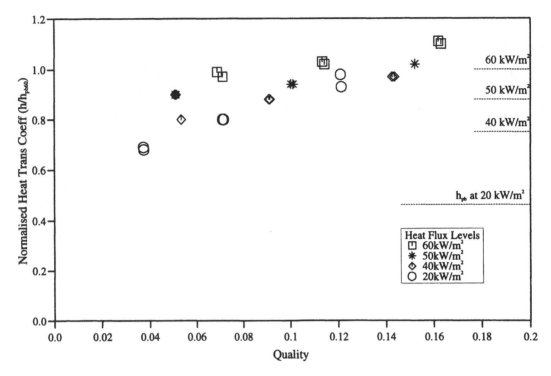

Fig.3. Effect of Quality on Flow Boiling Heat Transfer Coefficient; Re = 17,850, iso-octane, P = 1 atm
(Row 16)

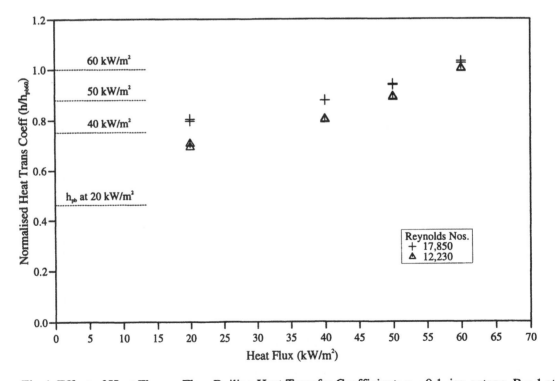

Fig.4. Effect of Heat Flux on Flow Boiling Heat Transfer Coefficient; x = 0.1, iso-octane, P = 1 atm

HEAT TRANSFER BEHAVIOUR OF AN ABSORBER TUBE WITH DIRECT STEAM GENERATION BY WATER INJECTION

O. Herbst, A. Fechner and W. Köhler
Thermal Hydraulics and Fluid Dynamics (NT31)
Siemens AG, Power Generation Group (KWU)
Erlangen, Germany

O. Goebel and B. Oberle
Institute for Technical Thermodynamics
DLR (German Aerospace Research Establishment)
Stuttgart, Germany

ABSTRACT

Two-phase flow investigations with special interest on the heat transfer behaviour are conducted on horizontal evaporator tubes in the diameter range between 50 and 85 mm. In view of the fact that future application of such horizontal tubes will be steam generation in the absorber tubes of parabolic troughs in solar power plants, the tubes are electrically heated on half the circumference alternately underneath or from the side. Heat fluxes up to 75 kW/m² - related to the inner wall of the tube - are realized. The effect of different types of injection nozzles is analyzed as well as the influence of different amounts and various conditions of the injected water itself. The pressure range is varied between 30 and 100 bar, and the mass flux between 50 and 500 kg/m² s.

1. INTRODUCTION

Today the parabolic trough solar power plants in California are the greatest commercially operating solar thermal plants in the world. 9 power blocks with a total electrical output of 354 MW_e have been in operation since the mid 80ies [PHARABOD]. Investment costs and complexity can be reduced by applying direct steam generation (DSG) in the absorber tubes [DAGAN] as an alternative to the existing two-circuit loop which uses a special high temperature oil in the primary loop. Two different DSG concepts are under investigation; the *once through-* and the *injection concept*.

Using the once through concept, the total amount of feed water is fed into the absorber tube. On its way along the tube, all water is evaporated, which results in steam qualities from 0 to 1. Flow instabilities can occur at certain combinations of mass flux and steam quality. Especially slug flow may cause damage to the tubes. Another critical aspect of this concept is the ability to control such a system.

Using the injection concept, the feed water is injected in sections by nozzles located along the absorber tube. The injection mass flow is controlled in such a way, that the injected water is evaporated before the next injection nozzle. In consequence the steam quality is always high in the absorber tube which avoids the danger of flow instability. The possibility to control the feed water seperately in each section allows the system to react fast to local irradiation disturbances, e.g. by clouds. For the realization of the injection concept a sufficient heat transfer has to be guaranted even for high steam qualities.

Siemens/KWU, the German Aerospace Research Establishment (DLR), the University of Munich and the Center for Solar Energy and Hydrogen Research (ZSW) are carrying out the project GUDE (Basic Investigations on Direct Steam Generation by Water Injection) which has been running since 1993 and will end in 1996. This project was initiated to demonstrate the feasibility of the injection concept and to generate a data basis for conception and design of such a collector field with DSG.

2. THE TEST FACILITY

The experiments are carried out at the BENSON test rig of Siemens/KWU in Erlangen [HEIN et al.], Fig. 1. The test loop is operated with demineralized and degasified water from a water treatment plant. The water flows through an electrical heater which can be controlled to vary the thermodynamic condition of the fluid as it enters the test section. The absorber tube is also electrically heated to simulate the solar radiation, reflected by the parabolic trough collector. The generated steam is condensed in a downstream spray-type condenser.

The tested absorber tube with a total length of 30 m is devided into three sections with one injection nozzle at the inlet of each. Electrical power up to 100 kW is available to each section for heating. The injected water is preheated in a recuperator and then conditioned to the desired injection temperature by a finely controllable electrical heater.

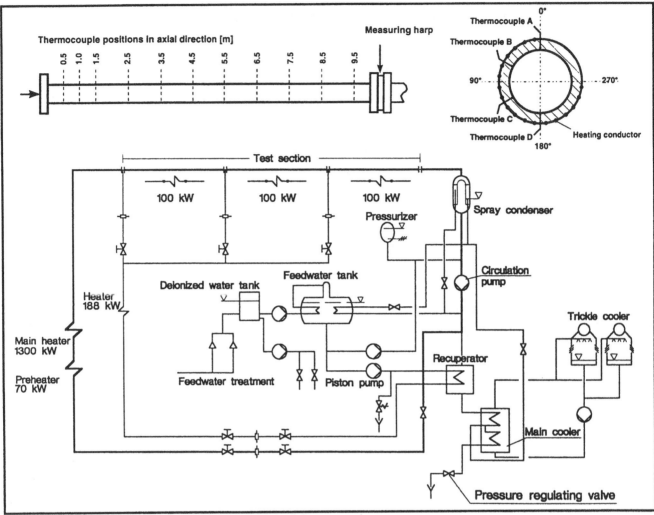

Figure 1: BENSON test facility and instrumentation of the test section

3. INSTRUMENTATION

The following measuring equipment is used to enable scientific conclusions:

- Thermocouples inside the tube wall
- Thermocouples in the fluid
- Pressure transducers for dynamic pressure in the flow
- Pressure transducers for pressure drop along the tube

Each tube is equipped with thermocouples at four lines (A,B,C,D) inside its wall, placed as shown in Fig.1.

A measuring harp (Figure 2) is located at the outlet of each tube. Each harp consists of 10 parallel tubes, which alternately contain a thermocouple or serve as a pitot tube. The thermocouples give information about the temperature profile in the flow, the pitot tubes about the pressure profile, and hence the velocity profile. The static pressure difference between section inlet and outlet is measured to obtain the pressure drop along the tube section.

Figure 2: Measuring harp

4. TEST MATRIX

The test matrix (Tab.1) covers a wide range of parameters which are relevant for a real solar power plant. The selected inner tube diameters correspond to the apertures of existing parabolic troughs. Up to now tests have been performed with an inner diameter of 65 mm only.

Parameter	Unit	Value
inner diameter	mm	50/65/85
type of nozzle		hollow cone/full cone/no nozzle
injection distance	m	10/30/(60)
heating profile		side/bottom (180°)
pressure level	bar	30/60/100
injection temperature	°C	Ts/Ts-30K/Ts+14K
mass flux	kg/m²s	50 to 500
heat flux	kW/m²	19/37/56/75

Table 1: Test matrix (Ts: Saturation temperature)

The type of injection nozzle is varied to evaluate its effect on the flow pattern and the heat transfer. The distance between two nozzles and the temperature of the injected water are varied for the same reason. If water is injected with a hollow conical spray pattern nozzle annular flow is expected with the inner surface of the tube wetted and therefore low temperature differences be present around its circumference. One object of these tests is to investigate the possibility of spraying technologies in maintaining the annular flow.

To simulate the solar radiation, both at noon and in the morning or evening, the absorber tube was heated from the bottom and from the side. Pressures between 30 and 100 bar are expected to be most interesting according to system analysis. The investigated mass fluxes cover the most interesting share of those which will occur in a real absorber tube, although the backmost regions can't be simulated due to energetic limitations of the steam producing test rig.

The simulated heat fluxes are close to those concentrated by the existing collectors with an additional margin. A heat flux of 75 kW/m² corresponds to 100% of the available electric power for each section while the heat flux of 56 kW/m² (75% power) is the nominal power of a parabolic trough collector system with an inner absorber tube diameter of 65mm.

5. TEST RESULTS

The temperature distribution around the circumference of the tube is the most interesting result of the tests. As a background information it should be made clear that if the difference between the hot and cold side of the tube exceeds 80 K, the tubes will not be able to withstand continuous operation because of excessive high thermal stresses (the tested tubes are qualified up to a temperature difference of 130 K). Too high thermal stresses may also cause defocussing of the absorber tubes. Since the absolute value of the temperature determines the losses of the absorber tube by back radiation, pressure is limited to 100 bar.

To investigate the heat transfer behaviour, experiments are performed as detailed below. In all of these experiments saturated steam generated in the heater enters the test section. Additional water is injected either into the first module or into all three modules.

Figure 3 shows the inside wall temperatures measured along the absorber tube at the instrumentated lines A to D. In the upper chart the tube is heated from the side (0°-180°). Stratified flow occurs with the inner surface wetted at line D and unwetted at line A to C. The temperatures at line B are higher than at line A because A is very close to the border of the nonheated region. Line C is close to the wetted region, which means that temperatures are lower than on line B.

The lower chart shows the tube heated from the bottom (90°-270°). Temperature differences are much lower than with the heating from the side. In this case stratified flow also occurs with most of the heated region wetted. The inner surface is also unwetted at lines A and B, but this region is not heated now.

The following figures concentrate on line B when the tube is heated from the side, because the highest wall temperatures have been observed there.

Figure 4 shows the influence of the mass flux on the heat transfer. With the wetted surface at line D which leads to inside wall temperatures close to the saturation temperature and the shown maximum values of line B, conclusion can be drawn that heat transfer is only sufficient for mass fluxes of 100 kg/m² s or higher. At the mass flux of 50 kg/m² s the temperature differences around the circumference exceed the limit of 80 K.

As the test results have shown there is no significant influence of the injection temperature on the heat transfer. No improvement was obtained with explosive atomization when superheated water was injected. In this case injection pressure was as high as a temperature increase of 14 K related to the saturation temperature (corresponding to the system pressure) did require.

The effect of different types of injection nozzles was also investigated. A comparison of temperature profiles obtained with a hollow conical spray pattern nozzle, a spray nozzle with full cone and without any nozzle has shown that a measurable improvement in heat transfer by spraying technologies is maintained only for the first two meters downstream. As expected the hollow cone nozzle leads to the best wetting characteristic in this section. Behind this short region no positive cooling effect of the different types of injection nozzles has been observed.

Figure 5 shows the differences between the wall temperatures inside the tube at line B and the saturation temperature in the pressure range between 30 and 100 bar. The system pressure of 60 bar leads to the highest values while the temperatures at 30 and 100 bar are approximately the same. This behaviour was almost identical for all other mass and heat fluxes when the absorber tube was heated from the side. Comparing the results at 60 bar with those at 30 or 100 bar, about 10 to 15 K higher temperatures were

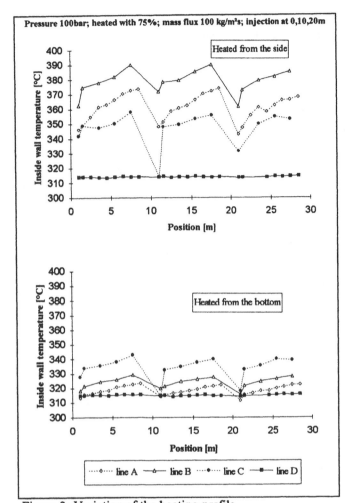

Figure 3: Variation of the heating profile

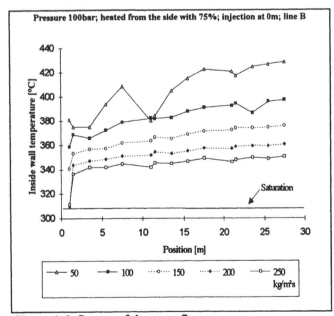

Figure 4: Influence of the mass flux

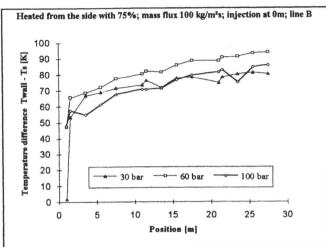

Figure 5: Influence of the system pressure

mcasurcd. In contrast when heating from the bottom the maximum temperature difference was not sensitive to system pressure.

In Figure 6 the temperature profiles at different heat fluxes are shown. As expected higher heat flux leads to higher wall temperatures. Since the differences between the single profiles are nearly equidistant, interpolations can be used to predict inside wall temperatures for other heat fluxes.

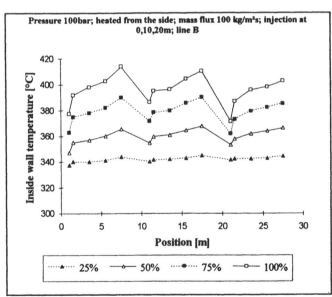

Figure 6: Influence of the heat flux

The effect of different distances between two injection nozzles is presented in Fig. 7. For both tests the mass flux of saturated steam at the tube inlet is 100 kg/m² s. The injected mass flow through the nozzles is adjusted in such way, that the injected water can be evaporated before it would reach the next nozzle downstream. Consequently the injected mass flow in the case with only one

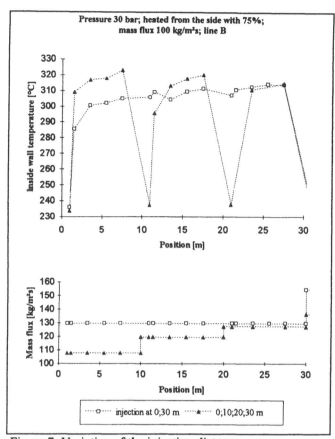

Figure 7: Variation of the injection distance

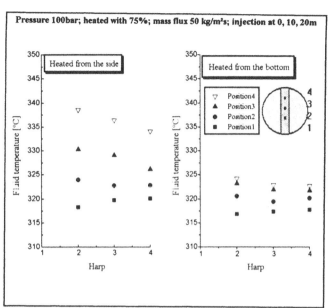

Figure 8: Thermal stratification

injection at the tube inlet is three times higher than the mass flow in the case with one injection each 10 m. It can be observed, that the wall temperatures in the case of one injection are lower than in the case of three injections considering the first two sections. This is due to the fact that the steam quality is lower corresponding to the larger water inventory in this case. The higher percentage of water in the flow causes a larger portion of the circumference of the tube to be wetted, consequently the wall temperatures are lower. This advantage of the case with only one injection disappeares downstream of the third injection position, because after it the steam quality is the same in both cases. Based on this consideration the conclusion can be drawn, that downstream of the short wetting zone behind the nozzle, the heat transfer is influenced by the local conditions only (heat flux, mass flux, pressure and steam quality).

Flow patterns and thermal stratification were detected with the measuring harps. Figure 8 shows the vertical temperature profiles in the fluid when the tube is heated from the side and from the bottom. The thermodynamic disequilibrium is higher when heating from the side because heat is transferred to the steam from the unwetted part of the inner surface. Not all of the injected water is evaporated. In contrast, thermodynamic disequilibrium is lower when heating from the bottom because most of the heat is transferred to the water. Only a part of the heated surface is unwetted and the effect of local superheating within the steam is therefore much lower.

Figure 9 shows the dynamic pressure profiles measured with the pitot tubes. In the upper chart water is injected in the first nozzle only. Since the highest heat flux is adjusted, the amount of injected water is high and can be detected at the outlet of the first section (Pitot tube 1 of harp 2). At this position only one third of the injected water is evaporated. The next third evaporates in the second section which results in a lower water level at the outlet of section 2 as the pitot tube 1 of harp 3 can detect. This means, that both harps 3 and 4 can measure "only" the velocity profile of the steam itself.

The lower chart shows the results when water is injected in each section. Since it is evaporated up to the next nozzle, all harps measure pure steam flow with increasing dynamic pressure along the absorber tube. The increasing flow velocity corresponds to the increase of the steam mass flux which is caused by the sectionwise injected and evaporated water.

All presented tests have been conducted with only half of the circumference heated. For a comparison with other applications it may be useful to give an estimation, how the wall temperatures would be in the case of an uniformly heated tube. Fig. 10 allows to make an estimation for this case. The dotted line shows the saturation temperature. The curve "T harp 4" shows the temperatures of the steam in the layer of position 4 (see Fig. 8). The curve "line B, superheating" shows the wall temperature as it would be if the steam would not be superheated, calculated with

$$Nu = 0.024 \cdot Re^{0.786} \cdot Pr^{0.45}. \qquad (1)$$

The curve "line B, calculated" shows the sum of steam superheating ("T harp 4" - T Saturation) and "line B, superheating". Above a mass flux of 100 kg/m² s the measured values ("line B, measured) are pretty close to the calculated ones. If the heat flux would be uniformly around the whole circumference, the wall superheating would not be different

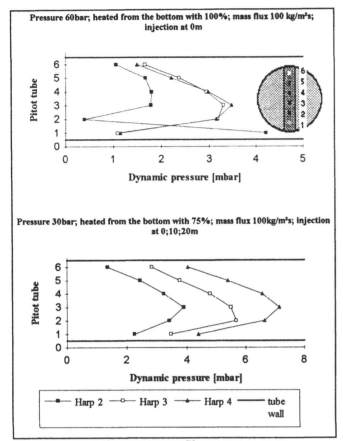

Figure 9: Dynamic pressure profiles

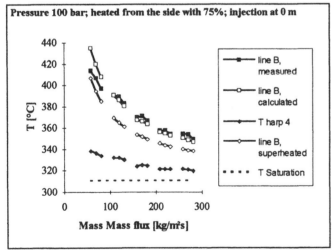

Fig 10: Prediction of wall temperatures at line B

6. CONCLUSIONS

It could be shown that the injection process offers a sufficient heat transfer behaviour for mass fluxes above 100 kg/m²s (this mass flux of steam is generated in an absorber tube with a length of approx. 100 m). This necessary minimum steam flow needs to be generated e.g. in a fossil fired boiler. Taking into account that an absorber tube of a real solar power plant will have a length of approx. 1000 m, the region with insufficient heat transfer (which needs to be supported by an additional steam flow) covers only 10 % of the whole tube length.

The shape of the injection nozzle and the temperature of the injected water don't have a sustainable influence on the flow pattern. About 2 m after the injection nozzles stratified flow was observed in all cases investigated so far. This result agrees with the predictions made with the Taitel/Dukler flow chart [TAITEL & DUKLER]. This result furthermore leads to the conclusion that there is no reason to place the injection nozzles close to each other. On the contrary, it could be shown that longer distances between the nozzles with higher injection mass flows lead to a better cooling of the tube wall. For the realization of the injection concept the injection distances have to be optimized to guaranty a reliable controllability of such a collector field.

After sunrise and before sunset when the tube is heated from the side, the highest temperature differences along the circumference occur. The heating from underneath (noon) is uncritical even for lower mass fluxes than 100 kg/m²s. The influence of the tube diameter is not yet investigated, but it is expected that smaller tube diameters will be less critical than larger ones. The whole test program will be completed in the end of 1995.

7. LITERATURE

Dagan, E., Müller, M., Lippke, F., "Direct Solar Steam Generation in Parabolic Trough Collectors", Plataforma Solar de Almeria (PSA), September 1992

Hein, D., Keil, H., Köhler, W., "The BENSON Test Rig", *VGB-Kraftwerkstechnik 57*, pp. 369-374, 1977

Pharabod, F.; Philiber, C., "LUZ Solar Power Plants", Deutsche Forschungsanstalt für Luft- und Raumfahrt (DLR) for IEA SSPS/Solar PACES, July 1991

Taitel, Y & Dukler, A. E. ,"A Model for Predicting Flow Regime Transitions in Horizontal and Near Horizontal Gas-Liquid Flow", *AIChE Jl 22*, pp. 47-55, 1976

because of the constant heat flux, but the steam temperature would well be different. The doubled heat flow per unit length would cause a doubled superheating of the steam. Consequently the estimation of the wall temperature at line B (uniformly heated tube) could be made with the measured steam superheating ("T harp 4" - T Saturation) added one more time to the measured temperature of line B (heated from the side).

FLOW PATTERNS IN AN ANNULAR GEOMETRY OF AN ABSORBER
PIPE WITH DIRECT STEAM GENERATION BY WATER INJECTION

P. Geskes
Institute of Technical Thermodynamics
DLR (German Aerospace Research Establishment)
Stuttgart, Germany

ABSTRACT

Results of investigated air-water two-phase flow behaviour in a horizontal pipe with a diameter of 70 mm and in annular geometries (diameter ratio: 0.71 and 0.86) with concentric and eccentric inner pipe are presented. The flow regimes, pressure losses and film thickness distributions in the pipe and the annuli are compared. Moreover, the first results of steam-water experiments in eccentric annuli (pipe diameter : 50 mm, diameter ratio : 0.68 and 0.86) are presented.

INTRODUCTION

The new generation of parabolic power plants is planned to work with direct steam production in the absorber pipe. Feed water can be supplied at the entrance of the absorber pipe (*once through concept*) or at different positions along the pipe (*injection concept*).

The regulation of the supplied water flow rate at transient insolation conditions is in the *once through concept* complex while the *injection concept* guarantees a quick regulation due to the low injected water flow rate at each injection flange.

Only in the initial section of the absorber pipe do critical thermodynamically problems arise. The low steam velocity in this first section leads on the one hand to a small heat transfer coefficient between steam and the wall and on the other hand to stratification without any droplets. Therefore the temperature gradient between the top and the bottom of the pipe exceeds the 100°C mark. In order to solve this problem of too high pipe tensions, the installation of a displacement body is highly suitable. The steam velocity increases and the flow pattern is influenced and changed remarkably, so that the pipe inner surface wetting is improved.

Literature reviews indicate that the studies of two-phase flow characteristics in horizontal annular geometries are rare. Osmusali et al. (1988) carried out experiments in horizontal annular geometries (pipe diameter: 5.08 cm; inner diameter: 2.5 cm). His model and experimentally identified flow regime map under atmospheric conditions showed no satisfactory conformity.

The beginning of slugging was detected at a superficial liquid velocity of nearly $U_{Ls} = 0.19\ m/s$. Weisman et al. (1984) diagnosed the transition at $U_{Ls} = 0.078\ m/s$ for nearly the same annular geometry. The marked difference between the results of both authors is not satisfactory, indicating that further investigations are necessary.

In order to optimize all parameter of the displacement body, experiments were carried out with air-water with a pressure of 2 bar. The results were helpful for the tests in heated pipes with steam-water flow with evaporation.

AIR-WATER TEST FACILITY

The experimental test loop consists of an open air and a recirculating water loop. Air is provided from two pressure tanks (3,9 m³) at pressure up to 40 bar which then passes a pressure reducer, pressure regulator and a flow meter. The air streams in the entrance of the pipe with a pressure of 2 bar. Behind the calm region of 2 m the water is injected through 6 bores distributed uniform over the circumference of the injection flange. The two-phase mixture flows through the test section with the whole length of 10 m and a diameter of 7 cm. Behind the test pipe the water is separated and repumped into the water tank. The influence of different kinds of air-water mixing disappears after $L/D = 40$, as tests with a nozzle, ring slot and the flange with 6 bores have demonstrated. In all following described experiments this flange is used. The annular geometries were created by installation of displacement bodies with a diameter of 5 cm ($d/D = 0.71$) and 6 cm ($d/D = 0.86$). In the following study with eccentric annulus a position is meant, where the displacement body is contacted directly with the bottom of the outer pipe.

In order to know the flow condition the flow rate and temperature of both phases are measured. Moreover, the pressure, the pressure loss over the test pipe and the required over-pressure for the water injection is registered. The film thickness is detected 4 m behind the injection position by stainless steel conductance probes. The potential difference across the probes (length: 8 mm, diameter: 0.5 mm, distance: 5 mm) is produced by a frequency generator (voltage: 3 V, frequency: 10 kHz). Contrary to all studies about the measurement of the film thickness the technology used here needs two pairs of probes.

While one pair is installed in the suction pipe between the tank and the pump the other is inserted in the acrylic test pipe. Both pairs are fed by the same frequency generator. In the transformer both signals are rectified, divided and amplified so that the output signal is 0-10 V. The advantage of this arrangement with two pairs of conductance probes is that a change in the ion concentration and the temperature of the water do not evoke a change in the signal. During the test period the film thickness is measured with a frequency of 1kHz and the flow pattern is filmed by a video camera.

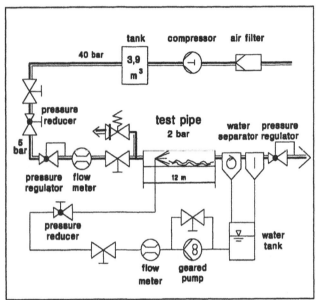

Figure 1: Air-water test facility

3. FLOW PATTERNS DEFINITION

The published flow regime maps for pipe two-phase flow of different authors exhibit crucial dissimilarities. The reason for this unsatisfactory fact is based on the slightly different definition of the flow patterns. This inadequacy makes an obvious definition necessary. The observed flow patterns in pipes and annuli at higher steam qualities are :

In **stratified smooth flow** both phases are separated and the two-phase interface is smooth while in **stratified wavy flow** the interface is characterized by ripple or roll waves, which create droplets at higher gas velocities. A thin unclosed film is produced, which grows in circumferential direction to the top by liquid or gas flow rate increase (**unclosed annular flow**). In **closed annular flow** the gas with droplets flows in the core, while the water film at the pipe inner surface is closed. Annular flow in concentric horizontal annuli show the characteristic disturbance waves, which are inclined and stable over a long period. Jayanti (1990) assumed that these disturbance waves are mainly responsible for the liquid transport to the top. In concentric annular geometries, where the inner pipe oppresses the transport mechanisms of entrainment and secondary flow, the disturbance waves seem to be the only possibility, of how the water film at the top of the pipe is replenished. The **wavy-annular flow**, sometimes called slug-annular flow, leads to a circumferential closed liquid film but the film is created by waves. In eccentric annular geometries the annular flow region is allways wavy-annular flow. In **slug flow**, waves become unstable and grow. Taitel and Dukler (1974) called slug and plug flow

"intermittent flow". The difference between both flow patterns is that in slug flow the slugs never close the whole pipe cross-section area. The crest of the slugs, which are never accelerated up to the gas velocity, are sprayed away before closing the pipe while in **plug flow** the waves reach the top of the pipe. The plugs are accelerated and pushed with high velocity through the pipe and can be dangerous in pipe bends.

SINGLE-PHASE PRESSURE LOSS

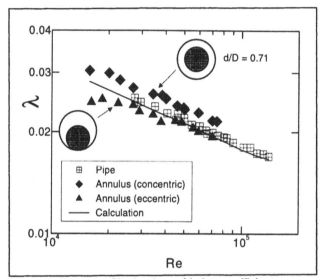

Figure 2 : Single-phase friction coefficient

Usually, the two-phase pressure drop is compared with the single-phase pressure loss by a friction multiplier. The pressure gradient in single-phase is given by

$$\Delta p = 0.3164 / \sqrt[4]{Re} \cdot (L / d_h) \cdot (\rho_G / 2) \cdot u_G^2 \qquad (1)$$

where L is the pipe length, ρ_G the gas density, u_G the gas velocity and Re the gas Reynolds number. In equation (1) the friction factor correlation of Blasius for smooth pipes is used. The hydraulic diameter for the annulus is $d_h = D - d$. The measured and calculated friction factor of gas flow is shown in figure 2. The experimental data of the pipe is consistent with the calculated curve. The pressure losses of the concentric annular geometry show a deviation of 10%. The measured friction factor of the pipe with concentric and eccentric inner pipe indicates a surprising difference of nearly 20 %.

TWO-PHASE PRESSURE LOSS

In many cases the applicability of theoretical models for the prediction of the two-phase pressure drop are limited to a narrow parameter range. Lockhart and Martinelli (1948) developed one of the earliest models by defining a two-phase friction multiplier Φ^2, which compares the single-phase with the two-phase pressure loss:

$$(\Delta p / \Delta L)_{2ph} = (\Delta p / \Delta L)_G \cdot \Phi_G^2 \qquad for \ X < 1 \qquad (2)$$

$$(\Delta p / \Delta L)_{2ph} = (\Delta p / \Delta L)_L \cdot \Phi_L^2 \qquad for \ X > 1 \qquad (3)$$

$$with \quad \Phi_G^2 = 1 + C \cdot X + X^2 \qquad (4)$$

$$or \quad \Phi_L^2 = 1 + C / X + 1 / X^2 \qquad (5)$$

In all experiments the Martinelli number X was less than 1.

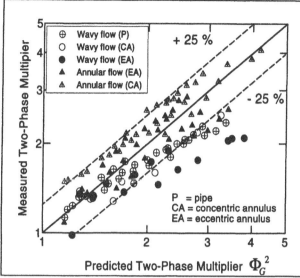

Figure 3 : Two-phase multiplier (X < 1)

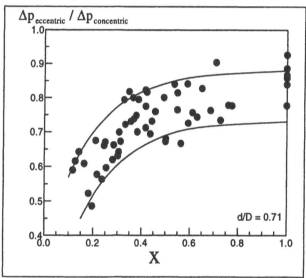

Figure 4 : Comparison of two-phase pressure loss in concentric and eccentric annulus

If both phases are flowing turbulently the Martinelli-number X can be calculated by

$$X_{tt} = \left(\rho_G / \rho_L \right)^{0.5} \left(\eta_L / \eta_G \right)^{0.1} \left((1-x) / x \right)^{0.9}. \qquad (6)$$

and C is equal to 20 in equation (4) and (5). In figure 3 the measured two-phase multiplier of the pipe and the annuli ($d/D = 0.71$) are shown for the turbulent-turbulent case. It can be recognized that the pressure loss is mainly overpredicted in stratified wavy flow and underpredicted in annular flow as well in the pipe as in the annuli. All experiments have shown that the pressure loss in a concentric annulus ($d/D = 0.71$) is 15 - 25 times higher than in the pipe. But in opposition to the concentric installation, the pressure losses are in the eccentric annulus 20 - 50 % less. In figure 4 the ratio of eccentric and concentric annulus pressure drop is shown. Obviously, the pressure drop differs extremely when x (gas quality) decreases. The reasons for the lower two-phase pressure loss in the eccentric annulus are:

- The pressure drop depends on the single-phase pressure drop, which is 20% less in the eccentric annulus
- In the concentric annulus disturbance waves are created in annular flow. On the other hand in the eccentric annulus annular flow is dominated by waves, which have the character of 'slugs'. At the same flow conditions the disturbance waves are sometimes 1.8 times faster (3-4 m/s) than the waves in the eccentric annulus. Based on this different liquid acceleration (momentum exchange) the pressure drop is higher in the concentric annulus.

FILM THICKNESS DISTRIBUTION

In heated pipes with water-steam flow the circumferential temperature distribution of the pipe contains all information about heat transfer coefficient and pipe cooling. In adiabatic two-phase flow the assessment of the flow is only possible if the circumferential film thickness distribution is detected. From figure 5 and 6 it can be noted that in the pipe the flow pattern

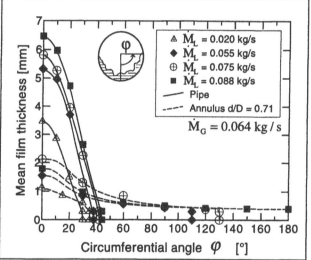

Figure 5 : Film thickness distribution at low gas flow rate in a pipe and a concentric annulus

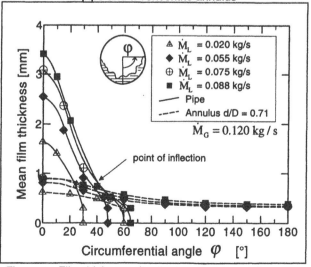

Figure 6 : Film thickness distribution at high gas flow rate in a pipe and a concentric annulus

and the angle of pipe wetting cannot be influenced significantly by increasing the liquid mass flow rate as well for low as for high gas mass flow rate. Stratified wavy flow dominates at low gas velocities, characterized by the typical film thickness distribution. At higher gas velocities unclosed annular flow occurs in the pipe (figure 6), which is characterized by a point of inflection in the curve. In the concentric annulus ($d/D = 0.71$) the flow changes at low gas flow rate from stratified wavy to unclosed annular and finally to wavy annular flow. At higher gas flow rate the flow in the concentric annulus transits from wavy flow to closed annular flow, which shows the known inclined disturbance waves.

The film thickness distributions manifest that the installation of a displacement body improves the pipe inner surface wetting remarkably and causes a high angle of wetness. This effect is based on an earlier alternation to unstable flow patterns (slug or annular flow), which theory is deduced in the following part.

FLOW REGIME MAP

Theoretical approach

There are a few models available, which are useful to predict the flow regime map in horizontal pipes. The Taitel-Dukler model is one of the theoretical models. In order to predict also the flow pattern in a horizontal annulus the Taitel-Dukler model must adapted to the new geometry (figure 7).

The momentum equation in stratified flow in an annulus can be calculated by

$$\tau_G \cdot \frac{(S_{Go} + S_{Gi})}{A_G} - \tau_L \cdot \frac{(S_{Lo} + S_{Li})}{A_L} + \tau_i \cdot S_i \cdot \left(\frac{1}{A_L} + \frac{1}{A_G}\right) = 0 \qquad (7)$$

with S_{Go}, S_{Gi} and S_{Lo}, S_{Li} for the perimeter on the outer and inner wall for the gas phase and the liquid phase, respectively.

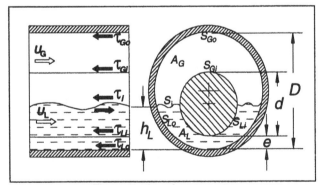

Figure 7 : Annular geometry

S_i is the width of the interface between gas and liquid. The shear stresses of both phases at the inner and outer wall are assumed to be equal ($\tau_G = \tau_{Go} = \tau_{Gi}$, $\tau_L = \tau_{Lo} = \tau_{Li}$). A_L and A_G are the cross-section areas, ρ_L and ρ_G the densities and U_L and U_G the mean velocities of the liquid and gas phase, respectively. The shear stresses are :

$$\tau_G = 0.5 \cdot f_G \cdot \rho_G \cdot U_G^2 \quad , \quad \tau_L = 0.5 \cdot f_L \cdot \rho_L \cdot U_L^2 \qquad (8), (9)$$

$$\tau_i = 0.5 \cdot f_i \cdot \rho_G \cdot (U_G - U_L)^2 \qquad (10)$$

with $f_L = C_L (D_L U_L / \nu_L)^{-n}$ and $f_G = C_G (D_G U_G / \nu_G)^{-m}$

The hydraulic diameter are evaluated as $D_L = 4A_L / S_L$ and $D_G = 4A_G / (S_G + S_i)$. The coefficients C_G and C_L equal 0.046 for turbulent flow and 16 for laminar flow, n and m take the value of 0.2 for turbulent flow and 1.0 for laminar flow. The interfacial friction factor f_i is assumed to be $5f_G$. The geometry parameters, which are only depending on the liquid height h_L can be computed by :

$$A_L = \frac{D^2}{4}\left[\pi - \cos^{-1}(a) + (a)\{1 - (a)^2\}^{0.5}\right] - k \cdot \frac{d^2}{4}\left[\pi - \cos^{-1}(b) + (b)\{1 - (b)^2\}^{0.5}\right]$$

$$A_G = \frac{\pi}{4}(D^2 - d^2) - A_L \qquad (11), (12)$$

$$S_i = D\left[1 - (a)^2\right]^{0.5} - k \cdot d\left[1 - (b)^2\right]^{0.5} \qquad (13)$$

$$S_{Lo} = D(\pi - \cos^{-1}(a)) \quad , \quad S_{Go} = \pi \cdot D - S_{Lo} \qquad (14), (15)$$

with $a = 2h_L / D - 1$ and $b = 2(h_L - e)/d - 1$
and separating the channel in 3 sections :

$h_L < e$	$e < h_L < d + e$	$h_L > d + e$
$k = 0$	$k = 1$	$k = 0$
$S_{Li} = 0$	$S_{Li} = d(\pi - \cos^{-1}(b))$	$S_{Li} = \pi d$
$S_{Gi} = \pi d$	$S_{Gi} = \pi d - S_{Li}$	$S_{Gi} = 0$

After calculating the theoretical liquid height, the condition of stability must be checked. Barnea (1985) gives the following equation for the stratified-intermittent or annular transition

$$(U_G - U_L) \leq \left(1 - \frac{h_L}{D}\right)\left[\left(\rho_L \frac{A_G}{A} + \rho_G \frac{A_L}{A}\right)\frac{\rho_L - \rho_G}{\rho_L \rho_G} g \frac{A}{dA_L / dh_L}\right]^{0.5} \qquad (16)$$

$$\text{with } \frac{dA_L}{dh_L} = D\left(1 - (a)^2\right)^{0.5} - k \cdot d\left(1 - (b)^2\right)^{0.5} \qquad (17)$$

Equation (16) and (7) must be solved iteratively to calculate the change between stable stratified and unstable slug or annular flow.

Experimental and theoretical flow regime map

The transition between different flow patterns is usually described by using flow regime maps. In figure 9 the total experimental flow map of a concentric annulus ($d/D = 0.71$) is shown. It can be recognized that the change of stratified to intermittent flow occurs at lower superficial liquid velocities than in the pipe. This phenomenon was also observed by Osmusali (1987) and Crawford (1984). In figure 9 the onset of atomisation curve is included, which improves the pipe wetting.

Several investigations were conducted to understand the influence of the inner pipe position. The most interesting transition line is that curve, which describes the change of the stable flow patterns (stratified flow) to an unstable region (slug or annular flow). Such an alternation would also mean a satisfactory pipe wetting and pipe cooling in absorber tubes.

Therefore figure 10 only contains this transition curve for different annular geometries. In figure 10 the calculated curves for the eccentric annulus with $d/D = 0.71$ and 0.86 ($e = 0$) is represented. The theoretical curves for the interesting eccentric annulus were computed by using the modified Taitel/Dukler-model, as presented before. The conformity between the model and the measured curve is quite good.

By comparing the transition lines of figure 9 and 10 it comes to light that the eccentric annulus creates a closed film at lower liquid flow rates than the concentric annulus does. The reason for this difference is the different flow behaviour as shown in figure 8 :

- In the concentric annulus the liquid film grows slowly from the bottom to the top of the annulus. The liquid film seems to be pulled up to the top. The replenish mechanisms of the liquid film are the same as in normal pipes (droplets, disturbance waves)

- In the eccentric annulus the known annular flow does not exist. The film in the annular region is created by intermittent waves (wavy-annular flow). If the waves are high enough with a high frequency a closed liquid film occurs. In wavy-annular flow the liquid transport to the top is nearly the same mechanism as in intermittent flow (plug and slug flow). Based on the different creation of the closed liquid film the transition 'region', as shown in figure 10, is reached earlier in the eccentric annulus.

- The disturbance waves are created in the concentric annulus and in the pipe at the bottom (Jayanti). Such a creation of disturbance waves is in the eccentric annulus impossible due to the narrow slot at the bottom.

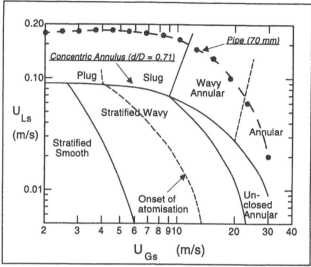

Figure 9 : Experimental flow regime map in an concentric annulus (d/D = 0.71)

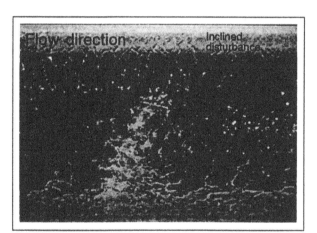

a) Concentric annulus

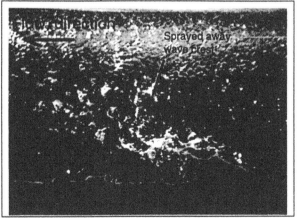

b) Eccentric annulus

Figure 8 : Flow patterns in eccentric and concentric annulus (U_{Gs} = 22.5 m/s, U_{Ls} = 0.05 m/s)

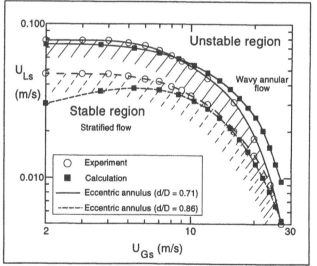

Figure 10 : Comparison of detected and calculated transition line in eccentric annuli

CONVECTIVE FLOW BOILING IN ANNULI

Based on the results of air-water experiments, tests in a 8 m steel pipe (inner diameter 49.5 mm) were carried out. The pressure was 30 bar and the diameter of the two tested displacement bodies were 33.7 mm (d/D = 0.68) and 42.3 mm (d/D = 0.86). The pipe was heated electrically from the side with 3.7 kW/m, in order to simulate the insolation in the morning and the afternoon and to investigate the more critical heating case. Grounded on the positive effect of an eccentric annulus in the air-water tests, the steel displacement bodies were also installed eccentrically, but the displacement body were not contacted directly with the bottom of the outer pipe. The distance was nearly e = 2-3 mm.

Classification of flow patterns in heated pipes

In heated pipes with high pressure it is not possible to detect the flow pattern by optical observations. Nevertheless, the transition between stratified and annular flow can be measured

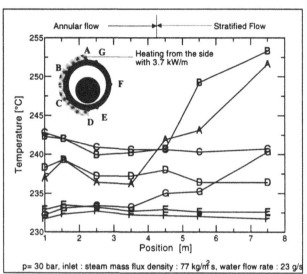

p= 30 bar, inlet : steam mass flux density : 77 kg/m² s, water flow rate : 23 g/s

Figure 11 : Temperature distribution in a heated annulus

by using the temperature distributions along the pipe and in circumference. In figure 11 the temperature distribution in axial direction of the annulus with $d/D = 0.86$ is shown.

The flow pattern changes during the evaporation from annular flow to stratified flow. In annular flow the temperature distribution shows a negative temperature gradient from the top to the bottom of the pipe. The liquid film is very thin at the top of the pipe and is responsible for the higher heat transfer coefficient at the top. In stratified flow the temperature increases from the bottom to the top of the pipe wall. If the pipe is heated from the side, the temperature distribution is influenced on the one hand from the heating profile and on the other hand from the heat transfer profile. Hence, the interpretation of the temperature distribution is more complicated than in uniform heated pipes. In order to differ the flow patterns the following criterions were used :

- Temperatures for Stratified flow : C < A < B
- Temperatures for Annular flow : C > B > A

Flow regime map in heated eccentric annuli

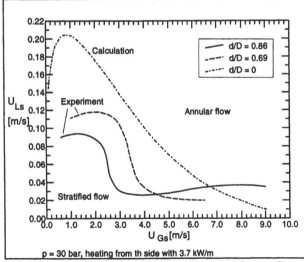

p = 30 bar, heating from th side with 3.7 kW/m

Figure 12 : Flow regime maps in eccentric annuli

From figure 12 it can be noted that the transition curve in eccentric annuli with steam-water flow with boiling is influenced remarkably by the diameter of the displacement body. Surprisingly, the curves do not touch the x-axis at high superficial steam velocities and low water velocities as it was observed in pipes. A minimum liquid flow rate has to be injected to realise a satisfactory pipe cooling. Probably, the inner displacement body catches the droplets and oppresses, that the droplets reach the top of the pipe. The sharp decrease of the curves may be caused by the distance between the displacement body and the pipe. It is also possible that waves, which do not reach the top but create enough droplets at the top of the pipe lead to a satisfactory pipe cooling. In air-water two phase flows only such a flow plattern was detected as annular flow, that created a closed water film.

CONCLUSIONS

Experiments in a horizontal pipe and different eccentric and concentric annular geometries were conducted with adiabatic air-water and heated steam-water two-phase flow. It was discovered that the pipe wetting can be improved significantly by the installation of a displacement. Surprisingly, the required liquid flow rate can moreover be reduced, if the displacement body is installed eccentrically. Moreover, the pressure drop is in the eccentric annulus 20-50% less. In steam-water tests with boiling the same flow behaviour was detected as in air-water flow. The effect of the displacement body cannot be explained by the reduction of the cross-section area. The required mass flux density for annular flow is in the annuli nearly 50 % less in comparison to the pipe. The total different flow behaviour in annuli is responsible for this effect. In order to calculate the flow regime map for annuli a modified Taitel/Dukler-model was presented.

ACKNOWLEDGEMENT

The author wish to express his sincere gratitude to Mr. Bissinger for valuable help in realising the film thickness measurement equipment and to the Solar Energy and Hydrogen Research Centre in Stuttgart, where all steam-water test were carried out.

REFERENCES

Barnea, D., Taitel, Y., 1994, "Interfacial and structural stability of separated flow," *Int. J. Multiphase Flow*, Vol. 20, Suppl., pp. 387-414.

Crawford, T., Weisman, J.,1984, "Two-phase Flow Pattern Transition in Ducts of Non-circular Cross-section and under Diabatic Conditions," *Int. J. Multiphase Flow*, Vol. 20, No. 3, pp. 385-391.

Jayanti, S., Hewitt, G. F., White, S. P., 1990, "Time-dependent behaviour of the liquid film in horizontal annular flow," *Int. J. Multiphase Flow*, Vol. 16, No. 6, pp. 1097-1116.

Osamusali, S. I., Chang, J. S., 1988, "Two phase flow regime transition in a horizontal pipe and annulus flow under gas-liquid two-phase flow," *Fund. of Gas-Liquid Flows*, Winter Annual Meeting of the ASME, Chicago, pp. 63-69.

Taitel, Y.,Dukler, A. E., 1976,"A Model for Predicting Flow Regime Transitions in Horizontal and Near Horizontal Gas-Liquid Flow," *AIChE Journal*, Vol. 22, No. 1, pp. 47-55

Martinelli, R.C., Nelson, D. B.,1948,"Prediction of Pressure Drop During Forced Circulating Boiling of Water," *American Society of Mechanical Engineers*, Transaction 70

PREDICTING BOILING HEAT TRANSFER IN A SMALL-DIAMETER ROUND TUBE USING AN ASYMPTOTIC METHOD

Jonathan P. Wattelet
Modine Manufacturing Company
Racine, Wisconsin

ABSTRACT

This work compared an asymptotic model for flow boiling heat transfer with experimental data for R-113 in a small-diameter round tube (2.92 mm) from Wambsganss et al. [1993]. The asymptotic model was developed using data from alternative refrigerant studies for larger tubes (7-11 mm) [Wattelet, 1994]. The small-diameter, round-tube experimental data were shown to be dependent on heat flux alone and not mass flux, indicating flow boiling behavior dominated by nucleate boiling effects. The asymptotic model predicted the data fairly well, with a mean deviation of 14.4%. This mean deviation was lower than several other correlations from the literature based on the superposition model [Wambsganss et al., 1993]. Experimental data for $Fr_l < 0.25$ were also well correlated by the asymptotic model (mean deviation: 14.8%), showing that a Froude number dependent term can be used to account for diminished convective boiling effects for low mass fluxes. However, better prediction of the small-diameter round-tube experimental data was obtained using nucleate boiling correlations, such as those of Lazarek and Black [1982] and Cooper [1984], which had mean deviations of 12.7% and 11.4%, respectively, when compared with the experimental data.

INTRODUCTION

Evaporators used in the automotive and stationary air conditioning industries are becoming more and more compact in nature. These compact heat exchangers have many attractive features, such as large heat transfer area per volume ratio, low weight per heat transfer duty, and reduced charge inventory [Wambsganss et al., 1992]. Recent studies on boiling heat transfer of R-113 in a horizontal, small-diameter (2.92 mm) round tube indicate that nucleate boiling is the dominant heat transfer mechanism due to the relatively high heat fluxes

and low mass fluxes inherent in small channels [Wambsganss et al., 1993]. This differs quite drastically from the convective boiling dominated data typically found during flow boiling in larger diameter tubes [Jung and Radermacher, 1989]. However, for low values of wall superheat ($\leq 3°C$), a convective dominant region was identified for small-diameter channels [Tran et al., 1994]. To adequately predict heat transfer for all of these situations, a correlation must be developed which has accurate terms for both convective and nucleate boiling. In addition, the form of the correlation must properly suppress either term when the other term dominates.

In this paper, flow boiling data for refrigerants in horizontal large diameter tubes (7-11 mm) will be presented and compared with an asymptotic model for flow boiling heat transfer developed by Wattelet [1994]. This data will validate the model for larger diameter tubes, where convective boiling typically dominates over nucleate boiling for low heat fluxes. Next, the asymptotic model will be compared with flow boiling data from Wambsganss et al. [1993] for small-diameter horizontal tubes. Comparisons will also be made between the superposition model, the asymptotic model, and the "greater of the two" model with regards to accurately predicting the small-diameter round-tube experimental flow boiling data.

FLOW BOILING CORRELATING FORMS

Correlations used to predict flow boiling data in larger diameter tubes have been based on three typical forms: the superposition model, the "greater of the two" model, and the asymptotic model.

The superposition model [e.g. Chen, 1966] is an additive model which typically combines a nucleate boiling term and a convective boiling term as follows:

$$h_{TP} = S h_{nb} + F h_l \qquad (1)$$

where S is a suppression factor for nucleate boiling, h_{nb} is a pool boiling heat transfer correlation used to account for nucleate boiling, F is a two-phase multiplier for convective heat transfer, and h_l is a single-phase convective heat transfer correlation such as the Dittus-Boelter correlation [1930].

The "greater of the two" model [e.g. Shah, 1982] evaluates expressions for convective and nucleate boiling and uses the larger of the values as follows:

$$h_{TP} = \varphi_{max} h_l \qquad (2)$$

$$\varphi_{max} = \max\left[\varphi_{nb}(Bo), \varphi_{cb}(Co)\right] \qquad (3)$$

where Bo is the boiling number, a dimensionless heat flux, and Co is the convection number, a modified form of the Lockhart-Martinelli parameter, X_{tt}.

The main feature of the asymptotic model is the "built in" suppression of the weaker component. The form of this correlation is given as

$$h_{TP} = \left[h_{nb}^n + h_{cb}^n\right]^{1/n} \qquad (4)$$

where $h_{cb} = Fh_l$. For n equal to 1, the form becomes that of the superposition model. As n approaches infinity, the form becomes the "greater of the two" model. For a large convective component and a small nucleate boiling component, the total two-phase heat transfer coefficient is made up almost entirely of the convective boiling component. For a mixed situation where both nucleate boiling and convective boiling occur, the total two-phase heat transfer coefficient is made up of a combination of the two components. For a nucleate boiling dominated situation, the total two-phase heat transfer coefficient is made up almost entirely by the nucleate boiling component.

A correlation developed to predict flow boiling heat transfer for alternative refrigerants in larger diameter tubes (7-11 mm) was recently developed which uses an asymptotic form to combine nucleate boiling and convective boiling terms [Wattelet, 1994]. It is listed as follows:

$$h_{TP} = \left[h_{nb}^n + h_{cb}^n\right]^{1/n}, \quad n = 2.5 \qquad (5)$$

$$h_{nb} = 55 q''^{0.67} M^{-0.5} P_r^{0.12} \left[-\log_{10} P_r\right]^{-0.55} \qquad (6)$$

$$h_{cb} = Fh_l R \qquad (7)$$

$$F = 1 + 1.925 X_{tt}^{-0.83} \qquad (8)$$

$$h_l = 0.023 \frac{k_l}{D} Re_l^{0.8} Pr_l^{0.4} \qquad (9)$$

$$R = 1.32 Fr_l^{0.2}, \text{ for } Fr_l < 0.25 \qquad (10)$$

$$R = 1, \text{ for } Fr_l \geq 0.25 \qquad (11)$$

Units of heat flux in Eq. 6 are W/m^2-K. The correlation was developed using data described in the next section and has been shown to be accurate over the following range of conditions: mass flux, 25-1000 kg/m^2-s; heat flux, 2-300 kW/m^2; quality, 5-95% [Wattelet, 1994].

LARGE TUBE FLOW BOILING DATA

An experimental study of flow boiling heat transfer of refrigerants in a horizontal-tube evaporator was conducted by Wattelet [1994]. A single-tube evaporation test facility was designed and developed to measure the evaporation characteristics of alternative refrigerants. Measurements were made in several instrumented, horizontal copper tubes with inside tube diameters ranging from 7 to 11 mm and lengths ranging from 1.22 to 2.44 m using R-12, R-22, R-134a, and a 60%/40% azeotropic mixture of R-32/R-125. Mass fluxes ranged between 25 and 1000 kg/m^2-s and heat fluxes ranged between 2 and 40 kW/m^2. The two main flow regimes found during objective and visual evaluation of the flow patterns during adiabatic and diabatic flow in smooth, horizontal tubes were wavy-stratified flow and annular flow.

Figure 1 shows the variation of heat transfer coefficient versus quality for annular flow tests with a fixed mass flux and varying heat flux for R-134a for a 10.92 mm i.d. As can be noted, for low heat fluxes, the heat transfer coefficient increased with quality. Intense evaporation at the liquid-vapor interface diminished the liquid film thickness, reducing the thermal resistance, which was associated with heat conduction across the film. Nucleate boiling appeared to be largely suppressed for these low heat flux cases. As heat flux increased, the heat transfer coefficients increased in the lower quality region and eventually merged at higher qualities with the heat transfer coefficients for low heat flux cases. Nucleate boiling at these lower qualities enhanced the heat transfer coefficient. At higher qualities, nucleate boiling was again largely suppressed due to significant surface cooling promoted by the thinning of the annular film. Wall superheats for this set of data ranged from 1 to 6°C.

Figure 2 shows the comparison of the R-12, R-134a, R-22, and R-32/R-125 experimental data of Wattelet [1994] with the asymptotic model, Eq. 5. Mean deviation of the asymptotic model, Eq. 5, compared with the R-12, R-134a, R-22, and R-32/R-125 experimental data was 11.8%.

SMALL TUBE FLOW BOILING DATA

A study of flow boiling heat transfer of R-113 in a horizontal small-diameter tube (2.92 mm) was recently conducted [Wambsganss et al., 1993]. Local heat transfer coefficients were reported for a range of heat fluxes between 9 kW/m^2 and 91 kW/m^2 and a range of mass fluxes between 50 and 300 kg/m^2-s. Wall superheats ranged from approximately 5 to 20°C. Due to the relatively high heat fluxes and low mass fluxes, high boiling numbers resulted. Based on flow pattern maps of adiabatic air-water mixture experiments in small-diameter tubes [Damianides and Westwater, 1988], the predominant flow pattern in the test tube was determined to be slug flow. As a result of the high heat fluxes and slug flow pattern, nucleate boiling was found to dominate as the heat transfer mechanism. Figure 3 shows the heat transfer coefficient versus quality for a typical set of tests. Heat transfer coefficients were a function of heat flux alone, with very little effect of quality, indicative of a nucleate boiling mechanism.

Figure 4 shows the comparison of the R-113 experimental data with the asymptotic model, Eq. 5. Mean deviation of the asymptotic model, Eq. 5, compared with the R-113 experimental data was 14.4%. This mean deviation was lower than other correlations which use superposition models [Wambsganss et al., 1993], such as the Jung and Radermacher correlation [1989] (16.0%), the Chen correlation [1966] (36.0%), and the Kandlikar correlation [1990] (22.0%). However, nucleate boiling only correlations, such as Lazarek and Black [1982] (12.7%) and Stephan and Abdelsalam [1980] (12.9%), predicted the data better.

Questions have arisen in the open literature whether a Froude number term attached to the convective boiling part of flow boiling correlations is an accurate way to account for low mass flux effects. Experimental data for $Fr_l < 0.25$ were also well correlated by the asymptotic model (mean deviation: 14.8%), showing that a Froude number dependent term can be used to account for diminished convective boiling effects for low mass fluxes.

Comparing the magnitudes of the convective and nucleate boiling terms in Eq. 5 using the small tube flow boiling conditions, the nucleate boiling term was larger in approximately 90% of the cases. However, the convective boiling term did contribute significantly to the predicted flow boiling heat transfer coefficient, with both mass flux and quality effects present. This indicated that the correlation did not completely follow the trends of the experimental data, which showed little effect of mass flux and quality.

A comparison was also made in Fig. 5 between the nucleate boiling term, Eq. 6, and the small-diameter, round-tube, flow boiling data. The nucleate boiling term is actually the Cooper correlation for pool boiling [Cooper, 1984]. The mean deviation of Eq. 6 with the small-diameter, round-tube, flow boiling data was 11.4%. This indicates that a "greater of the two" model does a better job of predicting this small-diameter data

than the asymptotic model. However, it should be noted that the large wall superheats (5 to 20°C) are probably more responsible for the nucleate boiling behavior than any small tube diameter effects such as surface tension. But it is the large heat fluxes and low mass fluxes inherent in small channels for typical small-tube applications that result in the large wall superheats.

CONCLUSIONS

An asymptotic model for correlating flow boiling heat transfer coefficients, developed using data from alternative refrigerant studies for larger tube data (7-11 mm), was compared with experimental data for R-113 in a small-diameter, horizontal round tube (2.92 mm). The small tube diameter flow boiling experimental data was dominated by heat flux effects, with very little variation of heat transfer coefficient with mass flux or quality. This type of behavior is indicative of flow boiling dominated by nucleate boiling effects. The asymptotic model predicted the experimental data fairly well, with a mean deviation of 14.4%. This mean deviation was lower than several other correlations based on the superposition model. Better prediction was found using only nucleate boiling correlations. Therefore, a "greater of the two" model with an accurate nucleate boiling term would do a better job of predicting this small-diameter data than an asymptotic model.

ACKNOWLEDGMENT

The author would like to thank David France of the University of Illinois at Chicago and Martin Wambsganss of Argonne National Laboratory for use of their experimental data on flow boiling in horizontal, small-diameter, round tubes.

NOMENCLATURE

Bo	boiling number
Co	convection number
D	diameter
F	two-phase multiplier for heat transfer
Fr_l	liquid Froude number
h_l	single-phase heat transfer coefficient
h_{cb}	convective boiling heat transfer coefficient
h_{nb}	nucleate boiling heat transfer coefficient
h_{TP}	two-phase heat transfer coefficient
k_l	liquid thermal conductivity
M	molecular weight

n	exponent in Eq. 5
P_r	reduced pressure
Pr_l	liquid Prandtl number
q''	heat flux
R	reduction factor in Eq. 5
Re_l	liquid alone Reynolds number
S	suppression factor in Eq. 1
X_{tt}	Lockhart-Martinelli parameter
ψ_{cb}	Shah two-phase convective boiling multiplier
ψ_{nb}	Shah two-phase nucleate boiling multiplier
ψ_{max}	maximum value of ψ_{cb}, ψ_{nb}

REFERENCES

Chen, J.C., 1966, "A Correlation for Boiling Heat Transfer to Saturated Fluids in Convective Flow," *I&EC Process Des. and Development*, Vol. 5 (3), pp. 322-329.

Cooper, M.G., 1984, "Saturation pool boiling--A simple correlation," International Chemical Engineering Symposium Series, 86, pp. 785-792.

Damianides, C.A. and Westwater, J.W., 1988, "Two-Phase Flow Patterns in a Compact Heat Exchanger and in Small Tubes," Proc. 2nd U.K. National Conference on Heat Transfer, Glasgow, Scotland, Vol. 2, 1257-1268.

Dittus, F.W. and L.M.K. Boelter, 1930, University of California Publications on Engineering, 2, 443.

Jung, D.S., McLinden, M., Radermacher, R., and Didion, D., 1989, "A Study of Flow Boiling Heat Transfer with Refrigerant Mixtures," *Int. J. of Heat and Mass Transfer*, Vol. 32(9), pp. 1751-1764.

Kandlikar, S.S., 1990, "A General Correlation for Predicting Two-Phase Flow Boiling Heat Transfer Coefficients in Horizontal and Vertical Tubes," *Heat Exchangers for Two-Phase Flow Applications*, Vol. 112, pp. 219-228.

Lazarek, G.M., and Black, S.H., 1982, "Evaporative Heat Transfer, Pressure Drop and Critical Heat Flux in a Small Vertical Tube with R-113," *Int. J Heat and Mass Transfer*, Vol. 25 (7), pp. 945-960.

Shah, M.M. 1982, "Chart Correlation for Saturated Boiling Heat Transfer: Equations and Further Study," *ASHRAE Transactions*, Vol. 82 (2), pp. 66-68.

Stephan, K, and Abdelsalam, M., 1980, "Heat Transfer Correlations for Natural Convection Boiling," *Int. J. of Heat and Mass Transfer*," Vol. 23, pp. 73-87.

Tran, T.N., Wambsganss, M.W., France, D.M., 1994, "Boiling Heat Transfer in Compact Heat Exchangers," 1994 AIChE Annual Meeting, session on Compact Heat Exchangers, San Francisco, CA.

Wambsganss, M.W., D.M. France, Jendrzejczyk, J.A., Tran, T.N., 1992, "Boiling Heat Transfer of Refrigerant R-113 in a Small-diameter, Horizontal Tube," Argonne National Laboratory, ANL-92/12.

Wambsganss, M.W., France, D.M, Jendrzejczyk, J.A., and Tran, T.N., 1993, "Boiling Heat Transfer in a Horizontal Small-Diameter Tube," *Journal of Heat Transfer*, Vol. 115, pp. 963-972.

Wattelet, J.P., 1994, "Heat Transfer Flow Regimes of Refrigerants in a Horizontal-Tube Evaporator", Ph.D. thesis, University of Illinois at Urbana-Champaign.

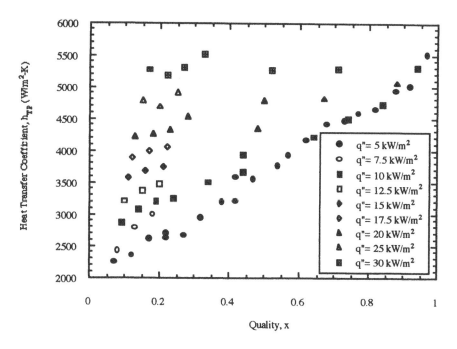

Figure 1. R-134a Heat Transfer Coefficient Versus Quality in Annular Flow in 10.92 mm i.d. Test Section. Mass Flux, 300 kg/m²-s; Saturation Temperature, 5°C.

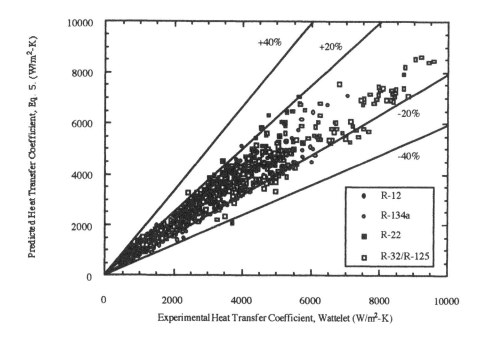

Figure 2. Predicted Heat Transfer Coefficient Using Eq. 5 Versus Experimental Heat Transfer Coefficient for R-12, R-134a, R-22, and R-32/R-125 from Wattelet [1994].

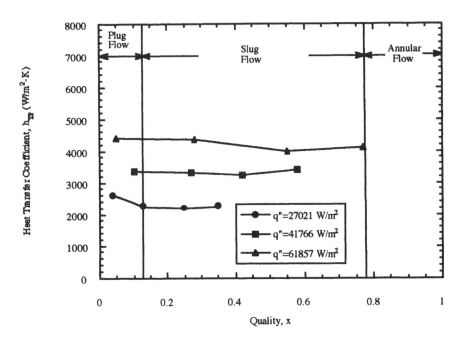

Figure 3. R-113 Heat Transfer Coefficient Versus Quality in 2.92 mm i.d. Test Section [Wambsganss et al., 1993]. Mass Flux, 200 kg/m^2-s.

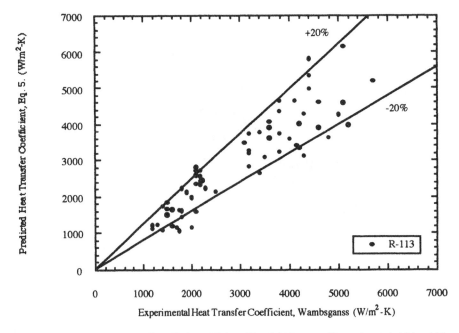

Figure 4. Predicted Heat Transfer Coefficient Using Eq. 5. Versus Experimental Heat Transfer Coefficient of R-113 from Wambsganss et al. [1993].

ANALYSIS OF PASSIVE DECAY POWER REMOVAL VIA TWO-PHASE BOILING-DRIVEN NATURAL CONVECTION IN THE NPR-HWR THIN RECTANGULAR FUEL CHANNELS

Christopher E. Henry
Nuclear Process Safety
Fauske & Associates, Inc.
Burr Ridge, Illinois

George Bankoff
Chemical Engineering
Northwestern University
Evanston, Illinois

Hans K. Fauske
Nuclear Process Safety
Fauske & Associates, Inc.
Burr Ridge, Illinois

ABSTRACT

As part of an assessment of severe accident prevention and mitigation in the NPR-HWR advanced production reactor, in-core, boiling-driven natural convection was identified as an effective mechanism for long-term, passive decay power removal. Henry (1993) conducted experiments which showed that decay power levels (1-4% of nominal full power) could be removed via latent heat on a long-term basis. Incipient fuel dryout and overheat would ensue only when the two-phase level within the fuel assemblies was insufficient to cover the heated length. Henry (1993) then developed a thermal hydraulic model for predicting threshold conditions spawning incipient dryout in the thin rectangular subchannels that comprise the coolant annuli within the fuel assembly. Such a prediction is necessary for determining the duration of successful power removal provided by this mechanism. The model provides good agreement with the noted data and, more importantly, with data from independent tests on a full-scale assembly. The noted mechanism is gravity-dominant two-phase natural circulation. Thus, this work is distinct from the large body of research in high-power removal limits from completely submerged channels, entailing a friction-/acceleration-dominant two-phase natural circulation. Cheng (1991) and Mishima (1987) are two of numerous contributors in this related field.

BACKGROUND

Figure 1 shows a radial cross section (≈ 0.1 m dia.) of a Mark 22 fuel assembly, which is used in the current generation of production reactors. While this differs in detail from that proposed for the NPR-HWR, the fundamental geometry is essentially unchanged. Concentric cylinders of fuel, which generate nearly all assembly power, and target material are bounded by coolant

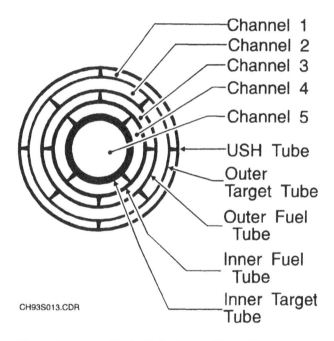

CH93S013.CDR

Figure 1   Mark 22 fuel assembly, with concentric fuel and target rings and coolant annuli

annuli. The axial spacers on the cylinders divide each annulus into four subchannels. Figure 2 shows an axial cross section of the in-core assembly (1 of ≈ 250). The key design aspect that fosters the noted passive mechanism is the coupled fuel-moderator flow circuit. The moderator coolant, which would be the sole heat sink for decay power removal, would provide the hydrostatic pressure driving the two-phase mixture, created by boiling within each assembly.

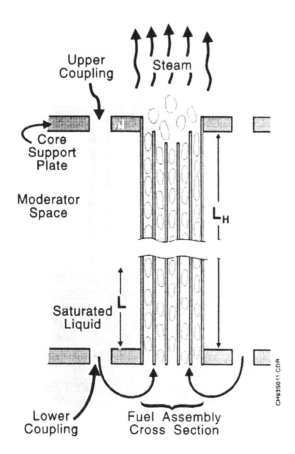

Figure 2   Boiling-driven natural convection via
the coupled fuel-moderator flow circuit

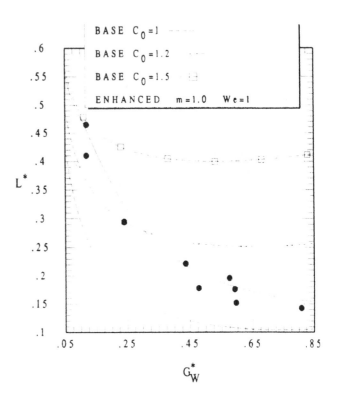

Figure 3   Benchmark of base and enhanced
thermal hydraulic models against
test data

EXPERIMENT DESCRIPTION

The test configuration was essentially that shown in
Figure 2, except only a half-length of the inner fuel ring
and one adjacent coolant annulus were simulated. These
were converted into a rectilinear (flat plate) geometry so
that heaters could be applied. Incipient dryout occurred
only when the moderator level was no longer sufficient to
sustain a two-phase level at or above the channel exit. As
shown by the data summary in Figure 3, the dimensionless
threshold collapsed level L^* was consistently below 50%.
With increasing power, the threshold L^* decreased, due to
a larger void fraction. Thus, in a LOCA scenario where
the core is initially submerged and coolant is removed only
by vaporization, the test results reveal that the moderator
would indeed serve as a large coolant inventory necessary
for long-term decay power removal. Finally, the two-
phase flow regime is observed to be churn-turbulent and
is characterized by thick, wavy liquid films in the narrow
edges of the rectangular subchannels. These films bridge
intermittently, forming gas and liquid slugs.

BASE THERMAL HYDRAULIC MODEL

Since the two-phase, churn-turbulent flow is gravity-

dominant, the momentum equation reduces to a simple
hydrostatic pressure balance between the collapsed moderator liquid and the boiled-up mixture within the channels,
as expressed in Eq. (1).

$$\rho_f \, g \, L = \int_0^{L_H} (\rho_g \, g \, \alpha + \rho_f \, g \, (1 - \alpha)) \, dx \qquad (1)$$

The symbols are detailed in the Nomenclature. Eq. (1)
is nondimensionalized via division by $\rho_f g L_H$, yielding:

$$L^* = 1 - (1 - \zeta^{-1}) \bar{\alpha} \qquad (2)$$

where $\zeta \equiv \rho_f / \rho_g$, $L^* \equiv L/L_H$ and the average void fraction $\bar{\alpha}$ is defined as:

$$\bar{\alpha} \equiv \int_0^1 \alpha(x^*) \, dx^* \qquad (3)$$

The axial void fraction profile is conventionally modeled
using the basic drift flux formulation along with continuity considerations. The nondimensional drift flux is
expressed as:

$$\alpha^{-1} \equiv \zeta^{1/2} \frac{U_\alpha^*}{j_g^*} + C_0 \left(1 + \zeta^{-1/2} \frac{j_f^*}{j_g^*} \right) \qquad (4)$$

where the terminal rise velocity, U_∞^*, and the superficial velocity of phase i, j_i^*, are defined as:

$$U_\infty^* \equiv \frac{\rho_f^{1/2} C_\infty U_\infty}{(\Delta \rho\, g\, w)^{1/2}} \tag{5}$$

$$j_i^* \equiv \frac{\rho_i^{1/2}\, j_i}{(\Delta \rho\, g\, w)^{1/2}} \tag{6}$$

where $C_\infty = 1.6$ and $U_\infty^* = 0.24$. Note also that the characteristic dimension for thin rectangular channels is the channel width, w. Lee's (1993) analysis of air-water flooding experiments in such channels provides sufficient justification for the use of width. He also cites other research that concurs with this assumption. The superficial phase velocities can be expressed as a function of axial position, x^*, via mass continuity for each phase and two prevailing boundary conditions:

(i)   The entry of saturated liquid into the channel so that thermal power is removed solely by latent heat of vaporization.

(ii)   All vapor exit from the channel, according to the imposed condition of incipient dryout.

This yields the superficial phase velocity profiles:

$$j_g^* = G_w^* x^* \quad ; \quad j_f^* = G_w^* \zeta^{1/2}(1 - x^*) \tag{7}$$

where the dimensionless mass flux, G_w^*, is defined as:

$$G_w^* \equiv \frac{G}{(\rho_g \Delta \rho\, g\, w)^{1/2}} \quad ; \quad G = \frac{Q}{A_c\, h_{fg}} \tag{8}$$

Substitution of velocity expressions from Eq. (7) into Eq. (4) yields the void fraction profile given by the basic drift flux formulation.

$$\alpha_{base}^{-1} = \left(\frac{\zeta^{-1/2} U_\infty^*}{G_w^*} + C_0 \zeta^{-1} \right)\frac{1}{x^*} + C_0 (1 - \zeta^{-1}) \tag{9}$$

Substitution of Eq. (9) for $\alpha(x^*)$ in Eq. (3) and integration over the channel length yields the average void fraction $\bar{\alpha}$.

$$\bar{\alpha} = K_2^{-2}[K_2 - K_1 \ln{(K_1 + K_2)} + K_1 \ln{K_1}] \tag{10}$$

where the following symbols have been defined to simplify notation:

$$K_1 \equiv \frac{\zeta^{-1/2} U_\infty^*}{G_w^*} + C_0 \zeta^{-1} \quad ; \quad K_2 \equiv C_0(1 - \zeta^{-1}) \tag{11}$$

Substitution of Eq. (10) into Eq. (2) comprises the base thermal-hydraulic model for predicting threshold conditions at incipient dryout of the heated channel.

Comparison of the model with threshold hydrostatic pressure (L^*) data from conducted experiments is shown in Figure 3.

Three curves are generated by varying the constant value of parameter C_0. It is termed the phase distribution parameter since it is intended to account for nonuniform, lateral (2-D) phase distribution in this 1-D analysis. A generally accepted empirical range exists from $C_0 = 1$ (uniform) to $C_0 = 1.5$ (skewed). As noted earlier, the two-phase regime is characterized by relatively thick, liquid films in the narrow edges of the channel cross section, creating a relatively skewed distribution. Thus, for small mass fluxes G_w^*, the good agreement between model and data for $C_0 \to 1.5$ is expected.

However, with increasing G_w^*, the data appears to migrate consistently toward $C_0 \to 1$. The trend toward a more uniform lateral phase distribution is not surprising in view of increasing vapor flux, which would tend to expand the two-phase flow and diminish the narrow edge films in the channel cross section. Thus, the existing model is deficient in the respect that, with a single constant C_0, it cannot adequately predict the threshold L^* over the entire gravity-dominant region.

ENHANCED THERMAL HYDRAULIC MODEL

The primary insight drawn from the base model benchmark is that the lateral phase distribution is not constant throughout the channel. Therefore, the fundamental assertion of the proposed enhanced thermal hydraulic model is that the phase distribution is a local quantity, capable of either constant or variable behavior. Furthermore, the local distribution is controlled by the behavior of the narrow edge liquid films. This approach is applied to the boiling channel, as shown in Figure 4.

The channel is divided into two regions: pre-flooding and flooding. In the pre-flooding region, the interfacial vapor velocity is not sufficient to flood the liquid films. Thus, these falling films are opposed only by wall shear stress, resulting in a relatively constant average film thickness δ and phase distribution throughout the region. However, in the flooding region, the films are perpetually flooded by the vapor. This condition, along with increasing vapor velocity, creates a decreasing film thickness in moving axially through the region, resulting in an asymptotic approach toward a more uniform phase distribution.

This conceptual description of the flow regime is supplemented by the graphical analog, given in Figure 5. The j_f^*-j_g^* plot presents three equations:

$$j_g^* = \psi \zeta^{1/2} j_f^* + \psi \zeta^{1/2} \frac{U_\infty^*}{C_0} \quad ; \tag{12}$$

$$\psi \equiv C_0 \left(\frac{\alpha}{1 - C_0 \alpha} \right)$$

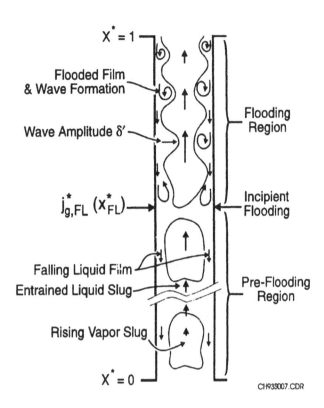

Figure 4   Division of subchannel into pre-
                  flooding, incipient flooding, and
                  flooding regions

$$j_g^{*1/2} + m\left(-j_f^*\right)^{1/2} = C_{Wa} \tag{13}$$

$$j_g^* = -\zeta^{1/2} j_f^* + G_w^* \tag{14}$$

Eq. (12) is a restatement of basic drift flux, Eq. (4).
Eq. (13) is the flooding correlation presented by Wallis
(1969), where m and C_{Wa} are empirical constants both
initially set to unity. Eq. (14) is simply a synthesis of
the phase continuity equations, created by eliminating
x^*.

Inspection of Figure 5 reveals that, in tracking
axially though the pre-flooding region, the drift flux, with
its constant $C_0 \approx 1.5$, creates a fixed x-intercept and a
fanning of lines with increasing α. This continues until the
drift flux becomes tangent to the Wallis flooding curve.
Since the drift flux equation cannot cross over this
boundary, it must remain tangent to the curve for the
remainder of the channel length. This initial tangency
corresponds to the flooding transition. Within the flooding
region, a modified drift flux equation prevails, one with a
continually varying lateral distribution, manifested in the
changing C_0 and x-intercept.

In Figure 5, the intersection of drift flux, Eq. (12),
with the continuity equation, Eq. (14), should yield the
superficial velocities at the transition, $j_{f,FL}^*$ and $j_{g,FL}^*$.

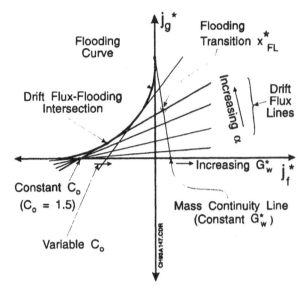

Figure 5   Flooding incipience marked by on-set
                  of tangency of drift flux with flooding
                  curve

However, a third equation is required for the third unknown, the void at transition, α_{FL}. The noted cross
sectional phase distribution in Figure 4 shows that the local
void can be expressed as a function of average film thickness.

$$\alpha = 1 - \frac{2\delta}{w} \tag{15}$$

Furthermore, Wallis (1969) showed that the thickness of
a turbulent falling film, opposed only by wall shear
(i.e., pre-flooding region), is given empirically by:

$$\frac{\delta}{w} = k_{film} \left(\frac{(-j_f^*)}{U_\infty^*} \right)^{2/3} \tag{16}$$

where $k_{film} = 0.12$. With substitution of Eq. (16) for
δ/w, Eq. (15), together with Eqs. (12) and (14), comprise a system of equations, the solution of which yields
$j_{g,FL}^*, j_{f,FL}^*$ and α_{FL}. Knowing $j_{g,FL}^*$, the axial location
of incipient flooding, x_{FL}^*, is readily determined from
Eq. (7).

Thus, the two-region character of the enhanced
model can be stated succinctly as:

$$\bar{\alpha} = \int_0^{x_{FL}^*} \alpha_{base}(x^*) dx^* + \int_{x_{FL}^*}^1 \alpha_{mod}(x^*) dx^* \tag{17}$$

The void profile from the base drift flux model, Eq. (9),
is integrated through the pre-flooding region. The void
profile from the modified drift flux formulation is then
integrated through the flooding region, and the resulting
summation is then the channel average void $\bar{\alpha}$ in Eq.

(2). However, the modified drift flux formulation remains to be determined.

It begins with the previous expression of local α in terms of the local δ.

$$\alpha_{mod} = 1 - \frac{2\delta}{w} \qquad (18)$$

Figure 4 shows that the flooding region is characterized by the formation of intermittently stable roll waves. Their quasi-stability is reflected in their occasional bridging of the channel core, forming a liquid slug. The flooding process can be viewed as a stagnation and "buckling" of the narrow edge falling film, forming the roll waves. Richter (1981) asserts that if a wave is to possess a stable, bounded growth, then a balance is required between the interfacial forces of frictional drag and restoring surface tension. This condition is expressed as:

$$\frac{1}{2}\rho_g u_g^2 \leq \frac{\sigma}{\delta'} \quad ; \quad \delta' \sim \delta \qquad (19)$$

As shown, the wave amplitude δ' scales with the film thickness δ. Richter's wave stability criterion is conventionally expressed in terms of the Weber number, We.

$$We = \text{constant} \quad ; \quad We \equiv \frac{\rho_g u_g^2 \delta}{\sigma} \qquad (20)$$

Determination of the constant value will be discussed later. Since the roll waves are considered intermittently stable, only the equality from Eq. (19) applies to Eq. (20). Replacing j_g/α for u_g in the We definition and substituting the result for δ in Eq. (18) yields a quadratic equation for α. The solution becomes the void fraction profile stemming from the modified drift flux formulation:

$$\alpha_{mod} = \left(\frac{(w \cdot G_w^* x^*)^4}{16\,We^2} + \frac{(w \cdot G_w^* x^*)^2}{2\,We} \right)^{1/2}$$
$$- \frac{(w \cdot G_w^* x^*)^2}{4\,We} \qquad (21)$$

$G_w^* x^*$ has replaced j_g^*. The value of We is determined at the flooding transition location, $x^* = x_{FL}^*$, where:

$$\alpha_{mod} = \alpha_{base} \qquad (22)$$

Equating the right-hand side of Eq. (9) with that of Eq. (21) yields an implicit expression for We, which is ≈ 1 for the current benchmark.

Eqs. (9) and (21) are then the respective integrands for the integrals in Eq. (17), both of which are solved analytically. This yields the final expression for $\bar{\alpha}$ in the enhanced model.

$$\bar{\alpha} = K_2^{-2}\left[K_2\, x_{FL}^* - K_1 \ln\,(K_1 + K_2\, x_{FL}^*) + K_1 \ln\,K_1\right]$$
$$+ \frac{K_3^{-2}}{12}\left[(K_3^2 + 8K_3)^{3/2} - (K_3^2\, x_{FL}^{*2} + 8K_3)^{3/2}\right] \qquad (23)$$
$$+ \frac{K_3}{12}\,(x_{FL}^{*3} - 1)$$

where $\quad K_3 \equiv \dfrac{(w \cdot G_w^*)^2}{We} \qquad (24)$

In addition to the gravitational term in Eq. (2), an additional term, modeling the friction and acceleration heads for a homogeneous mixture, is added to approximate these effects and complete the momentum equation.

$$L^* = 1 - (1 - \zeta^{-1})\bar{\alpha} + G_{41}^{*2}\left(1 + \frac{C_1\,L_H}{D_{hyd}}\right) \qquad (25)$$

Note, the latter term has only a minor influence that is restricted to the upper end of the G_w^* range investigated.

The enhanced model, Eq. (25), is added to the benchmark in Figure 3. Indeed, it seems to track the data quite well. In the case of low G_w^*, the pre-flooding region dominates, and the base drift flux, with $C_0 \approx 1.5$, prevails. However, with increasing G_w^*, the flooding region becomes significant and the modified drift flux prevails. This is manifested in the movement toward a more uniform phase distribution, which is consistent with data.

An additional benchmark (Babcock & Wilcox, 1992) in Figure 6 was performed against boil-up data at incipient dryout within a full-scale, electrically heated Mark 22 assembly, shown in Figure 1. Note, the multi-annulus geometry is more complex than the four-subchannel, one-annulus test section and, therefore, should be a definitive evaluation. Keeping the flooding parameters, m and C_{Wa}, at unity yields We ≈ 3. The model slightly underpredicts data. Setting m = 0.94 yields We ≈ 4 and provides good agreement. Sudo and Kaminaga (1989) demonstrate that such a value is justified for the generally large aspect ratio in the Mark 22 subchannels. This independent evaluation provides further evidence that the enhanced model correctly depicts the thermal hydraulics of boil-up in thin rectangular channels.

SUMMARY AND CONCLUSIONS

The fundamental objectives of this research were:

i) the demonstration of boiling-driven natural convection as an effective mechanism for long-term, in-vessel passive decay power removal within the NPR-HWR production reactor, and

ii) the development of a thermal hydraulic model that would predict threshold conditions incipient to dryout in the thin rectangular subchannels that comprise the coolant annuli within each fuel assembly.

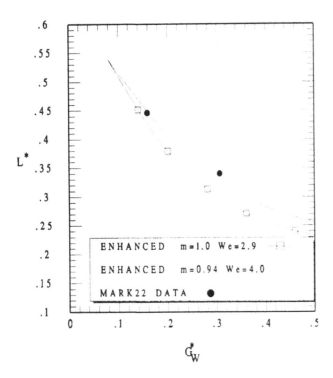

Figure 6   Benchmark of enhanced model against
                 Mark 22 boil-up data

Tests showed that decay power levels within the range investigated could be successfully removed via two-phase natural convection. Dryout and overheat occurred only when the hydrostatic head was no longer sufficient to sustain a two-phase level at or above the channel exit. An initial thermal hydraulic model, based upon the standard drift flux formulation for churn-turbulent flow, was developed. Comparison with data of the threshold hydrostatic head at dryout revealed that the model required refinement to accommodate the locally variable phase distribution throughout the channel. An enhanced thermal hydraulic model identified distinct pre-flooding and flooding regions within the channel. The pre-flooding region was modeled with the standard drift flux formulation, while the flooding region was modeled with a new, modified drift flux formulation. This enhanced model compared well with the noted data and with data from an independent test performed on a full-scale fuel assembly.

NOMENCLATURE

A_c cross sectional flow area

C_0 lateral phase distribution parameter

C_f wall friction coefficient

C_{Wa} flooding correlation parameter

D_{hyd} hydraulic diameter

G mass flux

g gravitational acceleration

j_i superficial velocity of phase i

K_i serial notation symbol

k_{film} empirical constant for film thickness correlation

L moderator collapsed liquid height

L^* dimensionless height ($=L/L_H$)

L_H heated length

m flooding correlation parameter

Q thermal power

u_i phase i real velocity

U_∞ terminal slug rise velocity

w subchannel width

x axial position from channel bottom

Subscripts/Superscripts

FL   incipient flooding

base   base drift flux formulation

mod   modified drift flux formulation

w   characteristic dimension (width)

L_H characteristic dimension (length)

*   dimensionless quantity

Greek Symbols

α void fraction

$\bar{\alpha}$ average void fraction

ρ_i phase i density

δ average film thickness

δ' wave amplitude

REFERENCES

Babcock & Wilcox 1992, New Production Reactor Flow Instability Experiments With Coolant Upflow, RDD:92:4024-01-01:01, Alliance Research Center, Alliance, OH.

Cheng, L. & Tichler, P. 1991, Critical Heat Flux for Free Convection Boiling in Thin Rectangular Channels, ANS Proceedings - 1991 National Heat Transfer Conf. v. 5 pp. 83-90.

Henry, C.E. 1993, Investigation of Buoyancy-Driven Flows, Ph.D. Thesis, Northwestern University, Evanston, IL.

Mishima, K. & Nishihara, H. 1987, The Effect of Flow Direction and Magnitude on CHF for Low Pressure Water in Thin Rectangular Channels, Nucl. Eng. and Design v. 86 pp. 165-181.

Richter, H.J. 1981, Flooding in Tubes and Annuli, Int. J. of Multiphase Flow, v. 7, no. 6, pp. 647-658.

Sudo, Y. & Kaminaga, M. 1989, A CHF Characteristic for Downward Flow in a Narrow Vertical Rectangular Channel Heated From Both Sides, Int. J. of Multiphase Flow, v. 15 no.5 pp. 755-766.

Wallis, G.B. 1969, One-Dimensional Two-Phase Flow, McGraw-Hill, New York.

INDEX

Printed and bound by CPI Group (UK) Ltd, Croydon, CR0 4YY

23/10/2024

01777686-0019